Springer Series on Biofilms

Volume 15

Springer Series on Biofilms presents carefully refereed volumes on selected topics on this field of research. All volumes reflect the latest findings and developments. Once anchored to a surface, biofilm microorganisms carry out a variety of detrimental (like the plaque on teeth) or beneficial (bioremediating hazardous waste sites) reactions. International experts will write, or edit, volumes on specific aspects of biofilms, like research methods, or the roles of biofilms in diseases. While based in microbiology, biofilms are of intense interest to many other scientists, because they affect whole ecosystems, and lie at the root of chronic bacterial infections.

Gordon Ramage • David Andes
Editors

Fungal Biofilms

Springer

Editors
Gordon Ramage
School of Health and Life Sciences
Glasgow Caledonian University
Glasgow, UK

David Andes
Department of Medicine
University of Wisconsin
Madison, WI, USA

ISSN 1863-9607 ISSN 1863-9615 (electronic)
Springer Series on Biofilms
ISBN 978-3-031-93314-1 ISBN 978-3-031-93315-8 (eBook)
https://doi.org/10.1007/978-3-031-93315-8

This Springer imprint is published by the registered company Springer Nature Switzerland AG
The registered company address is: Gewerbestrasse 11, 6330 Cham, Switzerland

Editorial Introduction

It is just over 40 years since Marrie and Costerton presented what may be the first published image of medical device infected by a fungal biofilm (PMID: 6429190). Since that time, we have witnessed skyrocketing concern over the medical impact of fungal biofilms. The magnitude of impact has increased alongside growth of the patient population. In addition, the working definition of a biofilm has broadened to include both surface-associated and non-surface-associated aggregates (PMID: 35922483), enabling the biofilm "paradigm" to inform analysis of diverse infection scenarios.

What ideas are central to the biofilm paradigm? How are biofilm populations distinct from planktonic populations? First and foremost is the concept of cell heterogeneity, a feature intrinsic to biofilms. Cells on an exposed surface encounter a different environment than those within the biofilm or attached to the surface (if present). These unique environments lead to unique phenotypes as cells adapt to local signals. Second is the concept of extracellular matrix accumulation and its biological impact. The matrix can act as a shield or sponge to help a biofilm control its own internal environment, and may include microbially synthesized molecules as well as host molecules in the infection environment. Third is the concept of hypoxic adaptation. Fungal pathogens can all grow in aerobic environments, and most biological analyses have been carried out under such conditions. But oxygen diffuses poorly in aqueous environments, so a submerged biofilm will mostly be in a low oxygen environment. That in turn impacts growth rate and metabolism, with downstream impact on such things as drug tolerance. Fourth is the concept of cell-cell communication and interaction. Perhaps the simplest specific idea in this regard is the production of and response to quorum sensing molecules, but such activities as mating or interspecies interactions, both positive and negative, fit the bill as well. Fifth is the important realization that not all biofilms are the same; different strains or biofilm context may yield quite different observations. The challenge is to identify the common features of, for example, a mucosal biofilm and an implanted device biofilm, which may yield the most promising therapeutic or diagnostic targets. Finally, there is a concept that represents an amalgam of all of the features of biofilm populations: biofilm cells are remarkably resistant to most antimicrobials that were developed using planktonic cell models. Ultimately, the most important driver of medical biofilm research is to identify therapeutic strategies that are effective against biofilm-based infections. This volume provides testament to the significant advances on all of these fronts in the area of fungal biofilms.

This book aims to explore fungal biofilms as a growth area of research. We have brought together a range of experts to cover a range of contemporary topics that will provide a valuable resource to those entering the field of fungal biofilms, but also a resource for those in the field. The book leads with a range of clinically focused chapters (oral cavity, airways, vaginosis, and wounds) that touches on complex interkingdom biofilms, and examines considerations for fungal biofilms role in infection transmission and control. The following chapters cover mechanistic insights into how *Candida albicans* biofilms behave, how these behaviors influence immunological control, and then explore novel antifungal approaches to improving anti-biofilm therapies. Finally, we look in more detail at the molds and present the latest insights into *Aspergillus* biofilm research, and look beyond the lung environment to fungal keratitis as another important reservoir for fungal biofilm mediated disease. Together, this book will provide a non-exhaustive resource of the latest development in fungal biofilm research.

University of Georgia
Athens, GA, USA — Aaron Mitchell

Unveristy of Wisconsin
Madison, WI, USA — David Andes

Glasgow Caledonian University
Glasgow, UK — Gordon Ramage

Contents

Chapter 1
Oral Candidiasis

Christopher Delaney, Hafsa Abduljalil, Suror Shaban, Muhanna Alshehri, and Jason L. Brown

Abstract Oral candidiasis is a common opportunistic fungal infection caused primarily by *Candida albicans*, affecting diverse populations, particularly immunocompromised individuals. This chapter provides an overview of the clinical spectrum of oral candidiasis, including acute and chronic manifestations of the disease, underlying risk factors, current treatment strategies, and emerging therapeutic options. Additionally, it explores experimental models that enhance our understanding of *Candida* pathogenesis and facilitate the development of innovative antifungal strategies both in vitro and in vivo. Understanding these aspects is critical for improving patient outcomes and addressing the growing challenge of antifungal resistance.

Introduction

Candida species form a component of the natural flora within the oral cavity and the oropharynx. Estimates of commensal carriage in healthy individuals vary among populations, but it is predicted that *Candida* species can be present in up to 80% of the healthy population (Akpan and Morgan 2002; Darwazeh and Darwazeh 2014; Zomorodian et al. 2016; Alves et al. 2023; Lu 2021). Infections are most associated with the dimorphic fungal species *Candida albicans* which form a biofilm on the mucosal surfaces of the tongue dorsum and palatal tissue. Non-*albicans* species have been associated with oral candidiasis in less frequent numbers, including *C. glabrata*, *C. krusei*, and *C. tropicalis* (Turner and Butler 2014). *C. albicans* is estimated to be the causative organism in up to 75% of oral *Candida*-related infections and more than 80% of denture stomatitis infections (Figueiral et al. 2007; Hu

C. Delaney (✉) · H. Abduljalil · S. Shaban · M. Alshehri · J. L. Brown (✉)
Oral Sciences Research Group, Glasgow Dental School, School of Medicine, College of Medical, Veterinary and Life Sciences, University of Glasgow, Glasgow, UK

Glasgow Biofilm Research Network, Glasgow, UK
e-mail: christopher.delaney@glasgow.ac.uk; jason.brown@glasgow.ac.uk

G. Ramage, D. Andes (eds.), *Fungal Biofilms*, Springer Series on Biofilms 15,
https://doi.org/10.1007/978-3-031-93315-8_1

et al. 2019a). Oral candidiasis is similar to other candidiasis in that it is an infection commonly seen in patients with other health conditions. Indeed, microbial dysbiosis in the natural flora is associated with poor oral hygiene, immunodeficient, or immunocompromised individuals and other predisposing risk factors (Baumgardner 2019). *Candida* infections are seen as opportunistic, with the organisms responsible exploiting reduced host defences. Yeast carriage in the oral cavity of asymptomatic healthy individuals further demonstrates that oral candidiasis is often seen as a "disease of the diseased," with underlying immune conditions strongly associated with its onset (Mun et al. 2016).

Opportunistic infection of the oral cavity can therefore be an indicator of other local and systemic health concerns (Ambe et al. 2020). Infections in immunocompromised and vulnerable patient groups can cause complications, including pain, and other quality-of-life impairments, such as difficulty swallowing and impaired speech. While uncommon, untreated oral biofilm infections in at-risk patient groups can lead to the dissemination of *Candida* (Mohamed et al. 2019). The purpose of this chapter is to outline the common presentations, risk factors, and treatments of oral candidiasis. The current therapeutic interventions, model systems used for studying disease pathogenesis, and emerging techniques for controlling this disease will additionally be explored.

Clinical Manifestations

For the majority of the clinical presentations of oral candidiasis, the dorsum aspect of the tongue is the starting point of infection, as it is the primary reservoir for the opportunistic *Candida* pathogens (Vila et al. 2020). Different studies have utilized several classification systems for the various clinical manifestations of oral candidiasis, and these clinical forms of oral disease can be detected in multiple sites. The disease can be generally classified into two main broad categories, namely, (1) acute manifestations and (2) chronic manifestations, both of which contain a variety of different sub-categories depending on clinical features. Despite *C. albicans* being the main causative agent in oral candidiasis, *C. albicans* and non-albicans oral manifestations are ultimately identical (Vila et al. 2020), which can make therapeutic interventions challenging for clinicians.

Acute Manifestations of Oral Candidiasis

Acute Pseudomembranous Candidiasis

This is the classical presentation of oral candidiasis and more commonly known as "thrush," often presenting as merging white plaques on the tissue surface of the tongue, buccal mucosa, hard and soft palate, and the oral pharynx. A distinctive clinical feature of pseudomembranous candidiasis is the ability to wipe off the associated plaques, which mainly consist of both host and microbial factors. These can include desquamated epithelium and immune cells together with the yeast and/or hyphal remnants of the *Candida* species, leaving behind an erythematous surface (Millsop and Fazel 2016). This form of oral candidiasis is often observed in newborns and the elderly, and diagnosis is typically based primarily on clinical presentation. It is also the most common variant of oral candidiasis observed in human immunodeficiency virus (HIV) patients (Thompson 3rd et al. 2010). Pseudomembranous candidiasis is often asymptomatic; however, some patients may suffer from a burning sensation in the oral cavity and/or difficulty in swallowing. The white lesions observed in oral thrush presentation have similar characteristics to those observed in lichen planus, lichenoid reaction, and leukoplakia (Fotos and Hellstein 1992).

Acute Atrophic Candidiasis

This form of oral candidiasis was previously referred to as "antibiotic sore mouth" or "burning mouth syndrome." This disease usually occurs as a consequence of a disturbed oral microbiome after prolonged use of a broad-spectrum antibiotic or corticosteroid inhalers, which ultimately results in an overgrowth of *Candida* species (Hellstein and Marek 2019). Acute atrophic candidiasis presents as painful erythematous lesions throughout the oral cavity, with the hard palate being mostly affected. Burning sensation of the tongue is usually reported by affected individuals if the mid-posterior dorsum of the tongue is involved (Patil et al. 2015). Diagnosis is usually made based on clinical presentation alongside a recent history of antibiotic or corticosteroid consumption, hence the alternative nomenclature of this classification. No therapeutic intervention is usually needed as cessation of antibiotics or the corticosteroid therapy will result in lesion resolution (Lewis and Williams 2017). Differential diagnosis includes erythroplakia, oral lichen planus, or allergic reactions (Shirasuna 2014).

Chronic Manifestations of Oral Candidiasis

Chronic Erythematous Atrophic Candidiasis

Commonly referred to as denture stomatitis, chronic erythematous atrophic candidiasis is the most prevalent chronic inflammatory oral disease among denture wearers, affecting more than 75% of edentulous patients (Ramage et al. 2004). The presence of a denture or orthodontic appliance is a prerequisite for developing this form of candidiasis, with other contributing factors including poor oral hygiene, continuous use of dentures (e.g., nocturnal use), and decreased salivary flow. Trauma that results from ill-fitting full or partial dentures can also predispose individuals to denture stomatitis (Hannah et al. 2017). Combined, these risk factors create a favorable environment for *C. albicans* to invade the oral tissues adjacent to the surface of the dental appliance, which subsequently initiates an inflammatory reaction. Based on the severity, denture stomatitis is classified into three types according to Newton's classification (Newton 1962): (1) a localized, simple inflammation or a pinpoint hyperemia; (2) a diffuse erythematous or generalized simple type involving the entire denture-covered mucosa; and (3) a granular type of inflammation (e.g., inflammatory papillary hyperplasia) involving the central part of the hard palate and the alveolar ridges (Arendorf and Walker 1987). Given the proximity of dentures to the respiratory tract, denture plaque is a potential reservoir of opportunistic respiratory pathogens. This can potentially increase the risk of developing aspiration pneumonia, particularly in elderly populations (O'donnell et al. 2016).

Chronic Hyperplastic Candidiasis

This type of oral candidiasis, sometimes referred to as Candidal leukoplakia, presents clinically as well-defined white plaques that cannot be scraped away. These are most commonly found at the buccal mucosa near the commissure of the mouth and also on the lateral surface of the tongue (Winocur-Arias et al. 2023). These lesions are often asymptomatic and present in higher prevalence among middle-aged male smokers (Arendorf et al. 1983). Chronic hyperplastic candidiasis is a rare subtype of oral candidiasis, and diagnosis is usually based on clinical symptoms in conjunction with a complete medical history. Additionally, swabs from the lesion can be obtained to confirm the presence of *Candida* species. Studies have shown an association between chronic hyperplastic candidiasis and oral squamous cell carcinoma. However, the role of *Candida* and its association with malignant transformation are unclear (Zhang et al. 2021).

Angular Cheilitis

Angular cheilitis is an inflammatory disease of variable etiology presented clinically as erythema, rhagades, ulcerations, and crusting of the lip corners and surrounding skin, either in one or both commissures (Cabras et al. 2020). Angular cheilitis is a multifactorial disease, and *C. albicans* is the most commonly isolated fungal pathogen from these lesions, with carriage being estimated as high as ~93% (Narayanan and Rogge 2024). In addition to the high prevalence of fungi, it is common for these lesions to be co-infected with *Staphylococcus aureus* and beta-hemolytic *Streptococci* (Chiriac et al. 2024). Reduced facial vertical dimension due to edentulism and ill-fitting dentures was reported to be the most common factor in initiating the disease. Other local contributing factors include xerostomia, poor oral hygiene, and cheilitis through habitual frequent lip biting, sialorrhea in babies, and smoking (Cabras et al. 2020). Angular cheilitis can present concomitantly with other variants of oral candidiasis, such as acute pseudomembranous candidiasis or denture stomatitis (Sharon and Fazel 2010).

Systemic factors that positively influence the development of angular cheilitis include reduced systemic levels of nutrients such as iron, vitamin B, and zinc. Nutritional deficiencies compromise the integrity of oral epithelium and diminish the immune defence mechanisms to resist and recover from microbial insults (Blagec et al. 2023). In addition, systemic conditions that can cause an impairment in immune function and inflammatory bowel diseases, such as Crohn's, can predispose an individual to developing the oral manifestations of this disease (Pandarathodiyil et al. 2021).

Angular cheilitis presents with unique clinical features that differ from other forms of oral candidiasis. Based on severity, Ohman has classified the disease into four subtypes (Ohman et al. 1986):

- *Type I:* small rhagades limited to the corner of the mouth, with slight involvement of adjacent skin.
- *Type II:* lesion with ragged border, with more extension in length and depth than a type I lesion.
- *Type III:* lesion consisting of several rhagades radiating from the corner of the mouth into the adjacent skin.
- *Type IV:* lesion presenting no rhagades but erythema of skin contagious to the vermilion border.

Accurate differentiation of angular cheilitis from other oral lesions is vital for proper management. Conditions that may mimic angular cheilitis, such as herpetic stomatitis, impetigo, or perioral atopic dermatitis, can be observed around the mouth, leading to misdiagnosis (Lugović-Mihić et al. 2018). Diagnosis is often made based on clinical features; however, in case of atypical or treatment-resistant cases, laboratory testing, such as a swab for microbial cultures or biopsy, may be required.

Median Rhomboid Glossitis

This variation of oral candidiasis is a rare oral developmental disorder of the oral cavity, sometimes known as posterior midline atrophic candidiasis. Median rhomboid glossitis is estimated to affect ~0.01% of the general population (John et al. 2023). Lesions are more predominant in males than females, usually affecting the central part of the dorsum surface of the tongue and presenting on clinical examination as a well-defined symmetric red oval or rhomboid lesion with an associated papillary atrophy. The infection is often asymptomatic, but some patients may complain of mild itching and irritation (Goregen et al. 2011). It was previously suggested that this lesion is a result of an embryologic defect caused by the failure of the tuberculum to fully unite with the lateral processes of the tongue, resulting in a smooth, depapillated erythematous area on the posterior dorsal surface. This theory was rejected later because of the rare occurrence of this lesion in children and increasing evidence of *C. albicans* infection as a causative agent for this lesion (Baughman 1971).

Cheilocandidiasis

Cheilocandidiasis is a newly recognized variant of oral candidiasis and clinically presents as an ulcerating lesion located on the lip, typically on the lower vermillion border (Samaranayake et al. 2009). It is a rare variant of oral candidiasis caused by the *C. albicans* pathogen. Personal habits, such as frequent lip-licking or thumb-sucking, are believed to promote moist environments ideal for *Candida* species to proliferate, leading to this form of the disease (Vila et al. 2020).

Chronic Mucocutaneous Candidiasis Syndromes

This chronic form of oral candidiasis is characterized by persistent or recurrent mucocutaneous candidiasis involving the skin, nails, and genital mucosa, with greater than 90% of patients presenting with oral involvement (Kirkpatrick 1994). These fungal lesions occur secondary to other medical conditions that involve abnormal immune function, such as HIV infection with T-cell deficiency, diabetes, and immunosuppressive therapy (Humbert et al. 2018). Patients with this form of candidiasis are often unresponsive to standard antifungal therapy and are at a great risk for developing oral squamous cell carcinoma (Rautemaa et al. 2007).

Risk Factors

A number of risk factors specific to the different forms of oral candidiasis have already been covered above. The following sections will outline the general predisposing factors of oral candidiasis related to immune suppression, both local and systemic, followed by other factors not involving the immune system. All risk factors are depicted schematically in Fig. 1.1.

Immune Suppression

Immune suppression is the major predisposing factor for oral candidiasis occurring in up to ~70% of individuals receiving immunosuppressive treatment and up to 90% of individuals with HIV (Dockrell et al. 2019; Lalla et al. 2010). Immunosuppressive regimes include corticosteroids, biologics, systemic disease, and chemotherapy, all of which can predispose and promote the invasion of pathogenic *Candida* species (Dropulic and Lederman 2016).

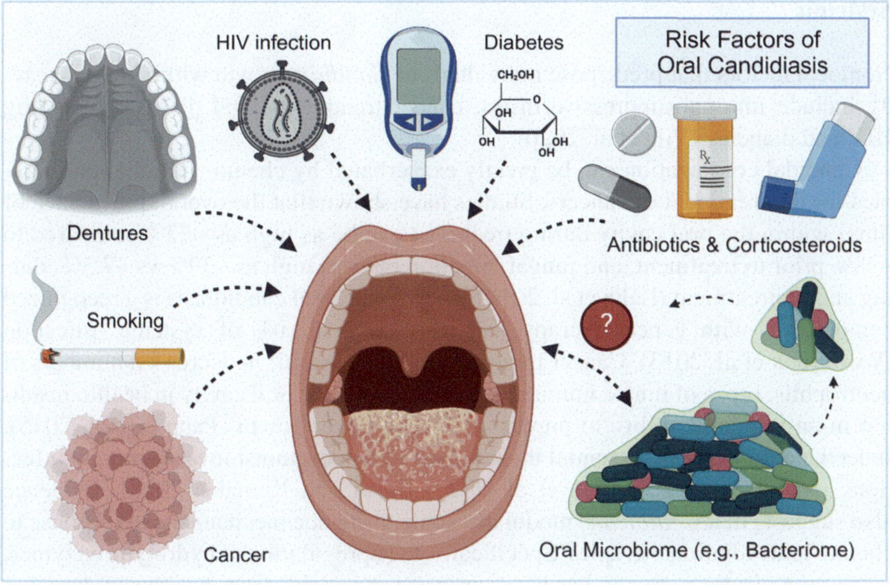

Fig. 1.1 Risk factors for oral candidiasis. Image created using Biorender

Local

Corticosteroids are typically used in the treatment of respiratory disorders such as asthma and chronic obstructive pulmonary disease. These interventions lead to a local reduction in the immune response in the oral cavity and upper respiratory systems (Dekhuijzen et al. 2016). A prescription sequence symmetry analysis identified the occurrence of oral candidiasis following corticosteroid usage. It demonstrated an increase in risk for oral candidiasis following the first 3 months after inhaled corticosteroid usage (Van Boven et al. 2013). Similarly, a study performed in 2016 indicated that there is a relationship between the incidence of candidiasis and the dosage or type of inhaled corticosteroid administered (Dekhuijzen et al. 2016). Early attempts to identify the odds of developing oral candidiasis following steroid usage indicated an odds ratio of 3.6, suggesting a strong link between corticosteroids and adverse events in the oropharynx (Rachelefsky et al. 2006). Corticosteroids are believed to induce the pathogenesis of *Candida* species through the reduced immune response and increasing the stimulation of saliva (Ellepola and Samaranayake 2001). Although the mechanisms are not fully understood, it is postulated that reduced immunoglobulin A production at the mucous membranes, in addition to an increase in salivary glucose levels, contributed to an ideal microenvironment for *Candida* species to flourish (Fukushima et al. 2005; Dekhuijzen et al. 2016).

Systemic

Nonlocal factors that predispose individuals to *Candida* growth within the oral cavity include immunosuppressive drugs, cancer treatments, and diseases including HIV and diabetes (Vila et al. 2020).

Candidal colonization can be greatly exacerbated by chemo- or radiation therapies for the treatment of cancers. Studies have shown that the overall prevalence of fungi within the oral cavity during treatment can be as high as ~72% compared to ~48% prior to treatment, and fungal infection rates as high as ~39% vs ~7.5%, during and pretreatment (Lalla et al. 2010). To this end, oral candidiasis is a recognized comorbidity with cancer therapy and also poses a risk of systemic infection (Westbrook et al. 2013). Cancer treatment regimens result in decreased numbers of neutrophils, types of innate immune cells that patrol the oral cavity in health, resulting in an impaired ability to prevent a *Candida* overgrowth (Pappas et al. 2015). Indeed, neutrophils are essential in early host defence against invasive fungal infections (Netea et al. 2015; He et al. 2022; Desai and Lionakis 2018). Evidence also suggests that *C. albicans* modulates some virulence mechanisms in response to chemo- and radiation therapies. Specifically, the upregulation of hydrolytic enzymes, a common virulence factor, has been observed in isolates from patients undergoing cancer therapy (Ramla et al. 2016).

Patients with HIV and Acquired Immunodeficiency Syndrome (HIV/AIDS) display increased susceptibility to numerous infections, including those from opportunistic commensal fungi. Notably, recent evidence suggests that there is an increasing level of antifungal resistance associated with oral candidiasis in patients with HIV infection (Keyvanfar et al. 2024). *C. albicans* is commonly the most prevalent species isolated from HIV patients (Rafat et al. 2021). Prior to highly active antiretroviral therapy, oral candidiasis was estimated to occur in up to 90% of HIV-infected individuals (Lortholary et al. 2012). Despite the reductions observed in incidence following the introduction of this antiviral therapy, oral candidiasis remains a concerning opportunistic pathogen in HIV/AIDs patients (Erfaninejad et al. 2022; Low et al. 2016).

HIV/AIDs diagnoses, alongside the opportunistic colonization of *Candida* species and the presentation of oral candidiasis, are associated directly with immunological suppression in HIV patients. A marked reduction in the T lymphocytes associated with HIV infection is the predominant risk factor. Specifically, a $CD4^+$ T-lymphocyte count <200 cells/mm^3 has previously been positively associated with the disease, similarly, a $CD4^+/CD8^+$ ratio <0.5 is also an indicator of susceptibility to developing oral candidiasis in HIV patients (Gaurav et al. 2011; Taverne-Ghadwal et al. 2022). $CD4^+$ T lymphocytes in healthy individuals are active in a localized manner against *Candida* within the oral mucosa and have been demonstrated to be necessary in host recovery from *Candida* infection (Farah et al. 2002). The importance of CD4+ T lymphocytes in antifungal immunity is highlighted in (Section "Disease Models") through the use of immunodeficient mice for studying disease pathogenesis.

Other Risk Factors

Known risk factors not directly associated with immune suppression include smoking and antibiotic use. Other predisposing factors can include the edentulous nature of individuals and diabetic patient groups.

Smoking is a predisposing risk factor for many diseases and has been associated with an increase in carriage of *Candida*, as well as being described as an independent risk factor in the development of oral candidiasis in HIV patients (Soysa and Ellepola 2005; Chattopadhyay and Patton 2013). The cellular processes associated with this association are not known; however, there is some evidence that cigarette smoke can promote adhesion and biofilm formation of the fungus on the mucosal surface (Semlali et al. 2014). Chemicals in smoke can also influence specific immune pathways; it has been shown through the negative regulation of Nrf2, a protein involved in oxidative stress and immunomodulation, leading to smoking-induced immunosuppression (Ye et al. 2021).

Broad-spectrum antibiotic use is associated with an increased risk of oral candidiasis in HIV/AIDs patients (Ambe et al. 2020). Antibiotic sore mouth, a specific type of candidiasis (Section "Acute Atrophic Candidiasis"), is most commonly associated with the broad class of antibiotics, tetracycline. Tetracycline has been associated with up to 3.5-fold increase in *Candida* carriage rates in hospitalized patients (Soysa et al. 2008). The reduction in competition for surfaces and nutrients from the natural microflora or bacteriome within the oral cavity has been suggested as an etiological factor in the overgrowth of *Candida* on the mucosa. Conversely, antibiotics have also been shown to induce salivary changes and have bystander immunomodulatory effects (Soysa et al. 2008).

Edentulous patients who wear removable appliances commonly develop denture stomatitis (Section "Chronic Erythematous Atrophic Candidiasis"). Denture stomatitis is associated with *Candida* infection of the appliance and inflammation of the contacted mucosa. The infection is typically triggered through a combination of the presence of *Candida* and an ill-fitting denture (Gendreau and Loewy 2011). Denture wearing itself is classified as a risk factor for oral candidiasis, but this variation is known to be exacerbated by many of the systemic predisposing factors described above (Perić et al. 2024). Additional risk factors outlined in a recent systematic review include salivary pH, age, overnight denture wearing, and sex of the patient (Perić et al. 2024). The incidence of denture stomatitis can reach up to ~75% in edentulous patients, with *C. albicans* being the most commonly associated pathogen (Sartawi et al. 2021; Ramage et al. 2004).

Carriage of *Candida* has been shown to be higher in diabetics compared to non-diabetic patients, with *C. glabrata* being shown to be the most common colonizer (Chouhan et al. 2019). Prevalence of oral candidiasis and/or denture stomatitis in type II diabetic patients varies, with fungal carriage reaching ~60–70% in some studies (Belazi et al. 2005; Martinez et al. 2013; Kumar et al. 2005). Nevertheless, appropriate control of diabetes can reduce the candidal load (Chouhan et al. 2019). Although direct causal links between diabetes and oral candidiasis are yet to be identified, causes potentially influencing this include higher salivary glucose levels, lower salivary pH, and insufficient host immunity to microbial infections (Rodrigues et al. 2019). Indeed, these links have been confirmed in published literature. For example, a higher frequency of *Candida* carriage overall has been associated with increases in salivary glucose concentrations (Darwazeh et al. 1991). Conversely, the raised salivary glucose levels as well as salivary flow changes associated with type 2 diabetes have been described as promoting the growth of *Candida* in the oral cavity (Mohammadi et al. 2016). Nevertheless, further studies are necessary to investigate the cellular mechanisms at play during these comorbidities.

The oral microbiome, or bacteriome, may also influence and facilitate the colonization of *C. albicans* and other fungal species of the oral cavity. Evidence suggests that higher levels of *Streptococcus* species, such as *Streptococcus oralis*, can drive oral dysbiosis in humans with hyper-IgE inflammatory syndrome-associated oropharyngeal candidiasis, resulting from excessively high levels of IgE antibody (Abusleme et al. 2018). Other publications are in agreement, whereby bacterial species such as *Rothia mucilaginosa* and *Streptococcus mitis* are more abundant in the saliva of patients with chronic erythematous candidiasis (Lyu et al. 2021). Indeed, a number of studies have shown that *Candida* species form unique interkingdom interactions with various oral bacteria, including both commensal and pathogenic microorganisms (Delaney et al. 2019; Fujinami et al. 2021; Lyu et al. 2021). However, these interactions may not influence microbial diversity or cause changes in fungal load. For example, with increased Candida load, the diversity of the salivary microbiome decreases, resulting in a composition dominant with saccharolytic and acidogenic bacteria (likely arising from preferential interactions with the fungus) (Kraneveld et al. 2012). Conversely, fungal load in denture swabs, mucosal swabs, and dental plaque samples from denture-wearing patients remained comparable regardless of microbiome diversity (Delaney et al. 2019). In agreement, others have shown through longitudinal studies that bacterial diversity remains the same regardless of candidiasis, pre- and post-antifungal intervention (Diaz et al. 2019). It is apparent from this that while intricate interactions occur between fungi and bacteria within the oral cavity, the evidence remains inconclusive on whether the oral microbiome, or bacteriome to be exact, influences the *Candida* pathogenesis or if the fungus merely serves as a passive bystander due to its structural capacity for bacterial adhesion. With the supporting evidence from various in vitro models, whereby *C. albicans* can alter the composition of bacterial biofilms (Young et al. 2020; Janus et al. 2017), this is perhaps suggestive of its role as a facilitator of bacterial pathogenesis, rather than the other way around. Further research is required.

Frontline and Novel Therapies

A successful treatment of oral candidiasis depends on an early and accurate diagnosis, eliminating the predisposing factors, and correcting the underlying diseases. It is also pertinent for clinicians, where possible, to determine the type of *Candida* infection to help inform the appropriate antifungal regimen. A conservative approach to controlling oral candidiasis includes brushing teeth, cleaning the tongue, buccal cavity, and dentures, using mouth washes such as chlorhexidine, and for denture wearers, dentures to be removed at night, sterilized, and submerged in a disinfectant solution. A reduction in the recurrence rate only by following this approach has been noted (Garcia-Cuesta et al. 2014). The pharmaceutical approach, on the other hand, involves using topical or systemic treatments. Most oral candidiasis cases can be successfully treated with topical application of antifungal agents, while the systemic administration is only confined to disseminated, secondary, or refractory infections. The antifungal agents, when applied locally, can quickly reach the therapeutic concentration at the infection site while limiting the incidence of adverse effects (Lu 2021).

Conventional Agents

The typical antifungal agents used in treating oral candidiasis fall into the three main classes: polyenes, azoles, and echinocandins. Polyenes, including nystatin and amphotericin B, have long been considered the first choice in treating oral candidiasis. Polyenes are fungicidal, and their main mechanism of action is interacting with ergosterol in the fungal cell membrane and forming transmembranous pores that result in increased permeability and leakage of intracellular components, leading to cellular death (Quindós et al. 2019; Xiao et al. 2022). Nystatin can be applied topically on the affected area or provided as pastilles, lozenges, or oral suspensions. Meta-analysis studies have shown that this antifungal is effective in treating denture stomatitis (Lyu et al. 2016; Rai et al. 2022). Amphotericin B, administered as a lozenge or oral suspension, is also an effective antifungal agent in treating oral candidiasis (Xiao et al. 2022).

Second-line antifungals used in oral candidiasis are azoles, which include imidazoles (miconazole, clotrimazole, and ketoconazole) or triazoles (fluconazole and itraconazole). Azoles are fungistatic, and their antifungal activity is mediated by inhibiting the biosynthesis of ergosterol, affecting the integrity and permeability of the cell wall and cell membrane (Ghannoum and Rice 1999). The clinical formulations of fluconazole include oral suspensions or in the form of a mouthwash. Miconazole, on the other hand, is available as an oral gel applied to the infected area or in the form of tablets. Alternatively, clotrimazole can be administered as a lozenge (Xiao et al. 2022). A meta-analysis systematic review has reported that fluconazole showed higher efficacy than nystatin in eradicating *Candida* biofilms (Lyu et al. 2016). Similarly, fluconazole has been shown to have similar efficacy to amphotericin B and a greater clinical response rate to clotrimazole in immunocompetent patients, with no significant differences in mycological cure rate (e.g., reduction in fungal load) or the incidence of adverse reactions (Xiao et al. 2022). This demonstrates that azoles may be useful in combating candidal infections, through the inhibition of growth and allowing for the restoration of a symbiotic oral microbiome.

Echinocandins (such as caspofungin) are also fungicidal agents, demonstrating a strong selectivity against fungi through inhibiting the biosynthesis of β-(1,3)-D-glucan; a specific component of the fungal cell wall (Ghannoum and Rice 1999). Early comparative studies have shown that caspofungin has a broad-spectrum antifungal activity similar to that of amphotericin B and is as effective in controlling localized and systemic candidiasis (Mccormack and Perry 2005; Arathoon et al. 2002; Mora-Duarte et al. 2002). However, more recently, caspofungin clinical use is limited to systemic or refractory cases due to the emergence of many echinocandin-resistant *Candida* species (Pristov and Ghannoum 2019).

Novel Interventions

Novel therapeutic alternatives are needed to address the limitations of the traditional antifungals in terms of drug resistance and systemic adverse effects. There have been extensive efforts in research to discover novel antifungal agents targeting key fungal virulence factors and other pathogenic traits within the context of oral candidiasis and associated infections. Alternative strategies investigated in clinical trials include probiotics, photodynamic therapy, naturally derived compounds, and immunotherapies (summarized in Table 1.1). It is noteworthy that a number of other novel interventions have been and are in the process of testing, but such work is limited to in vitro and in vivo. These include vaccinations, phytotherapy, repurposed drug libraries, herbal medicines, and naturally occurring bioactive compounds (Gharibpour et al. 2021; Romo et al. 2017; Pierce et al. 2015; Contaldo et al. 2023; Van De Veerdonk et al. 2010).

Of these clinical studies, probiotic therapy has shown arguably the most promising ability as an alternative to antifungal drug use. These microorganisms are able to restore the balance between the microbial communities in the oral cavity, while regulating innate and acquired immunity (Hu et al. 2019b). A wide variety of *Lactobacillus* species, such as *L. rhamnosus*, *L. acidophilus* and *L. reuteri,* and *Streptococcus salivarius,* have all been tested in clinical trials and have shown an inhibitory effect against Candidal load and amelioration of the signs and symptoms of the fungal infection (Hatakka et al. 2007; Kraft-Bodi et al. 2015; Ishikawa et al. 2015). An alternative technique that has shown similar promise is photodynamic therapy (PDT), a technique that uses light-activated photosensitizers applied to the diseased tissues. These compounds will result in photochemical reactions that generate reactive oxygen species, leading to cellular damage in the microbiota. The low cytotoxicity on healthy tissues, broad-spectrum activity, and short treatment duration make PDT a potential therapeutic option for oral fungal infections (Javed et al. 2014). Indeed, a recent meta-analysis systematic review highlighted a total of 11 studies investigating the prospect of PDT being used as an alternative for oral candidiasis or denture stomatitis. Observations from the study concluded that the

Table 1.1 Alternative therapies investigated in clinical trials for the treatment of oral candidiasis or denture stomatitis

Alternative therapy	Mode of action	References
Probiotics	Disruption of the pathogenic microbiome, including anti-inflammatory and/or immuno-modulatory properties to control the host response during infection	(Hatakka et al. 2007; Kraft-Bodi et al. 2015; Ishikawa et al. 2015; Li et al. 2014; Passariello et al. 2020; Lee et al. 2019) *(Hu et al. 2019b; Mundula et al. 2019)*
Photodynamic Therapy	Photo-sensitizers activated by light to generate reactive oxygen species that are antimicrobial	*(Javed et al. 2014, Hu et al. 2023)*
Naturally derived compounds *Chitosan* *Curcumin* *Fatty acids*	A number of natural products have been shown to have antimicrobial and/or anti-inflammatory activity *Chitosan*: Polysaccharides derived from chitin. When formulated into chitosan nanoparticles, these provide localized delivery of other antimicrobial agents *Curcumin*: Topical application of a polyphenolic compound from the spice, turmeric, with antimicrobial properties *Fatty acids*: Administered in a mouth rinse, evidence of antimicrobial properties	*(Inacio Silveira et al. 2023; Gharibpour et al. 2021)* (Mohammed and Fatalla 2020; Gamil et al. 2024; Namangkalakul et al. 2020) (Tatapudi et al. 2021) (Lee Muñoz et al. 2023)
Antibodies	Antibody opsonization of *Candida* through IgG administered in a mouthwash	(Pedraza-Sanchez et al. 2018)

*Asterisks denote meta-analysis systematic reviews reporting of multiple primary articles

efficacy of PDT was similar to that of fluconazole, providing a promising alternative to antifungal therapy (Hu et al. 2023).

Disease Models

Several different models exist for studying host–pathogen interactions, disease pathogenesis, and antimicrobial interventions in oral candidiasis. These can encompass in vitro and in vivo systems utilizing organotypic cells or animals. The following sub-sections will discuss these models, with a description of the model, advantages, and limitations summarized in Table 1.2 and depicted schematically in Fig. 1.2.

Table 1.2 Summary of in vitro and in vivo model systems used for assessing oral candidiasis. Descriptions of the models, advantages, and limitations associated with them are included

Model system		Description of model	Advantages	Limitations
In vitro *models*	*Two-dimensional organotypic co-culture systems*	Oral epithelial cells (e.g., keratinocytes) grown as monolayers exposed to *Candida* species (± bacterial species)	Allows for monitoring of fungal adhesion mechanisms and pro-inflammatory host responses	Limited host factors involved. Model utilizes single layers of epithelial cells or representative of in vivo
	Three-dimensional organotypic co-culture systems	In-house grown or commercially available tissue models consisting of epithelial cells grown at an air–liquid interface	Multi-layered structure allows for a more representative recapitulation of the oral mucosa to assess host–pathogen interactions	Troublesome to grow if created in-house (can be prone to contamination). Commercial tissue is expensive. Doesn't contain underlying connective tissue/ vasculature, thus limited in its applicability to in vivo tissue
	Biofilm model systems	Can be used in conjunction with tissue models for host–pathogen interactions or for antimicrobial testing—often utilized for complex microbial consortia pertaining to denture stomatitis and similar polymicrobial oral diseases	Can be complex containing different microbial consortia. Co-culture tissue models utilising such biofilm models allow for more accurate mimicry of host–biofilm interactions at the oral mucosa, than individual microbial species	Time consuming to grow, and even the most complex model still lacks true representation of the oral microbiome (>100 species)

(continued)

Table 1.2 (continued)

Model system		Description of model	Advantages	Limitations
In vivo *models*	*Immunosuppressed mice*	Cortisone acetate injections leads to immunosuppression, followed by infection with *C. albicans* or other *Candida* species	Microbial colonization and host tissue responses are similar to humans while utilizing a common risk factor of candidiasis (e.g., immunosuppression)	Requires infection with high concentrations of *Candida* species. Most infection models limited to acute phenotype due to ethical considerations for prolonged experimentation necessary for chronic manifestations of the disease.
	Transgenic mice	Created from mutations to key genes responsible for important immunological responses, followed by infection with *C. albicans* or other *Candida* species	Allows for the assessment of the role that key inflammatory markers and/or immunological cell lineages play in antifungal immunity	Immune responses are dampened, elevated and/or skewed, failing to recapitulate the same conditions in humans with similar genetic mutations. Additional ethical considerations needed for normal mouse models
	Rat	Used to assess microbial colonization of substrata leading to oral diseases such as denture stomatitis	Due to size, easier to handle than murine models, and allows for insertion of denture acrylic or implants to mimic infection arising from contaminated or colonized substrata	Similar to murine models, clear anatomical differences between humans and rats (e.g., salivary flow) that can influence microbial colonization and/or host response. Rats do not wear implanted dentition, thus lack a true representation of human disease

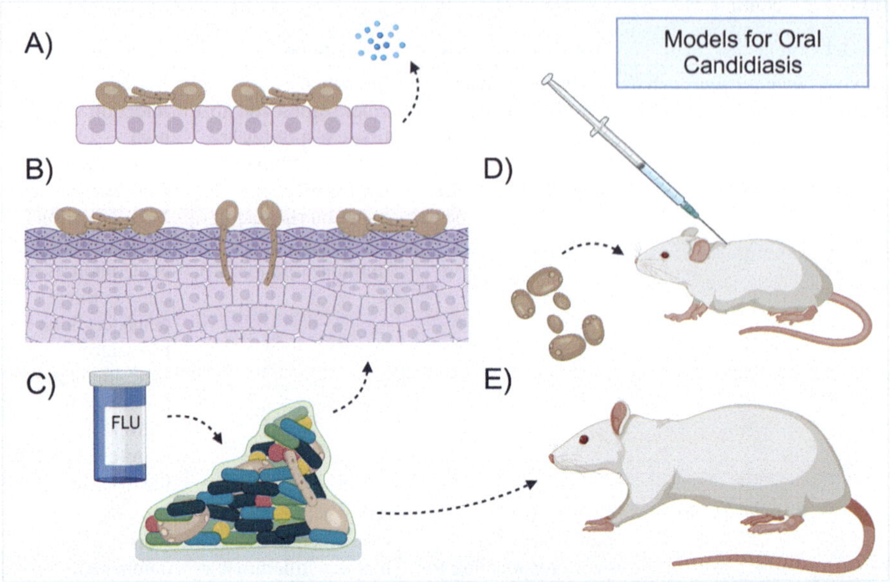

Fig. 1.2 In vitro and in vivo model systems utilized for studying oral candidiasis. (**a**) Epithelial cells grown as a monolayer and exposed to *Candida* species to assess microbial adhesion and host responses. (**b**) 3D organotypic models for investigating tissue invasion, often mediated by Candida hyphae formation. (**c**) Complex, polymicrobial biofilm models for assessing antimicrobial interventions (e.g., fluconazole; FLU) in denture stomatitis, models that can be utilized for host colonization investigations. (**d**) The most common mouse models involve immunosuppression with subcutaneous injections of cortisone acetate followed by oral inoculation with Candida species. (**e**) Rat models can be used for implantation with acyclic substrata inoculated with microorganisms (e.g., biofilms) to mimic edentulous patients in diseases such as denture stomatitis

In Vitro Models

In vitro model systems hold many advantages for understanding disease pathogenesis. These systems provide low-cost suitability for multiple screening assays and testing the efficacy of antifungals in a controlled manner. In vitro studies that exist for this disease have enhanced our understanding of surface adherence properties during biofilm formation on various substrata and following antifungal interventions, either in *Candida* only biofilms or when grown alongside bacterial species (Chevalier et al. 2018; Li et al. 2010; Lamfon et al. 2005). Conversely, the lack of representation of host factors that play a role in biofilm formation and disease pathogenesis is considered a major drawback in in vitro biofilm models (Last et al. 2021). To counter this, co-culture model systems now exist for investigating candidiasis pathogenesis in the oral cavity, from simple monolayers of epithelial cells to complex three-dimensional organotypic tissues that mimic in vivo tissue. These complex models can include commercially available tissue systems or in-house engineered tissue equivalents (Yadev et al. 2011).

In vitro monolayers of oral epithelial cells have been used to evaluate the role of *Candida* filamentation in disease pathogenesis (Fig. 1.2a). This has been achieved by assessing the virulence of *C. albicans* strains and their mutants through cellular adhesion and/or proinflammatory cytokine responses (Villar et al. 2004; Dongari-Bagtzoglou and Kashleva 2003). Others have reported unique cellular damage and transcriptional profiles in oral epithelial cells following exposure to *Candida* species, including *C. albicans*, *C. parapsilosis,* and *C. glabrata,* with the former being more inflammatory (Horvath et al. 2021; Schaller et al. 2002). This enhanced virulence of *C. albicans* in the oral cavity was thought to be attributable to its true hyphal form exhibited during tissue invasion, a unique phenotype reduced or absent from other oral-relevant *Candida* species (Turner and Butler 2014; Malinovska et al. 2023). Such pathogenicity has also been linked to the production of the fungal peptide toxin, Candidalysin, with mutant strains lacking the virulence factor incapable of activating or damaging epithelial cells in vitro and in vivo (Moyes et al. 2016). These aforementioned organotypic systems have allowed for assessing the potency of peptide orthologs of Candidalysin identified in other *Candida* species, and their effects on the host (Richardson et al. 2022).

Although extremely effective in preliminary studies, models of epithelial cells grown as monolayers, also termed two-dimensional systems, lack a true representation of the oral mucosa. Multi-cellular layered three-dimensional in vitro culture models have been developed (Fig. 1.2b) to assess the extent of fungal invasion in histologically relevant tissues created in vitro (Dongari-Bagtzoglou and Kashleva 2006; Moharamzadeh et al. 2008). Alternatively, commercially available reconstituted human oral epithelial (RHOE) models exist for investigations into the expression patterns of *C. albicans* genes during tissue invasion and to evaluate the consequences of genetic mutations on pathogenicity (Schaller and Weindl 2009; Schaller et al. 2006). From a host perspective, the transcriptional and proteomic inflammatory response in the RHOE tissue following infection has also been reported (Lu et al. 2006; Huang et al. 2019; Moyes et al. 2014).

A number of model systems also exist for studying different forms of oral candidiasis, including denture stomatitis. Given the complex microbiome of the denture surface in infected individuals (O'donnell et al. 2015; Morse et al. 2019a), most of these models are concerned with investigating microbial interactions and efficacy of antimicrobial therapies using in vitro interkingdom biofilms (Fig. 1.2c). Alongside *C. albicans*, these models can range from containing a few species to up to ten different bacteria (Morse et al. 2019b; Sherry et al. 2016; Brown et al. 2022; Delaney et al. 2019). Undefined biofilm communities developed from denture plaque or saliva also exist (Coulthwaite and Verran 2008; Lamfon et al. 2005). Such biofilms can also be applied to three-dimensional models studying host–pathogen interactions. A study by Morse et al. (2018) studied the effects of a polymicrobial biofilm containing *C. albicans* and four bacterial species (*Streptococcus sanguinis*, *Streptococcus gordonii*, *Actinomyces viscosus,* and *Actinomyces odontolyticus*) on various oral mucosal models, including the RHOE tissue (Morse et al. 2018). Although rudimentary in nature, such tissue models provide excellent testing platforms for preclinical investigations in oral candidiasis and associated Candidal diseases of the oral cavity.

In Vivo Models

Mouse models for oral candidiasis have been well validated and used to study oral candidiasis (Fig. 1.2d). In order to represent the infection, mice are rendered susceptible through immune suppression from subcutaneous cortisone acetate injections, followed by oral inoculation with *Candida* species (Kamai et al. 2001; Sofaer et al. 1982). The resulting infection is representative of acute pseudomembranous candidiasis (Section "Acute Pseudomembranous Candidiasis") seen in humans. This model allows for observation of white lesion characteristics through microscopic analysis, histological examination, and culturing to measure fungal burden in the infected tissues (Solis and Filler 2012). Alternative models exist for investigating oral candidiasis, such as immunodeficient mice lacking functional CD4 T-lymphocyte activity mimicking HIV infection (Vila et al. 2020). For example, the CD4/HIV MutA transgenic model identified the key role played by CD4+ T-lymphocyte subsets in host protection from the disease (De Repentigny et al. 2002). These models have also been effectively used to show that IL-17 and IL-22 are critical for mucosal host defence against oral fungal infections, whereby deficiency in IL-17 signaling and/or absence of TH17 cellular responses increased the susceptibility to oral candidiasis (Goupil et al. 2014; Conti et al. 2009). More recently, a TH17 inflammation model in Rag1−/− deficient mice was created for investigations into the host response during oral *Candida* infection of the lingual tissue. Bhaskaran et al. (2015) demonstrated in this methodology that control of the TH17 T-cell axis through the adoptive transfer of T regulatory cells reduced inflammation in infected mice (Bhaskaran et al. 2015). TH17 responses have since been highlighted as key immunological controllers at mucosal surfaces by maintaining barrier function and providing host immunity to commensal fungi such as *C. albicans* both in vitro and in vivo (Gaffen and Moutsopoulos 2020; Verma et al. 2017).

Mice models have also been used to study *Candida*-bacterial interactions in oral candidiasis. For example, it has been shown in immunocompetent mice that the frequency and the severity of oral thrush lesions were augmented during *C. albicans* and *S. oralis* coinfection when compared to mono-infections (Xu et al. 2014). It was later found that *S. oralis* promoted coaggregation in oral mucosal biofilms by inducing expression of key virulence factors such as *ALS1* genes in *C. albicans*, whilst increasing tissue invasion and dissemination (Xu et al. 2016, 2017). Alternatively, rat models are the recommended experimental animals for denture stomatitis, mainly due to the size difference within the murine models, allowing for the insertion of denture acrylic substrata into their oral cavities (Fig. 1.2e). An intraoral removable denture system that can be cultured with microorganisms has been developed to allow for multiple sampling during longitudinal studies (Johnson et al. 2012; Nett et al. 2010). Although the host response and histological features in the palatal mucosa are similar to humans in these models, they do not come without limitations as discussed in detail elsewhere (Sugio et al. 2020). One obvious limitation is that clinical manifestations in the murine models are associated with acute forms of the disease, not taking into account the chronic features of pathogenesis of various oral candidiasis disease forms. Nevertheless, in vivo and in vitro model systems are indispensable tools for unraveling the complex mechanisms of disease pathogenesis of this fungal disease. These models allow for controlled evaluations of conventional and novel therapeutic strategies, prior to applications in humans.

Final Discussion

In summary, oral candidiasis remains an extremely prevalent and multifaceted disease influenced by a combination of host factors and underlying health conditions. The range of clinical manifestations underpinning the different forms of the disease highlights the importance of early diagnosis and appropriate interventions for patients and clinicians alike. Understanding risk factors, including immunosuppression, denture use, and microbiome imbalances, highlights opportunities for targeted prevention, although most frontline therapies for combating the disease still involve conventional antifungal treatment. Although such therapies remain effective, emerging antifungal resistance and treatment failures call for the application of novel approaches like probiotics, immunotherapies, and biofilm-targeting strategies. Such therapeutics and strategies require further testing with the use of in vitro and in vivo model systems, which provide excellent testing platforms for disease pathogenesis and therapeutic activity. Continued research integrating these perspectives is essential for preclinical testing, with the hope of improving patient outcomes and addressing gaps in disease management following clinical trials.

References

Abusleme L, Diaz PI, Freeman AF, Greenwell-Wild T, Brenchley L, Desai JV, Ng WI, Holland SM, Lionakis MS, Segre JA, Kong HH, Moutsopoulos NM (2018) Human defects in STAT3 promote oral mucosal fungal and bacterial dysbiosis. JCI Insight 3:e122061

Akpan A, Morgan R (2002) Oral candidiasis. Postgrad Med J 78:455–459

Alves A, Lopes BO, Leite A, Cruz GS, Brito EHS, Lima LF, Cernakova L, Azevedo NF, Rodrigues CF (2023) Characterization of Oral Candida spp. biofilms in children and adults carriers from Eastern Europe and South America. Antibiotics (Basel) 12:797

Ambe NF, Longdoh NA, Tebid P, Bobga TP, Nkfusai CN, Ngwa SB, Nsai FS, Cumber SN (2020) The prevalence, risk factors and antifungal sensitivity pattern of oral candidiasis in HIV/AIDS patients in Kumba District Hospital, South West Region, Cameroon. Pan Afr Med J 36:23

Arathoon EG, Gotuzzo E, Noriega LM, Berman RS, Dinubile MJ, Sable CA (2002) Randomized, double-blind, multicenter study of caspofungin versus amphotericin B for treatment of oropharyngeal and esophageal candidiases. Antimicrob Agents Chemother 46:451–457

Arendorf TM, Walker DM (1987) Denture stomatitis: a review. J Oral Rehabil 14:217–227

Arendorf TM, Walker DM, Kingdom RJ, Roll JR, Newcombe RG (1983) Tobacco smoking and denture wearing in oral candidal leukoplakia. Br Dent J 155:340–343

Baughman R (1971) Median rhomboid glossitis: a developmental anomaly? Oral Surg Oral Med Oral Pathol 31:56–65

Baumgardner DJ (2019) Oral fungal microbiota: to thrush and beyond. J Patient Cent Res Rev 6:252–261

Belazi M, Velegraki A, Fleva A, Gidarakou I, Papanaum L, Baka D, Daniilidou N, Karamitsos D (2005) Candidal overgrowth in diabetic patients: potential predisposing factors. Mycoses 48:192–196

Bhaskaran N, Weinberg A, Pandiyan P (2015) Th17 inflammation model of oropharyngeal candidiasis in immunodeficient mice. J Vis Exp (96):52538

Blagec T, Glavina A, Spiljak B, Beslic I, Bulat V, Lugovic-Mihic L (2023) Cheilitis: a cross-sectional study-multiple factors involved in the aetiology and clinical features. Oral Dis 29:3360–3371

Brown JL, Young T, Mckloud E, Butcher MC, Bradshaw D, Pratten JR, Ramage G (2022) An in vitro evaluation of denture cleansing regimens against a polymicrobial denture biofilm model. Antibiotics (Basel) 11:113

Cabras M, Gambino A, Broccoletti R, Lodi G, Arduino PG (2020) Treatment of angular cheilitis: a narrative review and authors' clinical experience. Oral Dis 26:1107–1115

Chattopadhyay A, Patton LL (2013) Smoking as a risk factor for oral candidiasis in HIV-infected adults. J Oral Pathol Med 42:302–308

Chevalier M, Ranque S, Precheur I (2018) Oral fungal-bacterial biofilm models in vitro: a review. Med Mycol 56:653–667

Chiriac A, Chiriac AE, Pinteala T, Spinei A, Savin L, Zelenkova H, Wollina U (2024) Angular cheilitis-an oral disease with many facets. Wien Med Wochenschr 174(15):315–322

Chouhan S, Kallianpur S, Prabhu KT, Tijare M, Kasetty S, Gupta S (2019) Candidal prevalence in diabetics and its species identification. Int J Appl Basic Med Res 9:49–54

Contaldo M, Di Stasio D, Romano A, Fiori F, Della Vella F, Rupe C, Lajolo C, Petruzzi M, Serpico R, Lucchese A (2023) Oral candidiasis and novel therapeutic strategies: antifungals, phytotherapy, probiotics, and photodynamic therapy. Curr Drug Deliv 20:441–456

Conti HR, Shen F, Nayyar N, Stocum E, Sun JN, Lindemann MJ, Ho AW, Hai JH, Yu JJ, Jung JW, Filler SG, Masso-Welch P, Edgerton M, Gaffen SL (2009) Th17 cells and Il-17 receptor signaling are essential for mucosal host defense against oral candidiasis. J Exp Med 206:299–311

Coulthwaite L, Verran J (2008) Development of an in vitro denture plaque biofilm to model denture malodour. J Breath Res 2:017004

Darwazeh AMG, Darwazeh TA (2014) What makes oral candidiasis recurrent infection? A clinical view. J Mycol 2014:758394

Darwazeh AMG, Macfarlane TW, Mccuish A, Lamey P-J (1991) Mixed salivary glucose levels and candidal carriage in patients with diabetes mellitus. J Oral Pathol Med 20:280–283

De Repentigny L, Aumont F, Ripeau J-S, Fiorillo M, Kay DG, Hanna Z, Jolicoeur P (2002) Mucosal candidiasis in transgenic mice expressing human immunodeficiency virus type 1. J Infect Dis 185:1103–1114

Dekhuijzen PNR, Batsiou M, Bjermer L, Bosnic-Anticevich S, Chrystyn H, Papi A, Rodríguez-Roisin R, Fletcher M, Wood L, Cifra A, Soriano JB, Price DB (2016) Incidence of oral thrush in patients with COPD prescribed inhaled corticosteroids: Effect of drug, dose, and device. Respir Med 120:54–63

Delaney C, O'donnell LE, Kean R, Sherry L, Brown JL, Calvert G, Nile CJ, Cross L, Bradshaw DJ, Brandt BW, Robertson D, Ramage G (2019) Interkingdom interactions on the denture surface: Implications for oral hygiene. Biofilms 1:100002

Desai JV, Lionakis MS (2018) The role of neutrophils in host defense against invasive fungal infections. Curr Clin Microbiol Rep 5:181–189

Diaz PI, Hong BY, Dupuy AK, Choquette L, Thompson A, Salner AL, Schauer PK, Hegde U, Burleson JA, Strausbaugh LD, Peterson DE, Dongari-Bagtzoglou A (2019) Integrated Analysis of Clinical and Microbiome Risk Factors Associated with the Development of Oral Candidiasis during Cancer Chemotherapy. J Fungi (Basel) 13;5(2):49

Dockrell DH, O'shea D, Cartledge JD, Freedman AR (2019) British HIV Association guidelines on the management of opportunistic infection in people living with HIV: the clinical management of Candidiasis 2019. HIV Med 20(Suppl 8):2–24
Dongari-Bagtzoglou A, Kashleva H (2003) Candida albicans triggers interleukin-8 secretion by oral epithelial cells. Microb Pathog 34:169–177
Dongari-Bagtzoglou A, Kashleva H (2006) Development of a novel three-dimensional in vitro model of oral Candida infection. Microb Pathog 40:271–278
Dropulic LK, Lederman HM (2016) Overview of infections in the immunocompromised host. Microbiol Spectr 4. https://doi.org/10.1128/microbiolspec.DMIH2-0026-2016
Ellepola A, Samaranayake L (2001) Inhalational and topical steroids, and oral candidosis: a mini review. Oral Dis 7:211–216
Erfaninejad M, Zarei Mahmoudabadi A, Maraghi E, Hashemzadeh M, Fatahinia M (2022) Epidemiology, prevalence, and associated factors of oral candidiasis in HIV patients from Southwest Iran in post-highly active antiretroviral therapy era. Front Microbiol 13:983348
Farah CS, Elahi S, Drysdale K, Pang G, Gotjamanos T, Seymour GJ, Clancy RL, Ashman RB (2002) Primary role for CD4+ T lymphocytes in recovery from oropharyngeal candidiasis. Infect Immun 70:724–731
Figueiral MH, Azul A, Pinto E, Fonseca PA, Branco FM, Scully C (2007) Denture-related stomatitis: identification of aetiological and predisposing factors – a large cohort. J Oral Rehabil 34:448–455
Fotos PG, Hellstein JW (1992) Candida and candidosis. Epidemiology, diagnosis and therapeutic management. Dent Clin N Am 36:857–878
Fujinami W, Nishikawa K, Ozawa S, Hasegawa Y, Takebe J (2021) Correlation between the relative abundance of oral bacteria and Candida albicans in denture and dental plaques. J Oral Biosci 63:175–183
Fukushima C, Matsuse H, Saeki S, Kawano T, Machida I, Kondo Y, Kohno S (2005) Salivary IgA and oral candidiasis in asthmatic patients treated with inhaled corticosteroid. J Asthma 42:601–604
Gaffen SL, Moutsopoulos NM (2020) Regulation of host-microbe interactions at oral mucosal barriers by type 17 immunity. Sci Immunol 5:eaau4594
Gamil Y, Hamed MG, Elsayed M, Essawy A, Medhat S, Zayed SO, Ismail RM (2024) The antifungal effect of miconazole and miconazole-loaded chitosan nanoparticles gels in diabetic patients with Oral candidiasis-randomized control clinical trial and microbiological analysis. BMC Oral Health 24:196
Garcia-Cuesta C, Sarrion-Pérez M-G, Bagán JV (2014) Current treatment of oral candidiasis: a literature review. J Clin Exp Dent 6:e576
Gaurav S, Keerthilatha PM, Archna N (2011) Prevalence of oral manifestations and their association with CD4/CD8 Ratio and HIV viral load in South India. Int J Dent 2011:964278
Gendreau L, Loewy ZG (2011) Epidemiology and etiology of denture stomatitis. J Prosthodont 20:251–260
Ghannoum MA, Rice LB (1999) Antifungal agents: mode of action, mechanisms of resistance, and correlation of these mechanisms with bacterial resistance. Clin Microbiol Rev 12:501–517
Gharibpour F, Shirban F, Bagherniya M, Nosouhian M, Sathyapalan T, Sahebkar A (2021) The effects of nutraceuticals and herbal medicine on Candida albicans in oral candidiasis: a comprehensive review. Adv Exp Med Biol 1308:225–248
Goregen M, Miloglu O, Buyukkurt MC, Caglayan F, Aktas AE (2011) Median rhomboid glossitis: a clinical and microbiological study. Eur J Dent 5:367–372
Goupil M, Cousineau-Côté V, Aumont F, Sénéchal S, Gaboury L, Hanna Z, Jolicoeur P, De Repentigny L (2014) Defective Il-17-and Il-22-dependent mucosal host response to Candida albicans determines susceptibility to oral candidiasis in mice expressing the Hiv-1 transgene. BMC Immunol 15:1–13
Hannah VE, O'donnell L, Robertson D, Ramage G (2017) Denture stomatitis: causes, cures and prevention. Prim Dent J 6:46–51

Hatakka K, Ahola AJ, Yli-Knuuttila H, Richardson M, Poussa T, Meurman JH, Korpela R (2007) Probiotics reduce the prevalence of oral Candida in the elderly—a randomized controlled trial. J Dent Res 86:125–130

He Y, Liu J, Chen Y, Yan L, Wu J (2022) Neutrophil extracellular traps in Candida albicans infection. Front Immunol 13:913028

Hellstein JW, Marek CL (2019) Candidiasis: red and white manifestations in the oral cavity. Head Neck Pathol 13:25–32

Horvath M, Nagy G, Zsindely N, Bodai L, Horvath P, Vagvolgyi C, Nosanchuk JD, Toth R, Gacser A (2021) Oral epithelial cells distinguish between Candida species with high or low pathogenic potential through microRNA regulation. mSystems 6

Hu L, He C, Zhao C, Chen X, Hua H, Yan Z (2019a) Characterization of oral candidiasis and the Candida species profile in patients with oral mucosal diseases. Microb Pathog 134:103575

Hu L, Zhou M, Young A, Zhao W, Yan Z (2019b) In vivo effectiveness and safety of probiotics on prophylaxis and treatment of oral candidiasis: a systematic review and meta-analysis. BMC Oral Health 19:1–12

Hu Q, Li T, Yang J, Peng Y, Liu Q, Liu N (2023) Efficacy of photodynamic therapy in the treatment of oral candidiasis: a systematic review and meta-analysis. BMC Oral Health 23:802

Huang F, Song Y, Chen W, Liu Q, Wang Q, Liu W, Wang X, Wang W (2019) Effects of Candida albicans infection on defense effector secretion by human oral mucosal epithelial cells. Arch Oral Biol 103:55–61

Humbert L, Cornu M, Proust-Lemoine E, Bayry J, Wemeau JL, Vantyghem MC, Sendid B (2018) Chronic mucocutaneous candidiasis in autoimmune polyendocrine syndrome type 1. Front Immunol 9:2570

Inacio Silveira DQ, Lia EN, Massignan C, Stefani CM (2023) Natural products for the treatment of denture stomatitis: a systematic review. J Prosthet Dent 129:413–424

Ishikawa KH, Mayer MP, Miyazima TY, Matsubara VH, Silva EG, Paula CR, Campos TT, Nakamae AE (2015) A multispecies probiotic reduces oral Candida colonization in denture wearers. J Prosthodont 24:194–199

Janus MM, Crielaard W, Volgenant CM, Van Der Veen MH, Brandt BW, Krom BP (2017) Candida albicans alters the bacterial microbiome of early in vitro oral biofilms. J Oral Microbiol 9:1270613

Javed F, Samaranayake LP, Romanos GE (2014) Treatment of oral fungal infections using antimicrobial photodynamic therapy: a systematic review of currently available evidence. Photochem Photobiol Sci 13:726–734

John HA, Ahuja K, Dakhale R, Heda K, Sedani S (2023) Median rhomboid glossitis: a developmental disorder involving the central part of the tongue. Cureus 15:e48908

Johnson CC, Yu A, Lee H, Fidel PL Jr, Noverr MC (2012) Development of a contemporary animal model of Candida albicans-associated denture stomatitis using a novel intraoral denture system. Infect Immun 80:1736–1743

Kamai Y, Kubota M, Kamai Y, Hosokawa T, Fukuoka T, Filler SG (2001) New model of oropharyngeal candidiasis in mice. Antimicrob Agents Chemother 45:3195–3197

Keyvanfar A, Najafiarab H, Talebian N, Tafti MF, Adeli G, Ghasemi Z, Tehrani S (2024) Drug-resistant oral candidiasis in patients with Hiv infection: a systematic review and meta-analysis. BMC Infect Dis 24:546

Kirkpatrick CH (1994) Chronic mucocutaneous candidiasis. J Am Acad Dermatol 31:S14–S17

Kraft-Bodi E, Jørgensen M, Keller M, Kragelund C, Twetman S (2015) Effect of probiotic bacteria on oral Candida in frail elderly. J Dent Res 94:181S–186S

Kraneveld EA, Buijs MJ, Bonder MJ, Visser M, Keijser BJF, Crielaard W, Zaura E (2012) The relation between oral Candida load and bacterial microbiome profiles in Dutch older adults. PLoS One 7:e42770

Kumar BV, Padshetty NS, Bai KY, Rao MS (2005) Prevalence of Candida in the oral cavity of diabetic subjects. J Assoc Physicians India 53:599–602

Lalla RV, Latortue MC, Hong CH, Ariyawardana A, D'amato-Palumbo S, Fischer DJ, Martof A, Nicolatou-Galitis O, Patton LL, Elting LS, Spijkervet FK, Brennan MT (2010) A systematic review of oral fungal infections in patients receiving cancer therapy. Support Care Cancer 18:985–992

Lamfon H, Al-Karaawi Z, Mccullough M, Porter SR, Pratten J (2005) Composition of in vitro denture plaque biofilms and susceptibility to antifungals. FEMS Microbiol Lett 242:345–351

Last A, Maurer M, Mosig AS, Gresnigt MS, Hube B (2021) In vitro infection models to study fungal–host interactions. FEMS Microbiol Rev 45:fuab005

Lee Muñoz X, Vergara Núñez C, Jerez Fuentes J, Lozano CP (2023) Medium-chain fatty acids as an alternative treatment for denture stomatitis in older people. Clin Oral Investig 27:3713–3720

Lee X, Vergara C, Lozano CP (2019) Severity of Candida-associated denture stomatitis is improved in institutionalized elders who consume Lactobacillus rhamnosus Sp1. Aust Dent J 64:229–236

Lewis MAO, Williams DW (2017) Diagnosis and management of oral candidosis. Br Dent J 223:675–681

Li L, Finnegan MB, Ozkan S, Kim Y, Lillehoj PB, Ho CM, Lux R, Mito R, Loewy Z, Shi W (2010) In vitro study of biofilm formation and effectiveness of antimicrobial treatment on various dental material surfaces. Mol Oral Microbiol 25:384–390

Li D, Li Q, Liu C, Lin M, Li X, Xiao X, Zhu Z, Gong Q, Zhou H (2014) Efficacy and safety of probiotics in the treatment of C andida-associated stomatitis. Mycoses 57:141–146

Lortholary O, Petrikkos G, Akova M, Arendrup MC, Arikan-Akdagli S, Bassetti M, Bille J, Calandra T, Castagnola E, Cornely OA, Cuenca-Estrella M, Donnelly JP, Garbino J, Groll AH, Herbrecht R, Hope WW, Jensen HE, Kullberg BJ, Lass-Flörl C, Meersseman W, Richardson MD, Roilides E, Verweij PE, Viscoli C, Ullmann AJ (2012) ESCMID* *This guideline was presented in part at ECCMID 2011. European Society for Clinical Microbiology and Infectious Diseases. guideline for the diagnosis and management of Candida diseases 2012: patients with Hiv infection or Aids. Clin Microbiol Infect 18:68–77

Low A, Gavriilidis G, Larke N, B-Lajoie MR, Drouin O, Stover J, Muhe L, Easterbrook P (2016) Incidence of opportunistic infections and the impact of antiretroviral therapy among HIV-infected adults in low- and middle-income countries: a systematic review and meta-analysis. Clin Infect Dis 62:1595–1603

Lu SY (2021) Oral candidosis: pathophysiology and best practice for diagnosis, classification, and successful management. J Fungi (Basel) 7:555

Lu Q, Jayatilake JA, Samaranayake LP, Jin L (2006) Hyphal invasion of Candida albicans inhibits the expression of human beta-defensins in experimental oral candidiasis. J Invest Dermatol 126:2049–2056

Lugović-Mihić L, Pilipović K, Crnarić I, Šitum M, Duvančić T (2018) Differential diagnosis of cheilitis – how to classify cheilitis? Acta Clin Croat 57:342–351

Lyu X, Zhao C, Yan Z-M, Hua H (2016) Efficacy of nystatin for the treatment of oral candidiasis: a systematic review and meta-analysis. In: Drug Des Devel Ther. 2016 Mar 16:10:1161–71

Lyu X, Zheng H, Wang X, Zhang H, Gao L, Xun Z, Zhang Q, He X, Hua H, Yan Z, Chen F (2021) Oral microbiota composition and function changes during chronic erythematous candidiasis. Front Cell Infect Microbiol 11:691092

Malinovska Z, Conkova E, Vaczi P (2023) Biofilm formation in medically important Candida species. J Fungi (Basel) 9:955

Martinez RF, Jaimes-Aveldanez A, Hernandez-Perez F, Arenas R, Miguel GF (2013) Oral Candida spp. carriers: its prevalence in patients with type 2 diabetes mellitus. An Bras Dermatol 88:222–225

Mccormack PL, Perry CM (2005) Caspofungin: a review of its use in the treatment of fungal infections. Drugs 65:2049–2068

Millsop JW, Fazel N (2016) Oral candidiasis. Clin Dermatol 34:487–494

Mohamed AA, Lu X-L, Mounmin FA (2019) Diagnosis and treatment of esophageal candidiasis: current updates. Can J Gastroenterol Hepatol 2019:3585136

Mohammadi F, Javaheri MR, Nekoeian S, Dehghan P (2016) Identification of Candida species in the oral cavity of diabetic patients. Curr Med Mycol 2:1–7
Mohammed H, Fatalla AA (2020) The effectiveness of chitosan nano-particles addition into soft denture lining material on Candida albicans adherence. Pak J Med Health Sci 14:3
Moharamzadeh K, Brook I, Van Noort R, Scutt A, Smith K, Thornhill M (2008) Development, optimization and characterization of a full-thickness tissue engineered human oral mucosal model for biological assessment of dental biomaterials. J Mater Sci Mater Med 19:1793–1801
Mora-Duarte J, Betts R, Rotstein C, Colombo AL, Thompson-Moya L, Smietana J, Lupinacci R, Sable C, Kartsonis N, Perfect J, Caspofungin Invasive Candidiasis Study, G (2002) Comparison of caspofungin and amphotericin B for invasive candidiasis. N Engl J Med 347:2020–2029
Morse DJ, Wilson MJ, Wei X, Lewis MAO, Bradshaw DJ, Murdoch C, Williams DW (2018) Denture-associated biofilm infection in three-dimensional oral mucosal tissue models. J Med Microbiol 67:364–375
Morse DJ, Smith A, Wilson MJ, Marsh L, White L, Posso R, Bradshaw DJ, Wei X, Lewis MAO, Williams DW (2019a) Molecular community profiling of the bacterial microbiota associated with denture-related stomatitis. Sci Rep 9:10228
Morse DJ, Wilson MJ, Wei X, Bradshaw DJ, Lewis MAO, Williams DW (2019b) Modulation of Candida albicans virulence in in vitro biofilms by oral bacteria. Lett Appl Microbiol 68:337–343
Moyes DL, Shen C, Murciano C, Runglall M, Richardson JP, Arno M, Aldecoa-Otalora E, Naglik JR (2014) Protection against epithelial damage during Candida albicans infection is mediated by Pi3K/Akt and mammalian target of rapamycin signaling. J Infect Dis 209:1816–1826
Moyes DL, Wilson D, Richardson JP, Mogavero S, Tang SX, Wernecke J, Hofs S, Gratacap RL, Robbins J, Runglall M, Murciano C, Blagojevic M, Thavaraj S, Forster TM, Hebecker B, Kasper L, Vizcay G, Iancu SI, Kichik N, Hader A, Kurzai O, Luo T, Kruger T, Kniemeyer O, Cota E, Bader O, Wheeler RT, Gutsmann T, Hube B, Naglik JR (2016) Candidalysin is a fungal peptide toxin critical for mucosal infection. Nature 532:64–68
Mun M, Yap T, Alnuaimi A, Adams G, Mccullough M (2016) Oral candidal carriage in asymptomatic patients. Aust Dent J 61:190–195
Mundula T, Ricci F, Barbetta B, Baccini M, Amedei A (2019) Effect of probiotics on oral candidiasis: a systematic review and meta-analysis. Nutrients 11:2449
Namangkalakul W, Benjavongkulchai S, Pochana T, Promchai A, Satitviboon W, Howattanapanich S, Phuprasong R, Ungvijanpunya N, Supakanjanakanti D, Chaitrakoonthong T (2020) Activity of chitosan antifungal denture adhesive against common Candida species and Candida albicans adherence on denture base acrylic resin. J Prosthet Dent 123:181. e1–181. e7
Narayanan D, Rogge M (2024) Cheilitis: a diagnostic algorithm and review of underlying etiologies. Dermatitis 35(5):431–442
Netea MG, Joosten LA, Van Der Meer JW, Kullberg BJ, Van De Veerdonk FL (2015) Immune defence against Candida fungal infections. Nat Rev Immunol 15:630–642
Nett JE, Marchillo K, Spiegel CA, Andes DR (2010) Development and validation of an in vivo Candida albicans biofilm denture model. Infect Immun 78:3650–3659
Newton A (1962) Denture sore mouth. A possible etiology. Br Dent J 112:357–360
O'donnell LE, Robertson D, Nile CJ, Cross LJ, Riggio M, Sherriff A, Bradshaw D, Lambert M, Malcolm J, Buijs MJ, Zaura E, Crielaard W, Brandt BW, Ramage G (2015) The oral microbiome of denture wearers is influenced by levels of natural dentition. PLoS One 10:e0137717
O'donnell LE, Smith K, Williams C, Nile CJ, Lappin DF, Bradshaw D, Lambert M, Robertson DP, Bagg J, Hannah V, Ramage G (2016) Dentures are a reservoir for respiratory pathogens. J Prosthodont 25:99–104
Ohman SC, Dahlen G, Moller A, Ohman A (1986) Angular cheilitis: a clinical and microbial study. J Oral Pathol 15:213–217
Pandarathodiyil AK, Anil S, Vijayan SP (2021) Angular cheilitis-an updated overview of the etiology, diagnosis, and management. Int J Dent Oral Sci 8:1433–1438
Pappas PG, Kauffman CA, Andes DR, Clancy CJ, Marr KA, Ostrosky-Zeichner L, Reboli AC, Schuster MG, Vazquez JA, Walsh TJ, Zaoutis TE, Sobel JD (2015) Clinical practice guideline

for the management of candidiasis: 2016 update by the Infectious Diseases Society of America. Clin Infect Dis 62:e1–e50

Passariello C, Di Nardo F, Polimeni A, Di Nardo D, Testarelli L (2020) Probiotic Streptococcus salivarius Reduces Symptoms of Denture Stomatitis and Oral Colonization by Candida albicans. Appl Sci 10:3002

Patil S, Rao RS, Majumdar B, Anil S (2015) Clinical appearance of oral Candida infection and therapeutic strategies. Front Microbiol 6:1391

Pedraza-Sanchez S, Mendez-Leon JI, Gonzalez Y, Ventura-Ayala ML, Herrera MT, Lezana-Fernandez JL, Bellanti JA, Torres M (2018) Oral administration of human polyvalent IgG by mouthwash as an adjunctive treatment of chronic oral candidiasis. Front Immunol 9:2956

Perić M, Miličić B, Kuzmanović Pfićer J, Živković R, Arsić Arsenijević V (2024) A systematic review of denture stomatitis: predisposing factors, clinical features, etiology, and global Candida spp. Distribution. J Fungi 10:328

Pierce CG, Chaturvedi AK, Lazzell AL, Powell AT, Saville SP, Mchardy SF, Lopez-Ribot JL (2015) A novel small molecule inhibitor of Candida albicans biofilm formation, filamentation and virulence with low potential for the development of resistance. NPJ Biofilms Microbiomes 1:15012

Pristov K, Ghannoum M (2019) Resistance of Candida to azoles and echinocandins worldwide. Clin Microbiol Infect 25:792–798

Quindós G, Gil-Alonso S, Marcos-Arias C, Sevillano E, Mateo E, Jauregizar N, Eraso E (2019) Therapeutic tools for oral candidiasis: current and new antifungal drugs. Med Oral Patol Oral Cir Bucal 24:e172

Rachelefsky GS, Li H, Liao N, Liao Y, Combates NJ, Faruqi R, Varghese ST (2006) Inhaled corticosteroids and the risk of oropharyngeal adverse events: results from a meta-analysis. J Allergy Clin Immunol 117:S92

Rafat Z, Sasani E, Salimi Y, Hajimohammadi S, Shenagari M, Roostaei D (2021) The prevalence, etiological agents, clinical features, treatment, and diagnosis of HIV-associated oral candidiasis in pediatrics across the world: a systematic review and meta-analysis. Front Pediatr 9:805527

Rai A, Misra SR, Panda S, Sokolowski G, Mishra L, Das R, Lapinska B (2022) Nystatin effectiveness in oral candidiasis treatment: a systematic review & meta-analysis of clinical trials. Life (Basel) 12:1677

Ramage G, Tomsett K, Wickes BL, Lopez-Ribot JL, Redding SW (2004) Denture stomatitis: a role for Candida biofilms. Oral Surg Oral Med Oral Pathol Oral Radiol Endod 98:53–59

Ramla S, Sharma V, Patel M (2016) Influence of cancer treatment on the Candida albicans isolated from the oral cavities of cancer patients. Support Care Cancer 24:2429–2436

Rautemaa R, Hietanen J, Niissalo S, Pirinen S, Perheentupa J (2007) Oral and oesophageal squamous cell carcinoma—a complication or component of autoimmune polyendocrinopathy-candidiasis-ectodermal dystrophy (APECED, APS-I). Oral Oncol 43:607–613

Richardson JP, Brown R, Kichik N, Lee S, Priest E, Mogavero S, Maufrais C, Wickramasinghe DN, Tsavou A, Kotowicz NK, Hepworth OW, Gallego-Cortes A, Ponde NO, Ho J, Moyes DL, Wilson D, D'enfert C, Hube B, Naglik JR (2022) Candidalysins are a new family of cytolytic fungal peptide toxins. MBio 13:e0351021

Rodrigues CF, Rodrigues ME, Henriques M (2019) Candida sp. infections in patients with diabetes mellitus. J Clin Med 8:76

Romo JA, Pierce CG, Chaturvedi AK, Lazzell AL, Mchardy SF, Saville SP, Lopez-Ribot JL (2017) Development of anti-virulence approaches for candidiasis via a novel series of small-molecule inhibitors of Candida albicans filamentation. MBio 8

Samaranayake LP, Keung Leung W, Jin L (2009) Oral mucosal fungal infections. Periodontol 49:39–59

Sartawi SY, Abu-Hammad S, Salim NA, Al-Omoush S (2021) Denture stomatitis revisited: a summary of systematic reviews in the past decade and two case reports of papillary hyperplasia of unusual locations. Int J Dent 2021:7338143

Schaller M, Weindl G (2009) Models of oral and vaginal candidiasis based on in vitro reconstituted human epithelia for the study of host-pathogen interactions. Methods Mol Biol 470:327–345

Schaller M, Mailhammer R, Grassl G, Sander CA, Hube B, Korting HC (2002) Infection of human oral epithelia with Candida species induces cytokine expression correlated to the degree of virulence. J Invest Dermatol 118:652–657
Schaller M, Zakikhany K, Naglik JR, Weindl G, Hube B (2006) Models of oral and vaginal candidiasis based on in vitro reconstituted human epithelia. Nat Protoc 1:2767–2773
Semlali A, Killer K, Alanazi H, Chmielewski W, Rouabhia M (2014) Cigarette smoke condensate increases C. albicans adhesion, growth, biofilm formation, and EAP1, HWP1 and SAP2 gene expression. BMC Microbiol 14:61
Sharon V, Fazel N (2010) Oral candidiasis and angular cheilitis. Dermatol Ther 23:230–242
Sherry L, Lappin G, O'donnell LE, Millhouse E, Millington OR, Bradshaw DJ, Axe AS, Williams C, Nile CJ, Ramage G (2016) Viable compositional analysis of an eleven species oral polymicrobial biofilm. Front Microbiol 7:912
Shirasuna K (2014) Oral lichen planus: malignant potential and diagnosis. Oral Sci Int 11:1–7
Sofaer JA, Holbrook WP, Southam JC (1982) Experimental oral infection with the yeast Candida albicans in mice with or without inherited iron-deficiency anaemia (sla). Arch Oral Biol 27:497–503
Solis NV, Filler SG (2012) Mouse model of oropharyngeal candidiasis. Nat Protoc 7:637–642
Soysa N, Ellepola A (2005) The impact of cigarette/tobacco smoking on oral candidosis: an overview. Oral Dis 11:268–273
Soysa N, Samaranayake L, Ellepola A (2008) Antimicrobials as a contributory factor in oral candidosis – a brief overview. Oral Dis 14:138–143
Sugio CYC, Garcia A, Albach T, Moraes GS, Bonfante EA, Urban VM, Neppelenbroek KH (2020) Candida-associated denture stomatitis and murine models: what is the importance and scientific evidence? J Fungi (Basel) 6:70
Tatapudi R, Abdul Samad SK, Manyam R, Dasari D, Lakshmi RV (2021) Efficacy of curcumin in the treatment of denture stomatitis: a randomized double-blind study. J Oral Maxillofac Pathol 25:286–291
Taverne-Ghadwal L, Kuhns M, Buhl T, Schulze MH, Mbaitolum WJ, Kersch L, Weig M, Bader O, Groß U (2022) Epidemiology and prevalence of oral candidiasis in HIV patients from chad in the Post-Haart era. Front Microbiol 13:844069
Thompson GR 3rd, Patel PK, Kirkpatrick WR, Westbrook SD, Berg D, Erlandsen J, Redding SW, Patterson TF (2010) Oropharyngeal candidiasis in the era of antiretroviral therapy. Oral Surg Oral Med Oral Pathol Oral Radiol Endod 109:488–495
Turner SA, Butler G (2014) The Candida pathogenic species complex. Cold Spring Harb Perspect Med 4:a019778
Van Boven JFM, De Jong-Van Den Berg LTW, Vegter S (2013) Inhaled corticosteroids and the occurrence of oral candidiasis: a prescription sequence symmetry analysis. Drug Saf 36:231–236
Van De Veerdonk FL, Netea MG, Joosten LA, Van Der Meer JW, Kullberg BJ (2010) Novel strategies for the prevention and treatment of Candida infections: the potential of immunotherapy. FEMS Microbiol Rev 34:1063–1075
Verma AH, Richardson JP, Zhou C, Coleman BM, Moyes DL, Ho J, Huppler AR, Ramani K, Mcgeachy MJ, Mufazalov IA, Waisman A, Kane LP, Biswas PS, Hube B, Naglik JR, Gaffen SL (2017) Oral epithelial cells orchestrate innate type 17 responses to Candida albicans through the virulence factor candidalysin. Sci Immunol 2:eaam8834
Vila T, Sultan AS, Montelongo-Jauregui D, Jabra-Rizk MA (2020) Oral candidiasis: a disease of opportunity. J Fungi (Basel) 6:15
Villar C, Kashleva H, Dongari-Bagtzoglou A (2004) Role of Candida albicans polymorphism in interactions with oral epithelial cells. Oral Microbiol Immunol 19:262–269
Westbrook SD, Kirkpatrick WR, Wiederhold NP, Freytes CO, Toro JJ, Patterson TF, Redding SW (2013) Microbiology and epidemiology of oral yeast colonization in hemopoietic progenitor cell transplant recipients. Oral Surg Oral Med Oral Pathol Oral Radiol 115:354–358

Winocur-Arias O, Zlotogorski-Hurvitz A, Ben-Zvi Y, Chaushu G, Edel J, Vered M, Kaplan I (2023) The profile of chronic hyperplastic candidiasis: a clinico-pathological study. Virchows Arch 483:527–534

Xiao Y, Yuan P, Sun Y, Xu Y, Deng X, Wang X, Liu R, Chen Q, Jiang L (2022) Comparison of topical antifungal agents for oral candidiasis treatment: a systematic review and meta-analysis. Oral Surg Oral Med Oral Pathol Oral Radiol 133:282–291

Xu H, Sobue T, Thompson A, Xie Z, Poon K, Ricker A, Cervantes J, Diaz P, Dongari-Bagtzoglou A (2014) Streptococcal co-infection augments C andida pathogenicity by amplifying the mucosal inflammatory response. Cell Microbiol 16:214–231

Xu H, Sobue T, Bertolini M, Thompson A, Dongari-Bagtzoglou A (2016) Streptococcus oralis and Candida albicans synergistically activate mu-calpain to degrade E-cadherin from oral epithelial junctions. J Infect Dis 214:925–934

Xu H, Sobue T, Bertolini M, Thompson A, Vickerman M, Nobile CJ, Dongari-Bagtzoglou A (2017) S. oralis activates the Efg1 filamentation pathway in C. albicans to promote cross-kingdom interactions and mucosal biofilms. Virulence 8:1602–1617

Yadev NP, Murdoch C, Saville SP, Thornhill MH (2011) Evaluation of tissue engineered models of the oral mucosa to investigate oral candidiasis. Microb Pathog 50:278–285

Ye P, Chen W, Huang F, Liu Q, Zhu Y-N, Wang X, Han X-D, Wang W-M (2021) Smoking increases oral mucosa susceptibility to Candida albicans infection via the Nrf2 pathway: In vitro and animal studies. J Cell Mol Med 25:7948–7960

Young T, Alshanta OA, Kean R, Bradshaw D, Pratten J, Williams C, Woodall C, Ramage G, Brown JL (2020) Candida albicans as an essential "Keystone" component within polymicrobial oral biofilm models? Microorganisms 9:59

Zhang W, Wu S, Wang X, Gao Y, Yan Z (2021) Malignant transformation and treatment recommendations of chronic hyperplastic candidiasis-a six-year retrospective cohort study. Mycoses 64:1422–1428

Zomorodian K, Kavoosi F, Pishdad GR, Mehriar P, Ebrahimi H, Bandegani A, Pakshir K (2016) Prevalence of oral Candida colonization in patients with diabetes mellitus. J Mycol Médicale 26:103–110

Chapter 2
Cross-Kingdom Communication: The Interplay Between *Candida albicans* and *Streptococcus mutans*

Zhenting Xiang, Hyun Koo, and Yuan Liu

Abstract Microbial communities in the oral cavity exhibit complex interspecies interactions that significantly influence biofilm dynamics, pathogenesis, and therapeutic outcomes. Among these, the cross-kingdom interaction between *Candida albicans* and *Streptococcus mutans* has drawn increasing attention due to its role in exacerbating dental caries, particularly early childhood caries (ECC) and root caries. This chapter summarizes the synergistic interplay between these two oral pathogens, emphasizing their physical, metabolic, and molecular interactions. Given the intricate dynamics and clinical relevance of *C. albicans–S. mutans* interactions, we also explore emerging therapeutic strategies, ranging from natural compounds and nanomaterials to antimicrobial peptides, probiotics, and phototherapies. A deeper understanding of cross-kingdom communication offers novel insights into managing oral diseases and developing innovative interventions.

Introduction

Fungi are ubiquitous within the human body and are increasingly recognized as common members of the microbiome on nearly all mucosal surfaces (Limon et al. 2017). Within these diverse microbial consortia, fungi coexist with a range of bacterial species. Through these interactions, they modulate both themselves and their

Z. Xiang · Y. Liu (✉)
Oral Health Sciences, Kornberg School of Dentistry, Temple University, Philadelphia, PA, USA
e-mail: liuyuan1@upenn.edu

H. Koo
Biofilm Research Laboratories, Department of Orthodontics, Divisions of Community Oral Health & Pediatric Dentistry, School of Dental Medicine, University of Pennsylvania, Philadelphia, PA, USA

Center for Innovation & Precision Dentistry (CiPD), School of Dental Medicine, School of Engineering and Applied Sciences, University of Pennsylvania, Philadelphia, PA, USA

G. Ramage, D. Andes (eds.), *Fungal Biofilms*, Springer Series on Biofilms 15,
https://doi.org/10.1007/978-3-031-93315-8_2

bacterial counterparts, shaping the overall microbiome and influencing virulence (Peleg et al. 2010; Proctor et al. 2023). In the oral cavity, microbial communities form highly organized structures known as biofilms, which play a pivotal role in both health maintenance and disease pathogenesis (Li et al. 2022a). When ecological perturbations exceed a threshold, interspecies interactions that shape the biofilm community during health become altered, leading to dysbiosis and disease progression (Edlund et al. 2015). Among these interactions, recent studies have highlighted notable relationships between *Candida albicans*, the most abundant opportunistic fungus in the oral mycobiome, and *Streptococcus mutans*, a highly acidogenic and aciduric bacterium that is well established as the primary etiological agent of dental caries, particularly in the context of early childhood caries (ECC), a rampant form of tooth decay affecting children at a young age (Jin et al. 2024). Left untreated, ECC can progressively destroy the tooth's mineralized structures, often leading to pulp involvement, severe pain, and even systemic complications (Anil and Anand 2017).

Their interaction is not merely incidental; rather, it often intensifies cariogenic virulence and modulates tolerance to environmental stressors (Lu et al. 2023). These relationships are multifaceted, involving physical interactions by direct cell–cell contact, chemical interactions through the secretion of small molecules that are often involved in quorum sensing, environmental modifications such as pH changes, the use of metabolic by-products, and alterations in host responses (Peleg et al. 2010). The intricate relationship and unpredictable consequences of their interactions pose great therapeutic challenges in clinical settings, which demand the de novo emergence of potential antimicrobial therapies with multi-targets or combinatorial approaches (Li et al. 2023). This chapter presents a comprehensive summary of the current state of knowledge regarding the interactions between *C. albicans* and *S. mutans*, in terms of physical associations and molecular dialogues, especially considering the therapeutic aspect aimed at intervening in this connection.

Coexistence and Interactions in Oral Biofilms

Clinical and epidemiological evidence strongly supports the co-existence of *S. mutans* and *C. albicans*, particularly in the context of ECC and root caries (Du et al. 2022). Compared to children without caries, those with ECC present significantly higher levels of *C. albicans* in the dental plaque (de Carvalho et al. 2006; Xiao et al. 2018b). Notably, in severe ECC (S-ECC) cases, the elevated fungal load correlates with shifts in bacterial community composition, including an increased abundance of *S. mutans* (Xiao et al. 2018a). Moreover, a greater proportion of *S. mutans* is commonly detected in plaque samples where *C. albicans* is co-isolated in ECC lesions (Sridhar et al. 2020). A similar relationship extends to root caries, another form of severe carious disease. Longitudinal studies have demonstrated that elevated *Candida* levels in saliva are associated with the progression of root caries in older adults (Scheinin et al. 1992; Scheinin et al. 1994). Furthermore, clinical data indicate that individuals with root caries harbor a higher carriage rate of *C. albicans* in plaque, which positively correlates with increased *S. mutans* counts compared to those without root caries (Du et al. 2021).

Collectively, these findings suggest a synergistic relationship between *S. mutans* and *C. albicans*, wherein their co-presence fosters the formation of hypervirulent biofilms. This synergy likely contributes to more aggressive caries progression and underscores the clinical importance of targeting their interactions to prevent or mitigate severe oral disease.

Physical Interaction

The physical interactions between fungal and bacterial cells can range from seemingly disordered polymicrobial assemblages to highly specific, receptor-mediated symbiotic associations. *C. albicans* partners with mitis group streptococci (MGS) species, including *Streptococcus gordonii*, *Streptococcus sanguinis*, and *Streptococcus oralis*, through well-characterized cell wall surface proteins and receptors (Bamford et al. 2015; Bertolini and Dongari-Bagtzoglou 2019). This mode of interaction implies that *S. mutans*, also a member of the oral streptococci community, may employ similar mechanisms for fungal adhesion. Indeed, antigen I/II has been implicated as a key mediator of *S. mutans*–*C. albicans* attachment (Yang et al. 2018). The loss of the *spaP* gene, which codes for antigen I/II in *S. mutans*, markedly reduces fungal colonization in dual-species biofilms, both under in vitro conditions and in vivo models (Yang et al. 2018). Notably, while the interaction of *C. albicans* with *S. gordonii* involves binding of SspB to the fungal Als3 adhesin, Als3 is not strictly necessary for *C. albicans*–*S. mutans* biofilm formation (Silverman et al. 2010; Hoyer et al. 2014). Instead, alternative adhesins, such as Als1 and Hwp1, appear more critical for facilitating the *C. albicans*–*S. mutans* partnership, since dual-species biofilm formation was not enhanced when the ΔΔAls1 or ΔΔHwp1 strains were cultured with *S. mutans* in dual-species biofilms (Martorano-Fernandes et al. 2023). Recent studies have further refined this model by showing that ΔΔEfg1 and ΔΔAls1/ΔΔAls3 deletion led to a severe reduction in the coassembly between *C. albicans* and *S. mutans* and a significant disruption of surface colonization, whereas ΔΔHwp2, ΔΔHyr1, and ΔΔEap1 did not show a significant impact (Ren et al. 2022). It is worth noting that although filamentation and hyphal morphogenesis in *C. albicans* have been validated as crucial for mutualism effects with MGS, *C. albicans* displays distinctive adaptive mechanisms on the tooth surface and develops interactions with *S. mutans* regardless of fungal morphology (Xiang et al. 2023). These findings indicate that *C. albicans* possesses adaptive, multifactorial mechanisms for establishing stable associations with *S. mutans*, ensuring robust colonization and persistence in the complex and ever-changing oral environment.

In contrast to the mechanisms seen with MGS, *S. mutans* employs distinct strategies to associate with *C. albicans*, and this synergistic collaboration is

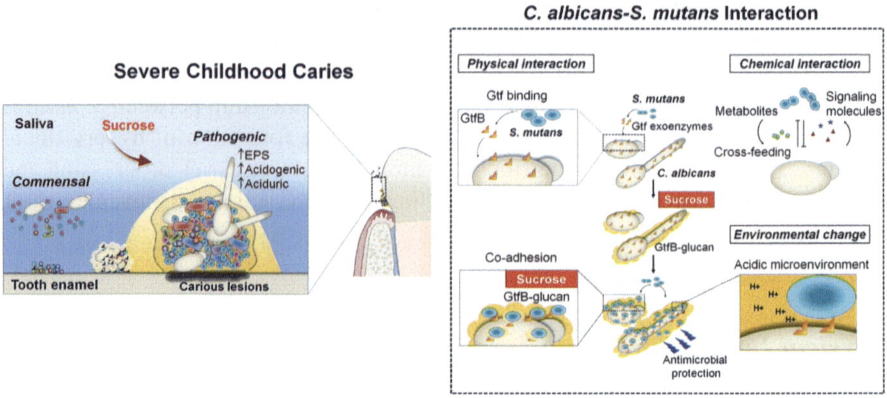

Fig. 2.1 Pathogenic mechanisms of *C. albicans*–*S. mutans* cross-kingdom interactions. (Adapted from Koo et al. (2018))
Host diet (dietary sugars, particularly sucrose) promotes the interactions between *C. albicans* and *S. mutans* by providing a substrate for EPS α-glucans production by streptococcal Gtfs that enhances coadhesion and bacterial–fungal tooth colonization, stimulating cross-kingdom biofilms. This interaction enhances the carriage of the cariogenic pathogen and acid production, while the presence of *C. albicans* increases EPS matrix production (via Gtf induction and fungal-derived EPS) and biofilm acidurícity, resulting in cariogenic conditions on the tooth surface. EPS exopolysaccharides, Gtf glucosyltransferase

predominantly driven by a sucrose-dependent partnership (Fig. 2.1). Glucosyltransferase (Gtf) exoenzymes secreted by *S. mutans* have been observed to bind to the surface of *C. albicans* cells and function to synthesize extracellular glucans. Among them, GtfB has demonstrated the highest affinity for *C. albicans* (Gregoire et al. 2011). Under sucrose-limited conditions, *C. albicans* exhibits nearly twice the adhesion strength to *S. gordonii* compared to *S. mutans*. However, upon glucan production, the binding force between *S. mutans* and *C. albicans* surfaces dramatically increases by approximately sixfold (Wan et al. 2021). Furthermore, atomic force microscopy measurements indicate that the adhesion of GtfB is 2.5 times stronger and 20 times more stable to *C. albicans* cell surfaces than to *S. mutans* (Hwang et al. 2015). This enhanced adhesion allows *C. albicans* cells to serve as critical physical scaffolds for the development of microcolonies of *S. mutans* embedded in the extracellular polysaccharide matrix. The Gtf-mediated binding also facilitates *C. albicans* recruitment into EPS-rich niches, thereby amplifying fungal colonization and contributing to the assembly of complex, three-dimensional inter-kingdom superstructures where *S. mutans* microcolonies intertwine with *C. albicans* yeasts, hyphae, and pseudohyphae (Gregoire et al. 2011; Falsetta et al. 2014; Lu et al. 2023). Likewise, *S. mutans* defective in GtfB is unable to form mixed-species biofilms with *C. albicans*.

While *S. mutans*-derived Gtfs are central to this interaction, fungal-derived β-1,3-glucans also contribute to the stability and architecture of the co-cultured biofilm matrix (Khoury et al. 2020). Moreover, the transcription factor Bcr1, which regulates key cell surface proteins such as Hwp1, Als1, and Als3 in *C. albicans*, is critical for biofilm development. Interestingly, GtfB augments the *C. albicans* counterpart in mixed-species biofilms through a Bcr1-independent mechanism (Ellepola et al. 2017). Further insights by Hwang et al. demonstrated that both the O- and N-terminal structures of mannans on the fungal cell wall play an important role in the direct binding with GtfB, as they observed *C. albicans* mutant strains (ΔΔOch1 and ΔΔPmt4, which exhibit functional defects in the biological synthesis of both N- and O-linked mannans) showed reduced GtfB binding and impaired mixed-biofilm formation with *S. mutans* (Hwang et al. 2017). Meanwhile, mannan-degrading endo- and exo-enzymes decrease the binding force between *S. mutans*-derived GtfB and *C. albicans* by roughly 15-fold, accompanied by modifications in biofilm formation and structure (Kim et al. 2021a).

Beyond simply serving as biological scaffolds for bacterial adhesion, recent evidence indicates that fungi and bacteria can cooperatively coordinate spatiotemporal behaviors that promote disease (Ren et al. 2022). Using multiscale real-time microscopy and computational analysis, researchers observed that aggregates of *C. albicans* and *S. mutans* exhibit emergent properties associated with remarkable group-level surface mobility and disease-promoting functions (Fig. 2.2). The mixed-species assemblages form a highly organized structure in which *S. mutans* clusters are embedded within an intricate network of fungal yeasts, hyphae, and exopolysaccharides. The resulting assemblages demonstrate enhanced surface colonization capabilities. A particularly noteworthy phenomenon is their "forward-leaping motion" directed by fungal filamentation. As hyphae elongate and contact the surface, they effectively "pull" the bacterial clusters along, enabling the bacteria to "hitchhike" and proliferate continuously during migration. Although the underlying molecular mechanism remains to be fully elucidated, these interkingdom assemblages show markedly increased adhesion and resilience, rendering them

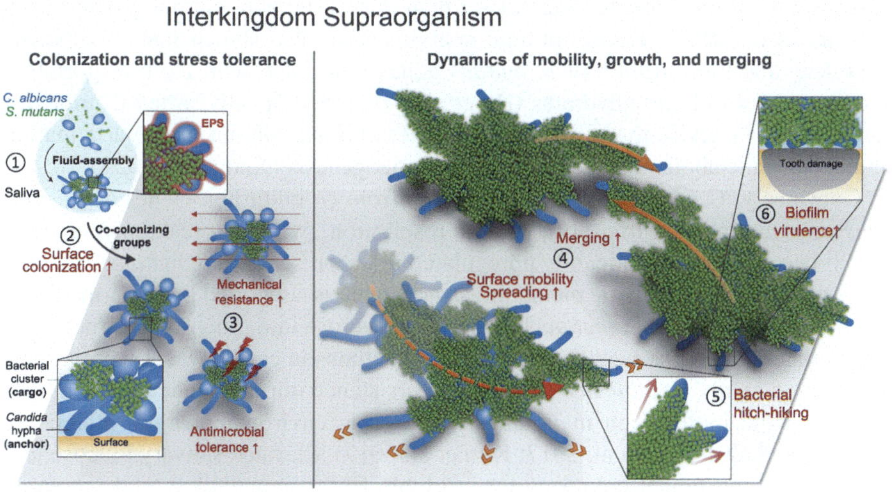

Fig. 2.2 *C. albicans* and *S. mutans* coassemble into interkingdom assemblages in human saliva behaving like supraorganisms with new functionalities and disease-promoting activity. (Adapted from Ren et al. (2022))

significantly more resistant to both mechanical disturbances and antimicrobial interventions (Ren et al. 2022).

Metabolic Interaction

Building upon their spatial and structural synergy, *C. albicans* and *S. mutans* also engage in intricate metabolic exchanges that enhance their collective cariogenic potential. Evidence from rodent models has proven that their co-colonization leads to enhanced bacterial–fungal carriage within plaque biofilm, resulting in the aggressive onset of tooth decay with rampant carious lesions similar to those found in ECC (Falsetta et al. 2014). Moreover, shotgun whole metagenome sequencing analyses reveal that the presence of *S. mutans* and *C. albicans* together induces not only a dysbiotic/cariogenic microbial community but also alters the functional landscape of the biofilm. Notably, pathways involved in amino acid biosynthesis and pyrimidine metabolism become more prominent under sucrose-rich conditions when *C. albicans* and *S. mutans* coexist (Liu et al. 2023). This suggests that interkingdom cross-feeding and metabolic cooperativity fundamentally influence biofilm growth patterns, proliferation rates, and virulence.

Clinical observations further support this notion. Studies have identified a positive correlation between *C. albicans* burden and caries severity in ECC, implying that the fungus may enhance the cariogenic potential of *S. mutans* (Xiao et al. 2018a). It has been shown by an in vitro study that conditioned medium from *S. mutans–C. albicans* biofilm favors its growth, Gtf activity, and microcolony development (Kim et al. 2017). Moreover, *C. albicans* promotes the carbohydrate intake of *S. mutans*, evidenced by the fact that the glucose metabolic rate of *S. mutans* co-cultured with *C. albicans* was higher than that of the pure culture group (Costa Oliveira et al. 2021). Transcriptomic analyses have also shown that *C. albicans* modulates the transcription of *S. mutans* genes associated with the transportation and metabolism of carbohydrates (He et al. 2017). Specifically, when *C. albicans* and *S. mutans* coexist, many upregulated genes of *S. mutans* are those participating in galactose metabolism and glycolysis/gluconeogenesis (Xiao et al. 2023).

Conversely, *C. albicans* similarly benefits from co-culturing. Genes of *C. albicans* involved in sugar transport, aerobic respiration, pyruvate breakdown, and the glyoxylate cycle are significantly upregulated in the presence of *S. mutans* (Ellepola et al. 2019). This mutualistic metabolic arrangement is supported by GtfB-mediated cross-feeding, wherein *S. mutans* breaks down sucrose into glucose and fructose—sugars that *C. albicans* readily metabolizes (Williamson et al. 1993; Ellepola et al. 2019). As a result, *C. albicans* growth and persistence improve under acidic conditions, and the biofilm can maintain its cariogenic environment more effectively (Sztajer et al. 2014; He et al. 2017; Kim et al. 2017). Elevated salivary glucose levels further drive fungal colonization, insoluble EPS accumulation, and increased protein and phospholipase activities, reinforcing the metabolic loop that supports both species (Brito et al. 2021).

Ultimately, this metabolic interplay between *S. mutans* and *C. albicans* leads to the production of lactate, along with small amounts of formate and fumarate as a result of carbohydrate fermentation, facilitating the growth of *S. mutans* and *C. albicans* within cariogenic biofilms, as both organisms are highly aciduric (He et al. 2017; Kim et al. 2017). Such acidogenic synergy exacerbates tooth demineralization and enamel erosion. Intriguingly, the presence of sucrose can also lessen the inhibitory effects of certain sugar alcohols, further favoring biofilm persistence and virulence (Chan et al. 2020). Thus, metabolic cooperation between *C. albicans* and *S. mutans* forms a self-reinforcing cycle, promoting hypervirulent biofilm formation and contributing significantly to the pathogenesis of dental caries.

Molecular Mechanisms

Having established that metabolic exchanges between *C. albicans* and *S. mutans* foster a hypervirulent biofilm environment, it is not surprising that their partnership is also regulated by intricate molecular dialogues. These interactions involve a complex network of extracellular signaling molecules and quorum sensing (QS) systems that can yield both synergistic and antagonistic outcomes (Fig. 2.3).

During the early exponential growth phase, *S. mutans* secretes competence-stimulating peptide (CSP) and the fatty acid signaling molecule trans-2-decenoic acid (SDSF), both of which influence *C. albicans* morphology by inhibiting germ

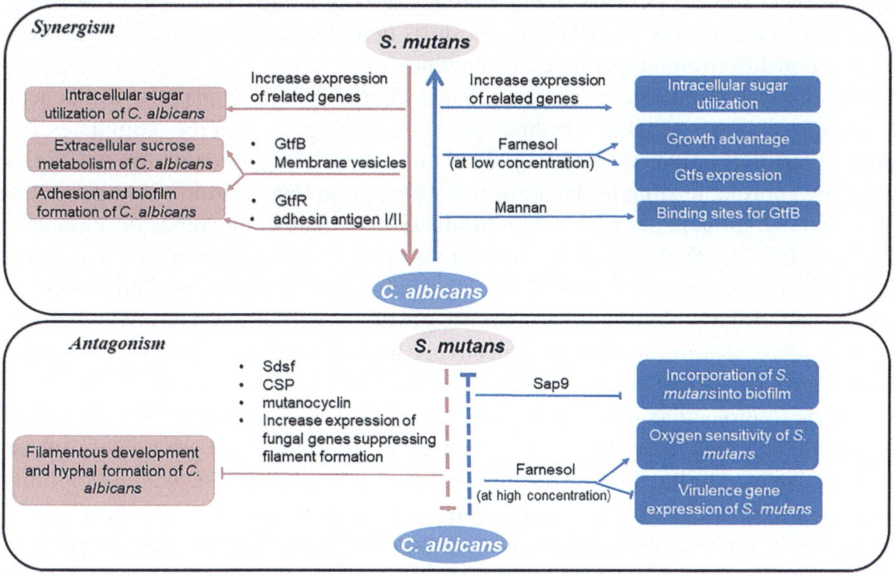

Fig. 2.3 Synergism and antagonism between *C. albicans* and *S. mutans*. (Adapted from Lu et al. (2023))

tube formation, thus favoring fungal proliferation in yeast form (Jarosz et al. 2009; Vílchez et al. 2010). Concomitantly, coculturing with *S. mutans* can upregulate *C. albicans* regulatory genes such as Tup1 and Nrg1, both of which are well-known repressors of filamentation, further modulating fungal morphology to suit the evolving biofilm environment (Jarosz et al. 2009). Beyond these initial inhibitory effects, the bacterial partner can exert additional subtle and context-dependent controls over fungal filamentous development. For example, *S. mutans* secretes mutanocyclin, a tetramic acid secondary metabolite that can alter fungal signaling pathways and morphology (Tao et al. 2022). Mutanocyclin can metabolize the subunit Tpk2 in a dose-dependent manner by regulating the protein kinase A (PKA) pathway to inhibit filamentous growth. By adjusting PKA activity, mutanocyclin effectively influences fungal morphogenesis, cell wall dynamics, and possibly stress responses—factors that collectively determine how *C. albicans* integrates into the polymicrobial community. Furthermore, the inactivation of Tpk2 heightens fungal sensitivity to mutanomycin. Additionally, mutanomycin shows a global impact on *C. albicans* transcriptional profile, primarily through the regulation of cell wall components via the Ras1-cAMP/PKA signaling pathway and the modulation of Efg1 and the subset of filamentous regulators controlled by Sfl1.

On the fungal side, farnesol, a key QS molecule in *C. albicans* that modulates the yeast-to-hyphae transition, adds another layer of complexity to this interspecies dialogue. Interestingly, a nonmonotonic response to farnesol concentration is observed in *S. mutans* growth. In *C. albicans*–*S. mutans* dual-species biofilms, low levels of farnesol (~25 μM) stimulate GtfB expression/activity and increase bacterial growth. Conversely, when farnesol levels exceed 100 μM, it suppresses *S. mutans* proliferation (Kim et al. 2017). This nonmonotonic response underscores the fine-tuned chemical interplay at work, where slight shifts in molecular concentration can pivot the relationship from cooperative to antagonistic.

The cross-communication is not unidirectional. *C. albicans* also influences *S. mutans* quorum sensing. In dual-species biofilms, *C. albicans* stimulates the expression of *S. mutans comS*, a gene that encodes the precursor for the SigX-inducing peptide, leading to the activation of the complete quorum sensing regulon of *S. mutans*, influencing bacteriocin production and late competence development (Sztajer et al. 2014). Additionally, membrane vesicles released by *S. mutans* offer a unique regulatory avenue: these vesicles can enhance *C. albicans* biofilm formation without altering its planktonic growth, suggesting that discrete nanoscale delivery systems exist within the biofilm, selectively transferring cargo that molds fungal behavior (Wu et al. 2020).

All these molecular interactions work in concert with the previously described metabolic and structural synergies, culminating in biofilms that are notably robust and persistent. During biofilm formation, mixed *C. albicans*–*S. mutans* biofilms display an abundance of extracellular matrix, elevated overall biomass, and a refined three-dimensional biofilm architecture compared to their monospecies counterparts (Falsetta et al. 2014; Kim et al. 2021b). Multiple studies have demonstrated that the presence of *C. albicans* enhances the expression and activity of *S. mutans gtfB*, leading to elevated EPS production and structurally more complex biofilms (Kim et al. 2017; Hwang et al. 2017; Lu et al. 2023). Meanwhile, *S. mutans* also enhances the expression of *C. albicans hwp1*, *als1*, and *als3*, which are critical for fungal adherence, filamentous growth, and biofilm formation (Ellepola et al. 2017). Moreover, pH monitoring reveals that *C. albicans–S. mutans* consortia rapidly establish and maintain an acidic microenvironment (~pH 4.3), a stark contrast to *S. mutans* single-species biofilms which cannot sustain such low pH conditions when cultured in saliva (Kim et al. 2021b). In this potent acidogenic condition of cross-kingdom biofilm, *S. mutans* increases the expression of *ldh* (associated with acid production) and *fabM* (involved in monounsaturated fatty acid synthesis necessary for acid tolerance), while simultaneously downregulating *ureC* and *arcA* (genes linked to acid neutralization) (Du et al. 2021; Li et al. 2023). In parallel, the resilience of *C. albicans* in these acidic biofilms depends on pH-adaptive genes, such as Phr2. Deletion of Phr2 dramatically reduces fungal cell numbers, underscoring the importance of acid adaptation for maintaining a robust fungal presence within the cariogenic biofilm milieu (Du et al. 2021).

Although many studies rely on laboratory strains to investigate this interkingdom interaction, there is evidence that strain variation plays a significant role. For instance, certain *S. mutans* strains isolated from children's saliva and dental plaque carry collagen-binding proteins (Cbp)—strongly linked to higher caries risk and severity. Notably, co-infection with *C. albicans* and these Cbp^+ *S. mutans* strains has been associated with caries recurrence, suggesting that they may exhibit distinct interactions compared to Cbp^- strains (Garcia et al. 2021a). Taking into account the genetic variability among clinical isolates, other investigations using clinical co-isolates of *S. mutans* and *C. albicans* have confirmed that *Candida*-mediated protection of *S. mutans* against H_2O_2 is not limited to laboratory strains (Katrak et al. 2023). Furthermore, genotypically and phenotypically diverse *C. albicans* strains form similarly robust interkingdom biofilms with *S. mutans*, exhibiting enhanced sugar metabolism irrespective of the fungus's filamentation state (Xiang et al. 2023).

Potential Therapeutic Strategies

Numerous clinical and experimental studies have conclusively demonstrated the cross-kingdom interaction between *S. mutans* and *C. albicans* in the pathogenesis of dental caries, particularly in ECC and root caries. Therefore, simultaneously regulating *S. mutans* and *C. albicans* in the oral cavity sheds new light on caries prevention. Recently, significant progress has been made to treat this notorious cross-kingdom biofilm by using novel nanomaterials, natural compounds, probiotics, photodynamic therapy, or combination therapy (Fig. 2.4).

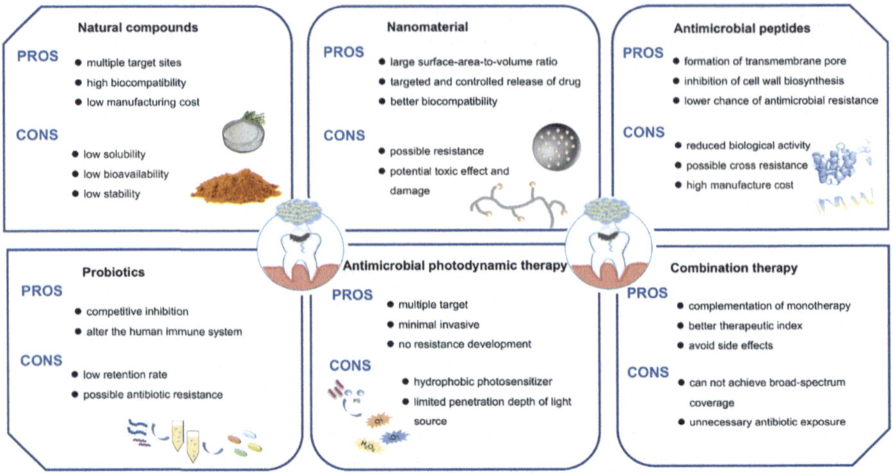

Fig. 2.4 Novel approaches for the treatment of cross-kingdom biofilm. (Adapted from Li et al. (2023))

Natural Compounds

Natural compounds, as a novel and more attractive option with good biocompatibility, are considered an alternative approach for controlling biofilm-forming pathogens (Mishra et al. 2020). A number of studies have been published showing that natural compounds have antimicrobial efficacy against oral biofilms (Karygianni et al. 2015). For example, natural compounds belonging to terpenoids or essential oils have been tested for their antibiofilm capacity on cross-kingdom biofilms. Caffeic acid phenethyl ester (CAPE), a classical type of terpenoids extracted from propolis, has been recognized for its biological and pharmacological properties, including anti-inflammatory, antioxidant, antibacterial, immunomodulatory, anticancer, and wound healing and repair properties (Olgierd et al. 2021). Based on the pleiotropic effects of CAPE, its potential utility in the prevention of oral diseases has been described in some reports, in which it displayed antibacterial activity against *S. mutans* (Niu et al. 2020) and antifungal activity against *C. albicans* (de Barros et al. 2021), and suppressed *S. mutans–C. albicans* cross-kingdom biofilm formation as well as extracellular polysaccharide (EPS) synthesis (Yin et al. 2022). The efficacy of CAPE is probably due to its downregulation of the expression level of glucosyltransferase genes. Another common type of natural compound is phenolics. As an example, curcumin, which is a natural polyphenol extracted from the root of turmeric, has been proven to inhibit *S. mutans–C. albicans* dual-species biofilm formation and decrease the amount of EPS in the biofilms (Li et al. 2019). This inhibitory effect occurs through downregulating *gtfB* and *gtfC* as well as the competence-stimulating peptide-quorum sensing (CSP-QS) system, including *comC*, *comD*, and *comE*. In addition, curcumin significantly reduced *als1* and *als3* gene expression, which belong to the agglutinin-like sequence (ALS) family and play an important role in *C. albicans* formation.

Although natural compounds have many advantages, such as high biocompatibility, diminished likelihood of resistance, and low cost, some drawbacks, like low solubility and stability, hinder the full development of their clinical applications. One feasible method of improving the efficient delivery of natural compounds is to use nanotechnology to overcome those limitations. For example, a recent study used polymeric nanoparticles as versatile carriers to efficiently deliver poorly soluble antimicrobial essential oils against cross-kingdom biofilms (Park et al. 2024). In this study, carvacrol-loaded biodegradable nanoemulsions (NEs) with systematically varied surface charges were investigated against *S. mutans–C. albicans* biofilms. Interestingly, the positively charged NEs exhibited superior diffusion into oral biofilms and yielded a faster and more effective eradication of both bacterial and fungal components simultaneously, providing a valuable framework for future advancements in this critical field with high clinical translational potential.

Nanomaterials

In recent years, there has been an emerging "outside of the box" approach for the treatment of recalcitrant bacterial and biofilm infections (Makabenta et al. 2021). The tunable properties of nanomaterials, particularly their surface functionalities, provide design spaces that can be fine-tuned to maximize the therapeutic effect while minimizing host toxicity (Makabenta et al. 2021). The array of sizes and shapes adopted by nanomaterials offers unique capabilities for targeting bacteria (Soenen et al. 2011). Nanomaterials can use multiple bactericidal mechanisms, including direct cell wall and/or membrane damage, generation of reactive oxygen species (ROS), and binding to intracellular components (Fig. 2.5). Furthermore, nanomaterials can be used as nanocarriers for delivery of therapeutic agents (Baptista et al. 2018; Chen et al. 2020).

Certain metal nanomaterials show their capacities in controlling the oral biofilm. Among them, silver nanoparticles (AgNPs) are the most effective against oral pathogens (Besinis et al. 2014). AgNPs have proven antimicrobial action against a broad spectrum of microorganisms such as bacteria (Pérez-Díaz et al. 2015; Perveen et al. 2018; Chittrarasu et al. 2021), fungi (Soares et al. 2018; Peralta et al. 2023), and viruses (Naumenko et al. 2023). The action of AgNPs is related to the interference of cell wall synthesis and also interferes in the replication and transcription process (Bapat et al. 2018). Recently, the synthesis process of metal nanoparticles using eco-friendly natural sources such as plant extracts, fruits, fungi, honey, and microorganisms (Philip 2010; Qamer et al. 2021) has received great attention, especially because they are less cytotoxic compared to conventional synthesis (Rajora et al. 2016). In a study, AgNPs were obtained by a "green" route using pomegranate extract and evaluated for biofilm killing. "Green" AgNPs significantly reduced the number of viable cells of the *S. mutans* biofilm. However, for *C. albicans* biofilm,

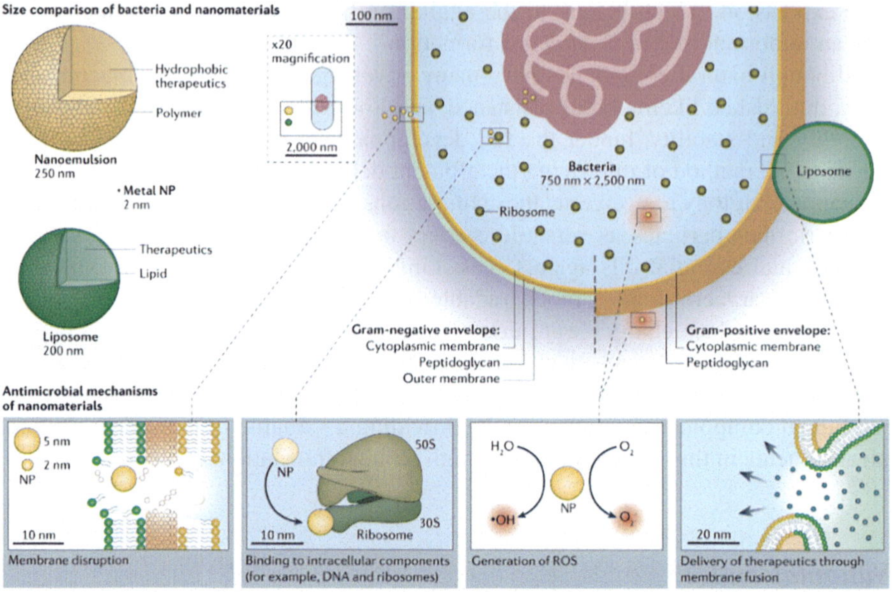

Fig. 2.5 Antimicrobial mechanisms of nanomaterials. (Adapted from Makabenta et al. (2021)) **Electrostatic interactions of nanomaterials with the negatively charged groups present on bacterial surfaces result in membrane damage and cytoplasmic leakage. Nanomaterials can bind various intracellular components, such as ribosomes, proteins, and/or DNA, disrupting their functions. Nanomaterials with catalytic activities increase the production of reactive oxygen species (ROS), such as hydroxyl radicals and superoxides, causing oxidative cellular stress. Nanomaterials can also be used for the delivery of therapeutic agents; some nanomaterials readily enter bacterial cells through membrane fusion, facilitating the delivery of their cargo. NP nanoparticle**

when AgNPs were coupled with tyrosol, a quorum-sensing molecule secreted by *C. albicans*, which plays a role in morphogenesis and biofilm development (Wongsuk et al. 2016), there was a significant reduction in the number of biofilm cells (Souza et al. 2024).

Chitosan is a nontoxic natural cationic polysaccharide with characteristics of adhesiveness, antimicrobial activity, biocompatibility, and biodegradability (Kong et al. 2010). Because of its poor solubility above pH 6.5, chitosan exhibits its antibacterial activity better under acidic conditions (Qi et al. 2004). Previous studies showed that chitosan inhibited the growth and adherence of *S. mutans* (Bae et al. 2006). The chitosan nanoparticles (CNPs), although smaller than chitosan, retain antimicrobial activity (Qi et al. 2004). Owing to a higher surface charge density, CNPs could interact with the negatively charged surface of bacterial cells, causing bacterial cell death (Shi et al. 2006). CNPs have been demonstrated to decrease the viability and the biofilm biomass of *C. albicans* and *S. mutans* biofilm after 18 h incubation (Ikono et al. 2019). However, CNPs showed a limited effect against established dual-species biofilms even treated for 3 h. Therefore, scientists have engineered CNPs as drug carriers to transport promising drugs into the infection site, with the end goal of making the drug more efficient. A representative example is the encapsulation of phloroglucinol into CNPs via the ionic gelification method (Khan et al. 2022). The chitosan loaded with phloroglucinol exhibited inhibitory effects on the cross-kingdom biofilm of *C. albicans* and other bacteria in a concentration-dependent manner. Another study showed that chitosan microparticles loaded with essential oils significantly reduced *S. mutans–C. albicans* cross-kingdom biofilm formation and metabolic activities while decreasing the cytotoxicity of essential oils (Garcia et al. 2023).

The "smart" drug delivery system is a delivery system of drugs by nanoparticles whose release is triggered by environmental stimuli such as pH, glucose, or bacterial products (Park 2014). Nanomaterials that are pH responsive are stable at physiologic pH levels but degrade or disrupt at acidic pH levels to release the active drug (Horev et al. 2015). Given that dental caries always occur at a persistent low pH site around 4.5 ~ 5.5 on teeth (Bowen 2013), where cariogenic organisms ferment sugar and create acidic niches, approaches to the control of cariogenic biofilms by targeting specific microenvironments have also been studied. Polymeric nanoparticles with tertiary cationic surface residues enable farnesol-loaded nanoparticles binding to pellicle and/or biofilms until the pH becomes acidic (below 5.5), and subsequently, the drug is released through protonation-mediated destabilization (Horev et al. 2015). This "smart release" system exerts antibiofilm activity by dual targeting *S. mutans* and *C. albicans*, suppressing cariogenic bacterial–fungal accumulation and preventing tooth enamel demineralization (Ito et al. 2022). Although effective and promising, the systemic safety and long-term effects of nanomaterials on the body are still among the major barriers to clinical use. Furthermore, the underlying antimicrobial mechanisms and potential for drug resistance of nanoparticles should be revealed in future work.

Antimicrobial Peptides (AMPs)

Antimicrobial peptides (AMPs) are small proteins with antibacterial, antiviral, and antifungal activities (Lazzaro et al. 2020). AMPs have been considered as potential therapeutic sources for future antibiotics because of their broad-spectrum activities, unique mechanisms of action, and the low possibility of inducing antimicrobial resistance (Nayab et al. 2022). The major mechanism of AMP action is to integrate into the bacterial cell membrane and disrupt its integrity, which could be described by some models like "barrel-stave" pore, "carpet," or "toroidal-pore" mechanism (Sanderson 2012) (Fig. 2.6), resulting directly or indirectly in cell lysis. However, AMPs may also have more complex activities, including metabolic and translational inhibition (Gagnon et al. 2016; Florin et al. 2017) and formation of nanonets (Chu et al. 2012; Loth et al. 2019). In addition to their antimicrobial function, AMPs also act as effector molecules in wound healing, inflammation, and immune activation (Nayab et al. 2022).

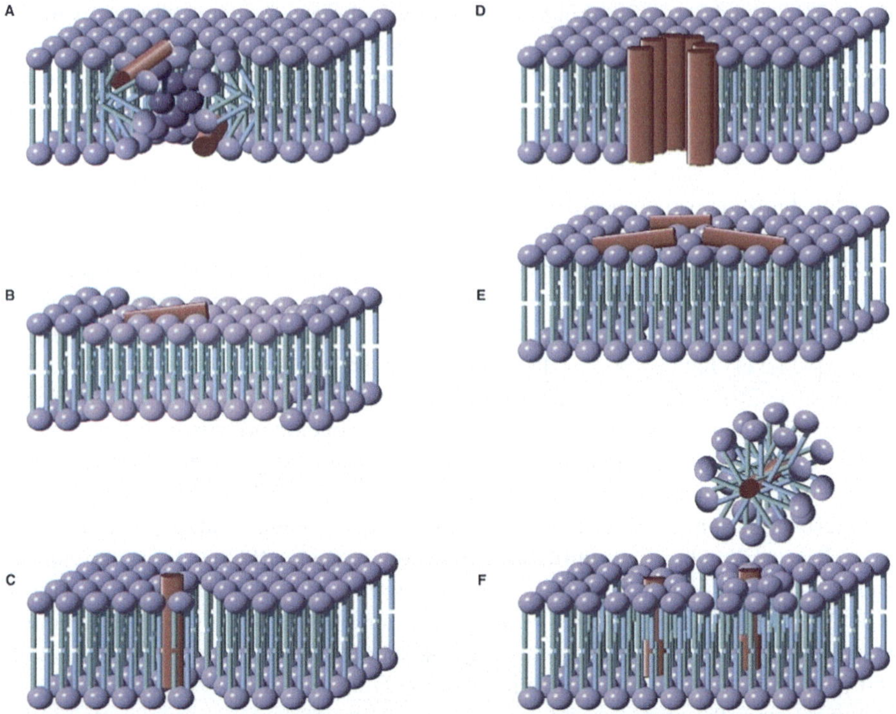

Fig. 2.6 Models for the mode of action of antimicrobial peptides. (Adapted from Sanderson (2012)) **(a) Toroidal pores, stabilized by the peptide, consisting of a lipid-lined pore and contiguous membrane leaflets; (b) membrane thinning, in which peptide binding leads to a reduction in the thickness of the bilayer aggregates: bilayer presents a reduced barrier to the flip-flop of lipids between leaflets, as well as to the formation of transient defects; (c) defect-mediated poration, in which the formation of transient defects (that are significantly less hydrated than toroidal pores) is promoted following peptide binding; (d) the barrel-stave model, consisting of a peptide-lined pore; (e) the carpet model, in which bilayer integrity is disrupted by the formation of peptide aggregates; (f) detergent models, in which membrane disruption occurs through the formation of peptide-lipid micelles that remove lipids from the membrane**

AMPs have been studied as a new strategy in oral disease treatment and prevention, providing a potential alternative to traditional antimicrobial therapy. As an example, a recent study developed a novel AMP using the antimicrobial domain of polyphemusin-I (PI), an AMP derived from horseshoe crabs, and the mineralizing domain of gallic acid (GA). They found that GAPI inhibited planktonic *S. mutans*, *L. casei*, and *C. albicans*, and demonstrated antibiofilm effect on multiple-species biofilms with more than tenfold reduction in viable cells (Zhang et al. 2023). In addition, GAPI reduced the demineralization of early enamel carious lesions. Simon and coworkers screened 75 synthetic cyclic dipeptides against oral pathogens and found that five new active compounds containing two aromatic amino acids could effectively inhibit the adhesion of *C. albicans* and *S. mutans* to the HA disk and prevent biofilm formation (Simon et al. 2019). GERM CLEAN is a newly synthesized antimicrobial peptide molecule with potent antibiofilm activity. Xiong et al.

evaluated the effects of GERM CLEAN on *S. mutans* and *C. albicans* dual-species biofilm formation and found it significantly reduced the adherence ability, EPS production, and total biomass of *S. mutans–C. albicans* biofilm. Furthermore, GERM CLEAN suppressed the acidogenicity of dual-species biofilm and influenced enamel demineralization (Xiong et al. 2022). The advent of AMP has increased the availability of treatment in *S. mutans* and *C. albicans* cross-kingdom biofilm; however, some challenges still exist in their therapeutic utility. As most experiments were performed in vitro without mimicking the oral environment, the long-term stability of AMP in the oral cavity is still questionable. Also, retention in the oral cavity and keeping biological activity under an acidic environment are critical for AMP's clinical translation. Moreover, although AMP has a lower chance of inducing antimicrobial resistance, some bacteria harbor sensors that could activate AMP resistance mechanisms. Therefore, AMPs still need to be further explored, such as the improvement of long-term stability, maintenance of bioactivity, and antimicrobial resistance.

Probiotics

Recently, probiotics have demonstrated their potential in treating oral disease by maintaining the delicate balance between host and microbiome, necessary for health. To date, the most commonly used probiotics include *Lactobacilli* and *Bifidobacterium*, both of which produce lactic acid and other bioactive substances, including hydrogen peroxide, carbon dioxide, bacteriocins, and adhesion inhibitors that could potentially affect the growth of cariogenic microorganisms (Meurman 2005). Zeng et al. assessed four probiotic *Lactobacillus* strains on *S. mutans* and *C. albicans* under high-risk clinical conditions, and their results showed that *L. plantarum* demonstrated superior inhibition on the growth of *C. albicans* and *S. mutans*, the disruption of virulent biofilm formation with reduced bacteria and EPS components, and a reduction in the formation of virulent microcolonies structures. Moreover, *L. plantarum* disrupted *S. mutans* and *C. albicans* cross-kingdom interactions. Genes of *S. mutans* and *C. albicans* involved in metabolic pathways (e.g., EPS formation, carbohydrate metabolism, glycan biosynthesis, and metabolism) were significantly downregulated. More significantly, genes related to *C. albicans* resistance to antifungal medication, fungal cell wall chitin remodeling, and resistance to oxidative stress were also significantly downregulated. In contrast, *Lactobacillus* genes that contribute to antimicrobial peptide plantaricin production were significantly upregulated (Zeng et al. 2022).

Different probiotic strains may prevent caries via different mechanisms, including the inhibition of cariogenic bacteria growth and pathogenesis, EPS production reduction, quorum sensing inhibition, and reduction of the expression of key metabolic and virulence genes. Using a rat model, Zhang et al. showed that *L. plantarum* CCFM8724 reduced the *S. mutans–C. albicans* dual-species biofilm mass and decreased the caries score (Zhang et al. 2020). The same group extended their research to investigate the disruption mechanisms of *L. plantarum* CCFM8724 by using metabolomics and transcriptomics analysis. They found that some carbohydrates related to biofilm formation were decreased, and virulence genes associated with adhesion were downregulated. In addition, *L. plantarum* supernatant had an active role in sugar metabolism during the formation of the *S. mutans–C. albicans* mixed-species biofilm, and was also related to carbohydrate utilization, glucan biosynthesis, and mycelium formation (Zhang et al. 2021). Despite their potential effects, probiotics also have some drawbacks that constrain their application as a drug. On the one hand, most probiotics present a low retention rate in the oral cavity or need high concentrations. Besides, the effects of probiotics on long-term changes in microbiome composition merit further investigation.

Phototherapies

Since the concept of light therapy was introduced in medicine more than a century ago, a diverse array of diagnostic and therapeutic devices utilizing light has been developed in clinical practice (Wei et al. 2020). Two forms of phototherapy have been evaluated for their direct effect on microbial viability: photodynamic and photothermal therapies, which exhibit distinct mechanisms of action. Photodynamic therapy kills bacteria via the generation of reactive oxygen species (ROS) when a photosensitizer is activated by light through either type I mechanisms (electron transfer) or type II mechanisms (energy transfer) (Fig. 2.7a). In contrast, photothermal therapy damages bacteria via plasmonic heating, electron-hole generation, non-radiative relaxation, or thermal vibration of molecules (Fig. 2.7b) (Huang et al. 2023).

Phototherapies have been utilized in dentistry to treat oral infectious diseases. For example, photodynamic therapy (PDT) can effectively inhibit the growth of a variety of microorganisms associated with cariogenic biofilms, such as *S. mutans*, *Lactobacillus*, and *C. albicans* (Garcia et al. 2021b). Other articles have also investigated the antimicrobial effects of photosensitizers such as methylene blue and curcumin against *S. mutans* and *C. albicans* dual-species biofilm after irradiation (Pereira et al. 2011; Quishida et al. 2016). The antimicrobial effectiveness of PDT is dependent on multiple factors, including the characteristics of photosensitizers, oxygen concentration, light dose, and so on. PDT is currently only proposed as an adjunct disinfection method in dental applications, mainly due to the EPS matrix and efflux pumps of bacteria that hinder the binding capacity of photosensitizers and thus affect the antibacterial effect. Moreover, the generated ROS has a short lifespan and diffusion length in the biological environment.

Another form of light therapy is photobiomodulation therapy (PBMT), a non-pharmacologic form of host-modulatory therapy. PBMT is not inherently bacteriocidal; rather, it enhances antimicrobial immunity via effects on a variety of different host immune and stromal cells (Tanum et al. 2024). Using an in vitro model that simulates direct interactions between host cells and intact pathogens, Tanum et al. demonstrated that PBMT ameliorated the proinflammatory response of human gingival keratinocytes challenged with pathogenic oral microbes, including *S. oralis* and *C. albicans*, while enhancing the ability of the cells to mount a direct

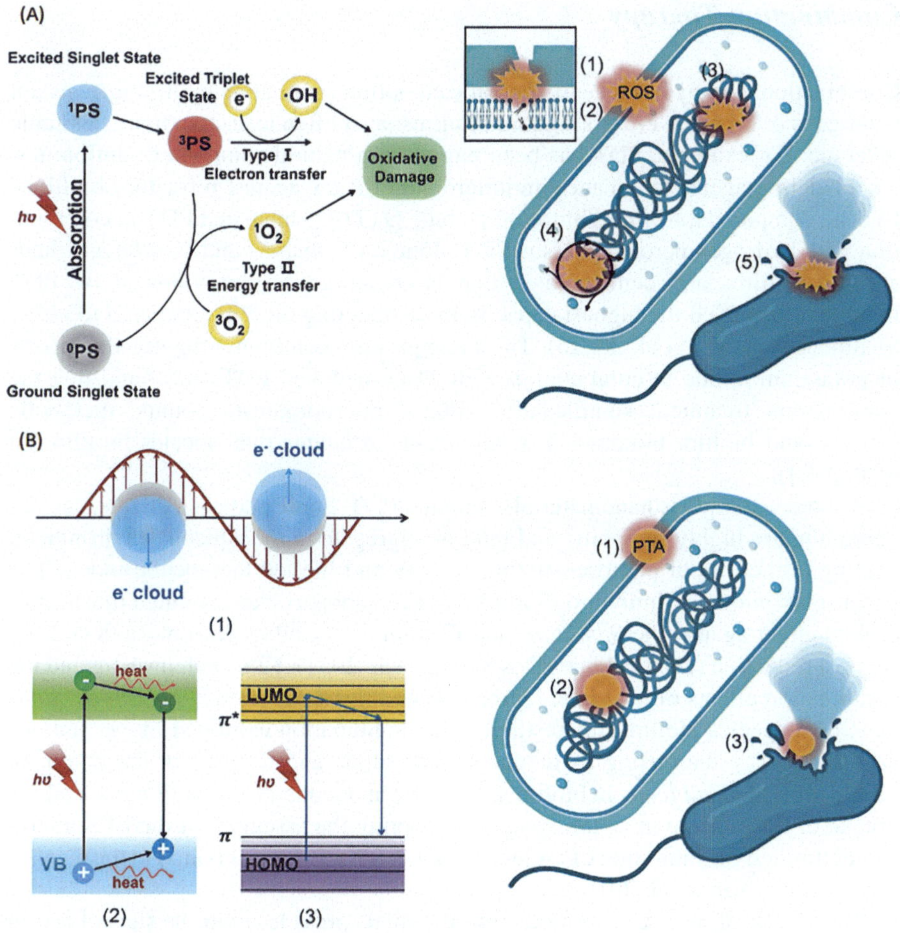

Fig. 2.7 Mechanisms of phototherapies. (Adapted from Huang et al. (2023))
Mechanism of (a) photodynamic therapy and its functions on bacteria illustrating (1) alteration of bacterial outer membrane permeability, (2) oxidation of lipids, (3) degradation of protein or DNA, (4) interference with bacterial metabolism, and (5) permanent damage of bacterial envelope and (b) photothermal therapy and its functions on bacteria illustrating (1) alteration of bacterial outer membrane permeability, (2) denaturation of proteins, and (3) permanent damage of bacterial envelope

antibacterial response (Tanum et al. 2024). While the findings of this study exhibit the potent benefit of using PBMT, further studies are warranted to enhance its clinical efficacy.

Combination Therapy

Combination therapy has recently become an option for treating multidrug-resistant pathogens or complex cross-kingdom biofilms, as it often leads to a more desirable outcome. For example, PDT has been paired with natural compounds, antibiotics, and chelating agents to enhance antimicrobial efficacy against refractory biofilms. Li et al. compared the effects of toluidine blue O (TBO)-mediated PDT in combination with hydrogen peroxide versus PDT alone on *S. mutans* and *C. albicans* dual-species biofilms and demonstrated that H_2O_2 administration followed by PDT treatment displayed the highest capacity in disinfecting biofilm compared to either treatment alone (Li et al. 2022b). The same group recently investigated the effects of potassium iodide in combination with TBO-mediated PDT and found that the combination treatment significantly reduced microorganism counts, metabolic activity, and biofilm biomass of *S. mutans*–*C. albicans* dual-species biofilm (Li et al. 2024).

Another example is nanomaterial-mediated PDT. Some photosensitizers, such as curcumin, are highly lipophilic and tend to aggregate in the aqueous environment, leading to the loss of photosensitizing activity and limited therapeutic index. The encapsulation of curcumin into Pluronics®-127 nanoparticles exhibited the antimicrobial effect against the *S. mutans* and *C. albicans* biofilms with more than 3 log steps of reduction (Dantas Lopes Dos Santos et al. 2021). Kim et al. investigated the combination effects of povidone iodine (PI) and fluconazole on *S. mutans*–*C. albicans* dual-species biofilm. Interestingly, the combination exhibited higher antimicrobial efficacy than using PI or fluconazole alone and resulted in the complete elimination of *C. albicans* in biofilms. It has been documented that PI enhanced the uptake of fluconazole in *C. albicans* by disrupting the assembly of the EPS matrix shield through the inhibition of a-glucan synthesis by *S. mutans* bound to the *C. albicans* surface (Kim et al. 2018).

The combination of two or more antimicrobials provides multiple sites of action to complement monotherapy, which is beneficial for the treatment of cross-kingdom biofilms formed by different structured microorganisms. Besides, the combination leads to a decrease in the antimicrobial concentration used and avoids some side effects.

Conclusions and Future Perspectives

The *S. mutans* and *C. albicans* cross-kingdom biofilm has been demonstrated to orchestrate the development and progression of dental caries. Most studies focus on gene regulation, adhesion, and metabolism interactions between these two microbes using laboratory strains. Moreover, the cross-kingdom interactions under complex biological conditions, such as carbohydrate consumption, remain unclear. In addition, examining physical interactions alone is insufficient, given the imminent threat of antibiotic resistance faced by both species. With the innovation and development of technology, further exploration of *S. mutans–C. albicans* interactions employing clinical isolates and environmental factors will expand the current knowledge of caries mechanisms and provide broader perspectives on microbial interaction within various niches.

To date, numerous approaches have been developed to target cross-kingdom biofilm interactions in oral diseases. Non-antibiotic therapeutic approaches, especially natural extracts and nanomaterials, have shown significant potential in inhibiting microorganisms and eliminating biofilm. Among the strategies reviewed here, treating the cross-kingdom biofilm by combinatory therapy offers a better choice to target different virulence factors. However, there are considerable uncertainties about these newly emerging treatments, which require further work to be done before they can be implemented in the clinic.

References

Anil S, Anand PS (2017) Early childhood caries: prevalence, risk factors, and prevention. Front Pediatr 5:157. https://doi.org/10.3389/fped.2017.00157

Bae K, Jun EJ, Lee SM, Paik DI, Kim JB (2006) Effect of water-soluble reduced chitosan on Streptococcus mutans, plaque regrowth and biofilm vitality. Clin Oral Investig 10:102–107. https://doi.org/10.1007/s00784-006-0038-3

Bamford CV, Nobbs AH, Barbour ME, Lamont RJ, Jenkinson HF (2015) Functional regions of Candida albicans hyphal cell wall protein Als3 that determine interaction with the oral bacterium Streptococcus gordonii. Microbiology (Reading) 161:18–29. https://doi.org/10.1099/mic.0.083378-0

Bapat RA, Chaubal TV, Joshi CP, Bapat PR, Choudhury H, Pandey M, Gorain B, Kesharwani P (2018) An overview of application of silver nanoparticles for biomaterials in dentistry. Mater Sci Eng C Mater Biol Appl 91:881–898. https://doi.org/10.1016/j.msec.2018.05.069

Baptista PV, McCusker MP, Carvalho A, Ferreira DA, Mohan NM, Martins M, Fernandes AR (2018) Nano-strategies to fight multidrug resistant bacteria-"a Battle of the titans". Front Microbiol 9:1441. https://doi.org/10.3389/fmicb.2018.01441

Bertolini M, Dongari-Bagtzoglou A (2019) The relationship of Candida albicans with the Oral bacterial microbiome in health and disease. Adv Exp Med Biol 1197:69–78. https://doi.org/10.1007/978-3-030-28524-1_6

Besinis A, De Peralta T, Handy RD (2014) The antibacterial effects of silver, titanium dioxide and silica dioxide nanoparticles compared to the dental disinfectant chlorhexidine on Streptococcus mutans using a suite of bioassays. Nanotoxicology 8:1–16. https://doi.org/10.3109/17435390.2012.742935

Bowen WH (2013) The Stephan curve revisited. Odontology 101:2–8. https://doi.org/10.1007/s10266-012-0092-z

Brito ACM, Bezerra IM, Borges MHDS, Cavalcanti YW, Almeida LDFDD (2021) Effect of different salivary glucose concentrations on dual-species biofilms of Candida albicans and Streptococcus mutans. Biofouling 37:615–625. https://doi.org/10.1080/08927014.2021.1946519

Chan A, Ellepola K, Truong T, Balan P, Koo H, Seneviratne CJ (2020) Inhibitory effects of xylitol and sorbitol on Streptococcus mutans and Candida albicans biofilms are repressed by the presence of sucrose. Arch Oral Biol 119:104886. https://doi.org/10.1016/j.archoralbio.2020.104886

Chen H, Gu L, Liao B, Zhou X, Cheng L, Ren B (2020) Advances of caries nanomaterials. Molecules 25:5047. https://doi.org/10.3390/molecules25215047

Chittrarasu M, Ahamed AS, Ravi V (2021) Antimicrobial efficacy of green synthesis of silver nanoparticles against cariogenic pathogens - an in vitro study. J Pharm Bioallied Sci 13:S1188–S1192. https://doi.org/10.4103/jpbs.jpbs_338_21

Chu H, Pazgier M, Jung G, Nuccio S-P, Castillo PA, de Jong MF, Winter MG, Winter SE, Wehkamp J, Shen B, Salzman NH, Underwood MA, Tsolis RM, Young GM, Lu W, Lehrer RI, Bäumler AJ, Bevins CL (2012) Human α-defensin 6 promotes mucosal innate immunity through self-assembled peptide nanonets. Science 337:477–481. https://doi.org/10.1126/science.1218831

Costa Oliveira BE, Ricomini Filho AP, Burne RA, Zeng L (2021) The route of sucrose utilization by Streptococcus mutans affects intracellular polysaccharide metabolism. Front Microbiol 12:636684. https://doi.org/10.3389/fmicb.2021.636684

Dantas Lopes Dos Santos D, Besegato JF, de Melo PBG, Oshiro Junior JA, Chorilli M, Deng D, Bagnato VS, Rastelli ANDS (2021) Curcumin-loaded Pluronic® F-127 micelles as a drug delivery system for curcumin-mediated photodynamic therapy for oral application. Photochem Photobiol 97:1072–1088. https://doi.org/10.1111/php.13433

de Barros PP, Rossoni RD, Garcia MT, Kaminski VDL, Loures FV, Fuchs BB, Mylonakis E, Junqueira JC (2021) The anti-biofilm efficacy of Caffeic Acid Phenethyl Ester (CAPE) in vitro and a murine model of oral candidiasis. Front Cell Infect Microbiol 11:700305. https://doi.org/10.3389/fcimb.2021.700305

de Carvalho FG, Silva DS, Hebling J, Spolidorio LC, Spolidorio DMP (2006) Presence of mutans streptococci and Candida spp. in dental plaque/dentine of carious teeth and early childhood caries. Arch Oral Biol 51:1024–1028. https://doi.org/10.1016/j.archoralbio.2006.06.001

Du Q, Ren B, He J, Peng X, Guo Q, Zheng L, Li J, Dai H, Chen V, Zhang L, Zhou X, Xu X (2021) Candida albicans promotes tooth decay by inducing oral microbial dysbiosis. ISME J 15:894–908. https://doi.org/10.1038/s41396-020-00823-8

Du Q, Ren B, Zhou X, Zhang L, Xu X (2022) Cross-kingdom interaction between Candida albicans and oral bacteria. Front Microbiol 13:911623. https://doi.org/10.3389/fmicb.2022.911623

Edlund A, Yang Y, Yooseph S, Hall AP, Nguyen DD, Dorrestein PC, Nelson KE, He X, Lux R, Shi W, McLean JS (2015) Meta-omics uncover temporal regulation of pathways across oral microbiome genera during in vitro sugar metabolism. ISME J 9:2605–2619. https://doi.org/10.1038/ismej.2015.72

Ellepola K, Liu Y, Cao T, Koo H, Seneviratne CJ (2017) Bacterial GtfB augments Candida albicans accumulation in cross-kingdom biofilms. J Dent Res 96:1129–1135. https://doi.org/10.1177/0022034517714414

Ellepola K, Truong T, Liu Y, Lin Q, Lim TK, Lee YM, Cao T, Koo H, Seneviratne CJ (2019) Multi-omics analyses reveal synergistic carbohydrate metabolism in Streptococcus mutans-Candida albicans mixed-species biofilms. Infect Immun 87:e00339–e00319. https://doi.org/10.1128/IAI.00339-19

Falsetta ML, Klein MI, Colonne PM, Scott-Anne K, Gregoire S, Pai C-H, Gonzalez-Begne M, Watson G, Krysan DJ, Bowen WH, Koo H (2014) Symbiotic relationship between Streptococcus mutans and Candida albicans synergizes virulence of plaque biofilms in vivo. Infect Immun 82:1968–1981. https://doi.org/10.1128/IAI.00087-14

Florin T, Maracci C, Graf M, Karki P, Klepacki D, Berninghausen O, Beckmann R, Vázquez-Laslop N, Wilson DN, Rodnina MV, Mankin AS (2017) An antimicrobial peptide that inhibits translation by trapping release factors on the ribosome. Nat Struct Mol Biol 24:752–757. https://doi.org/10.1038/nsmb.3439

Gagnon MG, Roy RN, Lomakin IB, Florin T, Mankin AS, Steitz TA (2016) Structures of proline-rich peptides bound to the ribosome reveal a common mechanism of protein synthesis inhibition. Nucleic Acids Res 44:2439–2450. https://doi.org/10.1093/nar/gkw018

Garcia BA, Acosta NC, Tomar SL, Roesch LFW, Lemos JA, Mugayar LRF, Abranches J (2021a) Association of Candida albicans and Cbp+ Streptococcus mutans with early childhood caries recurrence. Sci Rep 11:10802. https://doi.org/10.1038/s41598-021-90198-3

Garcia MT, Ward RADC, NMF G, Pedroso LLC, Neto JVDS, Strixino JF, Junqueira JC (2021b) Susceptibility of dental caries microcosm biofilms to photodynamic therapy mediated by Fotoenticine. Pharmaceutics 13:1907. https://doi.org/10.3390/pharmaceutics13111907

Garcia LGS, da Rocha MG, Freire RS, Nunes PIG, Nunes JVS, Fernandes MR, Pereira-Neto WA, Sidrim JJC, Santos FA, Rocha MFG, Rodrigues LKA, Vieira RS, Brilhante RSN (2023) Chitosan microparticles loaded with essential oils inhibit duo-biofilms of Candida albicans and Streptococcus mutans. J Appl Oral Sci 31:e20230146. https://doi.org/10.1590/1678-7757-2023-0146

Gregoire S, Xiao J, Silva BB, Gonzalez I, Agidi PS, Klein MI, Ambatipudi KS, Rosalen PL, Bauserman R, Waugh RE, Koo H (2011) Role of glucosyltransferase B in interactions of Candida albicans with Streptococcus mutans and with an experimental pellicle on hydroxyapatite surfaces. Appl Environ Microbiol 77:6357–6367. https://doi.org/10.1128/AEM.05203-11

He J, Kim D, Zhou X, Ahn S-J, Burne RA, Richards VP, Koo H (2017) RNA-Seq reveals enhanced sugar metabolism in Streptococcus mutans co-cultured with Candida albicans within mixed-species biofilms. Front Microbiol 8:1036. https://doi.org/10.3389/fmicb.2017.01036

Horev B, Klein MI, Hwang G, Li Y, Kim D, Koo H, Benoit DSW (2015) pH-activated nanoparticles for controlled topical delivery of farnesol to disrupt oral biofilm virulence. ACS Nano 9:2390–2404. https://doi.org/10.1021/nn507170s

Hoyer LL, Oh S-H, Jones R, Cota E (2014) A proposed mechanism for the interaction between the Candida albicans Als3 adhesin and streptococcal cell wall proteins. Front Microbiol 5:564. https://doi.org/10.3389/fmicb.2014.00564

Huang S, Qi M, Chen Y (2023) Photonics-based treatments: mechanisms and applications in oral infectious diseases. Front Microbiol 14:948092. https://doi.org/10.3389/fmicb.2023.948092

Hwang G, Marsh G, Gao L, Waugh R, Koo H (2015) Binding force dynamics of Streptococcus mutans-glucosyltransferase B to Candida albicans. J Dent Res 94:1310–1317. https://doi.org/10.1177/0022034515592859

Hwang G, Liu Y, Kim D, Li Y, Krysan DJ, Koo H (2017) Candida albicans mannans mediate Streptococcus mutans exoenzyme GtfB binding to modulate cross-kingdom biofilm development in vivo. PLoS Pathog 13:e1006407. https://doi.org/10.1371/journal.ppat.1006407

Ikono R, Vibriani A, Wibowo I, Saputro KE, Muliawan W, Bachtiar BM, Mardliyati E, Bachtiar EW, Rochman NT, Kagami H, Xianqi L, Nagamura-Inoue T, Tojo A (2019) Nanochitosan antimicrobial activity against Streptococcus mutans and Candida albicans dual-species biofilms. BMC Res Notes 12:383. https://doi.org/10.1186/s13104-019-4422-x

Ito T, Sims KR, Liu Y, Xiang Z, Arthur RA, Hara AT, Koo H, Benoit DSW, Klein MI (2022) Farnesol delivery via polymeric nanoparticle carriers inhibits cariogenic cross-kingdom biofilms and prevents enamel demineralization. Mol Oral Microbiol 37:218–228. https://doi.org/10.1111/omi.12379

Jarosz LM, Deng DM, van der Mei HC, Crielaard W, Krom BP (2009) Streptococcus mutans competence-stimulating peptide inhibits Candida albicans hypha formation. Eukaryot Cell 8:1658–1664. https://doi.org/10.1128/EC.00070-09

Jin P, Wang L, Chen D, Chen Y (2024) Unveiling the complexity of early childhood caries: Candida albicans and Streptococcus mutans cooperative strategies in carbohydrate metabolism and virulence. J Oral Microbiol 16:2339161. https://doi.org/10.1080/20002297.2024.2339161

Karygianni L, Al-Ahmad A, Argyropoulou A, Hellwig E, Anderson AC, Skaltsounis AL (2015) Natural antimicrobials and oral microorganisms: a systematic review on herbal interventions for the eradication of multispecies oral biofilms. Front Microbiol 6:1529. https://doi.org/10.3389/fmicb.2015.01529

Katrak C, Garcia BA, Dornelas-Figueira LM, Nguyen M, Williams RB, Lorenz MC, Abranches J (2023) Catalase produced by Candida albicans protects Streptococcus mutans from H2O2 stress—one more piece in the cross-kingdom synergism puzzle. mSphere 8:e00295–23. https://doi.org/10.1128/msphere.00295-23

Khan F, Oh D, Chandika P, Jo D-M, Bamunarachchi NI, Jung W-K, Kim Y-M (2022) Inhibitory activities of phloroglucinol-chitosan nanoparticles on mono- and dual-species biofilms of Candida albicans and bacteria. Colloids Surf B Biointerfaces 211:112307. https://doi.org/10.1016/j.colsurfb.2021.112307

Khoury ZH, Vila T, Puthran TR, Sultan AS, Montelongo-Jauregui D, Melo MAS, Jabra-Rizk MA (2020) The role of Candida albicans secreted polysaccharides in augmenting Streptococcus mutans adherence and mixed biofilm formation: in vitro and in vivo studies. Front Microbiol 11:307. https://doi.org/10.3389/fmicb.2020.00307

Kim D, Sengupta A, Niepa THR, Lee B-H, Weljie A, Freitas-Blanco VS, Murata RM, Stebe KJ, Lee D, Koo H (2017) Candida albicans stimulates Streptococcus mutans microcolony development via cross-kingdom biofilm-derived metabolites. Sci Rep 7:41332. https://doi.org/10.1038/srep41332

Kim D, Liu Y, Benhamou RI, Sanchez H, Simón-Soro Á, Li Y, Hwang G, Fridman M, Andes DR, Koo H (2018) Bacterial-derived exopolysaccharides enhance antifungal drug tolerance in a cross-kingdom oral biofilm. ISME J 12:1427–1442. https://doi.org/10.1038/s41396-018-0113-1

Kim HE, Dhall A, Liu Y, Bawazir M, Koo H, Hwang G (2021a) Intervening in symbiotic cross-kingdom biofilm interactions: a binding mechanism-based nonmicrobicidal approach. mBio 12:e00651–21. https://doi.org/10.1128/mBio.00651-21

Kim H-E, Liu Y, Dhall A, Bawazir M, Koo H, Hwang G (2021b) Synergism of Streptococcus mutans and Candida albicans reinforces biofilm maturation and acidogenicity in saliva: an in vitro study. Front Cell Infect Microbiol 10. https://doi.org/10.3389/fcimb.2020.623980

Kong M, Chen XG, Xing K, Park HJ (2010) Antimicrobial properties of chitosan and mode of action: a state of the art review. Int J Food Microbiol 144:51–63. https://doi.org/10.1016/j.ijfoodmicro.2010.09.012

Koo H, Andes DR, Krysan DJ (2018) Candida–streptococcal interactions in biofilm-associated oral diseases. PLoS Pathog 14:e1007342. https://doi.org/10.1371/journal.ppat.1007342

Lazzaro BP, Zasloff M, Rolff J (2020) Antimicrobial peptides: application informed by evolution. Science 368:eaau5480. https://doi.org/10.1126/science.aau5480

Li X, Yin L, Ramage G, Li B, Tao Y, Zhi Q, Lin H, Zhou Y (2019) Assessing the impact of curcumin on dual-species biofilms formed by Streptococcus mutans and Candida albicans. Microbiology 8:e937. https://doi.org/10.1002/mbo3.937

Li X, Liu Y, Yang X, Li C, Song Z (2022a) The oral microbiota: community composition, influencing factors, pathogenesis, and interventions. Front Microbiol 13:895537. https://doi.org/10.3389/fmicb.2022.895537

Li Y, Du J, Huang S, Wang S, Wang Y, Cai Z, Lei L, Huang X (2022b) Hydrogen peroxide potentiates antimicrobial photodynamic therapy in eliminating Candida albicans and Streptococcus mutans dual-species biofilm from denture base. Photodiagn Photodyn Ther 37:102691. https://doi.org/10.1016/j.pdpdt.2021.102691

Li Y, Huang S, Du J, Wu M, Huang X (2023) Current and prospective therapeutic strategies: tackling Candida albicans and Streptococcus mutans cross-kingdom biofilm. Front Cell Infect Microbiol 13:1106231. https://doi.org/10.3389/fcimb.2023.1106231

Li Y, Huang S, Du J, Wang S, Cai Z, Huang X (2024) Deciphering the killing mechanisms of potassium iodide in combination with antimicrobial photodynamic therapy against cross-kingdom biofilm. Front Cell Infect Microbiol 14:1444764. https://doi.org/10.3389/fcimb.2024.1444764

Limon JJ, Skalski JH, Underhill DM (2017) Commensal fungi in health and disease. Cell Host Microbe 22:156–165. https://doi.org/10.1016/j.chom.2017.07.002

Liu Y, Daniel SG, Kim H-E, Koo H, Korostoff J, Teles F, Bittinger K, Hwang G (2023) Addition of cariogenic pathogens to complex oral microflora drives significant changes in biofilm compositions and functionalities. Microbiome 11:123. https://doi.org/10.1186/s40168-023-01561-7

Loth K, Vergnes A, Barreto C, Voisin SN, Meudal H, Da Silva J, Bressan A, Belmadi N, Bachère E, Aucagne V, Cazevielle C, Marchandin H, Rosa RD, Bulet P, Touqui L, Delmas AF, Destoumieux-Garzón D (2019) The ancestral N-terminal domain of big defensins drives bacterially triggered assembly into antimicrobial nanonets. mBio 10:e01821–19. https://doi.org/10.1128/mBio.01821-19

Lu Y, Lin Y, Li M, He J (2023) Roles of Streptococcus mutans-Candida albicans interaction in early childhood caries: a literature review. Front Cell Infect Microbiol 13:1151532. https://doi.org/10.3389/fcimb.2023.1151532

Makabenta JMV, Nabawy A, Li C-H, Schmidt-Malan S, Patel R, Rotello VM (2021) Nanomaterial-based therapeutics for antibiotic-resistant bacterial infections. Nat Rev Microbiol 19:23–36. https://doi.org/10.1038/s41579-020-0420-1

Martorano-Fernandes L, Goodwine JS, Ricomini-Filho AP, Nobile CJ, Del Bel Cury AA (2023) Candida albicans adhesins Als1 and Hwp1 modulate interactions with Streptococcus mutans. Microorganisms 11:1391. https://doi.org/10.3390/microorganisms11061391
Meurman JH (2005) Probiotics: do they have a role in oral medicine and dentistry? Eur J Oral Sci 113:188–196. https://doi.org/10.1111/j.1600-0722.2005.00191.x
Mishra R, Panda AK, De Mandal S, Shakeel M, Bisht SS, Khan J (2020) Natural anti-biofilm agents: strategies to control biofilm-forming pathogens. Front Microbiol 11:566325. https://doi.org/10.3389/fmicb.2020.566325
Naumenko K, Zahorodnia S, Pop CV, Rizun N (2023) Antiviral activity of silver nanoparticles against the influenza A virus. J Virus Erad 9:100330. https://doi.org/10.1016/j.jve.2023.100330
Nayab S, Aslam MA, Rahman SU, Sindhu ZUD, Sajid S, Zafar N, Razaq M, Kanwar R, Amanullah (2022) A review of antimicrobial peptides: its function, mode of action and therapeutic potential. Int J Pept Res Ther 28:46. https://doi.org/10.1007/s10989-021-10325-6
Niu Y, Wang K, Zheng S, Wang Y, Ren Q, Li H, Ding L, Li W, Zhang L (2020) Antibacterial effect of caffeic acid phenethyl ester on cariogenic bacteria and Streptococcus mutans biofilms. Antimicrob Agents Chemother 64:e00251–20. https://doi.org/10.1128/AAC.00251-20
Olgierd B, Kamila Ż, Anna B, Emilia M (2021) The pluripotent activities of caffeic acid phenethyl ester. Molecules 26:1335. https://doi.org/10.3390/molecules26051335
Park K (2014) Controlled drug delivery systems: past forward and future back. J Control Release 190:3–8. https://doi.org/10.1016/j.jconrel.2014.03.054
Park J, Xiang Z, Liu Y, Li C-H, Chen C, Nagaraj H, Nguyen T, Nabawy A, Koo H, Rotello VM (2024) Surface-charge tuned polymeric nanoemulsions for carvacrol delivery in interkingdom biofilms. ACS Appl Mater Interfaces 16:37613–37622. https://doi.org/10.1021/acsami.4c06618
Peleg AY, Hogan DA, Mylonakis E (2010) Medically important bacterial-fungal interactions. Nat Rev Microbiol 8:340–349. https://doi.org/10.1038/nrmicro2313
Peralta LCF, Almeida NLM, Pontes FML, Rinaldo D, Carneiro CA, Neppelenbroek KH, Lara VS, Porto VC (2023) Silver nanoparticles in denture adhesive: an antimicrobial approach against Candida albicans. J Dent 131:104445. https://doi.org/10.1016/j.jdent.2023.104445
Pereira CA, Romeiro RL, Costa ACBP, Machado AKS, Junqueira JC, Jorge AOC (2011) Susceptibility of Candida albicans, Staphylococcus aureus, and Streptococcus mutans biofilms to photodynamic inactivation: an in vitro study. Lasers Med Sci 26:341–348. https://doi.org/10.1007/s10103-010-0852-3
Pérez-Díaz MA, Boegli L, James G, Velasquillo C, Sánchez-Sánchez R, Martínez-Martínez R-E, Martínez-Castañón GA, Martinez-Gutierrez F (2015) Silver nanoparticles with antimicrobial activities against Streptococcus mutans and their cytotoxic effect. Mater Sci Eng C Mater Biol Appl 55:360–366. https://doi.org/10.1016/j.msec.2015.05.036
Perveen S, Safdar N, Chaudhry G-E-S, Yasmin A (2018) Antibacterial evaluation of silver nanoparticles synthesized from lychee peel: individual versus antibiotic conjugated effects. World J Microbiol Biotechnol 34:118. https://doi.org/10.1007/s11274-018-2500-1
Philip D (2010) Rapid green synthesis of spherical gold nanoparticles using Mangifera indica leaf. Spectrochim Acta A Mol Biomol Spectrosc 77:807–810. https://doi.org/10.1016/j.saa.2010.08.008
Proctor DM, Drummond RA, Lionakis MS, Segre JA (2023) One population, multiple lifestyles: commensalism and pathogenesis in the human mycobiome. Cell Host Microbe 31:539–553. https://doi.org/10.1016/j.chom.2023.02.010
Qamer S, Romli MH, Che-Hamzah F, Misni N, Joseph NMS, Al-Haj NA, Amin-Nordin S (2021) Systematic review on biosynthesis of silver nanoparticles and antibacterial activities: application and theoretical perspectives. Molecules 26:5057. https://doi.org/10.3390/molecules26165057
Qi L, Xu Z, Jiang X, Hu C, Zou X (2004) Preparation and antibacterial activity of chitosan nanoparticles. Carbohydr Res 339:2693–2700. https://doi.org/10.1016/j.carres.2004.09.007
Quishida CCC, De Oliveira Mima EG, Jorge JH, Vergani CE, Bagnato VS, Pavarina AC (2016) Photodynamic inactivation of a multispecies biofilm using curcumin and LED light. Lasers Med Sci 31:997–1009. https://doi.org/10.1007/s10103-016-1942-7

Rajora N, Kaushik S, Jyoti A, Kothari SL (2016) Rapid synthesis of silver nanoparticles by pseudomonas stutzeri isolated from textile soil under optimised conditions and evaluation of their antimicrobial and cytotoxicity properties. IET Nanobiotechnol 10:367–373. https://doi.org/10.1049/iet-nbt.2015.0107

Ren Z, Jeckel H, Simon-Soro A, Xiang Z, Liu Y, Cavalcanti IM, Xiao J, Tin N-N, Hara A, Drescher K, Koo H (2022) Interkingdom assemblages in human saliva display group-level surface mobility and disease-promoting emergent functions. Proc Natl Acad Sci USA 119:e2209699119. https://doi.org/10.1073/pnas.2209699119

Sanderson JM (2012) Resolving the kinetics of lipid, protein and peptide diffusion in membranes. Mol Membr Biol 29:118–143. https://doi.org/10.3109/09687688.2012.678018

Scheinin A, Pienihäkkinen K, Tiekso J, Holmberg S (1992) Multifactorial modeling for root caries prediction. Community Dent Oral Epidemiol 20:35–37. https://doi.org/10.1111/j.1600-0528.1992.tb00670.x

Scheinin A, Pienihäkkinen K, Tiekso J, Holmberg S, Fukuda M, Suzuki A (1994) Multifactorial modeling for root caries prediction: 3-year follow-up results. Community Dent Oral Epidemiol 22:126–129. https://doi.org/10.1111/j.1600-0528.1994.tb01587.x

Shi Z, Neoh KG, Kang ET, Wang W (2006) Antibacterial and mechanical properties of bone cement impregnated with chitosan nanoparticles. Biomaterials 27:2440–2449. https://doi.org/10.1016/j.biomaterials.2005.11.036

Silverman RJ, Nobbs AH, Vickerman MM, Barbour ME, Jenkinson HF (2010) Interaction of Candida albicans cell wall Als3 protein with Streptococcus gordonii SspB adhesin promotes development of mixed-species communities. Infect Immun 78:4644–4652. https://doi.org/10.1128/IAI.00685-10

Simon G, Bérubé C, Voyer N, Grenier D (2019) Anti-biofilm and anti-adherence properties of novel cyclic dipeptides against oral pathogens. Bioorg Med Chem 27:2323–2331. https://doi.org/10.1016/j.bmc.2018.11.042

Soares MRPS, Corrêa RO, Stroppa PHF, Marques FC, Andrade GFS, Corrêa CC, Brandão MAF, Raposo NRB (2018) Biosynthesis of silver nanoparticles using Caesalpinia ferrea (Tul.) Martius extract: physicochemical characterization, antifungal activity and cytotoxicity. PeerJ 6:e4361. https://doi.org/10.7717/peerj.4361

Soenen SJ, Rivera-Gil P, Montenegro J-M, Parak WJ, De Smedt SC, Braeckmans K (2011) Cellular toxicity of inorganic nanoparticles: common aspects and guidelines for improved nanotoxicity evaluation. Nano Today 6:446–465. https://doi.org/10.1016/j.nantod.2011.08.001

Souza JAS, do Amaral JG, Monteiro DR, Fernandes RA, Fernandes GL, Gorup LF, de Souza Neto FN, de Camargo ER, Agostinho AM, Barbosa DB, Delbem ACB (2024) "Green" silver nanoparticles combined with tyrosol as potential oral antimicrobial therapy. J Dent 143:104867. https://doi.org/10.1016/j.jdent.2024.104867

Sridhar S, Suprabha BS, Shenoy R, Suman E, Rao A (2020) Association of Streptococcus mutans, Candida albicans and oral health practices with activity status of caries lesions among 5-year-old children with early childhood caries. Oral Health Prev Dent 18:911–919. https://doi.org/10.3290/j.ohpd.a45411

Sztajer H, Szafranski SP, Tomasch J, Reck M, Nimtz M, Rohde M, Wagner-Döbler I (2014) Cross-feeding and interkingdom communication in dual-species biofilms of Streptococcus mutans and Candida albicans. ISME J 8:2256–2271. https://doi.org/10.1038/ismej.2014.73

Tanum J, Kim HE, Lee SM, Kim A, Korostoff J, Hwang G (2024) Photobiomodulation of gingival cells challenged with viable oral microbes. J Dent Res 103:745–754. https://doi.org/10.1177/00220345241246529

Tao L, Wang M, Guan G, Zhang Y, Hao T, Li C, Li S, Chen Y, Huang G (2022) Streptococcus mutans suppresses filamentous growth of Candida albicans through secreting mutanocyclin, an unacylated tetramic acid. Virulence 13:542–557. https://doi.org/10.1080/21505594.2022.2046952

Vílchez R, Lemme A, Ballhausen B, Thiel V, Schulz S, Jansen R, Sztajer H, Wagner-Döbler I (2010) Streptococcus mutans inhibits Candida albicans hyphal formation by the fatty acid signaling molecule trans-2-decenoic acid (SDSF). Chembiochem 11:1552–1562. https://doi.org/10.1002/cbic.201000086

Wan SX, Tian J, Liu Y, Dhall A, Koo H, Hwang G (2021) Cross-kingdom cell-to-cell interactions in cariogenic biofilm initiation. J Dent Res 100:74–81. https://doi.org/10.1177/0022034520950286

Wei G, Yang G, Wang Y, Jiang H, Fu Y, Yue G, Ju R (2020) Phototherapy-based combination strategies for bacterial infection treatment. Theranostics 10:12241–12262. https://doi.org/10.7150/thno.52729

Williamson PR, Huber MA, Bennett JE (1993) Role of maltase in the utilization of sucrose by Candida albicans. Biochem J 291(Pt 3):765–771. https://doi.org/10.1042/bj2910765

Wongsuk T, Pumeesat P, Luplertlop N (2016) Fungal quorum sensing molecules: role in fungal morphogenesis and pathogenicity. J Basic Microbiol 56:440–447. https://doi.org/10.1002/jobm.201500759

Wu R, Tao Y, Cao Y, Zhou Y, Lin H (2020) Streptococcus mutans membrane vesicles harboring glucosyltransferases augment Candida albicans biofilm development. Front Microbiol 11:581184. https://doi.org/10.3389/fmicb.2020.581184

Xiang Z, Wakade RS, Ribeiro AA, Hu W, Bittinger K, Simon-Soro A, Kim D, Li J, Krysan DJ, Liu Y, Koo H (2023) Human tooth as a fungal niche: Candida albicans traits in dental plaque isolates. MBio 14:e0276922. https://doi.org/10.1128/mbio.02769-22

Xiao J, Grier A, Faustoferri RC, Alzoubi S, Gill AL, Feng C, Liu Y, Quivey RG, Kopycka-Kedzierawski DT, Koo H, Gill SR (2018a) Association between oral Candida and bacteriome in children with severe ECC. J Dent Res 97:1468–1476. https://doi.org/10.1177/0022034518790941

Xiao J, Huang X, Alkhers N, Alzamil H, Alzoubi S, Wu TT, Castillo DA, Campbell F, Davis J, Herzog K, Billings R, Kopycka-Kedzierawski DT, Hajishengallis E, Koo H (2018b) Candida albicans and early childhood caries: a systematic review and meta-analysis. Caries Res 52:102–112. https://doi.org/10.1159/000481833

Xiao J, Zeng Y, Rustchenko E, Huang X, Wu TT, Falsetta ML (2023) Dual transcriptome of Streptococcus mutans and Candida albicans interplay in biofilms. J Oral Microbiol 15:2144047. https://doi.org/10.1080/20002297.2022.2144047

Xiong K, Chen X, Zhu H, Ji M, Zou L (2022) Anticaries activity of GERM CLEAN in Streptococcus mutans and Candida albicans dual-species biofilm. Oral Dis 28:829–839. https://doi.org/10.1111/odi.13799

Yang C, Scoffield J, Wu R, Deivanayagam C, Zou J, Wu H (2018) Antigen I/II mediates interactions between Streptococcus mutans and Candida albicans. Mol Oral Microbiol 33:283–291. https://doi.org/10.1111/omi.12223

Yin W, Zhang Z, Shuai X, Zhou X, Yin D (2022) Caffeic Acid Phenethyl Ester (CAPE) inhibits cross-kingdom biofilm formation of Streptococcus mutans and Candida albicans. Microbiol Spectr 10:e0157822. https://doi.org/10.1128/spectrum.01578-22

Zeng Y, Fadaak A, Alomeir N, Wu TT, Rustchenko E, Qing S, Bao J, Gilbert C, Xiao J (2022) Lactobacillus plantarum disrupts S. mutans-C. albicans cross-kingdom biofilms. Front Cell Infect Microbiol 12:872012. https://doi.org/10.3389/fcimb.2022.872012

Zhang Q, Qin S, Xu X, Zhao J, Zhang H, Liu Z, Chen W (2020) Inhibitory effect of Lactobacillus plantarum CCFM8724 towards Streptococcus mutans- and Candida albicans-induced caries in rats. Oxidative Med Cell Longev 2020:4345804. https://doi.org/10.1155/2020/4345804

Zhang Q, Li J, Lu W, Zhao J, Zhang H, Chen W (2021) Multi-omics reveals the inhibition of Lactiplantibacillus plantarum CCFM8724 in Streptococcus mutans-Candida albicans mixed-species biofilms. Microorganisms 9:2368. https://doi.org/10.3390/microorganisms9112368

Zhang OL, Niu JY, Yu OY, Mei ML, Jakubovics NS, Chu CH (2023) Development of a novel peptide with antimicrobial and Mineralising properties for caries management. Pharmaceutics 15:2560. https://doi.org/10.3390/pharmaceutics15112560

Chapter 3
Formation and Prevention of Biofilms on Airway Management Devices

Ed Deshmukh-Reeves, Felicity Ryan, and Campbell W. Gourlay

Abstract Polymicrobial biofilm formation on airway management devices such as endotracheal and tracheostomy tubing is a significant risk factor for the onset of ventilator-associated pneumonia. The most common colonising microbes found within biofilms are reported to be similar for both endotracheal and tracheostomy tubing, with the most common bacteria including *P. aeruginosa, Klebsiella* spp., *S. aureus* and coagulase-negative *Staphylococci*, and the most common fungal pathogen being *C. albicans.* Characterisation of polymicrobial biofilms on tracheal tubing is generally limited to microbial culture-based analysis, and we postulate this may have underestimated the role that fungi play. We propose that the interplay between *Candida* species and co-colonising bacteria is critical to the development of biofilms that form on airway management devices. This review also argues for increased consideration of host factors and conditioning films in the characterisation of tracheal tube biofilms and the development of prevention strategies. Host factors such as mucin and bile found on tracheostomy tubing can significantly alter microbial behaviour, but we have much to learn as to their role in biofilm formation and maturation. Significant efforts have been made to determine the composition of polymicrobial biofilms; however, much remains to be determined if more effective management strategies are to reach the clinic and improve patient outcomes.

E. Deshmukh-Reeves · C. W. Gourlay (✉)
Kent Fungal Group, School of Biosciences, University of Kent, Canterbury, Kent, UK
e-mail: C.W.Gourlay@kent.ac.uk

F. Ryan
Microbiology, Pathology Department, East Kent Hospitals University NHS Foundation Trust, Kent, UK

G. Ramage, D. Andes (eds.), *Fungal Biofilms*, Springer Series on Biofilms 15,
https://doi.org/10.1007/978-3-031-93315-8_3

Introduction

The trachea provides the critical passage of air between the larynx and the bronchi. In cases where the trachea has become blocked, compromised or that a patient requires mechanical ventilation, medical devices must be utilised to restore and support the airway (Terragni et al. 2014). Devices remain within the patient for extended periods of time and are bathed in bodily secretions that may contain both resident and environment-derived microbes. This will inevitably lead to microbial colonisation and biofilm formation on the medical device, which poses an infection risk (El Cheikh et al. 2018). Biofilms can be particularly problematic when formed on airway management devices, as this can lead to partial or complete occlusion of the device, leading to reduced or total loss of function and danger to life (Aguilera Xiol et al. 2015). Biofilm formation on tracheal tubing is also the primary risk factor for the development of ventilator-associated pneumonia (VAP), a severe nosocomial infection reported to occur in 5–40% of mechanically ventilated patients with an attributable mortality of approximately 10% (Papazian et al. 2020).

Here, we briefly summarise airway management devices, their use in clinical practice and their susceptibility to biofilm formation. We comment on the study aims and methods and the focus of species identification, with particular attention paid to whether fungi have been considered. Furthermore, we describe the importance of host factors that are present in the upper airway and how these may influence the formation of biofilms on medical devices. Lastly, we discuss current strategies for the prevention of biofilm formation.

Airway Management Devices

Airway management devices are employed for the maintenance and restoration of the respiratory system in patients who may be unable to breathe. Such devices are often formed of specialised tubing, designed to bypass obstructions and provide a secure pathway for air to travel in and out of the lungs. In cases where the act of breathing requires support, tubing can be connected to an oxygen supply for mechanical ventilation (Stocker and Biro 2005).

Mechanical Ventilation Respiratory failure is caused by many factors, including, but not limited to, trauma, infection/sepsis, cardiac failure or neuromuscular disorders. In these cases, patients require mechanical ventilation to either support or replace their respiratory function. Mechanical ventilation works by using positive pressure to drive air into the patient's lungs. This process is tightly regulated based on air composition, pressure, flow rate and volume. Following delivery of the air, the innate elasticity of the lungs passively expires the delivered air. Mechanical ventilation may be used with varying degrees of assistance. Patients capable of breathing spontaneously can use the ventilator in such a manner that it delivers air with each patient-controlled breath, thus acting supportively rather than by total respiratory replacement (Hickey et al. 2024).

Supraglottic Airway Devices Supraglottic airway devices function above the epiglottis. Often, supraglottic airway devices are used transiently in emergency situations and out-of-hospital environments. The two most commonly used supraglottic airway devices are the laryngeal mask airway devices and laryngeal tubes. Laryngeal mask airway devices are used to temporarily maintain an open airway. They are used in emergency situations or to administer anaesthesia. The tip of the device has a large ovoid inflatable cuff that forms an airtight seal just above the epiglottis (Brain 1983). A laryngeal tube serves a similar purpose to the laryngeal mask airway but with a few differences in structure. The laryngeal tube has both a proximal and distal inflatable cuff. The proximal cuff is large and inflates at the root of the tongue to isolate the oropharynx, allowing for positive pressure ventilation. The distal cuff inflates in the oesophagus to prevent aspiration (Saracoglu et al. 2016).

Subglottic Airway Devices: Endotracheal and Tracheostomy Tubing Subglottic airway devices extend through the vocal cords and into the trachea itself. The most frequently used subglottic airway devices are endotracheal tubes (ETTs) and tracheostomy tubes (TTs) (Fig. 3.1). Both forms of tracheal tubing provide a secure passage of air from the trachea to the environment, allowing the patient to breathe through the tubing. Both can also be attached to a ventilator. As subglottic airway devices do not allow air to pass through the vocal cords, patient speech is not possible when using these devices (Gaszyński et al. 2023).

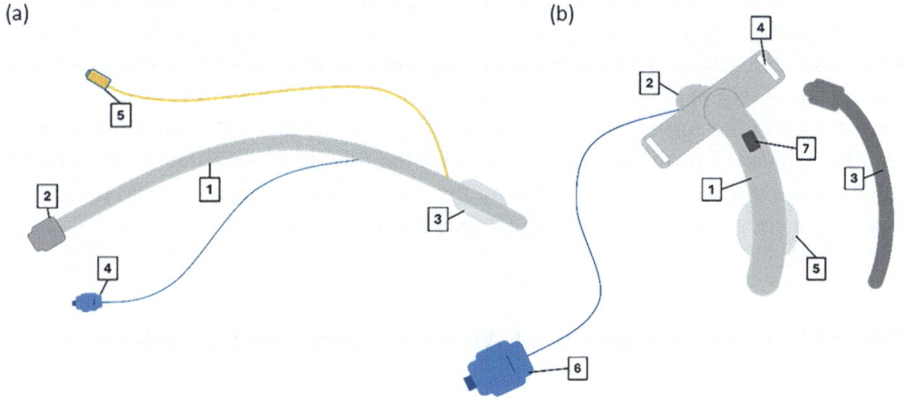

Fig. 3.1 Illustration of endotracheal and tracheostomy tubing including optional variations. (**a**) Endotracheal tube. (a1) Tube main body. (a2) Standardised adaptor. (a3) Optional inflatable cuff. (a4) Inflation port and line (present on cuffed tubes). (a5) Optional suction line for the removal of secretions pooling above the cuff. (**b**) Tracheostomy tube. (b1) Tube main body. (b2) Standardised adaptor. (b3) Tubing obturator; typically made of inflexible plastic for easier insertion of the tube. Removed after tube insertion. (b4) Flange with loops for straps to secure the tube to the patient. (b5) Optional inflatable cuff. (b6) Inflation port and line (present on cuffed tubes). (b7) Optional fenestration; allows air to be partially exhaled through the oral/nasal route and vocal cords facilitating speech and coughing

ETTs are inserted through the oral cavity and towards the trachea. They are made of a variety of materials including polyvinyl chloride (PVC), silicone and specialty polymers (Alves et al. 2023). ETTs can include an inflatable cuff at the distal tip, which provides an airtight seal against the tracheal wall and prevents aspiration of secretions from above the cuff. Some ETTs include a port just above the cuff, which is connected to a secondary line that is used to suction pooled subglottic secretions (Yuzkat and Demir 2019). To insert an ETT, clinicians use a laryngoscope to manipulate the epiglottis and visualise the vocal cords. The ETT tip is then inserted through the lumen of the vocal cords and into the trachea (Haas et al. 2014).

To use a TT, a tracheostomy must first be performed. This is the surgical generation of a stoma in the neck, extending to the trachea. This surgery can be performed as either an open or percutaneous tracheostomy. Open tracheostomy is performed in an operating theatre and involves the manual dissection of the tissues just below the cricoid cartilage to the trachea. A window is then made in the trachea, and a TT is inserted. The percutaneous tracheostomy is performed at the bedside and uses a needle and guidewire to puncture the trachea. A series of dilators is then used to expand the opening and insert a TT (Jackson 1909; Shelden 1957; Ciaglia et al. 1985).

TTs are made of similar materials to ETTs but are shorter in length. Often, TTs include an obturator, a rigid piece of plastic that fits in the lumen of the tubing to prevent the tube from bending during insertion. As with ETTs, TTs can include an inflatable cuff (Ng et al. 2024). While it is less common for TTs to include a built-in suction port, a line can be inserted through the lumen of the tube for routine suctioning of tracheal secretions (Buglass 1999). TTs are sometimes fitted with a removable inner cannula that can be cleaned or replaced easily without having to replace the full tube (Ng et al. 2024). An additional inclusion on some TTs is fenestration on the apex of the curve. If a patient does not require mechanical ventilation, a valve can be connected to the standard adaptor of the tube that allows air to be exhaled through the oral/nasal route, including the vocal cords. This allows the patient to speak and cough more functionally (Regan and Hunt 2008).

Microbial Contamination of Airway Management Devices

Of the airway management devices mentioned above, ETTs and TTs are particularly susceptible to microbial contamination since the tubes may remain in the patient for extended periods of time. Upon insertion, tracheal tubing is exposed to bodily secretions such as mucus and saliva (Winklmaier et al. 2006; Coppadoro et al. 2014). Potentially pathogenic microorganisms originating from these secretions, or the wider environment, adhere to and colonise the tubing (El Cheikh et al. 2018). If allowed to mature, these colonising pathogens organise into a biofilm, a structured colony of microorganisms that become encased within an extracellular matrix to protect themselves from external stresses and treatments (Fig. 3.2) (Limoli et al. 2015). Tubes that are contaminated by a biofilm are commonly changed to

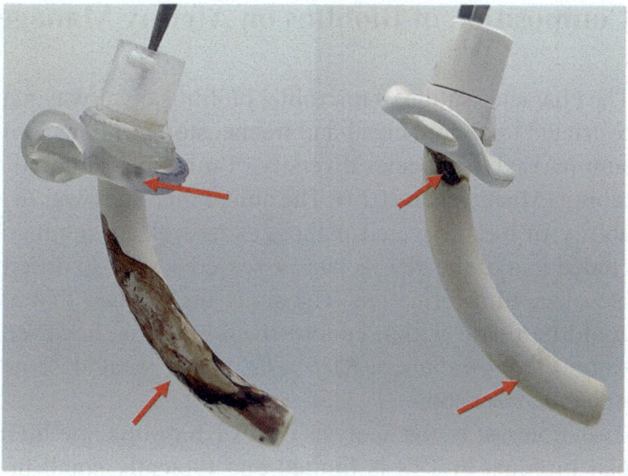

Fig. 3.2 Extracted tracheostomy tubes with visible biofilms. Red arrows indicate locations of biofilm

reduce the risk of infection and maintain the functionality of the device (Diaconu et al. 2018).

Biofilm Formation Biofilm formation begins with the colonisation of a surface by planktonic microorganism(s). Initially cells will form weak, non-specific associations with the surface. This allows for the formation of specific attachments by receptor–ligand interactions. Once adhered, biofilm-forming microorganisms undergo proliferation and organisation to form a community (Sharma et al. 2023). In this community, nutrients are exchanged to form local microbial economies that collectively benefit the colony (Arjes et al. 2021).

Once adhered, cells can begin to encapsulate themselves within an extracellular matrix (ECM). The composition of this ECM is highly diverse, as components can originate from both the colonising microorganism, such as secreted polysaccharides, as well as those scavenged from the environment, such as host proteins and cellular matter (Limoli et al. 2015; Nett et al. 2015). Once formed, the ECM functions as a protective barrier between the colonising cells and environmental stresses. Most notably, the presence of ECM provides increased tolerance to antimicrobial therapies, with minimum inhibitory concentrations often hundreds of times higher for biofilms, compared to planktonic cells of the same species (Liu et al. 2022).

Following the establishment of the biofilm and its extracellular matrix, planktonic cells can be released from the biofilm in a process termed dispersal. These cells can establish new biofilms at secondary sites of infection (Rumbaugh and Sauer 2020). In addition to progressing the spread and dissemination of the infection, cells dispersed from the primary biofilm are demonstrated to possess increased virulence traits compared to their primary colonising counterparts (Chua et al. 2014).

Another key characteristic of biofilms is their persistence. Following the formation of a biofilm, some cells enter a quiescent, dormant state, deeper within the architecture of the biofilm. The reactivation of these quiescent, persister cells is thought to be highly relevant to chronic and recurrent infections (Costerton et al. 1999).

Composition of Biofilms on Airway Management Devices

The characterisation of microbial biofilms on airway management devices is largely restricted to endotracheal and tracheostomy tubing. A recent systematic review has summarised the species diversity of microbes cultured from removed endotracheal tubing (Mishra et al. 2024). The aim of this study was to identify the biofilm-forming ability of the recovered pathogens as well as summarise prevalence data. It concluded that of the 981 isolates assessed across 20 different studies, 72% were capable of biofilm formation. The study highlighted that a diverse range of pathogens could be found within endotracheal biofilms, however, the most common species were *P. aeruginosa, Klebsiella* spp., coagulase-negative *Staphylococci* and *Acinetobacter* spp.

In contrast to the analysis of ETT biofilms, the literature to characterise microbial biofilms in TTs is limited and typically small in sample size (Koirala et al. 2010; Acharya et al. 2015; Cader et al. 2020; Ochońska et al. 2021; Raveendra et al. 2022). Here, we summarise the culture-based analysis of microorganisms recovered from TTs to date (Table 3.1)(28). As was found with endotracheal tubing, biofilms formed on TTs displayed a wide range of bacterial species, but were also primarily colonised by *P. aeruginosa, K. pneumoniae, A. baumannii* and *Staphylococcus* spp.

While bacterial species identification was conducted in all studies published to date, only three of the seven studies described a methodology that was inclusive for the detection of fungi. In these, all reported the presence of *Candida* spp. Further inspection of the 14 analyses of ETTs that aimed to characterise the microbial profile of the biofilms (Table 3.2) showed that only eight conducted fungal

Table 3.1 Summary of culture-based analyses of tracheostomy tube biofilms. (*) *Mammaliicoccus sciuri* formerly *Staphylococcus sciuri*, *Pichia kudriavzevii* formerly *Candida krusei* (**) Tracheal mucosa was swabbed directly following tube insertion and 7 days post intubation. This table includes data only from the second swab

Location of study, *n*, citation	Sample collection method	Organism identified	Incidence (%)	Methods/include detection of fungi?
India, *n* = 100, (Saravanam et al. 2022)	Swab taken from cuff (if present) or outer aspect of the maximum curvature	*Pseudomonas aeruginosa*	28	Yes
		Coagulase-negative *Staphylococcus*	27	
		Streptococcus spp.	26	
		Escherichia coli	14	
		Unspecified fungi	8	
		Klebsiella pneumoniae	3	
		Acinetobacter baumannii	1	
		Citrobacter spp.	1	

(continued)

Table 3.1 (continued)

Location of study, *n*, citation	Sample collection method	Organism identified	Incidence (%)	Methods/include detection of fungi?
Poland, *n* = 45 (Ścibik et al. 2022)	Tube removed and vortexed to collect biofilm material	*Staphylococcus aureus*	46.7	Yes
		Klebsiella pneumoniae	28.9	
		Staphylococcus epidermidis	28.9	
		Candida albicans	26.7	
		Viridans streptococci	22.2	
		Escherichia coli	15.6	
		Enterobacter cloacae	15.6	
		Pseudomonas aeruginosa	13.3	
		Corynebacterium macginleyi	6.7	
		Stenotrophomonas maltophilia	6.7	
		Proteus spp.	6.7	
		Corynebacterium spp.	6.7	
		Propionibacterium acnes	4.4	
		Saccharomyces spp.	4.4	
		Enterococcus faecalis	4.4	
		Streptococcus agalactiae	4.4	
		Neisseria spp.	4.4	
		Serratia liquefaciens	4.4	
		*Mammaliicoccus sciuri**	2.2	
		Candida parapsilosis	2.2	
		Streptococcus salivarius	2.2	
		Moraxella catarrhalis	2.2	
		*Pichia kudriavzevii**	2.2	
		Serratia marcescens	2.2	

(continued)

Table 3.1 (continued)

Location of study, *n*, citation	Sample collection method	Organism identified	Incidence (%)	Methods/include detection of fungi?
Raveendra, *n* = 35, (Raveendra et al. 2022)	Internal aspect of tubing swabbed 7 days post-intubation	*Klebsiella pneumoniae*	28.5	Precise culture methods not specified
		Acinetobacter baumannii	25.7	
		Pseudomonas aeruginosa	20.0	
		Staphylococcus aureus	14.3	
		Escherichia coli	11.4	
Oman, *n* = 108, (Cader et al. 2020)	Swab of the distal tip post-extubation	*Pseudomonas aeruginosa*	36.1	Precise culture methods not specified
		Acinetobacter baumannii	25.9	
		Klebsiella spp.	10.2	
		Coagulase-negative *Staphylococcus*	5.6	
		Proteus spp.	2.8	
		Escherichia coli	2.8	
		Staphylococcus aureus	2.8	
		Streptococcus spp.	1.9	
		Enterobacter spp.	1.9	
		Stenotrophomonas maltophilia	1.9	
		Providencia spp.	1.9	
		Unspecified coliform bacteria	1.9	
Nepal, *n* = 30, (Acharya et al. 2015)	Swab of tracheal mucosa(**)	*Pseudomonas aeruginosa*	56.7	Precise culture methods not specified
		Acinetobacter anitratus	40.0	
		Staphylococcus aureus	16.7	
		Enterococcus spp.	10.0	
		Klebsiella pneumoniae	6.7	

(continued)

Table 3.1 (continued)

Location of study, *n*, citation	Sample collection method	Organism identified	Incidence (%)	Methods/include detection of fungi?
Nepal, *n* = 50, (Koirala et al. 2010)	Tracheal aspiration (luken trap)	*Pseudomonas aeruginosa*	54.0	Culture methods inclusive of fungal growth but study aims only inclusive of bacterial isolates
		Klebsiella pneumoniae	22.0	
		Escherichia coli	18.0	
		Staphylococcus aureus	14.0	
		Klebsiella oxytoca	12.0	
		Unspecified Gram-negative bacteria	8.0	
		Viridans streptococci	4.0	
		Enterobacter cloacae	2.0	
USA, *n* = 21, (Solomon et al. 2009)	Whole tube collected. Outside of tubing sterilised. Contents of inner lumen collected by sonication	Coagulase-negative *Staphylococcus*	23.8	Yes
		Citrobacter koseri	19.0	
		Staphylococcus aureus	9.5	
		Pseudomonas aeruginosa	9.5	
		Escherichia coli	9.5	
		B-haemolytic group-b streptococci	9.5	
		Serratia marcescens	4.8	
		Stenotrophomonas maltophilia	4.8	
		Diptheroid bacteria	4.8	
		Proteus vulgaris	4.8	
		Candida albicans	4.8	
		Enterococcus faecalis	4.8	
		Streptococcus mitis	4.8	

identification (Mishra et al. 2024). Of these, six were positive for *Candida* spp., *one* for *A. fumigatus*, and one was negative for fungi. These data demonstrate that fungal pathogens, particularly *Candida* spp., are prevalent on TT and ETT biofilms but are often overlooked due to exclusive culture practices and study design. This is of added importance when considering that colonisation by *Candida* spp. in VAP patients is associated with longer ICU stay and higher mortality (Huang et al. 2020).

In summary, the biofilms formed on TT and ETT are reported to comprise similar prevalent species when common microbiology culture practices are used.

Table 3.2 Summary of studies included in Mishra et al. systematic review and whether fungi were included in the analyses. (*) Study describes culture methods that would result in fungal growth but the study aims were to identify bacterial isolates only

Location of study, *n*, citation	Inclusive of fungi?	Fungi detected?
Nepal, *n* = 70 (Baidya et al. 2021)	Yes	Yes, *Candida* spp.
UK, *n* = 20 (Cairns et al. 2011)	Yes	Yes, *Candida* spp.
Belgium, *n* = 31 (van Charante et al. 2022)	Yes	Yes, *Candida* spp.
France, *n* = 22 (Danin et al. 2015)	Yes	Yes, *Candida* spp.
Iran, *n* = NA (Dargahi et al. 2022)	No	No
Indonesia, *n* = 37 (Dewi et al. 2021)	No	No
Brazil, *n* = 50 (Ferreira et al. 2016)	No*	No
India, *n* = 87 (Gupta et al. 2015)	Yes	Yes, *Aspergillus* spp.
France, *n* = 64 (Maldiney et al. 2022)	Yes	Yes *Candida* spp.
Iran, *n* = 160 (Shahrokhi et al. 2018)	No	No
India, *n* = 15 (Singhai et al. 2012)	No*	No
India, *n* = 42 (Summaiya and Unrmi 2012)	Yes	No
India, *n* = 110 (Uppe et al. 2019)	Yes	Yes *Candida* spp.
India, *n* = 95 (Vinodkumar et al. 2017)	No*	No

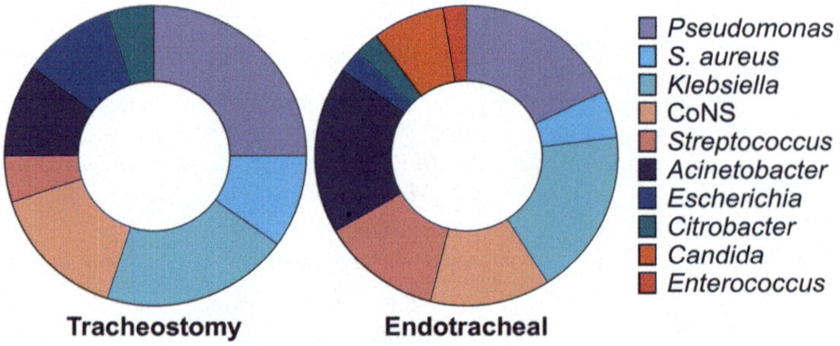

Fig. 3.3 Organisms most frequently isolated from tracheal tubing. Pie charts depicting the frequency of which organisms appear in the top 3 most prevalent organisms per study

P. aeruginosa and *Klebsiella* spp. are the most frequently identified, followed by *Acinetobacter* spp. and coagulase-negative *Staphylococci* (Fig. 3.3). *Candida* spp. are the most common fungal pathogens, but current methodologies may understate their significance and prevalence during polymicrobial biofilm formation.

Role of Host Factors in Biofilm Formation in the Airway

Role of Mucus in Biofilm Formation in the Airway When considering microbial interactions with airway management devices, it is essential to consider mucus, the resulting hydrogel formed by the polymerisation of mucin proteins. Mucus acts as a lubricant for the tissues and a barrier to external chemical and biological stresses (Bansil and Turner 2018). As the primary respiratory secretion, indwelling tracheal tubing is exposed to and coated with mucus rapidly following insertion (Markovetz et al. 2019). This leads to a change in surface properties of the tubing and its ability to become colonised by biofilm-forming microbes. Here, we summarise the composition of airway mucus, how it varies in disease phenotypes, and how microbes utilise mucus to establish a biofilm.

There are as many as 12 types of mucins documented to be found in the airway, but two, MUC5AC and MUC5B, are the most abundant (Rose and Voynow 2006). MUC5B is expressed in superficial epithelia as well as submucosal glands. MUC5B is known to play a vital role in mucociliary clearance and is therefore required for the prevention of infection (Bonser et al. 2016). MUC5AC is produced exclusively in the superficial epithelia and not deeper in the respiratory tract (Okuda et al. 2019). While not required for mucociliary clearance, MUC5AC is involved in the prevention of infection, but the precise mechanics are not well characterised (Bonser et al. 2016). To organise into mucus hydrogels, MUC5AC and MUC5B glycoproteins polymerise. Polymers are formed of each exclusive mucin type; MUC5AC forms highly branched polymers, whereas MUC5B polymers are typically linear. Large polymers are then cross-linked to form a gel (Fahy and Dickey 2010). The proportion and hydration of this gel play an important role in the physical properties of the mucus, the competency of the mucociliary escalator, and lung function. This is evident in muco-obstructive diseases such as chronic obstructive pulmonary disease, cystic fibrosis and bronchiectasis, in which mucin production and mucus hydration are both dysregulated (Henke et al. 2007; Ramsey et al. 2020; Shah et al. 2023). Those with underlying respiratory conditions are therefore more likely to have increased mucus production. They are also more likely to suffer from pulmonary inflammation, another key driver of mucin secretion, with early response cytokines such as IL-1β and TNF-α associated with MUC5AC overproduction (Ordoñez et al. 2001; Koo et al. 2002; Gray et al. 2004).

Intubation by endotracheal or tracheostomy tubing can cause local inflammation that subsequently drives mucin secretion and can affect its composition (Puyo et al. 2006). In addition, bypassing of the glottis following intubation results in an impaired cough, which reduces mucociliary clearance and causes the accumulation of mucus on the tubing surface, particularly around the distal tip and inflatable cuff (if present)(Goetz et al. 2022). All these variables considered, the ETT environment is one inundated with mucus. This is such a remarkable phenomenon that removing ETTs has been recommended as a method to sample airway mucus from patients, as it is easily obtained in high volumes and has representative rheological properties (Markovetz et al. 2019).

Many of the pathogens found to commonly colonise tracheal tubing have evolved specific responses to detect and respond to mucin to promote biofilm formation and virulence. Exposure to mucin has been shown to increase the virulence of *A. baumannii, S. agalactiae* and *Neisseria meningitidis* in a murine infection model. In these cases, it was concluded that increased virulence was due to the presence of iron in the purified porcine mucin used in these studies, not the mucin itself. While free iron levels are kept very low in the healthy airway, cystic fibrosis mucus has been shown to have unusually high levels of free iron and this is utilised by bacterial pathogens, particularly *P. aeruginosa* (Reid et al. 2007). In these investigations, iron was shown to enhance infection phenotypes, but other research has shown a clear link between iron availability and biofilm formation. Free iron available in cystic fibrosis mucus is therefore stipulated to assist in driving biofilm formation (Moreau-Marquis et al. 2015).

In addition to promoting general growth, mucin is also described as a biofilm dispersal-promoting signal in *P. aeruginosa*, where detection of MUC5AC led to active cell escape via flagellum-based motility (Co et al. 2018). The authors suggest that mucins may be used as a potential therapeutic for the disassembly of *P. aeruginosa* biofilms; however, the use of dispersal approaches is challenging as dispersed cells may relocate to secondary locations and promote infection (Thorarinsdottir et al. 2020). In *Staphylococcus* spp. growth on a mucin-containing substrate resulted in surface spreading motility by dendritic growth as governed by the *agr* quorum circuit (Jacob et al. 2024). Here, the authors present mucin-dependent motility as a mechanism by which *Staphylococcus* spp. travel over mucosal surfaces, but this same mechanism can be applied to a biofilm formed on tracheal tubing. Dendritic growth on the tracheal tube may promote growth across a wider area of the tubing increasing the risk of spread to the tracheal wall (tracheitis) as well as VAP (Ranjan et al. 2014; Blot et al. 2017).

Mucins have also been described to be detrimental to biofilm formation, notably by decreasing the ability of cells to adhere to the abiotic surface. This has been described in *C. albicans*, *P. aeruginosa*, *S. aureus* and *S. epidermidis*, in which surfaces coated with mucins lower total fungal/bacterial adhesion (Shi et al. 2000; Kavanaugh et al. 2014; Co et al. 2018). Interestingly, for *C. albicans* mucin can further inhibit biofilm formation by preventing the yeast-hyphal transition that is critical for successful *C. albicans* biofilm formation (Takagi et al. 2022).

Role of CO_2 in Biofilm Formation in the Airway Colonising organisms are exposed to the elevated CO_2 (5% compared to 0.04% found in the atmosphere) found in exhaled breath. It would appear that *C. albicans* has evolved to utilise this elevation in CO_2 as it promotes hyphal transition and filamentous growth (Mock et al. 1990; Klengel et al. 2005; Hall et al. 2010). Mechanistically, soluble CO_2 is hydrated to bicarbonate by a carbonic anhydrase. This can then bind to the adenylyl cyclase Cyr1p, which activates downstream hypha-specific genes (Hall et al. 2010). This transition drives invasive, virulent growth in host tissues but is also a key determinant of biofilm formation. Growth of *C. albicans* biofilms in the presence of 5% CO_2 resulted in increased metabolic activity, dry biomass and thickness. Increased biofilm formation was driven by an upregulation in nutrient scavenging genes, particularly that of the key iron homeostasis regulator Sef1 (Pentland et al. 2021).

The increase in *C. albicans* biofilm formation may have a profound effect on the recruitment of bacterial species. As *C. albicans* biofilms are composed of many intertwined strands of hyphae, these may behave like a scaffold to which many bacterial species can adhere (Brand et al. 2008; Harriott and Noverr 2009). While the relationships between *C. albicans* and bacterial pathogens are reported as both synergistic and antagonistic, the binding of bacterial cells to the *C. albicans* hyphae surface is observed across a broad range of pathogens relevant to the respiratory tract (Jenkinson et al. 1990; Brand et al. 2008; Harriott and Noverr 2009; de Jongh et al. 2024).

Role of Bile in Biofilm Formation in the Airway Bile is a digestive secretion produced by the liver. It is a mixture of water, bile salts, cholesterol, phospholipids and bilirubin. The role of bile is to emulsify lipids to aid in breakdown and absorption (Fuchs and Trauner 2022). As a digestive secretion, it is unsurprising that bile is not typically found in the airway. However, patients with co-morbidities that cause gastro-oesophageal reflux can aspirate bile, leading to airway inflammation (Hunt et al. 2018).

In the digestive tract, bile has antimicrobial properties, inhibiting the growth of both bacterial and fungal species, including *C. albicans* (Guinan et al. 2018; An et al. 2022). However, there is now growing evidence suggesting bile can have pro-biofilm effects on some strains of respiratory pathogens. Bile-induced biofilm formation has been documented in *P. aeruginosa*, *S. aureus* and *K. pneumoniae* (Chen et al. 2014b; Ulluwishewa et al. 2016; Flynn et al. 2019; Bayat et al. 2023). The impact of bile on *P. aeruginosa* is of critical importance, as reflux is a common symptom in cystic fibrosis patients (Bongiovanni et al. 2020). An estimated 60% of adult cystic fibrosis patients suffer from chronic *P. aeruginosa* infection, making them particularly susceptible to bile-induced biofilm formation (Maqbool and Pauwels 2017). This is further supported by the finding that long-term exposure to bile selects for hyper-biofilm forming strains and upregulates antibiotic tolerance in *P. aeruginosa*, exacerbating an already highly virulent infection (Reen et al. 2016; Flynn et al. 2019).

Research into the interaction between *C. albicans* and bile is limited to the digestive tract rather than the airway/biofilms, and reports mixed outcomes. One study reported that the secondary bile acids lithocholic and deoxycholic acid were able to inhibit *C. albicans* growth in vitro, suggesting an evolved mechanism to prevent *C. albicans* colonisation (Guinan et al. 2018). However, a more recent study documented the effect of taurocholic acid in a murine infection model. They showed that administering taurocholic acid induced *C. albicans* infection. Additionally, an increase in fungal load was detected in the faeces, implying increased dissemination (Thangamani et al. 2021). Furthermore, the fatty acid component of bile has been shown to reduce the efficacy of antifungals by encapsulating drugs such as fluconazole in micelles, preventing them from reaching the fungi (Hsieh and Brock 2017). These findings demonstrate that the regulation of individual bile components can significantly alter the microbial constituency and phenotypes during colonisation and infection.

Regarding tracheal tubing, there is limited documentation describing interactions with bile. If a tracheal tube is wrongly inserted into the oesophagus, bile can be aspirated, but this is temporary and conspicuous to the clinician. It is, however, well documented that tracheal tubing is exposed to large volumes of mucus, and that the tracheal aspirates of those suffering from VAP are more likely to contain detectable quantities of bile (Wu et al. 2009). While the precise role of bile in tracheal tube biofilms is yet to be uncovered, exposure to bile has the potential to promote biofilm formation and progress disease.

Conditioning Films as Mediators of Biofilm Formation on Airway Management Devices Conditioning films are formed from the deposition of biological molecules on the surface of an implanted device. The process of conditioning film formation, termed conditioning, occurs prior to biofilm formation and begins when a surface is exposed to an aqueous environment (Bhagwat et al. 2021). For indwelling medical devices, conditioning films are composed of human proteins, lipids, polysaccharides and nucleic acids (Bhagwat et al. 2021). The precise composition of conditioning films at different sites of the body is not well characterised but is hypothesised to vary depending on the availability of the surrounding biomolecules (Raya et al. 2022). For example, the conditioning films of urinary catheters contain large amounts of common urine markers such as Tamm-Horsfall protein and human serum albumin (Santin et al. 1999). Conditioning films are best understood in the oral cavity, specifically the dental pellicle, a conditioning film that forms on the enamel. As the quality and composition of the dental pellicle are closely associated with the formation of dental caries, it has been thoroughly analysed and characterised (Enax et al. 2023). From the understanding of the dental pellicle, the composition of the tracheal tube conditioning film can be speculated. Common proteins shared between tracheal secretions and saliva are core components of the dental pellicle, such as lysozyme, lactoferrin and, notably, mucin (Enax et al. 2023). As discussed above, mucins and mucus are abnormally abundant in intubated patients and are therefore likely to be a central component of the tracheal tube conditioning film.

Conditioning films are important in terms of biofilm formation as they provide an alternative site of adhesion for colonising pathogens. This is notable as many pathogens lack the ability to adhere to some abiotic surfaces but will be adept at binding human matrix proteins, such as fibrinogen, which are common in conditioning films (Busscher and Weerkamp 1987; Herrmann et al. 1988).

The study of how conditioning films influence biofilm formation on medical devices is surprisingly limited and often focuses on how single bacterial species adhere. For example, coating a plastic substrate with fibrinogen, fibronectin or laminin can increase the adhesion of coagulase-negative *Staphylococci* or *Corynebacterium* (Herrmann et al. 1988; Souza et al. 2015), but coating with bovine serum albumin can decrease the adhesion of *P. aeruginosa* (Hwang et al. 2013). While these findings are key building blocks in the process of understanding the precise role of conditioning films, it is unlikely that a film of pure fibrinogen exists in nature. Assessing the role of conditioning films in this manner is not recommended when considering the findings from dental research, which demonstrate that even small modulations in the constituency of the dental pellicle (over 100 components (Siqueira et al. 2007)) can result in the promotion or reduction of pathogenic colonisation (Siqueira et al. 2012).

Although the mechanisms relating conditioning and pathogenic colonisation remain complex, the relationship between conditioning and abiotic surfaces is better understood. Of note is the impact conditioning has on the efficacy of engineered antimicrobial surfaces. When conditioned with whole blood, a zirconium-nitride silver surface, documented to have strong antimicrobial properties when uncoated, had markedly reduced efficacy. This highlights the importance of conditioning in producing antimicrobial surfaces, and how engineering a surface to be tolerant of pathogenic colonisation may not be sufficient to prevent biofilm formation (Slate et al. 2019).

Approaches to the Prevention of Biofilm Formation on Airway Management Devices

To limit biofilm formation on airway management devices, numerous antifouling strategies have been developed. Here, we briefly summarise the most prominent approaches to antifouling that could be used on airway management devices.

Metal Cation Coatings Metal coatings are a commonly used antimicrobial strategy, relying on the ability of metals to ionically bind the surface of microorganisms and cause cell lysis by modifying membrane permeability (Tincu et al. 2022). Silver-based and zinc-based coatings are frequently included in novel coatings in the form of nanoparticles. There are numerous studies highlighting the efficacy of both silver and zinc nanoparticles against a wide range of pathogens in in vitro, in vivo and clinical studies (Seil and Webster 2011; Webster and Geilich 2013; Jiang et al. 2017; Zhang et al. 2017; Cruz-Pacheco et al. 2019; Bechtold et al. 2020; Tincu et al. 2022). A notable application of a silver-based antimicrobial that is used in clinical practice is the silver-gold-palladium coating produced by Bactiguard® Infection Protocol. Coating ETTs with this novel antimicrobial decreased the incidence of VAP in a 180-patient cohort by 16% (Tincu et al. 2022). Less commonly used metal coatings include selenium and titanium, which have smaller bodies of evidence to support their usage but still demonstrate efficacy against various bacterial pathogens (Tran and Webster 2013; Deng et al. 2016; Caratto et al. 2017).

Bio-inspired Coatings Bio-inspired coatings include all forms of antimicrobial peptides and biological approaches. These are typically positively charged polyamino acids that adhere to cell membranes and form pores, providing an often broad range of antimicrobial effects (Huang 2006). Another class of bio-inspired solutions is the ceragenins, also termed cationic steroid antibiotics. These are synthetic sterols modified with various amines. The mechanism of action of these molecules is thought to be similar to antimicrobial peptides in that their positive charge is attracted to the negatively charged cell membrane of bacterial cells (Savage et al. 2002). A more complex approach is the treatment and coating of surfaces with phages. Phages can be used to selectively inhibit the growth of target microbes. For example, treatment of an ETT surface with *P. aeruginosa* targeting phages removed the biofilm of multidrug-resistant strain (Amankwah et al. 2022).

Disinfectant Impregnation Polymers can also be impregnated with disinfectants such as chlorhexidine or gendine. Impregnation with such additives can be as potent as silver coating, and effective against a similarly broad range of pathogens (Raad et al. 2011).

Hydrogels, Nanopillars and Brush Coatings As an alternative to antimicrobials, the modification of the surface properties of the medical device is another viable option for the prevention of microbial colonisation. Surface modifications can be broadly categorised into three groups: hydrogel barriers, nanopillars and polymer brush coatings.

Synthetic hydrogels act as a barrier to medical devices, as mucus does to mucosal tissues. The composition of this hydrogel can be precisely tailored and optimised for the target system by modifying its initial formulation and inclusions. Hydrogels have been described as broadly anti-adhesive, but can also be co-formulated with the antimicrobial systems summarised above (Sanjanwala et al. 2024). Regarding tracheal tubing, coating ETTs with a silver-containing hydrogel has shown good efficacy in preventing device-related respiratory infections in a rabbit model (Hu et al. 2023). Another benefit of some hydrogel coatings is their ability to prevent protein adsorption to the device surface. As detailed above, the development of proteinaceous conditioning films can significantly increase bacterial adhesion and biofilm formation while also decreasing the efficacy of impregnated antimicrobial coatings, making synthetic hydrogels an attractive candidate (Slate et al. 2019; Chen et al. 2021).

Nanopillar surfaces are ultra-precise modifications to a surface to build spike-like projections smaller than that of a bacterial cell. Based on surfaces found in nature, including insect wings and shark skin, adsorption of the cell to the nanopillars results in cell surface stress and further membrane disruption and lysis (Bhadra et al. 2015; Ishak et al. 2021; Rostami et al. 2022; Ewunkem et al. 2023). Despite their intriguing mechanism of action, nanopillar-based antimicrobials have not been widely implemented within medical devices to date.

This summary has highlighted a number of antimicrobial strategies that could be used on tracheal tubing. However, despite many approaches demonstrating efficacy in vitro, most fail to proceed beyond the pre-clinical stage of development. This limitation may be due to the high cost of progressing to clinical trials, but the unpredictability and complexity of in vivo systems are also common challenges (Alves et al. 2023).

Perspectives and Concluding Remarks

Host Factors Have a Remarkable Impact on Biofilm Formation The importance of host factors in modelling infection is well understood. However, research examining biofilm formation often excludes host factors, focusing more on inter-microbial interactions and those between microbes and surfaces. Research aimed at characterising the role of host factors in biofilm formation is generally limited, often investigating the role of a single host protein and its impact on biofilm formation as determined by in vitro microtitre plate assays. While these investigations are important and provide an initial indication of specific proteins' involvement in biofilm formation, they reveal little information about the host factor's true role in a clinical biofilm that includes all associated host factors. For example, mucus and bile are two bodily secretions composed of a cocktail of biological molecules (Bansil and Turner 2018; Fuchs and Trauner 2022). Supplementation of biofilms with individual components of each of these mixtures can result in the promotion or reduction of biofilm formation in various microorganisms, as described above. Given that individual components of these host factors can have opposing effects on biofilm formation, it is difficult to conclude from in vitro studies the net impact of these secretions on biofilm formation. From clinical data, it is possible to infer the importance of host factors. For example, the regular suctioning of tracheal tubing to remove pooled secretions results in a significantly lower rate of biofilm formation (Buglass 1999; Aguilera Xiol et al. 2015), but this does not provide details of mechanisms.

In order to effectively model biofilm formation at a given location in the body, consideration of the host factors is critical. An in vivo investigation into device-associated biofilms demonstrated that local proteins may form a structural component of the biofilm itself, rather than just behave as an initial conditioning film, and that this is governed by the colonising microbes (Nett et al. 2015). Prior to colonisation by *C. albicans*, vascular catheters contained a diverse range of serum proteins, but post-colonisation, haemoglobins were by far the dominant protein. In urinary catheters, *C. albicans* colonisation selected for keratin and fibrinogen, and in denture biofilms amylase was selected. This study provides evidence of a single microbe specifically manipulating host factors at various anatomical locations. As clinical biofilms are typically polymicrobial, the manipulation of host factors is likely highly complex and dependent on the microbial profile.

Role of Fungi in Biofilm Formation on Tracheal Tubing is Largely Overlooked When observing the culture-based species analysis of endotracheal and tracheostomy tubing, a selection of pathogens dominate the tracheal tube biofilms, *P. aeruginosa*, *Staphylococcus* spp., *A. baumannii* and *K. pneumoniae* are by far the most ubiquitous members as reported by culture-based analyses. Perhaps overlooked in this list of universally prevalent microorganisms are *Candida* spp. While not present in all the culture-based analyses detailed above, many of these investigations aimed only to quantify prokaryotic presence and largely excluded fungi from their analysis. Of the studies that did include the detection of fungi in their analysis, *Candida* spp. were isolated in all but two (Solomon et al. 2009; Cairns et al. 2011; Summaiya and Unrmi 2012; Gupta et al. 2015; Danin et al. 2015; Uppe et al. 2019; Baidya et al. 2021; Ścibik et al. 2022; Saravanam et al. 2022; van Charante et al. 2022; Maldiney et al. 2022). Also worth noting is the chronology of biofilm formation and the suitability of culture-based analyses for investigating key pathogens in biofilm formation. For example, *P. aeruginosa* is documented to adhere to the hyphae of *C. albicans* with high affinity. Following adherence, *P. aeruginosa* secretes several virulence factors that kill the fungi and allow succession as the dominant organism in the polymicrobial biofilm (Fourie et al. 2016). Here, initial colonisation is mediated by *C. albicans*, which then facilitates the rapid colonisation by *P. aeruginosa*. As the fungi are generally outcompeted by *P. aeruginosa* in terms of direct killing and nutrient scavenging (Fourie et al. 2018), *C. albicans* may not be detected in the culture-based analysis, which provides only an end-point snapshot of the polymicrobial biofilm. To combat this problem, a DNA sequencing-based approach may be more appropriate as it can capture a fingerprint of all organisms that were present from initial colonisation to the end of the device's lifespan.

Presence of *C. albicans* is Associated with Colonisation by the Bacterial Causes of VAP The importance of *C. albicans* in biofilm formation on tracheal tubes is highlighted by its synergy with key bacterial pathogens. For each of the most common bacterial pathogens described above, there is literature describing a mechanism by which *C. albicans* can enhance their colonisation and/or pathogenesis (Table 3.3).

Further supporting the role of *C. albicans* in airway device colonisation is a clinical investigation into *Candida* bloodstream infections in which bacterial pathogens are co-isolated (Zhong et al. 2020). Here, it was shown that *P. aeruginosa*, *K. pneumoniae*, *Staphylococcus* spp. and *A. baumannii* made up 80% of the species co-isolated with *C. albicans*, aligning with the culture-based analyses of tracheal tube biofilms. Furthermore, of the patients who presented with *C. albicans* and more than one bacterial pathogen, the most common baseline characteristics noted were central venous catheter (95.8%); urethral catheter (87.5%); total parenteral nutrition (83.3%); and invasive mechanical ventilation (81.8%); all of which are closely associated with biofilm formation and device-related infection, particularly that of *C. albicans* (Swindell et al. 2009; Rishpana 2015; Timsit et al. 2019; Jung et al. 2020).

In the airway, although not established, it may be the case that the bacterial pathogens commonly found on tracheal tubing, which are established causes of VAP, take advantage of *C. albicans'* ability to utilise elevated CO_2 to grow large biofilms rapidly. If this is the case, the prevention of *C. albicans* colonisation may

Table 3.3 *C. albicans* synergy with key respiratory pathogens

Organism	Description	Citation
P. aeruginosa	*C. albicans* produced ethanol stimulates matrix deposition and adhesion in *P. aeruginosa*, promoting a biofilm phenotype	Chen et al. (2014a)
A. baumannii	Presence of *C. albicans* resulted in more frequent *A. baumannii* colonisation. Antifungal treatment lowered the incidence of bacterial infection	Tan et al. (2016)
Staphylococcus spp.	*S. aureus* binds to *C. albicans* hyphal scaffold to enhance adhesion, then uses *C. albicans* extracellular matrix to increase tolerance to vancomycin	Kong et al. (2016)
K. pneumoniae	*K. pneumoniae* infection is associated with co-isolation of *Candida* spp. possibly originating from parenteral feeding or TT	Papadimitriou-Olivgeris et al. (2014)

offer an alternative prophylactic strategy to prevent VAP in intubated patients. While this approach has not yet been attempted in airway management devices, it has been proven in the prevention of biofilm formation on voice prostheses (Pentland et al. 2020). It was initially noted that voice prostheses were rapidly colonised by yeast biofilms and needed frequent replacement. A microbiological analysis of the biofilms found that the *C. albicans* and *S. aureus* were the dominant species in this system. As it was understood in the literature that *S. aureus* can use *C. albicans* to enhance its own attachment and biofilm formation (Harriott and Noverr 2009; Peters et al. 2012), the investigators developed a prophylactic cleaning regime by repurposing the antifungal drug nystatin. The regimen was designed to be performed at home without the need of a clinician and would prevent fungal growth as well as remove residual conditioning. The targeted antifungal treatment was highly successful, extending device lifespan by an average of 270%, and is now the recommended practice by a range of healthcare providers (Pentland et al. 2020).

In summary, future characterisation of biofilm formation on airway management devices should attempt to understand fungal and bacterial pathogens, and the chronology of their colonisation, as well as the complex interface between the microbes and host factors. A fuller comprehension of these variables will assist significantly in designing targeted approaches to prevent biofilm formation and improve patient outcomes.

References

Acharya R, Wakode PT, Bhandary S et al (2015) Colonization and infection in tracheostomized patients at tertiary care hospital in Eastern Nepal. Health Renaissance 12:68–73. https://doi.org/10.3126/hren.v12i2.14102

Aguilera Xiol E, Li Bassi G, Wyncoll D et al (2015) Tracheal tube biofilm removal through a novel closed-suctioning system: an experimental study. Br J Anaesth 115:775–783. https://doi.org/10.1093/bja/aev340

Alves D, Grainha T, Pereira MO, Lopes SP (2023) Antimicrobial materials for endotracheal tubes: a review on the last two decades of technological progress. Acta Biomater 158:32–55. https://doi.org/10.1016/j.actbio.2023.01.001

Amankwah S, Adisu M, Gorems K et al (2022) Assessment of phage-mediated inhibition and removal of multidrug-resistant Pseudomonas aeruginosa biofilm on medical implants. Infect Drug Resist 15:2797–2811. https://doi.org/10.2147/IDR.S367460

An C, Chon H, Ku W et al (2022) Bile acids: major regulator of the gut microbiome. Microorganisms 10:1792. https://doi.org/10.3390/microorganisms10091792

Arjes HA, Willis L, Gui H et al (2021) Three-dimensional biofilm colony growth supports a mutualism involving matrix and nutrient sharing. elife 10. https://doi.org/10.7554/eLife.64145

Baidya S, Sharma S, Mishra SK et al (2021) Biofilm formation by pathogens causing ventilator-associated pneumonia at intensive care units in a tertiary care hospital: an armor for refuge. Biomed Res Int 2021:1–10. https://doi.org/10.1155/2021/8817700

Bansil R, Turner BS (2018) The biology of mucus: composition, synthesis and organization. Adv Drug Deliv Rev 124:3–15. https://doi.org/10.1016/j.addr.2017.09.023

Bayat M, Nahand JS, Farsad-Akhatr N, Memar MY (2023) Bile effects on the Pseudomonas aeruginosa pathogenesis in cystic fibrosis patients with gastroesophageal reflux. Heliyon 9:e22111. https://doi.org/10.1016/j.heliyon.2023.e22111

Bechtold M, Valério A, Ulson de Souza AA et al (2020) Synthesis and application of silver nanoparticles as biocidal agent in polyurethane coating. J Coat Technol Res 17:613–620. https://doi.org/10.1007/s11998-019-00297-0

Bhadra CM, Khanh Truong V, Pham VTH et al (2015) Antibacterial titanium nano-patterned arrays inspired by dragonfly wings. Sci Rep 5:16817. https://doi.org/10.1038/srep16817

Bhagwat G, O'Connor W, Grainge I, Palanisami T (2021) Understanding the fundamental basis for biofilm formation on plastic surfaces: role of conditioning films. Front Microbiol 12. https://doi.org/10.3389/fmicb.2021.687118

Blot M, Bonniaud-Blot P, Favrolt N et al (2017) Update on childhood and adult infectious tracheitis. Med Mal Infect 47:443–452. https://doi.org/10.1016/j.medmal.2017.06.006

Bongiovanni A, Manti S, Parisi GF et al (2020) Focus on gastroesophageal reflux disease in patients with cystic fibrosis. World J Gastroenterol 26:6322–6334. https://doi.org/10.3748/wjg.v26.i41.6322

Bonser LR, Zlock L, Finkbeiner W, Erle DJ (2016) Epithelial tethering of MUC5AC-rich mucus impairs mucociliary transport in asthma. J Clin Invest 126:2367–2371. https://doi.org/10.1172/JCI84910

Brain AI (1983) The laryngeal mask-A new concept in airway management. Br J Anaesth 55:801–805. https://doi.org/10.1093/bja/55.8.801

Brand A, Barnes JD, Mackenzie KS et al (2008) Cell wall glycans and soluble factors determine the interactions between the hyphae of Candida albicans and Pseudomonas aeruginosa. FEMS Microbiol Lett 287:48–55. https://doi.org/10.1111/j.1574-6968.2008.01301.x

Buglass E (1999) Tracheostomy care: tracheal suctioning and humidification. Br J Nurs 8:500–504. https://doi.org/10.12968/bjon.1999.8.8.6627

Busscher HJ, Weerkamp AH (1987) Specific and non-specific interactions in bacterial adhesion to solid substrata. FEMS Microbiol Lett 46:165–173. https://doi.org/10.1111/j.1574-6968.1987.tb02457.x

Cader SHA, Shah FA, Nair SKGR (2020) Tracheostomy colonisation and microbiological isolates of patients in intensive care units-a retrospective study. World J Otorhinolaryngol Head Neck Surg 6:49–52. https://doi.org/10.1016/j.wjorl.2019.04.002

Cairns S, Thomas JG, Hooper SJ et al (2011) Molecular analysis of microbial communities in endotracheal tube biofilms. PLoS One 6:e14759. https://doi.org/10.1371/journal.pone.0014759

Caratto V, Ball L, Sanguineti E et al (2017) Antibacterial activity of standard and N-doped titanium dioxide-coated endotracheal tubes: an in vitro study. Rev Bras Ter Intensiva 29:55. https://doi.org/10.5935/0103-507X.20170009

Chen AI, Dolben EF, Okegbe C et al (2014a) Candida albicans ethanol stimulates Pseudomonas aeruginosa WspR-controlled biofilm formation as part of a cyclic relationship involving Phenazines. PLoS Pathog 10:e1004480. https://doi.org/10.1371/journal.ppat.1004480

Chen K-M, Chiang M-K, Wang M et al (2014b) The role of pgaC in Klebsiella pneumoniae virulence and biofilm formation. Microb Pathog 77:89–99. https://doi.org/10.1016/j.micpath.2014.11.005

Chen Q, Zhang D, Gu J et al (2021) The impact of antifouling layers in fabricating bioactive surfaces. Acta Biomater 126:45–62. https://doi.org/10.1016/j.actbio.2021.03.022

Chua SL, Liu Y, Yam JKH et al (2014) Dispersed cells represent a distinct stage in the transition from bacterial biofilm to planktonic lifestyles. Nat Commun 5:4462. https://doi.org/10.1038/ncomms5462

Ciaglia P, Firsching R, Syniec C (1985) Elective percutaneous dilatational tracheostomy. Chest 87:715–719. https://doi.org/10.1378/chest.87.6.715

Co JY, Cárcamo-Oyarce G, Billings N et al (2018) Mucins trigger dispersal of Pseudomonas aeruginosa biofilms. NPJ Biofilms Microbiomes 4:23. https://doi.org/10.1038/s41522-018-0067-0

Coppadoro A, Bellani G, Bronco A et al (2014) Measurement of endotracheal tube secretions volume by micro computed tomography (MicroCT) scan: an experimental and clinical study. BMC Anesthesiol 14:22. https://doi.org/10.1186/1471-2253-14-22

Costerton JW, Stewart PS, Greenberg EP (1999) Bacterial biofilms: a common cause of persistent infections. Science (1979) 284:1318–1322. https://doi.org/10.1126/science.284.5418.1318

Cruz-Pacheco A, Muñoz-Castiblanco D, Gómez Cuaspud J et al (2019) Coating of polyetheretherketone films with silver nanoparticles by a simple chemical reduction method and their antibacterial activity. Coatings 9:91. https://doi.org/10.3390/coatings9020091

Danin P-E, Girou E, Legrand P et al (2015) Description and microbiology of endotracheal tube biofilm in mechanically ventilated subjects. Respir Care 60:21–29. https://doi.org/10.4187/respcare.02722

Dargahi Z, Hamad AA, Sheikh AF et al (2022) The biofilm formation and antibiotic resistance of bacterial profile from endotracheal tube of patients admitted to intensive care unit in southwest of Iran. PLoS One 17:e0277329. https://doi.org/10.1371/journal.pone.0277329

de Jongh CA, Bikker FJ, de Vries TJ et al (2024) Porphyromonas gingivalis interaction with Candida albicans allows for aerobic escape, virulence and adherence. Biofilms 7:100172. https://doi.org/10.1016/j.bioflm.2023.100172

Deng W, Ning S, Lin Q et al (2016) I-TiO2/PVC film with highly photocatalytic antibacterial activity under visible light. Colloids Surf B Biointerfaces 144:196–202. https://doi.org/10.1016/j.colsurfb.2016.03.085

Dewi FH, Purwanto B, Wasita B (2021) Quantitative biofilm for bacterial pathogens of ventilator-associated pneumonia. Anaesth Pain Intensive Care 25. https://doi.org/10.35975/apic.v25i2.1468

Diaconu O, Siriopol I, Poloşanu LI, Grigoraş I (2018) Endotracheal tube biofilm and its impact on the pathogenesis of ventilator-associated pneumonia. J Crit Care Med 4:50–55. https://doi.org/10.2478/jccm-2018-0011

El Cheikh M, Barbosa J, Caixêta J, Avelino M (2018) Microbiology of tracheal secretions: what to expect with children and adolescents with tracheostomies. Int Arch Otorhinolaryngol 22:050–054. https://doi.org/10.1055/s-0037-1601403

Enax J, Ganss B, Amaechi BT et al (2023) The composition of the dental pellicle: an updated literature review. Front Oral Health 4. https://doi.org/10.3389/froh.2023.1260442

Ewunkem AJ, Beard AF, Justice BL et al (2023) Honeybee wings hold antibiofouling and antimicrobial clues for improved applications in health care and industries. AIMS Microbiol 9:332–345. https://doi.org/10.3934/microbiol.2023018

Fahy JV, Dickey BF (2010) Airway mucus function and dysfunction. N Engl J Med 363:2233–2247. https://doi.org/10.1056/NEJMra0910061

Ferreira TDO, Koto RY, Leite GFDC et al (2016) Microbial investigation of biofilms recovered from endotracheal tubes using sonication in intensive care unit pediatric patients. Braz J Infect Dis 20:468–475. https://doi.org/10.1016/j.bjid.2016.07.003

Flynn S, Reen FJ, O'Gara F (2019) Exposure to bile leads to the emergence of adaptive signaling variants in the opportunistic pathogen Pseudomonas aeruginosa. Front Microbiol 10. https://doi.org/10.3389/fmicb.2019.02013

Fourie R, Ells R, Swart CW et al (2016) Candida albicans and Pseudomonas aeruginosa interaction, with focus on the role of eicosanoids. Front Physiol 7. https://doi.org/10.3389/fphys.2016.00064

Fourie R, Kuloyo OO, Mochochoko BM et al (2018) Iron at the Centre of Candida albicans interactions. Front Cell Infect Microbiol 8. https://doi.org/10.3389/fcimb.2018.00185

Fuchs CD, Trauner M (2022) Role of bile acids and their receptors in gastrointestinal and hepatic pathophysiology. Nat Rev Gastroenterol Hepatol 19:432–450. https://doi.org/10.1038/s41575-021-00566-7

Gaszyński T, Gómez-Ríos MÁ, Serrano-Moraza A et al (2023) New devices, innovative technologies, and non-standard techniques for airway management: a narrative review. Healthcare 11:2468. https://doi.org/10.3390/healthcare11182468

Goetz RL, Vijaykumar K, Solomon GM (2022) Mucus clearance strategies in mechanically ventilated patients. Front Physiol 13. https://doi.org/10.3389/fphys.2022.834716

Gray T, Coakley R, Hirsh A et al (2004) Regulation of MUC5AC mucin secretion and airway surface liquid metabolism by IL-1β in human bronchial epithelia. Am J Phys Lung Cell Mol Phys 286:L320–L330. https://doi.org/10.1152/ajplung.00440.2002

Guinan J, Villa P, Thangamani S (2018) Secondary bile acids inhibit Candida albicans growth and morphogenesis. Pathog Dis 76. https://doi.org/10.1093/femspd/fty038

Gupta R, Malik A, Rizvi M, Adhmed M (2015) Biofilm producing multidrug and extensive drug resistant bacterial pathogens from tracheal aspirates of intensive care unit patients—a threat to combat. Int J Curr Microbiol Appl Sci:1–9

Haas CF, Eakin RM, Konkle MA, Blank R (2014) Endotracheal tubes: old and new discussion. Respir Care 59:933–955. https://doi.org/10.4187/respcare.02868

Hall RA, De Sordi L, MacCallum DM et al (2010) CO2 acts as a signalling molecule in populations of the fungal pathogen Candida albicans. PLoS Pathog 6:e1001193. https://doi.org/10.1371/journal.ppat.1001193

Harriott MM, Noverr MC (2009) Candida albicans and Staphylococcus aureus form polymicrobial biofilms: effects on antimicrobial resistance. Antimicrob Agents Chemother 53:3914–3922. https://doi.org/10.1128/AAC.00657-09

Henke MO, John G, Germann M et al (2007) MUC5AC and MUC5B mucins increase in cystic fibrosis airway secretions during pulmonary exacerbation. Am J Respir Crit Care Med 175:816–821. https://doi.org/10.1164/rccm.200607-1011OC

Herrmann M, Vaudaux PE, Pittet D et al (1988) Fibronectin, fibrinogen, and laminin act as mediators of adherence of clinical staphylococcal isolates to foreign material. J Infect Dis 158:693–701. https://doi.org/10.1093/infdis/158.4.693

Hickey SM, Sankari A, Giwa AO (2024) Mechanical ventilation. In: StatPearls. https://www.ncbi.nlm.nih.gov/books/NBK539742/. Accessed 6 Aug 2024

Hsieh S-H, Brock M (2017) Lipid components of bile increase the protective effect of conjugated bile salts against antifungal drugs. Fungal Biol 121:929–938. https://doi.org/10.1016/j.funbio.2017.08.002

Hu C, Yu Y, Qian H et al (2023) Antibacterial endotracheal tube with silver-containing double-network hydrogel coating. Colloid Interface Sci Commun 55:100724. https://doi.org/10.1016/j.colcom.2023.100724

Huang HW (2006) Molecular mechanism of antimicrobial peptides: the origin of cooperativity. Biochim Biophys Acta (BBA) – Biomembranes 1758:1292–1302. https://doi.org/10.1016/j.bbamem.2006.02.001

Huang D, Qi M, Hu Y et al (2020) The impact of Candida spp airway colonization on clinical outcomes in patients with ventilator-associated pneumonia: a systematic review and meta-analysis. Am J Infect Control 48:695–701. https://doi.org/10.1016/j.ajic.2019.11.002

Hunt EB, Sullivan A, Galvin J et al (2018) Gastric aspiration and its role in airway inflammation. Open Respir Med J 12:1–10. https://doi.org/10.2174/1874306401812010001

Hwang G, Liang J, Kang S et al (2013) The role of conditioning film formation in Pseudomonas aeruginosa PAO1 adhesion to inert surfaces in aquatic environments. Biochem Eng J 76:90–98. https://doi.org/10.1016/j.bej.2013.03.024

Ishak MI, Jenkins J, Kulkarni S et al (2021) Insights into complex nanopillar-bacteria interactions: roles of nanotopography and bacterial surface proteins. J Colloid Interface Sci 604:91–103. https://doi.org/10.1016/j.jcis.2021.06.173

Jackson C (1909) Tracheotomy. Laryngoscope 19:285–290. https://doi.org/10.1288/00005537-190904000-00003

Jacob KM, Hernández-Villamizar S, Hammer ND, Reguera G (2024) Mucin-induced surface dispersal of Staphylococcus aureus and Staphylococcus epidermidis via quorum-sensing dependent and independent mechanisms. MBio 15:e0156224. https://doi.org/10.1128/mbio.01562-24

Jenkinson HF, Lala HC, Shepherd MG (1990) Coaggregation of Streptococcus sanguis and other streptococci with Candida albicans. Infect Immun 58:1429–1436. https://doi.org/10.1128/iai.58.5.1429-1436.1990

Jiang X, Lv B, Shen Q, Wang X (2017) Preparation of silicon-modified antimicrobial polyethylene endotracheal tubes. J Biomed Mater Res B Appl Biomater 105:91–98. https://doi.org/10.1002/jbm.b.33530

Jung P, Mischo CE, Gunaratnam G et al (2020) Candida albicans adhesion to central venous catheters: impact of blood plasma-driven germ tube formation and pathogen-derived adhesins. Virulence 11:1453–1465. https://doi.org/10.1080/21505594.2020.1836902

Kavanaugh NL, Zhang AQ, Nobile CJ et al (2014) Mucins suppress virulence traits of Candida albicans. mBio 5:e01911. https://doi.org/10.1128/mBio.01911-14

Klengel T, Liang W-J, Chaloupka J et al (2005) Fungal adenylyl cyclase integrates CO2 sensing with cAMP signaling and virulence. Curr Biol 15:2021–2026. https://doi.org/10.1016/j.cub.2005.10.040

Koirala P, Bhatta DR, Ghimire P et al (2010) Bacteriological profile of tracheal aspirates of the patients attending neuro-hospital Nepal. Int J Life Sci 4. https://doi.org/10.3126/ijls.v4i0.3496

Kong EF, Tsui C, Kucharíková S et al (2016) Commensal protection of Staphylococcus aureus against antimicrobials by Candida albicans biofilm matrix. mBio 7. https://doi.org/10.1128/mBio.01365-16

Koo JS, Kim Y-D, Jetten AM et al (2002) Overexpression of mucin genes induced by interleukin 1-1β, tumor necrosis factor-α, lipopolysaccharide, and neutrophil elastase is inhibited by a retinoic acid receptor α antagonist. Exp Lung Res 28:315–332. https://doi.org/10.1080/01902140252964393

Limoli DH, Jones CJ, Wozniak DJ (2015) Bacterial extracellular polysaccharides in biofilm formation and function. Microbiol Spectr 3. https://doi.org/10.1128/microbiolspec.MB-0011-2014

Liu D, Huang Q, Gu W, Zeng X-A (2022) A review of bacterial biofilm control by physical strategies. Crit Rev Food Sci Nutr 62:3453–3470. https://doi.org/10.1080/10408398.2020.1865872

Maldiney T, Pineau V, Neuwirth C et al (2022) Endotracheal tube biofilm in critically ill patients during the COVID-19 pandemic: description of an underestimated microbiological compartment. Sci Rep 12:22389. https://doi.org/10.1038/s41598-022-26560-w

Maqbool A, Pauwels A (2017) Cystic fibrosis and gastroesophageal reflux disease. J Cyst Fibros 16:S2–S13. https://doi.org/10.1016/j.jcf.2017.07.007

Markovetz MR, Subramani DB, Kissner WJ et al (2019) Endotracheal tube mucus as a source of airway mucus for rheological study. Am J Phys Lung Cell Mol Phys 317:L498–L509. https://doi.org/10.1152/ajplung.00238.2019

Mishra SK, Baidya S, Bhattarai A et al (2024) Bacteriology of endotracheal tube biofilms and antibiotic resistance: a systematic review. J Hosp Infect 147:146–157. https://doi.org/10.1016/j.jhin.2024.03.004

Mock RC, Pollack JH, Hashimoto T (1990) Carbon dioxide induces endotrophic germ tube formation in Candida albicans. Can J Microbiol 36:249–253. https://doi.org/10.1139/m90-043

Moreau-Marquis S, Coutermarsh B, Stanton BA (2015) Combination of hypothiocyanite and lactoferrin (ALX-109) enhances the ability of tobramycin and aztreonam to eliminate Pseudomonas aeruginosa biofilms growing on cystic fibrosis airway epithelial cells. J Antimicrob Chemother 70:160–166. https://doi.org/10.1093/jac/dku357

Nett JE, Zarnowski R, Cabezas-Olcoz J et al (2015) Host contributions to construction of three device-associated Candida albicans biofilms. Infect Immun 83:4630–4638. https://doi.org/10.1128/IAI.00931-15

Ng J, Hamrang-Yousefi S, Hohman M, et al (2024) Tracheostomy tube change. In: StatPearls. https://www.ncbi.nlm.nih.gov/books/NBK555919/. Accessed 31 July 2024

Ochońska D, Ścibik Ł, Brzychczy-Włoch M (2021) Biofilm formation of clinical Klebsiella pneumoniae strains isolated from tracheostomy tubes and their association with antimicrobial resistance, virulence and genetic diversity. Pathogens 10:1345. https://doi.org/10.3390/pathogens10101345

Okuda K, Chen G, Subramani DB et al (2019) Localization of secretory mucins MUC5AC and MUC5B in normal/healthy human airways. Am J Respir Crit Care Med 199:715–727. https://doi.org/10.1164/rccm.201804-0734OC

Ordoñez CL, Khashayar R, Wong HH et al (2001) Mild and moderate asthma is associated with airway goblet cell hyperplasia and abnormalities in mucin gene expression. Am J Respir Crit Care Med 163:517–523. https://doi.org/10.1164/ajrccm.163.2.2004039

Papadimitriou-Olivgeris M, Spiliopoulou A, Fligou F et al (2014) Association of KPC-producing Klebsiella pneumoniae colonization or infection with Candida isolation and selection of non-albicans species. Diagn Microbiol Infect Dis 80:227–232. https://doi.org/10.1016/j.diagmicrobio.2014.07.012

Papazian L, Klompas M, Luyt C-E (2020) Ventilator-associated pneumonia in adults: a narrative review. Intensive Care Med 46:888–906. https://doi.org/10.1007/s00134-020-05980-0

Pentland DR, Stevens S, Williams L et al (2020) Precision antifungal treatment significantly extends voice prosthesis lifespan in patients following total laryngectomy. Front Microbiol 11. https://doi.org/10.3389/fmicb.2020.00975

Pentland DR, Davis J, Mühlschlegel FA, Gourlay CW (2021) CO2 enhances the formation, nutrient scavenging and drug resistance properties of C. albicans biofilms. NPJ Biofilms Microbiomes 7:67. https://doi.org/10.1038/s41522-021-00238-z

Peters BM, Ovchinnikova ES, Krom BP et al (2012) Staphylococcus aureus adherence to Candida albicans hyphae is mediated by the hyphal adhesin Als3p. Microbiology (N Y) 158:2975–2986. https://doi.org/10.1099/mic.0.062109-0

Puyo CA, Dahms T, Tricomi S (2006) Trachea inflammation during endotracheal intubation. Chest 130:228S. https://doi.org/10.1378/chest.130.4_MeetingAbstracts.228S-c

Raad II, Mohamed JA, Reitzel RA et al (2011) The prevention of biofilm colonization by multidrug-resistant pathogens that cause ventilator-associated pneumonia with antimicrobial-coated endotracheal tubes. Biomaterials 32:2689–2694. https://doi.org/10.1016/j.biomaterials.2010.12.015

Ramsey KA, Chen ACH, Radicioni G et al (2020) Airway mucus hyperconcentration in non–cystic fibrosis bronchiectasis. Am J Respir Crit Care Med 201:661–670. https://doi.org/10.1164/rccm.201906-1219OC

Ranjan N, Chaudhary U, Chaudhry D, Ranjan KP (2014) Ventilator-associated pneumonia in a tertiary care intensive care unit: analysis of incidence, risk factors and mortality. Indian J Crit Care Med 18:200–204. https://doi.org/10.4103/0972-5229.130570

Raveendra N, Rathnakara SH, Haswani N, Subramaniam V (2022) Bacterial biofilms on tracheostomy tubes. Indian J Otolaryngol Head Neck Surg 74:4995–4999. https://doi.org/10.1007/s12070-021-02598-6

Raya D, Shreya A, Kumar A et al (2022) Molecular regulation of conditioning film formation and quorum quenching in sulfate reducing bacteria. Front Microbiol 13. https://doi.org/10.3389/fmicb.2022.1008536

Reen FJ, Flynn S, Woods DF et al (2016) Bile signalling promotes chronic respiratory infections and antibiotic tolerance. Sci Rep 6:29768. https://doi.org/10.1038/srep29768

Regan K, Hunt K (2008) Tracheostomy management. Continuing Educ Anaesth Crit Care Pain 8:31–35. https://doi.org/10.1093/bjaceaccp/mkm049

Reid DW, Carroll V, O'May C et al (2007) Increased airway iron as a potential factor in the persistence of Pseudomonas aeruginosa infection in cystic fibrosis. Eur Respir J 30:286–292. https://doi.org/10.1183/09031936.00154006

Rishpana MS (2015) Candiduria in catheter associated urinary tract infection with special reference to biofilm production. J Clin Diagn Res 9:DC11. https://doi.org/10.7860/JCDR/2015/13910.6690

Rose MC, Voynow JA (2006) Respiratory tract mucin genes and mucin glycoproteins in health and disease. Physiol Rev 86:245–278. https://doi.org/10.1152/physrev.00010.2005

Rostami S, Tekkeşin AI, Ercan UK, Garipcan B (2022) Biomimetic sharkskin surfaces with antibacterial, cytocompatible, and drug delivery properties. Biomater Adv 134:112565. https://doi.org/10.1016/j.msec.2021.112565

Rumbaugh KP, Sauer K (2020) Biofilm dispersion. Nat Rev Microbiol 18:571–586. https://doi.org/10.1038/s41579-020-0385-0

Sanjanwala D, Londhe V, Trivedi R et al (2024) Polysaccharide-based hydrogels for medical devices, implants and tissue engineering: a review. Int J Biol Macromol 256:128488. https://doi.org/10.1016/j.ijbiomac.2023.128488

Santin M, Motta A, Denyer SP, Cannas M (1999) Effect of the urine conditioning film on ureteral stent encrustation and characterization of its protein composition. Biomaterials 20:1245–1251. https://doi.org/10.1016/S0142-9612(99)00026-5

Saracoglu KT, Saracoglu A, Kafali H (2016) Basic airway equipments in pediatric cardiac arrest management. Trends Anaesth Crit Care 7–8:54–58. https://doi.org/10.1016/j.tacc.2016.03.006

Saravanam PK, Jayagandhi S, Shajahan S (2022) Microbial profile in tracheostomy tube and tracheostoma: a prospective study. Indian J Otolaryngol Head Neck Surg 74:1740–1743. https://doi.org/10.1007/s12070-019-01743-6

Savage PB, Li C, Taotafa U et al (2002) Antibacterial properties of cationic steroid antibiotics. FEMS Microbiol Lett 217:1–7. https://doi.org/10.1111/j.1574-6968.2002.tb11448.x

Ścibik Ł, Ochońska D, Golda-Cepa M et al (2022) Microbiological analysis of tracheostomy tube biofilms and antibiotic resistance profiles of potentially pathogenic microorganisms. Otolaryngol Pol 76:8–21. https://doi.org/10.5604/01.3001.0015.8827

Seil JT, Webster TJ (2011) Reduced Staphylococcus aureus proliferation and biofilm formation on zinc oxide nanoparticle PVC composite surfaces. Acta Biomater 7:2579–2584. https://doi.org/10.1016/j.actbio.2011.03.018

Shah BK, Singh B, Wang Y et al (2023) Mucus hypersecretion in chronic obstructive pulmonary disease and its treatment. Mediat Inflamm 2023:1–15. https://doi.org/10.1155/2023/8840594

Shahrokhi E, Hasani A, Ansarin K et al (2018) Bacterial biofilm in ventilator-associated pneumonia: a clinical concern. J Res Med Dent Sci 6:46–51

Sharma S, Mohler J, Mahajan SD et al (2023) Microbial biofilm: a review on formation, infection, antibiotic resistance, control measures, and innovative treatment. Microorganisms 11:1614. https://doi.org/10.3390/microorganisms11061614

Shelden CH (1957) Percutaneous tracheotomy. JAMA J Am Med Assoc 165:2068. https://doi.org/10.1001/jama.1957.02980340034009

Shi L, Ardehali R, Caldwell KD, Valint P (2000) Mucin coating on polymeric material surfaces to suppress bacterial adhesion. Colloids Surf B: Biointerfaces 17(4):229–239

Singhai M, Malik A, Shahid M et al (2012) A study on device-related infections with special reference to biofilm production and antibiotic resistance. J Glob Infect Dis 4:193. https://doi.org/10.4103/0974-777X.103896

Siqueira WL, Zhang W, Helmerhorst EJ et al (2007) Identification of protein components in in vivo human acquired enamel pellicle using LC−ESI−MS/MS. J Proteome Res 6:2152–2160. https://doi.org/10.1021/pr060580k

Siqueira WL, Custodio W, McDonald EE (2012) New insights into the composition and functions of the acquired enamel pellicle. J Dent Res 91:1110–1118. https://doi.org/10.1177/0022034512462578

Slate AJ, Wickens D, Wilson-Nieuwenhuis J et al (2019) The effects of blood conditioning films on the antimicrobial and retention properties of zirconium-nitride silver surfaces. Colloids Surf B Biointerfaces 173:303–311. https://doi.org/10.1016/j.colsurfb.2018.09.060

Solomon DH, Wobb J, Buttaro BA et al (2009) Characterization of bacterial biofilms on tracheostomy tubes. Laryngoscope 119:1633–1638. https://doi.org/10.1002/lary.20249

Souza MC, dos Santos LS, Sousa LP et al (2015) Biofilm formation and fibrinogen and fibronectin binding activities by Corynebacterium pseudodiphtheriticum invasive strains. Antonie Van Leeuwenhoek 107:1387–1399. https://doi.org/10.1007/s10482-015-0433-3

Stocker R, Biro P (2005) Airway management and artificial ventilation in intensive care. Curr Opin Anaesthesiol 18:35–45. https://doi.org/10.1097/00001503-200502000-00007

Summaiya M, Unrmi NJ (2012) Assessment of biofilm formation by the causative organisms of ventilator associated pneumonia at intensive care unit of a tertiary care hospital. Natl J Med Res 2:15–18

Swindell K, Lattif AA, Chandra J et al (2009) Parenteral lipid emulsion induces germination of Candida albicans and increases biofilm formation on medical catheter surfaces. J Infect Dis 200:473–480. https://doi.org/10.1086/600106

Takagi J, Aoki K, Turner BS et al (2022) Mucin O-glycans are natural inhibitors of Candida albicans pathogenicity. Nat Chem Biol 18:762–773. https://doi.org/10.1038/s41589-022-01035-1

Tan X, Chen R, Zhu S et al (2016) Candida albicans airway colonization facilitates subsequent Acinetobacter baumannii pneumonia in a rat model. Antimicrob Agents Chemother 60:3348–3354. https://doi.org/10.1128/AAC.02180-15

Terragni P, Faggiano C, Martin E, Ranieri V (2014) Tracheostomy in mechanical ventilation. Semin Respir Crit Care Med 35:482–491. https://doi.org/10.1055/s-0034-1383862

Thangamani S, Monasky R, Lee JK et al (2021) Bile acid regulates the colonization and dissemination of Candida albicans from the gastrointestinal tract by controlling host defense system and microbiota. J Fungi 7:1030. https://doi.org/10.3390/jof7121030

Thorarinsdottir HR, Kander T, Holmberg A et al (2020) Biofilm formation on three different endotracheal tubes: a prospective clinical trial. Crit Care 24:382. https://doi.org/10.1186/s13054-020-03092-1

Timsit J-F, Schwebel C, Styfalova L et al (2019) Impact of bronchial colonization with Candida spp. on the risk of bacterial ventilator-associated pneumonia in the ICU: the FUNGIBACT prospective cohort study. Intensive Care Med 45:834–843. https://doi.org/10.1007/s00134-019-05622-0

Tincu RC, Cobilinschi C, Tincu IF, Macovei RA (2022) Efficacy of noble metal–alloy endotracheal tubes in ventilator-associated pneumonia prevention: a randomized clinical trial. Balkan Med J 39:167. https://doi.org/10.4274/balkanmedj.galenos.2021.2021-7-86

Tran PA, Webster TJ (2013) Antimicrobial selenium nanoparticle coatings on polymeric medical devices. Nanotechnology 24:155101. https://doi.org/10.1088/0957-4484/24/15/155101

Ulluwishewa D, Wang L, Pereira C et al (2016) Dissecting the regulation of bile-induced biofilm formation in Staphylococcus aureus. Microbiology (N Y) 162:1398–1406. https://doi.org/10.1099/mic.0.000317

Uppe A, Gupta D, Sawant S et al (2019) Incidence of biofilm formation in ET tube and correlation with occurrence of VAP in a tertiary care ICU in India. Am J Infect Dis Microbiol 7:65–72

van Charante F, Wieme A, Rigole P et al (2022) Microbial diversity and antimicrobial susceptibility in endotracheal tube biofilms recovered from mechanically ventilated COVID-19 patients. Biofilms 4:100079. https://doi.org/10.1016/j.bioflm.2022.100079

Vinodkumar CS, Patil SS, Kumar AA et al (2017) Repercussion of biofilm and antibiotic resistance in ventilator associated pneumonia. Int J Res Med Sci 5:4419. https://doi.org/10.18203/2320-6012.ijrms20174571

Webster TJ, Geilich (2013) Reduced adhesion of Staphylococcus aureus to ZnO/PVC nanocomposites. Int J Nanomedicine 8:1177. https://doi.org/10.2147/IJN.S42010

Winklmaier U, Wüst K, Schiller S, Wallner F (2006) Leakage of fluid in different types of tracheal tubes. Dysphagia 21:237–242. https://doi.org/10.1007/s00455-006-9047-2

Wu Y-C, Hsu P-K, Su K-C et al (2009) Bile acid aspiration in suspected ventilator-associated pneumonia. Chest 136:118–124. https://doi.org/10.1378/chest.08-2668

Yuzkat N, Demir CY (2019) Effect of using the suction above cuff endotracheal tube (SACETT) on postoperative respiratory complications in rhinoplasty: a randomized prospective controlled trial. Ther Clin Risk Manag 15:571–577. https://doi.org/10.2147/TCRM.S200662

Zhang Y, Nguyen KC, Caldwell D et al (2017) Immune responses during single and repeated murine endotracheal exposures of zinc oxide nanoparticles. NanoImpact 7:54–65. https://doi.org/10.1016/j.impact.2017.06.003

Zhong L, Zhang S, Tang K et al (2020) Clinical characteristics, risk factors and outcomes of mixed Candida albicans/bacterial bloodstream infections. BMC Infect Dis 20:810. https://doi.org/10.1186/s12879-020-05536-z

Chapter 4
Exploring the Oral Mycobiome: Biofilms and Inter-kingdom Interactions in Health and Disease

Kassapa Ellepola and Chaminda Jayampath Seneviratne

Abstract In microbiome research, the mycobiome component remains relatively understudied. The oral cavity harbors a diverse range of fungal species, many of which are implicated in disease. Among them, *Candida albicans* has been abundantly detected in microbiome studies, with its interspecies interactions with bacterial species extensively documented. This chapter provides a comprehensive overview of current research on the oral mycobiome and its association with various oral and systemic diseases. Additionally, it highlights the role of fungal species in the oral cavity, with particular emphasis on *Candida albicans* as a key contributor to oral biofilm formation and its interactions with bacterial communities.

Introduction

The term "microbiome" can be confusing for readers. To the general public, "microbiome" often refers to all microorganisms living in a particular environment. For example, the popular term "gut microbiome" generally implies all microorganisms residing in the human gut. However, the term "microbiome" is more commonly associated with the bacterial component of resident microorganisms, as bacteria are the most extensively studied and well-characterized group of the human microbiome (Hopson et al. 2020). This focus has led many studies and related media reports to overlook the fungal component, or mycobiome.

Research has shown that, similar to commensal bacteria, there are resident fungal species that coexist harmoniously with the human host. The term "mycobiome"

K. Ellepola (✉)
Department of Oral Biology, College of Dentistry, University of Illinois Chicago, Chicago, IL, USA
e-mail: kassapa@uic.edu

C. J. Seneviratne (✉)
School of Dentistry, The University of Queensland, Brisbane, QLD, Australia
e-mail: jaya.seneviratne@uq.edu.au

G. Ramage, D. Andes (eds.), *Fungal Biofilms*, Springer Series on Biofilms 15,
https://doi.org/10.1007/978-3-031-93315-8_4

specifically refers to these fungal species in a given environment (Cui et al. 2013). In broader terms, however, it also encompasses the interactions of fungi with other microorganisms and the host, explored at multiple omics levels (genomics, transcriptomics, proteomics, metabolomics). Unlike bacteria, the fungal species within the mycobiome are eukaryotic, a key distinction from the bacteriome. In this chapter, "microbiome" will refer to the entirety of microbial residents in an environment, including microeukaryotes like fungi, amoebas, and flagellates, as well as prokaryotes such as bacteria, archaea, and viruses.

A significant limitation in defining the human mycobiome is the lack of studies identifying the fungal component in the microbiome. Most past research has relied on culture-dependent methods, which identified only a limited range of fungal species. Recently, however, sequencing-based studies have revealed a large number of fungal species colonizing various body sites (Cui et al. 2013; Jiang et al. 2022). These include the oral cavity, nasal cavity, lungs, gut, skin, vagina, and conjunctiva, which are areas previously studied primarily for their bacterial populations.

While fungi are typically found in smaller quantities during routine assessments (Diaz and Dongari-Bagtzoglou 2021), sequencing-based microbiome studies have revealed the presence of more than 100 fungal genera in the oral cavity (Diaz and Dongari-Bagtzoglou 2021; Ghannoum et al. 2010; Diaz et al. 2017). The oral mycobiome is primarily dominated by either *Candida* (Gabaldón et al. 2013; Turner and Butler 2014; Ahmad et al. 2014; Hong et al. 2020; Dupuy et al. 2014) or *Malassezia* (Diaz and Dongari-Bagtzoglou 2021; Hong et al. 2020; Dupuy et al. 2014) species. These fungi are believed to play distinct ecological roles based on their differing nutritional preferences where *Candida* tends to feed on sugars, whereas *Malassezia* typically utilizes lipids (Hong et al. 2020). In a mycobiome analysis of healthy individuals, *Candida* species were the most commonly identified fungi, found in 75% of participants. This was followed by *Cladosporium* at 65%, *Aureobasidium* and Saccharomycetales at 50% each, *Aspergillus* at 35%, *Fusarium* at 30%, and *Cryptococcus* at 20% (Ghannoum et al. 2010). Table 4.1 provides a summary of mycobiome studies associated with the oral cavity, carried out in the past decade.

Oral Mycobiome Associated with Oral Diseases

Although limited, several studies have examined the role of the mycobiome in oral diseases. In contrast to the healthy mycobiome community, which is highly variable, a dysbiotic mycobiome associated with oral diseases is often characterized by the overgrowth of *Candida* species (Abusleme et al. 2018). Few examples are discussed below.

In a study of the oral mycobiome in a cohort of patients with nonoral solid tumors undergoing chemotherapy, two distinct fungal communities, or "mycotypes," were observed, dominated by *Candida* and *Malassezia* species, respectively (Hong et al. 2020). Interestingly, the *Candida* mycotype showed a positive correlation with cancer occurrence and a concurrent enrichment of aciduric bacterial species,

Table 4.1 A compilation of research on the oral mycobiome (2014–2024)

Study aim(s)	Method(s)	Sample size	Reference
Exploring the characteristics of oral bacteria and fungi in idiopathic membranous nephropathy (IMN) patients with different tongue coatings, and clarify the therapeutic effect of Moshen Fuyuan Formula (MSFY) based on oral microbiome	16S rRNA and ITS gene sequencing	24 patients and 10 healthy individuals	Wang et al. (2024)
Characterizing the oral mycobiome of Portuguese children and young adults with allergic rhinitis alone or with asthma, asthmatics, and healthy controls.	ITS1 and ITS2 gene sequencing	36 adults and 313 children	Pérez-Losada et al. (2024)
Identifying indigenous fungal communities and environmental fungi present in the oral cavity and evaluate their role in oral dysbiosis	ITS2 gene sequencing	88 participants	Acosta-Pagán et al. (2024)
Studying dysbiosis and interactions of the mycobiome and bacteriome in mucosal lesions of erosive and non-erosive oral lichen planus patients	16S rRNA and ITS gene sequencing	49 patients and 10 healthy individuals	Beibei et al. (2024)
Studying oral fungal dysbiosis and systemic immune dysfunction in Chinese patients with schizophrenia	ITS1 gene sequencing	118 patients	Liu et al. (2024)
Identifying bacterial and fungal species in adult asthma patients and the effect of inhaled corticosteroid use on microbiome changes	16S rRNA and "panfungal" 5.8S rRNA gene sequencing	28 asthmatic patients on inhaler steroid therapy, 25 steroid-naive asthmatics, and 24 healthy controls.	Oğuzülgen et al. (2024)
Studying the changes of microbial domains across body sites (including oral cavity) in females from prepartum to postpartum phases	16S rRNA and ITS gene sequencing	30 pregnant women, and 29 nonpregnant women	Neumann et al. (2024)
Studying the alterations of the oral and gut mycobiome and cytokines during long-term follow-up of COVID-19 convalescents	ITS gene sequencing	35 COVID-19 confirmed patients	Ren et al. (2023)
Exploring the role of oral fungal communities in nasopharyngeal carcinoma (NPC) carcinogenesis	ITS2 gene sequencing	476 untreated incident NPC patients and 537 population-based controls	Chen et al. (2023a)

(continued)

Table 4.1 (continued)

Study aim(s)	Method(s)	Sample size	Reference
Characterizing the oral mycobiome in COVID-19 and recovered patients	ITS gene sequencing	71 COVID-19 patients, 36 suspected cases, 22 recovered COVID-19 patients, 36 suspected cases who recovered, and 132 controls	Hu et al. (2023)
Mapping the human oral and gut fungal microbiota in patients with metabolic dysfunction-associated fatty liver disease	Metagenomic sequencing	21 patients and 20 healthy controls	Niu et al. (2023)
Studying the characteristics and correlations of the oral and gut fungal microbiome with hypertension	Metagenomic sequencing	36 participants with hypertension and 24 healthy controls	Chen et al. (2023b)
Characterization of oral bacterial and fungal microbiome in recovered COVID-19 patients	16S rRNA and the ITS2 gene sequencing	23 COVID-19 recovered patients and 29 healthy individuals	Wei et al. (2023)
Exploration of the oral microbiome in key mucosal areas of the oral cavity and assessing the microbiome variations in premalignant and oral squamous cell carcinoma samples from both users and non-users of Toombak (smokeless tobacco)	16S rRNA and the ITS gene sequencing	47 Toombak smokeless tobacco users and 31 non-users	Sami et al. (2023)
Evaluating the effects of HIV and antiretroviral therapy on the oral microbiome	16S rRNA and the ITS gene sequencing	470 subjects	Beall et al. (2023)
Evaluating the extent of oral-gut transmission of bacterial and fungal microbiota in healthy Chinese adults	16S rRNA V3-V4 and ITS1 gene sequencing	470 healthy Chinese adults	Cheung et al. (2023)
Evaluating the supragingival mycobiome in determining the caries experience of HIV-exposed but uninfected children	ITS gene sequencing	568 children (189 HIV-infected, 189 HIV-exposed but uninfected and 190 controls)	O'Connell et al. (2023)
Examining oral microbiome changes and interplay with immune response following prenatal total oral rehabilitation	Metagenomic sequencing	15 pregnant women	Wu et al. (2023)
Analyzing the buccal mucosal fungal microbiome in oral lichen planus patients	ITS1 and ITS2 gene sequencing	20 patients and 10 healthy controls	Du et al. (2023)
Evaluating the effect of argon plasma pre-treatment of healing abutments on peri-implant microbiome	Next-generation DNA sequencing	25 patients	Canullo et al. (2023)

(continued)

Table 4.1 (continued)

Study aim(s)	Method(s)	Sample size	Reference
Cataloguing microbial communities in the human oral cavity by profiling saliva, gingival plaque, and mucosal samples at different stages of oral carcinogenesis	16S rRNA gene and ITS sequencing	90 healthy controls, 96 patients with oral premalignant lesions, 87 from OSCC subjects	Heng et al. (2022)
Evaluating smokeless tobacco consumption on the dysbiosis of oral mycobiome	ITS1 gene sequencing	40 participants	Sajid et al. (2022)
Understanding the altered landscape of the oral microbiome and mycobiome in SARS-CoV-2 infected patients and its correlation with risk factors compared to non-infected individuals	16S rRNA and ITS gene sequencing	30 patients, 24 non-infected individuals	Gupta et al. (2022)
Characterizing the oral bacterial and fungal microbiomes in a large cohort of systemically healthy Chinese adults	16S rRNA and the ITS gene sequencing	1469 participants	Cheung et al. (2022)
Evaluating correlation between core members of oral mycobiome and early childhood caries (ECC) progression	V3-V4 16S rRNA and ITS1 gene sequencing	18 CF, 15 ECC, and 22 S-ECC participants	Tu et al. (2022)
Evaluating oral mycobiome differences in various spatial niches with and without severe early childhood caries	ITS2 gene sequencing	8 S-ECC children and 8 CF participants	Cui et al. (2021)
Determining the effects of HIV and several other clinical variables on the oral fungal communities in a large cohort of HIV-positive and -negative individuals	ITS2 gene sequencing	149 HIV-positive and 88 HIV-negative subjects	Fidel Jr et al. (2021)
Analyzing the salivary mycobiome in a cohort of oral squamous cell carcinoma patients from Sudan	Next-generation sequencing (NGS)-based methodology by amplifying the ITS2 region	59 patients diagnosed with OSCC and 13 non-OSCC control volunteers	Mohamed et al. (2021)
Analyzing the oral microbial diversity among atrophic glossitis patients and healthy individuals	16S rRNA and ITS gene sequencing	50 participants with atrophic glossitis and 50 healthy participants	Li et al. (2021)
Analyzing the salivary mycobiome diversity in HIV-infected patients	ITS gene sequencing	30 HIV patients and 30 healthy controls	Chang et al. (2021)
Evaluating the keystone salivary mycobiome in postpartum period in health and disease conditions	ITS gene sequencing	3 postpartum females along with 1 healthy non-pregnant female	Khadija et al. (2021)

(continued)

Table 4.1 (continued)

Study aim(s)	Method(s)	Sample size	Reference
Characterization of supragingival plaque and oral swab microbiomes in children with severe early childhood caries	V4-16S rRNA and ITS1 rRNA gene sequencing	40 S-ECC and 40 CF	de Jesus et al. (2021)
Studying the oral microbiome in cystic fibrosis patients and matched controls	16S rRNA sequencing, and culture and proteomics-based identification of fungi	31 cystic fibrosis patients and matched controls	Willis et al. (2021)
A comparative study of oral bacteria and fungi microbiota in Tibetan and Chinese Han living at different altitude	16S rRNA and ITS1 rRNA genes sequencing	90 healthy native Tibetan and 35 healthy Han adults	Dong et al. (2021)
Characterizing the oral mycobiome, including mycobiome-bacteriome interactions, in the setting of remission-induction chemotherapy for acute myeloid leukemia	16S rRNA gene (V4) and ITS2 sequencing	39 patients	Robinson et al. (2020)
Generating mycobiome taxonomic profiles from site-specific supragingival plaque samples in children with different caries status	ITS1 gene sequencing	33 participants	O'Connell et al. (2020)
Studying the salivary mycobiome in non-cancer and patients scheduled to receive cancer chemotherapy and evaluate the utility of mycotypes as biomarkers for oral diseases	ITS1 gene sequencing	36 cancer patients and 23 noncancer individuals	Hong et al. (2020)
Analyzing potential sex-based differences in the supragingival plaque microbiota (bacteriome and mycobiome) of young children with severe early childhood caries and those caries-free	V4-16S rRNA and ITS1 rRNA gene sequencing	40 with S-ECC and 40 CF	de Jesus et al. (2020)
Determining the role of oral mycobiome in dental caries and its interaction with caries-associated bacteria	ITS2 gene sequencing	30 children with no caries, early caries, or advanced caries	Baraniya et al. (2020)
Determining the relationship between oral and gut microbiome diversity and chronic systemic inflammation in antiretroviral therapy-treated people living with HIV with prevalent severe periodontitis	16S rRNA and ITS gene sequencing	52 primarily postmenopausal women	Annavajhala et al. (2020)

(continued)

Table 4.1 (continued)

Study aim(s)	Method(s)	Sample size	Reference
Sequencing the salivary mycobiome and bacteriome associated with oral lichen planus (OLP)	Illumina MiSeq sequencing	18 healthy subjects, 17 reticular OLP patients, and 18 erosive OLP patients	Li et al. (2019)
Assessment of the oral mycobiome in childhood, to determine if the composition of fungi varies between children with and without caries	Illumina MiSeq to sequence the ITS2 gene	17 participants	Fechney et al. (2019)
Studying the longitudinal changes in oral and gastrointestinal bacteriome and mycobiome in a patient population with hematopoietic cell transplantation exposed to high-dose melphalan and antimicrobials	16S rRNA and ITS gene sequencing	15 participants	El Jurdi et al. (2019)
Characterizing the skin, oral, and anal mycobiomes of infants over the first month of life and the anal and vaginal mycobiomes of mothers	ITS2 gene sequencing	17 Infants, 16 mothers	Ward et al. (2018)
Assessing the relationship of microbial dysbiosis with smoking and markers of HIV disease	Ion-Torrent sequencing platform	48 HIV-infected smokers, 24 HIV-infected non-smokers, 24 HIV-uninfected smokers	Mukherjee et al. (2018)
Evaluating the consequences of STAT3 deficiency (patients with autosomal-dominant hyper IgE syndrome) on the oral commensal microbiome	16S rRNA and ITS1 gene sequencing	36 participants	Abusleme et al. (2018)
To describe and compare the oral mycobiome of subjects with and without periodontal disease	ITS gene sequencing	15 patients with periodontal disease, 15 healthy	Peters et al. (2017)
Evaluating the oral mycobiome at baseline and determining whether changes occur after admission to the intensive care unit	ITS1 gene sequencing	10 participants	Watkins et al. (2017)
To examine the oral fungal biodiversity of patients with pseudomembranous oral candidiasis compared to healthy individuals	Next-generation sequencing	27 patients and 66 healthy controls	Imabayashi et al. (2016)
Characterizing the infecting microbe, host, micro-, and mycobiomes leading up to infection onset in a leukemia patient that developed invasive mucormycosis	16S rRNA gene and ITS gene sequencing	1 participant	Shelburne et al. (2015)

(continued)

Table 4.1 (continued)

Study aim(s)	Method(s)	Sample size	Reference
Characterizing the mycobiome at different respiratory tract levels (including oral washes) in persons with and without HIV infection and in HIV-infected individuals with chronic obstructive pulmonary disease	18S and ITS gene sequencing	56 participants	Cui et al. (2015)
Characterizing the oral bacteriome and mycobiome of HIV-infected patients and uninfected individuals	Multitag 454 pyrosequencing (MTPS) technique	12 HIV-infected patients and 12 uninfected controls	Mukherjee et al. (2014)
Examining the fungal flora in patients with oral candidiasis and investigating changes in the flora after antifungal treatment	Length heterogeneity-polymerization chain reaction (LH-PCR) analysis of ITS regions	52 patients with oral candidiasis and 30 healthy controls	Ieda et al. (2014)
Revising current practices in sequence-based taxonomy assignments and employment of improvements to address the question of the fungal genera present in the healthy human mouth	ITS1 gene sequencing	6 participants	Dupuy et al. (2014)

ITS internal transcribed spacer, *CF* caries free, *ECC* early childhood caries, S-ECC severe early childhood caries, OSCC oral squamous cell carcinoma, *HIV* human immunodeficiency virus

suggesting that the oral mycobiome may also influence the oral bacteriome, contributing to cancer pathogenesis in the oral cavity. In another study, the relative abundance of *C. albicans* was found to be higher in oral squamous cell carcinoma (OSCC), whereas noncancerous healthy tissues were dominated by *Malassezia* (Krüger et al. 2019). While more data is needed to confirm the potential role of mycobiome diversity in oral cancer, it is possible that dysbiosis involving both fungal and bacterial communities may contribute. Other oral diseases have also been explored for potential mycobiome involvement. For example, a study on the oral mycobiome of patients with oral lichen planus revealed a variety of fungal species, predominantly *Candida* species, along with other fungi such as *Alternaria*, *Aspergillus*, *Cladosporium/Davidella*, *Saccharomyces*, *Phoma*, and *Malassezia* species (Heng et al. 2022).

However, there remains a significant research gap in understanding the relationship between the oral mycobiome and oral diseases, highlighting an urgent need for further studies in this field. Both traditional culture-based techniques and novel sequencing-based methods (Table 4.1) have highlighted the role of *Candida* species in the pathogenesis of various oral diseases. *Candida* often forms mixed-microbial communities with other oral pathogens. Therefore, later in the chapter, we will discuss a few examples of mixed *Candida*-bacterial infections in the context of the oral mycobiome and oral microbiome.

Prevalence of *Candida albicans*

C. albicans can form surface-attached communities called "biofilms," which are challenging to eliminate due to their heightened antifungal resistance (Seneviratne et al. 2008). While it typically exists as a commensal organism in the human body, *C. albicans* can lead to infections under certain conditions, such as in immunocompromised hosts (Mayer et al. 2013). *C. albicans* is the most frequently identified pathogen in fatal bloodstream infections (Seneviratne et al. 2011). Of the many *Candida* species identified, five are linked to 90% of human infectious diseases: *C. albicans*, *Candida tropicalis*, *Candida glabrata*, *Candida parapsilosis*, and *Candida krusei*. Among these, *C. albicans* is the most prevalent pathogenic fungus associated with candidiasis, accounting for 66% of *Candida*-related infections (Pfaller and Diekema 2007).

Candida albicans *Oral Infections*

Candida species are commonly isolated from the oral cavity of about 30–50% of humans (Odds 1988). Among these species, *C. albicans* is the most frequently identified fungus (Zegarelli 1993). Candidiasis can present in various oral forms, including the pseudomembranous variant known as thrush, the erythematous (atopic) variant, and the hyperplastic variant. It can also lead to *Candida*-associated conditions such as denture stomatitis, angular cheilitis or angular stomatitis, median rhomboid glossitis, and linear gingival erythema (Williams and Lewis 2011).

Candida-*Bacterial Interactions in the Oral Cavity*

Candida albicans is often isolated and is associated with polymicrobial interactions, forming mixed-species biofilms with both gram-positive and gram-negative bacteria. These complex interactions can significantly influence their physiological functions, metabolic activities, and potential to cause disease, underscoring the clinical relevance of polymicrobial infections. Recent research is deepening our understanding of the molecular mechanisms underlying these relationships and their effects on oral health. Notably, significant clinical implications arise from findings that show physical interactions between *C. albicans* and oral bacteria can enhance the tolerance of polymicrobial biofilms to antimicrobial agents and increase their resilience to physical disruption (Jenkinson and Lamont 2005).

Candida albicans and *Streptococcus mutans* Interactions

Clinical studies have shown that the fungus *C. albicans* is often found in high numbers in the dental plaque biofilms of toddlers with early childhood caries (ECC) (de Carvalho et al. 2006; Raja et al. 2010; Yang et al. 2012; Jenkinson et al. 1990; Qiu et al. 2015; Xiao et al. 2016). Studies have determined that the coaggregation of *C. albicans* and *S. mutans* leads to early childhood caries (Salehi et al. 2020). Estimates suggest that *C. albicans* is present in 62.3% of carious lesions in preschool children and 71.4% in school-aged children (Rozkiewicz et al. 2006). Additionally, it has been demonstrated that oral biofilms in toddlers with dental caries have a significantly higher prevalence of *C. albicans* compared to those without caries (de Carvalho et al. 2006).

Synergistic interactions between *C. albicans* and *S. mutans* have been documented, as demonstrated by their repeated coisolation from denture plaques and dental caries (de Carvalho et al. 2006). Intrinsic bacterial factors, such as the antigen I/II polypeptides in the cell wall and the glucosyltransferase B (GtfB) activity of *S. mutans* in the presence of sucrose, promote the association between *C. albicans* and *S. mutans*. This interaction enhances hyphal formation, increases cell density, boosts acid production, and improves surface attachment, thereby fostering cariogenic growth (Shirtliff et al. 2009; Hwang et al. 2017; Koo et al. 2018; Yang et al. 2018). In addition, there was an upregulation of extracellular DNA, glucan exopolysaccharides, and various quorum-sensing molecules, during *C. albicans*–*S. mutans* interactions (Koo et al. 2018; Falsetta et al. 2014; Sztajer et al. 2014; Klein et al. 2015; Wu et al. 2020). Figure 4.1 provides an overview of the interactions between species in the mixed-species biofilm of *S. mutans* and *C. albicans*.

Biofilm Formation, Extracellular Glucan Production and Carbohydrate Metabolism

The presence of sucrose is a crucial environmental factor that significantly enhances adhesion among certain organisms (Pereira-Cenci et al. 2008; Falsetta et al. 2014; Metwalli et al. 2013). *S. mutans*-derived Gtfs strongly bind to the surface of *C. albicans*, showing high affinity for N- and O-linked mannans, and produce substantial amounts of glucans when sucrose is available (Hwang et al. 2017; Gregoire et al. 2011). The presence of both *C. albicans* and *S. mutans* significantly promotes the formation of an EPS-rich matrix, resulting in the development of biofilms that are larger and thicker than those produced by either species alone (Falsetta et al. 2014). The EPS made up of glucan serves as an intermediary matrix between yeast and bacterial cells. In addition to binding to *C. albicans* yeast cells, these glucans increase the available binding sites for *S. mutans* (Gregoire et al. 2011; Branting et al. 1989), resulting in a highly pathogenic mixed-species biofilm (Falsetta et al. 2014). The EPS matrix of mixed biofilms also imparts a drug resistance mechanism to both fungal and bacterial cells by inhibiting drug uptake via entrapment of drugs in the matrix (Kim et al. 2018).

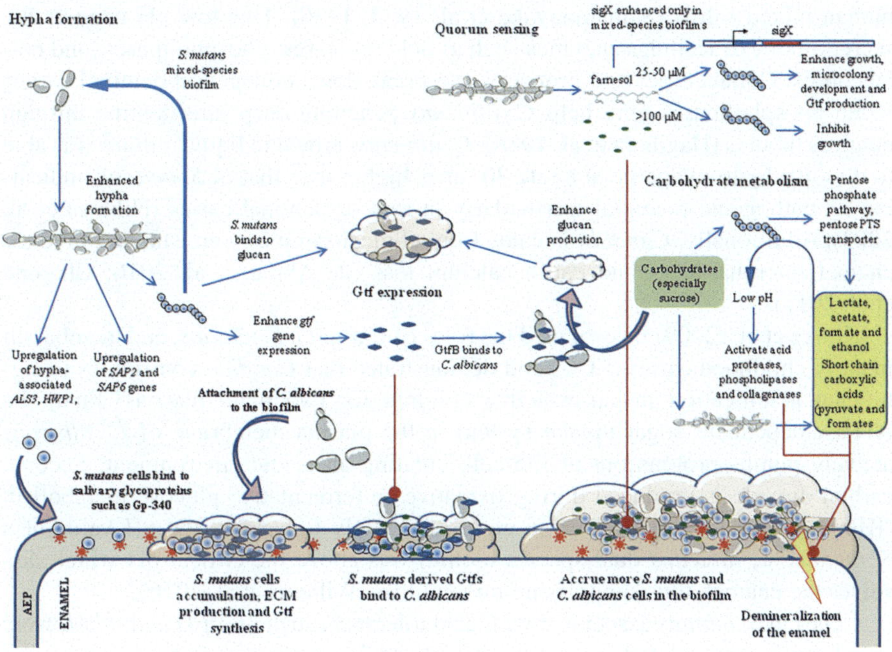

Fig. 4.1 Interspecies interactions in *Streptococcus mutans–Candida albicans* mixed-species biofilm. (Adapted from (Ellepola 2018))

The role of carbohydrate metabolism in *C. albicans–S. mutans* mixed-species biofilms linked to dental caries has been widely investigated. Studies utilizing RNA sequencing revealed that *S. mutans* exhibits increased sugar metabolism, particularly in pyruvate and galactose metabolism, when cocultured with *C. albicans* in these biofilms (He et al. 2017). Sztajer et al. (2014) proposed the occurrence of cross-feeding of carbohydrates in *S. mutans–C. albicans* mixed-species biofilms (Sztajer et al. 2014). While *C. albicans* has been shown to be inefficient at metabolizing sucrose directly (Williamson et al. 1993), it can utilize the breakdown products of sucrose, such as glucose and fructose, that are produced by *S. mutans* within the biofilm (Sztajer et al. 2014; Ellepola et al. 2019).

Acid Production and Tolerance

C. albicans can generate significant amounts of acids from carbohydrates (Parahitiyawa et al. 2006; Thein et al. 2007). In the presence of dietary sugars like glucose and sucrose, it produces short-chain carboxylic acids, including pyruvate and formate (Samaranayake et al. 1982, 1983). The aciduric capability of *C. albicans* was evident when the yeast survived for up to 72 hours in a highly acidic environment and thrived at a very low pH of 3.0 in glucose- or sucrose-supplemented human mixed saliva (Samaranayake et al. 1983, 1986). This low pH triggers the secretion of extracellular enzymes such as acid proteases, phospholipases, and collagenases. Collagenases and proteases can break down collagen in dentinal caries, while phospholipases may help *C. albicans* penetrate deep into dentine through dentinal tubules (Hagihara et al. 1988). *C. albicans* grown in liquid culture was able to dissolve hydroxyapatite at a rate 20 times higher than that of *S. mutans*, indicating its pathogenic potential, particularly in root or dentinal caries (Nikawa et al. 2003). Additionally, *Candida* isolates from clinical samples were shown to reduce enamel microhardness and cause calcium loss (de Abreu et al. 2016; Charone et al. 2017).

Klinke et al. (2009) investigated the roles of *Candida*, *S. mutans*, and lactobacilli in early childhood caries (ECC) and demonstrated that *Candida* contributes to significant acidification in caries-active children for two main reasons. First, the ATPase-dependent sugar uptake system in the plasma membrane of *C. albicans* actively pumps protons out of the cell, creating an acidic environment. Second, carbon dioxide is produced during oxidative or fermentative glucose metabolism (Klinke et al. 2009). However, in contrast, a study found that the pH value of a *S. mutans–C. albicans* dual-species biofilm was above the critical pH which dissolves the enamel, thus suppressing mineral loss (Willems et al. 2016).

Genes in *S. mutans* associated with acid tolerance, such as *atpD* and *fabM*, were significantly upregulated in mature mixed-species biofilms compared to single-species biofilms (Falsetta et al. 2014). The presence of *C. albicans* enhances the transcriptional activation of these genes, improving *S. mutans'* ability to survive and thrive in an acidic environment. This coexistence with *C. albicans* promotes the expression of key virulence genes in *S. mutans*, which are essential for its persistence in biofilms and its role in disease. Furthermore, a recent study supported the idea that *C. albicans* contributes to dental caries through its strong capacity to produce and tolerate acids (Klinke et al. 2011). Coinfection of rats with *S. mutans* and *C. albicans* enhanced the colonization and carriage of both organisms in vivo and significantly enhanced the virulence of plaque biofilms, resulting in the development of rampant carious lesions (Falsetta et al. 2014).

Physical Interactions Between *C. albicans* and *S. mutans*

Several in vitro studies have demonstrated that *C. albicans* promotes the accumulation of *S. mutans* in mixed-species biofilms. This suggests a potential facilitation mechanism where *C. albicans* cells serve as a support for bacterial adherence (Falsetta et al. 2014; Gregoire et al. 2011). While *S. mutans* boosts the expression of *gtf* genes during biofilm development, the presence of *C. albicans* seems to further elevate the expression of these genes during the early stages of mixed-species biofilm formation and in mature biofilms compared to *S. mutans* single-species biofilms (Falsetta et al. 2014). This expression pattern reinforces the crucial roles of GtfB and GtfC-derived polysaccharides in forming *S. mutans-C. albicans* biofilms. Notably, Hwang et al. (2015) found that GtfB, when associated with *C. albicans*, exhibits significantly higher activity and generates more glucans compared to when it is bound to *S. mutans* surfaces (Hwang et al. 2015). Research has demonstrated that *S. mutans* GtfB can enhance the expression of *HWP1*, *ALS1*, and *ALS3* in *C. albicans* (Ellepola et al. 2017). Additionally, *C. albicans* coaggregates with bacterial cells through the adhesive N-terminal domain of agglutinin-like sequence (ALS) proteins, which are various glycoproteins encoded by *ALS* genes on its surface (Hoyer et al. 2008; Silverman et al. 2010). In contrast, studies using an experimental *Galleria mellonella* candidiasis model have shown that *S. mutans* can secrete byproducts that inhibit biofilm formation, hyphal growth, and pathogenicity in *C. albicans* (Barbosa et al. 2016). Secretory molecules like trans-2-decanoic acid, mutanobactin A, and CSP produced by *S. mutans* play a role in the antagonistic interaction between *C. albicans* and *S. mutans* (Jarosz et al. 2009; Vílchez et al. 2010).

Quorum-Sensing-Mediated Interaction in the Mixed-Species Biofilm

The production of GtfB, virulence genes, and competence in *S. mutans* is activated by quorum-sensing signaling molecules like farnesol from *C. albicans* (Kim et al. 2017). Farnesol, an acyclic alcohol, was the first quorum-sensing molecule identified in *C. albicans* (Hornby et al. 2001). It plays a role in regulating morphogenesis by preventing the transformation from yeast to hyphae (Ramage et al. 2002). Numerous studies have demonstrated the broad-spectrum inhibitory effects of farnesol against *C. albicans* and various bacterial species (Jabra-Rizk et al. 2006). Farnesol has been shown to reduce the number of *C. albicans* and *S. mutans* cells in both single and mixed-species biofilms (Fernandes et al. 2016). Metabolite profiling and chromatographic analyses of the supernatant from *S. mutans-C. albicans* mixed-species biofilms also revealed increased levels of farnesol (Kim et al. 2017). Interestingly, farnesol concentrations between 25 and 50 μM detected in the supernatant promoted *S. mutans* cell growth, microcolony development, and Gtfs. However, higher concentrations (>100 μM) of farnesol inhibited *S. mutans* growth. Thus, farnesol may serve as a key modulator in this cross-kingdom interaction, with *S. mutans* growth responding nonmonotonically to varying farnesol concentrations. In experiments with an *S. mutans* strain that had a transcriptional fusion between a green fluorescent protein gene and the promoter for *sigX* (an alternative sigma factor induced by quorum sensing signals), *sigX* expression was strongly activated in dual-species biofilms but not in single-species biofilms (Sztajer et al. 2014). This further suggests the ability of *C. albicans* to stimulate the QS system of *S. mutans*.

Transcriptomic and Proteomic Changes During *Candida albicans–Streptococcus mutans* Mixed-Species Biofilms

He et al. (2017) and Ellepola et al. (2020) analyzed the impact on the whole transcriptome of *S. mutans* in the presence of *C. albicans* using RNA-Seq and systems analysis (He et al. 2017; Ellepola et al. 2019). *C. albicans* significantly influenced the transcriptome of *S. mutans*. In mixed-species biofilms, there was an increase in the expression of genes related to pyruvate metabolism, carbohydrate metabolism, and GtfB compared to single-species biofilms. Additionally, *C. albicans* altered the transcription of two-component signal transduction systems, which are crucial for the fitness and sucrose-dependent biofilm formation in *S. mutans*. Conversely, the production of a bacteriocin called mutacin was downregulated, potentially impacting the bacterial composition of biofilms in the presence of *C. albicans* (He et al. 2017). RNA sequencing and iTRAQ (isobaric tags for relative and absolute quantitation)-based quantitative proteomics revealed a significant increase in *C. albicans* genes and proteins associated with carbohydrate metabolism, including those involved in sugar transport, aerobic respiration, pyruvate breakdown, and the glyoxylate cycle. Additionally, other genes and proteins related to cell morphogenesis and cell wall components, such as mannan and glucan, were also upregulated, suggesting enhanced fungal activity in mixed-species biofilms (Ellepola et al. 2019).

Interaction of *C. albicans* and Mitis Group Streptococci

Mitis group streptococci (MGS), including *S. gordonii*, *S. oralis*, and *S. sanguinis*, have cell wall receptors that facilitate synergistic interactions by binding to *C. albicans* ALS proteins (Xu et al. 2014; O'Donnell et al. 2015). For instance, the adhesive N-terminal domains of Als3 and SspB proteins play a role in physical interactions with *S. gordonii* (Silverman et al. 2010). Souza et al. (2020) recently examined the formation of mixed biofilms by *C. albicans* and MGS on a titanium surface, finding that *C. albicans* promotes MGS biofilm development (Souza et al. 2020). Similarly, MGS exhibits a synergistic interaction with *C. albicans* by enhancing the expression of *efg1*, a hyphal-associated gene, which increases virulence. One study showed that, compared to single-species *C. albicans* biofilms, there was a significant ($P < 0.05$) upregulation of *ALS3*, *HWP1*, *SAP2*, and *SAP6* genes, along with increased hyphal production in biofilms containing *S. mutans* and *S. sanguinis* (Cavalcanti et al. 2016a). The mutualistic interaction between *C. albicans* and MGS on titanium surfaces in an in vitro infection model (titanium–mucosal interface) was also shown to increase tissue damage (Souza et al. 2020).

C. albicans and *S. gordonii*: Montelongo-Jauregui et al. (2016) utilized an artificial saliva medium that simulated the physiological conditions of the oral cavity in vitro to investigate the relationship and biofilm development between *S. gordonii* and *C. albicans (*Montelongo-Jauregui et al. 2016). Their findings revealed a synergistic interaction between *C. albicans* and *S. gordonii*, which enhanced the formation of polymicrobial biofilms and demonstrated a high level of antimicrobial resistance (Montelongo-Jauregui et al. 2016). Similarly, the synergistic relationship between *C. albicans* and *S. gordonii* has been explored on the surfaces of materials used in the biomedical industry, such as titanium (Montelongo-Jauregui et al. 2018). A transcriptomics study revealed that 75 genes in *C. albicans* related to responses to chemical stimuli, regulation, homeostasis, protein modification, and the cell cycle were significantly upregulated, while 36 genes primarily associated with transport and translation were downregulated (Dutton et al. 2016). The upregulation of filamentation-associated genes *TEC1*, *FGR42*, and *ALS1* supports previous findings that *S. gordonii* promotes the transition of *C. albicans* from yeast to hyphae. Increased expression of genes related to arginine biosynthesis in *C. albicans* indicates a response to oxidative stress. In contrast, the transcriptional response of *S. gordonii* was limited to just eight significantly upregulated genes (*glpK*, *rplO*, *celB*, *rplN*, *rplB*, *rpsE*, *ciaR*, and *gat*). These transcriptomic patterns suggest that signals from *S. gordonii* induce a positive filamentation response in *C. albicans*, while *S. gordonii* appears to be less affected by *C. albicans*. Research has also shown that *S. gordonii* aids the survival and escape of *C. albicans* from macrophage phagosomes (Salvatori et al. 2020).

C. albicans and *S. oralis*: While historically regarded as commensal organisms, *Streptococcus oralis* has been shown to work in concert with *C. albicans*, increasing the severity of oral candidiasis (Diaz et al. 2014; Cavalcanti et al. 2016b, 2017). The surface adhesin ALS1, regulated by the Efg1 master regulator of *C. albicans*, plays a role in the development of mixed biofilms formed by *C. albicans* and *S. oralis* (Xu et al. 2017). *S. oralis* and *C. albicans* work synergistically to elevate levels of μ-calpain, a proteolytic enzyme that degrades E-cadherin at the junctions of oral epithelial cells (Xu et al. 2016). This increased structural breakdown enhances *C. albicans'* ability to invade the oral and esophageal mucosa during coinfection with *S. oralis* (Diaz et al. 2012).

Interaction with *C. albicans* and Other Oral Bacteria

Several other oral bacteria, including *Porphyromonas gingivalis*, *Actinomyces* species (such as *A. viscosus*, *A. naeslundii*, *A. odontolyticus*, and *A. oris*), *Fusobacterium nucleatum*, *Rothia dentocariosa*, *Aggregatibacter actinomycetemcomitans*, and *S. salivarius*, have been reported to interact with *C. albicans* in the oral cavity (Montelongo-Jauregui and Lopez-Ribot 2018). Toxins, peptides, and autoinducer-2 secreted by *Aggregatibacter actinomycetemcomitans*, *S. mutans*, and *S. gordonii*, respectively, have been found to inhibit *C. albicans* filamentation (Jarosz et al. 2009; Brusca et al. 2013; Bachtiar et al. 2014; Jack et al. 2015).

C. albicans and *F. nucleatum:* Recent research demonstrated that a direct physical interaction between *C. albicans* and *F. nucleatum* diminished both the growth of *C. albicans* and its ability to produce hyphae, which are more associated with invasive candidal outgrowth than the yeast form (Bor et al. 2016). Although other species have been shown to inhibit the yeast–hyphae transition in *C. albicans*, this study is the first to describe a specific contact-dependent mechanism (Bor et al. 2016). The reduction of *C. albicans* hyphae resulting from this interaction enhanced the survival of *C. albicans* in the presence of macrophages and decreased the production of proinflammatory cytokines by coinfected macrophages. Thus, an intriguing working hypothesis suggests that *C. albicans* and *F. nucleatum* collaborate to minimize their outgrowth and avoid attracting excessive attention from the host immune system, promoting a commensal lifestyle (Bor et al. 2016).

*Interactions of Non-*albicans Candida *and Other Fungi with Oral Bacteria*

C. guilliermondii and *C. dubliniensis* have been isolated from the oral cavities of HIV-infected patients (Lasker et al. 2001), while *C. rugose* and *C. lusitaniae* have been found in the oral cavities of other patients (Pires-Gonçalves et al. 2007; Sardi et al. 2013). Oral bacteria such as *F. nucleatum* interact with *C. dubliniensis* (Hsu et al. 1990), and *F. nucleatum*, *A. viscosus*, oral streptococci, and *L. amylovorus* have been shown to coaggregate with *C. tropicalis* (Jabra-Rizk et al. 1999). Mixed biofilms of *S. epidermidis* and *S. xylosus* with *C. albicans* and non-*Candida albicans* species, such as *C. parapsilosis* and *C. famata*, have been isolated from facial prostheses (Ariani et al. 2015). Recent studies have also investigated the microbiome associated with oral cancers, revealing that a dysbiotic mycobiome enriched in *C. albicans* was common among patients with oral squamous cell carcinoma (OSCC) (Perera et al. 2017; Mohamed et al. 2021).

Conclusions

Growing evidence underscores the significant role of the mycobiome in oral and systemic health. The oral cavity hosts a diverse array of fungal species, with *Candida albicans* emerging as a key player due to its ability to form biofilms and interact with bacterial communities. Understanding these complex interspecies dynamics is crucial for advancing therapeutic strategies aimed at mitigating fungal-associated diseases. Continued research into the oral mycobiome will provide deeper insights into its contributions to health and disease.

References

Abusleme L et al (2018) Human defects in STAT3 promote oral mucosal fungal and bacterial dysbiosis. JCI Insight 3(17):e122061

Acosta-Pagán K et al (2024) Ecological competition in the oral mycobiome of Hispanic adults living in Puerto Rico associates with periodontitis. J Oral Microbiol 16(1):2316485

Ahmad KM et al (2014) Genome structure and dynamics of the yeast pathogen Candida glabrata. FEMS Yeast Res 14(4):529–535

Annavajhala MK et al (2020) Oral and gut microbial diversity and immune regulation in patients with HIV on antiretroviral therapy. mSphere. 5(1):e00798-19

Ariani N et al (2015) Efficacy of cleansing agents in killing microorganisms in mixed species biofilms present on silicone facial prostheses – an in vitro study. Clin Oral Investig 19(9):2285–2293

Bachtiar EW et al (2014) AI-2 of Aggregatibacter actinomycetemcomitans inhibits Candida albicans biofilm formation. Front Cell Infect Microbiol 4:94

Baraniya D et al (2020) Supragingival mycobiome and inter-kingdom interactions in dental caries. J Oral Microbiol 12(1):1729305

Barbosa JO et al (2016) Streptococcus mutans can modulate biofilm formation and attenuate the virulence of Candida albicans. PLoS One 11(3):e0150457

Beall CJ et al (2023) Independent effects of HIV and antiretroviral therapy on the oral microbiome identified by multivariate analyses. MBio 14(3):e0040923

Beibei L et al (2024) Dysbiosis and interactions of the mycobiome and bacteriome in mucosal lesions of erosive and non-erosive oral lichen planus patients. J Oral Microbiol 16(1):2374639

Bor B et al (2016) Morphological and physiological changes induced by contact-dependent interaction between Candida albicans and Fusobacterium nucleatum. Sci Rep 6:27956

Branting C, Sund M-L, Linder LE (1989) The influence of Streptococcus mutans on adhesion of Candida albicans to acrylic surfaces in vitro. Archs Oral Biol 34(5):347–353

Brusca MI et al (2013) Mechanisms of interaction between Candida albicans and Streptococcus mutans: an experimental and mathematical modelling study. Acta Odontol Scand 71(3–4):416–423

Canullo L et al (2023) Effect of argon plasma pre-treatment of healing abutments on peri-implant microbiome and soft tissue integration: a proof-of-concept randomized study. BMC Oral Health 23(1):27

Cavalcanti YW et al (2016a) Modulation of Candida albicans virulence by bacterial biofilms on titanium surfaces. Biofouling 32(2):123–134

Cavalcanti IM et al (2016b) Interkingdom cooperation between Candida albicans, Streptococcus oralis and Actinomyces oris modulates early biofilm development on denture material. Pathog Dis 74(3):ftw002

Cavalcanti IM et al (2017) Interactions between Streptococcus oralis, Actinomyces oris, and Candida albicans in the development of multispecies oral microbial biofilms on salivary pellicle. Mol Oral Microbiol 32(1):60–73

Chang S et al (2021) Comparative analysis of salivary mycobiome diversity in human immunodeficiency virus-infected patients. Front Cell Infect Microbiol 11:781246

Charone S et al (2017) Role of Candida species from HIV infected children in enamel caries lesions: an in vitro study. J Appl Oral Sci 25(1):53–60

Chen Y et al (2023a) Oral fungal profiling and risk of nasopharyngeal carcinoma: a population-based case-control study. EBioMedicine 96:104813

Chen BY et al (2023b) Characteristics and correlations of the oral and gut fungal microbiome with hypertension. Microbiol Spectr 11(1):e0195622

Cheung MK et al (2022) Determinants and interactions of oral bacterial and fungal microbiota in healthy Chinese adults. Microbiol Spectr 10(1):e0241021

Cheung MK et al (2023) Extent of oral-gut transmission of bacterial and fungal microbiota in healthy Chinese adults. Microbiol Spectr 11(1):e0281422
Cui L, Morris A, Ghedin E (2013) The human mycobiome in health and disease. Genome Med 5(7):63
Cui L et al (2015) Topographic diversity of the respiratory tract mycobiome and alteration in HIV and lung disease. Am J Respir Crit Care Med 191(8):932–942
Cui Y et al (2021) Oral mycobiome differences in various spatial niches with and without severe early childhood caries. Front Pediatr 9:748656
de Abreu C, Brandi T et al (2016) Demineralizing potential of dental biofilm added with Candida albicans and Candida parapsilosis isolated from preschool children with and without caries. Microb Pathog 100:51–55
de Carvalho FG et al (2006) Presence of mutans streptococci and Candida spp. in dental plaque/dentine of carious teeth and early childhood caries. Arch Oral Biol 51(11):1024–1028
de Jesus VC et al (2020) Sex-based diverse plaque microbiota in children with severe caries. J Dent Res 99(6):703–712
de Jesus VC et al (2021) Characterization of supragingival plaque and oral swab microbiomes in children with severe early childhood caries. Front Microbiol 12:683685
Diaz PI, Dongari-Bagtzoglou A (2021) Critically appraising the significance of the oral mycobiome. J Dent Res 100(2):133–140
Diaz PI et al (2012) Synergistic interaction between Candida albicans and commensal oral streptococci in a novel in vitro mucosal model. Infect Immun 80(2):620–632
Diaz PI, Strausbaugh LD, Dongari-Bagtzoglou A (2014) Fungal-bacterial interactions and their relevance to oral health: linking the clinic and the bench. Front Cell Infect Microbiol 4:101
Diaz PI et al (2017) Mining the oral mycobiome: methods, components, and meaning. Virulence 8(3):313–323
Dong K et al (2021) Comparative study of oral bacteria and fungi microbiota in Tibetan and Chinese Han living at different altitude. Tohoku J Exp Med 254(2):129–139
Du G et al (2023) Preliminary analysis of the buccal mucosal fungal microbiome in oral lichen planus patients. J Oral Pathol Med 52(7):680–690
Dupuy AK et al (2014) Redefining the human oral mycobiome with improved practices in amplicon-based taxonomy: discovery of Malassezia as a prominent commensal. PLoS One 9(3):e90899
Dutton LC et al (2016) Transcriptional landscape of trans-kingdom communication between Candida albicans and Streptococcus gordonii. Mol Oral Microbiol 31(2):136–161
El Jurdi N et al (2019) Gastrointestinal microbiome and mycobiome changes during autologous transplantation for multiple myeloma: results of a prospective pilot study. Biol Blood Marrow Transplant 25(8):1511–1519
Ellepola KJB (2018) Molecular microbiology of bacterial-fungal mixed-species biofilms and clinical implications. National University of Singapore, PhD dissertation.
Ellepola K et al (2017) Bacterial GtfB augments Candida albicans accumulation in cross-kingdom biofilms. J Dent Res 96(10):1129–1135
Ellepola K et al (2019) Multi-omics analyses reveal synergistic carbohydrate metabolism in Streptococcus mutans-Candida albicans mixed-species biofilms. Infect Immun 87(10):e00339-19.
Falsetta ML, Klein MI, Colonne PM, Scott-Anne K, Gregoire S, Pai CH, Gonzalez-Begne M, Watson G, Krysan DJ, Bowen WH, Koo H (2014) Symbiotic relationship between Streptococcus mutans and Candida albicans synergizes virulence of plaque biofilms in vivo. Infect Immun 82(5):1968–1981
Fechney JM et al (2019) Preliminary study of the oral mycobiome of children with and without dental caries. J Oral Microbiol 11(1):1536182
Fernandes RA et al (2016) Biofilm formation by Candida albicans and Streptococcus mutans in the presence of farnesol: a quantitative evaluation. Biofouling 32(3):329–338

Fidel PL Jr et al (2021) Effect of HIV/HAART and other clinical variables on the oral mycobiome using multivariate analyses. MBio. 12(2):e00294-21.
Gabaldón T et al (2013) Comparative genomics of emerging pathogens in the Candida glabrata clade. BMC Genomics 14:623
Ghannoum MA et al (2010) Characterization of the oral fungal microbiome (mycobiome) in healthy individuals. PLoS Pathog 6(1):e1000713
Gregoire S et al (2011) Role of glucosyltransferase B in interactions of Candida albicans with Streptococcus mutans and with an experimental pellicle on hydroxyapatite surfaces. Appl Environ Microbiol 77(18):6357–6367
Gupta A et al (2022) Oral dysbiosis and its linkage with SARS-CoV-2 infection. Microbiol Res 261:127055
Hagihara Y et al (1988) Degradation of human dentine collagen by an enzyme produced by the yeast Candida albicans. Arch Oral Biol 33(8):617–619
He J et al (2017) RNA-Seq reveals enhanced sugar metabolism in Streptococcus mutans co-cultured with Candida albicans within mixed-species biofilms. Front Microbiol 8:1036
Heng W et al (2022) Oral bacteriome and mycobiome across stages of oral carcinogenesis. Microbiol Spectr 10(6):e0273722
Hong BY et al (2020) The salivary mycobiome contains 2 ecologically distinct mycotypes. J Dent Res 99(6):730–738
Hopson LM et al (2020) Chapter five – Bioinformatics and machine learning in gastrointestinal microbiome research and clinical application. In: Kasselman LJ (ed) Progress in molecular biology and translational science. Academic Press, pp 141–178
Hornby JM et al (2001) Quorum sensing in the dimorphic fungus Candida albicans is mediated by farnesol. Appl Environ Microbiol 67(7):2982–2992
Hoyer LL et al (2008) Discovering the secrets of the Candida albicans agglutinin-like sequence (ALS) gene family – a sticky pursuit. Med Mycol 46(1):1–15
Hsu LY et al (1990) Coaggregation of oral Candida isolates with bacteria from bone marrow transplant recipients. J Clin Microbiol 28(12):2621–2626
Hu X et al (2023) Oral fungal alterations in patients with COVID-19 and recovered patients. Adv Sci (Weinh) 10(19):e2205058
Hwang G et al (2015) Binding force dynamics of Streptococcus mutans-glucosyltransferase B to Candida albicans. J Dent Res 94(9):1310–1317
Hwang G et al (2017) Candida albicans mannans mediate Streptococcus mutans exoenzyme GtfB binding to modulate cross-kingdom biofilm development in vivo. PLoS Pathog 13(6):e1006407
Ieda S et al (2014) Molecular analysis of fungal populations in patients with oral candidiasis using internal transcribed spacer region. PLoS One 9(6):e101156
Imabayashi Y et al (2016) Molecular analysis of fungal populations in patients with oral candidiasis using next-generation sequencing. Sci Rep 6:28110
Jabra-Rizk MA et al (1999) Coaggregation of Candida dubliniensis with Fusobacterium nucleatum. J Clin Microbiol 37(5):1464–1468
Jabra-Rizk MA et al (2006) Effect of farnesol on Staphylococcus aureus biofilm formation and antimicrobial susceptibility. Antimicrob Agents Chemother 50(4):1463–1469
Jack AA et al (2015) Streptococcus gordonii comCDE (competence) operon modulates biofilm formation with Candida albicans. Microbiology (Reading) 161(Pt 2):411–421
Jarosz LM et al (2009) Streptococcus mutans competence-stimulating peptide inhibits Candida albicans hypha formation. Eukaryot Cell 8(11):1658–1664
Jenkinson HF, Lamont RJ (2005) Oral microbial communities in sickness and in health. Trends Microbiol 13(12):589–595
Jenkinson HF, Lala HC, Shepherd MG (1990) Coaggregation of Streptococcus sanguis and other streptococci with Candida albicans. Infect Immun 58(5):1429–1436
Jiang S et al (2022) Next-generation sequencing applications for the study of fungal pathogens. Microorganisms 10(10):1882

Khadija B, Imran M, Faryal R (2021) Keystone salivary mycobiome in postpartum period in health and disease conditions. J Mycol Med 31(1):101101
Kim D et al (2017) Candida albicans stimulates Streptococcus mutans microcolony development via cross-kingdom biofilm-derived metabolites. Sci Rep 7:41332
Kim D et al (2018) Bacterial-derived exopolysaccharides enhance antifungal drug tolerance in a cross-kingdom oral biofilm. ISME J 12(6):1427–1442
Klein MI et al (2015) Streptococcus mutans-derived extracellular matrix in cariogenic oral biofilms. Front Cell Infect Microbiol 5:10
Klinke T et al (2009) Acid production by oral strains of Candida albicans and lactobacilli. Caries Res 43(2):83–91
Klinke T et al (2011) Dental caries in rats associated with <i>Candida albicans</i>. Caries Res 45(2):100–106
Koo H, Andes DR, Krysan DJ (2018) Candida-streptococcal interactions in biofilm-associated oral diseases. PLoS Pathog 14(12):e1007342
Krüger W et al (2019) Fungal-bacterial interactions in health and disease. Pathogens 8(2):70
Lasker BA et al (2001) Molecular epidemiology of Candida albicans strains isolated from the oropharynx of HIV-positive patients at successive clinic visits. Med Mycol 39(4):341–352
Li Y et al (2019) Salivary mycobiome dysbiosis and its potential impact on bacteriome shifts and host immunity in oral lichen planus. Int J Oral Sci 11(2):13
Li H et al (2021) Oral microbial diversity analysis among atrophic glossitis patients and healthy individuals. J Oral Microbiol 13(1):1984063
Liu X et al (2024) Oral fungal dysbiosis and systemic immune dysfunction in Chinese patients with schizophrenia. Transl Psychiatry 14(1):475
Mayer FL, Wilson D, Hube B (2013) Candida albicans pathogenicity mechanisms. Virulence 4(2):119–128
Metwalli KH et al (2013) Streptococcus mutans, Candida albicans, and the human mouth: a sticky situation. PLoS Pathog 9(10):e1003616
Mohamed N et al (2021) Analysis of salivary mycobiome in a cohort of oral squamous cell carcinoma patients from Sudan identifies higher salivary carriage of Malassezia as an independent and favorable predictor of overall survival. Front Cell Infect Microbiol 11:673465
Montelongo-Jauregui D, Lopez-Ribot JL (2018) Candida interactions with the oral bacterial microbiota. J Fungi (Basel) 4(4):122
Montelongo-Jauregui D et al (2016) An in vitro model for oral mixed biofilms of Candida albicans and Streptococcus gordonii in synthetic saliva. Front Microbiol 7:686
Montelongo-Jauregui D et al (2018) An in vitro model for Candida albicans-Streptococcus gordonii biofilms on titanium surfaces. J Fungi (Basel) 4(2):66
Mukherjee PK et al (2014) Oral mycobiome analysis of HIV-infected patients: identification of Pichia as an antagonist of opportunistic fungi. PLoS Pathog 10(3):e1003996
Mukherjee PK et al (2018) Dysbiosis in the oral bacterial and fungal microbiome of HIV-infected subjects is associated with clinical and immunologic variables of HIV infection. PLoS One 13(7):e0200285
Neumann CJ et al (2024) The dynamics of the female microbiome: unveiling abrupt changes of microbial domains across body sites from prepartum to postpartum phases. Microbiol Spectr 12(8):e0014724
Nikawa H et al (2003) In vitro cariogenic potential of Candida albicans. Mycoses 46(11–12):471–478
Niu C et al (2023) Mapping the human oral and gut fungal microbiota in patients with metabolic dysfunction-associated fatty liver disease. Front Cell Infect Microbiol 13:1157368
O'Connell LM et al (2020) Site-specific profiling of the dental mycobiome reveals strong taxonomic shifts during progression of early-childhood caries. Appl Environ Microbiol 86(7):e02825-19
O'Connell LM et al (2023) Supragingival mycobiome of HIV-exposed-but-uninfected children reflects a stronger correlation with caries-free-associated taxa compared to HIV-infected or uninfected children. Microbiol Spectr 11(6):e0149123

Odds FC (1988) Candida and Candidosis. Elsevier – Health Sciences Division
O'Donnell LE et al (2015) Polymicrobial Candida biofilms: friends and foe in the oral cavity. FEMS Yeast Res 15(7):fov077
Oğuzülgen K et al (2024) Inhaler steroid use changes oral and airway bacterial and fungal microbiome profile in asthma patients. Int Arch Allergy Immunol 185(1):10–19
Parahitiyawa NB et al (2006) Interspecies variation in Candida biofilm formation studied using the Calgary biofilm device. APMIS 114(4):298–306
Pereira-Cenci T et al (2008) The effect of Streptococcus mutans and Candida glabrata on Candida albicans biofilms formed on different surfaces. Arch Oral Biol 53(8):755–764
Perera M et al (2017) A dysbiotic mycobiome dominated by Candida albicans is identified within oral squamous-cell carcinomas. J Oral Microbiol 9(1):1385369
Pérez-Losada M et al (2024) Characterization of the oral mycobiome of Portuguese with allergic rhinitis and asthma. Curr Res Microb Sci 7:100300
Peters BA et al (2017) The oral fungal mycobiome: characteristics and relation to periodontitis in a pilot study. BMC Microbiol 17(1):157
Pfaller MA, Diekema DJ (2007) Epidemiology of invasive candidiasis: a persistent public health problem. Clin Microbiol Rev 20(1):133–163
Pires-Gonçalves RH et al (2007) Genetic relatedness of commensal strains of Candida albicans carried in the oral cavity of patients' dental prosthesis users in Brazil. Mycopathologia 164(6):255–263
Qiu R et al (2015) Genotypic diversity and cariogenicity of Candida albicans from children with early childhood caries and caries-free children. BMC Oral Health 15:144
Raja M, Hannan A, Ali K (2010) Association of oral candidal carriage with dental caries in children. Caries Res 44(3):272–276
Ramage G et al (2002) Inhibition of Candida albicans biofilm formation by farnesol, a quorum-sensing molecule. Appl Environ Microbiol 68(11):5459–5463
Ren Z et al (2023) Alterations of the oral and gut mycobiome and cytokines during long-term follow-up of COVID-19 convalescents. Signal Transduct Target Ther 8(1):166
Robinson S et al (2020) Observational cohort study of oral mycobiome and Interkingdom interactions over the course of induction therapy for leukemia. mSphere. 5(2):e00048-20.
Rozkiewicz D et al (2006) Oral Candida albicans carriage in healthy preschool and school children. Adv Med Sci 51(Suppl 1):187–190
Sajid M et al (2022) Smokeless tobacco consumption induces dysbiosis of oral mycobiome: a pilot study. Appl Microbiol Biotechnol 106(17):5643–5657
Salehi B et al (2020) Management of Streptococcus mutans-Candida spp. oral biofilms' infections: paving the way for effective clinical interventions. J Clin Med 9(2):517
Salvatori O et al (2020) Bacteria modify Candida albicans hypha formation, microcolony properties, and survival within macrophages. mSphere. 5(4):e00689-20.
Samaranayake LP et al (1982) Isotachophoretic analysis of short-chain carboxylic acids produced by Candida albicans. Microbios 35(140):91–98
Samaranayake LP et al (1983) Growth and acid production of Candida albicans in carbohydrate supplemented media. Microbios 37(148):105–115
Samaranayake LP et al (1986) Growth and acid production of Candida species in human saliva supplemented with glucose. J Oral Pathol 15(5):251–254
Sami A et al (2023) Altered oral microbiome in Sudanese Toombak smokeless tobacco users carries a newly emerging risk of squamous cell carcinoma development and progression. Sci Rep 13(1):6645
Sardi JCO et al (2013) Candida species: current epidemiology, pathogenicity, biofilm formation, natural antifungal products and new therapeutic options. J Med Microbiol 62(Pt 1):10–24
Seneviratne CJ et al (2008) Cell density and cell aging as factors modulating antifungal resistance of Candida albicans biofilms. Antimicrob Agents Chemother 52(9):3259–3266
Seneviratne CJ et al (2011) Antifungal susceptibility and virulence attributes of bloodstream isolates of Candida from Hong Kong and Finland. Mycopathologia 172(5):389–395

Shelburne SA et al (2015) Implementation of a pan-genomic approach to investigate holobiont-infecting microbe interaction: a case report of a leukemic patient with invasive mucormycosis. PLoS One 10(11):e0139851
Shirtliff ME, Peters BM, Jabra-Rizk MA (2009) Cross-kingdom interactions: Candida albicans and bacteria. FEMS Microbiol Lett 299(1):1–8
Silverman RJ et al (2010) Interaction of Candida albicans cell wall Als3 protein with Streptococcus gordonii SspB adhesin promotes development of mixed-species communities. Infect Immun 78(11):4644–4652
Souza JGS et al (2020) Biofilm interactions of Candida albicans and mitis group streptococci in a titanium-mucosal Interface model. Appl Environ Microbiol 86(9):e02950-19
Sztajer H et al (2014) Cross-feeding and interkingdom communication in dual-species biofilms of Streptococcus mutans and Candida albicans. ISME J 8(11):2256–2271
Thein ZM, Smaranayake YH, Smaranayake LP (2007) Dietary sugars, serum and the biocide chlorhexidine digluconate modify the population and structural dynamics of mixed Candida albicans and Escherichia coli biofilms. APMIS 115(11):1241–1251
Tu Y et al (2022) The crosstalk between saliva bacteria and fungi in early childhood caries. Front Cell Infect Microbiol 12:845738
Turner SA, Butler G (2014) The Candida pathogenic species complex. Cold Spring Harb Perspect Med 4(9):a019778
Vílchez R et al (2010) Streptococcus mutans inhibits Candida albicans hyphal formation by the fatty acid signaling molecule trans-2-decenoic acid (SDSF). Chembiochem 11(11):1552–1562
Wang XH et al (2024) Oral bacteriome and mycobiome of patients with idiopathic membranous nephropathy with different tongue coatings treated with a Chinese herbal formula. J Ethnopharmacol 331:118233
Ward TL et al (2018) Development of the human mycobiome over the first month of life and across body sites. mSystems 3(3)
Watkins RR et al (2017) Admission to the intensive care unit is associated with changes in the oral mycobiome. J Intensive Care Med 32(4):278–282
Wei N et al (2023) Characterization of oral bacterial and fungal microbiome in recovered COVID-19 patients. BMC Microbiol 23(1):123
Willems HM et al (2016) Candida albicans in oral biofilms could prevent caries. Pathog Dis 74(5):ftw039
Williams D, Lewis M (2011) Pathogenesis and treatment of oral candidosis. J Oral Microbiol 3:5771. https://doi.org/10.3402/jom.v3i0.5771
Williamson PR, Huber MA, Bennett JE (1993) Role of maltase in the utilization of sucrose by Candida albicans. Biochem J 291(Pt 3):765–771
Willis JR et al (2021) Citizen-science based study of the oral microbiome in cystic fibrosis and matched controls reveals major differences in diversity and abundance of bacterial and fungal species. J Oral Microbiol 13(1):1897328
Wu R et al (2020) Streptococcus mutans membrane vesicles harboring glucosyltransferases augment Candida albicans biofilm development. Front Microbiol 11:581184
Wu TT et al (2023) Metagenomic analysis examines oral microbiome changes and interplay with immune response following prenatal total oral rehabilitation. J Transl Med 21(1):172
Xiao J et al (2016) Candida albicans carriage in children with Severe Early Childhood Caries (S-ECC) and maternal relatedness. PLoS One 11(10):e0164242
Xu H, Jenkinson HF, Dongari-Bagtzoglou A (2014) Innocent until proven guilty: mechanisms and roles of Streptococcus-Candida interactions in oral health and disease. Mol Oral Microbiol 29(3):99–116
Xu H et al (2016) Streptococcus oralis and Candida albicans synergistically activate μ-Calpain to degrade E-cadherin from Oral epithelial junctions. J Infect Dis 214(6):925–934

Xu H et al (2017) S. oralis activates the Efg1 filamentation pathway in C. albicans to promote cross-kingdom interactions and mucosal biofilms. Virulence 8(8):1602–1617
Yang XQ et al (2012) Genotypic distribution of Candida albicans in dental biofilm of Chinese children associated with severe early childhood caries. Arch Oral Biol 57(8):1048–1053
Yang C et al (2018) Antigen I/II mediates interactions between Streptococcus mutans and Candida albicans. Mol Oral Microbiol 33(4):283–291
Zegarelli DJ (1993) Fungal infections of the oral cavity. Otolaryngol Clin N Am 26(6):1069–1089

Chapter 5
Vulvovaginal Candidiasis: Pathophysiology, the Role of Biofilms, and Emerging Therapeutics

Pranitha Murali, Anthony J. Slate, Ryan Kean, and William Johnston

Abstract Vulvovaginal candidiasis (VVC) is a prevalent fungal infection of the female genital tract that impacts millions of women worldwide, representing a considerable burden on health and quality of life. Despite comprehensive knowledge of the causative species (*Candida albicans* and non-*albicans Candida*), approximately 8% of women experience recurrent VVC (RVVC), characterised by more frequent, persistent infections. Fungal biofilm formation has been proposed as one possible mechanism underlying this response; however, this remains a topic of debate. In this chapter, we review the current understanding of R/VVC pathophysiology, clinical presentation and diagnostic approaches. We further discuss the evidence linking biofilms to RVVC and highlight emerging therapeutic strategies aimed at better managing this challenging condition.

Introduction

Vulvovaginal candidiasis (VVC), colloquially termed vaginal thrush, is a common fungal condition of the female genital tract. It is often reported that the majority of women (>70%) will experience at least one VVC episode in their lifetime (Bongomin

P. Murali · R. Kean · W. Johnston (✉)
Department of Biological and Biomedical Sciences, Glasgow Caledonian University, Glasgow, UK

Safeguarding Health through Infection Prevention Research Group, Research Centre for Health (ReaCH), Glasgow Caledonian University, Glasgow, UK
e-mail: Pranitha.Murali@gcu.ac.uk; Ryan.Kean@gcu.ac.uk; William.Johnston2@gcu.ac.uk

A. J. Slate
Department of Biological and Biomedical Sciences, Glasgow Caledonian University, Glasgow, UK

Department of Physical and Life Sciences, School of Applied Sciences, University of Huddersfield, Queensgate Campus, Huddersfield, UK
e-mail: A.J.Slate@hud.ac.uk

G. Ramage, D. Andes (eds.), *Fungal Biofilms*, Springer Series on Biofilms 15,
https://doi.org/10.1007/978-3-031-93315-8_5

et al. 2017). In addition, up to 8% go on to develop recurrent VVC (RVVC), usually defined as three or more cases within 1 year (Donders et al. 2022; Workowski 2021) However, the accuracy of these figures is disputed and the exact prevalence remains difficult to ascertain, due to high levels of self-diagnosis and self-treatment (Benedict et al. 2022b; Rathod and Buffler 2014). Despite this, VVC and RVVC represent a clear health and economic burden, with an estimated $14 billion (USD) associated with lost productivity in high-income countries (Denning et al. 2018). VVC results in vaginal inflammation, causing irritation, itching and burning (Jeanmonod et al. 2024; Nyirjesy et al. 2022). Furthermore, RVVC has been shown to significantly reduce quality of life and mental wellbeing among patients (Donders et al. 2022; Irving et al. 1998; Lietz et al. 2023).

The primary causative agent of VVC is *Candida albicans*, accounting for approximately 85–90% of cases (Martin Lopez 2015), which may fluctuate depending on geographical location (Sherry et al. 2017). Interestingly, this fungal species is also found in 10–30% of healthy asymptomatic nonpregnant women (Achkar and Fries 2010; Chatzivasileiou and Vyzantiadis 2019; McKloud et al. 2021), indicating its mere presence is not always indicative of VVC (Farhan et al. 2019; Jeanmonod et al. 2024). The exact cause of VVC is yet to be elucidated; however, it is proposed to be linked to disturbances in the vaginal microenvironment, which may be induced by treatment including antibiotics and hormone replacement therapy (HRT), as well as pregnancy and genetic predispositions (Gonçalves et al. 2016; Jeanmonod et al. 2024; Pereira et al. 2021; Sobel 2007). Although VVC is not considered a sexually transmitted infection (STI), it has been found to be more common among sexually active women (Jeanmonod et al. 2024; Kombade et al. 2021).

It is widely known that *C. albicans*, as well as non-albicans *Candida* (NAC) species, form fungal biofilms, which contribute to various human diseases (Cavalheiro and Teixeira 2018; Nobile and Johnson 2015; Ramage et al. 2005). Despite this, and in contrast to other biofilm-associated vaginal conditions such as bacterial vaginosis (BV) (Johnston et al. 2023; Machado et al. 2016; Sousa et al. 2023), the role of biofilms in VVC remains a topic of debate (McKloud et al. 2021; Noverr and Fidel 2019; Sobel 2015). For example, some studies have failed to identify *Candida* biofilms in vaginal tissue biopsies (Swidsinski et al. 2019), questioning the biofilm-mediated hypothesis of these conditions. In contrast, other studies have demonstrated the presence of such biofilms in both murine models (Harriott et al. 2010) and in human RVVC (Pan et al. 2023), supporting a role for *Candida* biofilms in recurrent cases.

The following chapter provides an overview of the current understanding of VVC and RVVC aetiology, including the debated role of microbial biofilms. It will explore the various factors contributing to the development of these conditions, considerations for accurate diagnosis and emerging therapeutic strategies.

Aetiology and Pathogenesis

The Vaginal Microbiome

The vaginal microbiome was initially investigated over 100 years ago by Albert S Döderlein, who recognised the importance of *Lactobacilli* spp., within this environment. Since then, it has become apparent that the exact composition of the vaginal microbiome is dependent on various factors including age, ethnicity, geographic location and physiological factors including hormonal fluctuations (Lehtoranta et al. 2022). However, as originally suggested, it has been found that in the majority of healthy premenopausal women, this environment is dominated by different lactobacilli species (Fig. 5.1a).

These organisms are present at approximately 1×10^7–1×10^8 CFU/gram of vaginal fluid (Borges et al. 2014) and play a crucial role in maintaining vaginal health through inhibition of competing microorganisms (McKloud et al. 2021). This is achieved through the production of lactic acid, hydrogen peroxide and bacteriocin-like substances which maintain a protective, acidic environment (pH <4.5) (Borges et al. 2014; Boris and Barbés 2000). In addition, these species directly compete for nutrient acquisition and block epithelial binding sites from potential pathogens (Ferrer 2000). Extensive community profiling has allowed researchers to designate community state types (CSTs) depending on the dominant lactobacilli species (Chee et al. 2020; Ravel et al. 2011). CST-I, II, III and V are dominated by *L. crispatus*, *L. gasseri*, *L. iners* and *L. jensenii*, respectively, while CST-IV represents a diverse microbial community with no lactobacilli dominance (Ravel et al. 2011).

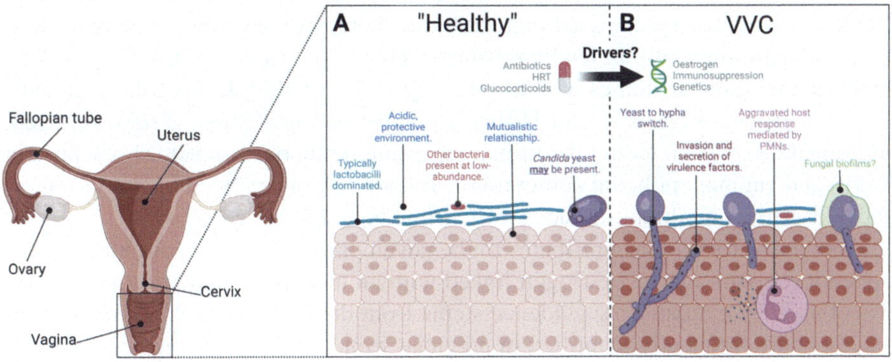

Fig. 5.1 Current understanding of R/VVC, displaying the vaginal microenvironment in healthy conditions* (**a**) and VVC (**b**). In "healthy" conditions, the vaginal microbiome is typically dominated by lactobacilli species. Other bacterial species are present at low abundance and *Candida* species may be present in yeast form. In VVC, *Candida* hyphal formation occurs promoting tissue invasion and secretion of virulence factors. An aggravated innate response is mediated by polymorphonuclear neutrophils (PMNs) which drives hyperinflammation and symptom onset. Fungal biofilms may contribute to recurrent disease and/or failed clinical treatment. Potential and suspected drivers of VVC are illustrated. *In this figure, "healthy" refers to asymptomatic women. (Created in BioRender (https://BioRender.com/m47u651))

Despite being the primary causative agent of VVC, fungal species including *C. albicans* may also be found in asymptomatic women, forming an often overlooked part of the microbiome (Achkar and Fries 2010; McKloud et al. 2021). *Candida* species are isolated in approximately 10–30% of asymptomatic nonpregnant women at any given time (Achkar and Fries 2010; Bauters et al. 2002; Chatzivasileiou and Vyzantiadis 2019), and over 70% of women show positive yeast colonisation when sampled longitudinally over a 1-year period (Beigi et al. 2004). In some patients, this appears to be transient to some degree, as the same study observed that only 4% of women were colonised with *Candida* spp., at every visit.

The Onset of VVC

The majority of VVC cases have no identifiable trigger; however, several factors are linked with an increased likelihood of developing infection (Farr et al. 2021; Rosati et al. 2020). As aforementioned, a large proportion of women will carry *Candida* species in the vaginal microenvironment at some point during their lifetime (Beigi et al. 2004). In otherwise healthy women, local defence mechanisms, including the protective actions of vaginal lactobacilli, and the host immune response, are believed to attenuate fungal adhesion and prevent overgrowth (Ferrer 2000). As such, alternations in the vaginal/vulvar environment may disrupt this balance and promote the onset of VVC (Gonçalves et al. 2016; Rosati et al. 2020).

One of the most notable associations stems from antibacterial therapy (i.e. antibiotics), leading to the so-called "antibiotic-induced VVC" (Sobel 2016; Xu et al. 2008). It is widely hypothesised that antibiotic therapy disrupts the protective vaginal lactobacilli populations, leading to the onset of VVC. However, this has not been supported by clinical studies quantifying vaginal lactobacilli levels following antibacterial exposure (Agnew and Hillier 1995; Shukla and Sobel 2019). As such, additional theories including the impact of antibiotic therapy on heat shock protein (HSP) and vaginal epithelial cell cytokine release have been hypothesised as alternative drivers of antibiotic-induced VVC (Shukla and Sobel 2019).

Another factor which contributes to the development of VVC is circulating oestrogen level, with higher oestrogen linked with an increased susceptibility to VVC (He et al. 2022). Supporting evidence stems from the finding that *Candida* species may be more frequently identified in pregnant women where oestrogen is elevated (Dias et al. 2011; Leli et al. 2013; Sampson and George 2019), and in women receiving hormone replacement therapy (HRT) or oral contraceptives (Fischer and Bradford 2011; Hillier and Lau 1997; Spinillo et al. 1995). One proposed mechanism behind this response is that oestrogen has been shown to increase the virulence of *C. albicans*, evidenced through increased fungal survival, growth and HSP90 expression following treatment with 17-β-oestradiol (Zhang et al. 2000). In addition, oestrogen exerts direct impacts on the vaginal epithelium which may promote fungal adhesion (He et al. 2022; Luan et al. 2020; Tarry et al. 2005). An example of this is through the elevated production of glycogen, which forms a suitable substrate for *C. albicans* binding (Dennerstein and Ellis 2001).

In contrast to other *Candida* infections of mucosal sites such as oropharyngeal candidiasis (OPC) which are rare in immunocompetent individuals (Jabra-Rizk et al. 2016), VVC occurs frequently in both immunocompetent and immunosuppressed women (Jacobsen 2023; Sobel 2007). Additionally, several studies demonstrate that defective adaptive immune responses do not predispose to VVC (Gow et al. 2011). Despite this, glucocorticoid use has been linked to VVC, thought to be linked to their immunosuppressive actions combined with the induction of a hyperglycaemic environment, which favours fungal overgrowth (Ferris and Kahn 2012; Gonçalves et al. 2016; Mohammed et al. 2021). In line with this, it has been demonstrated that VVC is more common among individuals with diabetes mellitus (DM), likely driven by similar factors (Berbudi et al. 2020; Mohammed et al. 2021; Sobel 2007).

Irrespective of the risk factors mentioned above, the majority of VVC cases occur in their absence. As such, a potential genetic predisposition to VVC has been suggested by several researchers (Bradford et al. 2013). In line with this theory, there is some evidence suggesting RVVC may be linked to polymorphisms in mannose-binding lectin gene (Babula et al. 2003) and Lewis blood group antigens (Chaim et al. 1997). Other theories include shifts in the vaginal microbiota across differing demographics and ethnicities, such as a reduction of lactobacilli which may predispose to VVC (Antonio et al. 1999; Gonçalves et al. 2016). However, it should be noted that the weight of influence that these factors have on VVC onset remains to be elucidated.

Causative Species

VVC is categorised as uncomplicated or complicated depending on the frequency of infection, causative fungal species and host immune function (Nyirjesy et al. 2022; Sobel et al. 1998). Uncomplicated VVC refers to cases caused by *C. albicans*, causing sporadic episodes which present as mild symptoms in otherwise healthy individuals. In contrast, complicated cases are those which are more severe and/or recurrent, caused by a NAC species, or present in immunocompromised women (Nyirjesy et al. 2022). As previously mentioned, *C. albicans* causes the vast majority of VVC cases (Martin Lopez 2015; Sobel et al. 1998), with NAC species accounting for 10–20% (McKloud et al. 2021). Common causative NAC species include *Nakaseomyces glabratus* (formerly *C. glabrata*), *C. parapsilosis, C. dubliniensis, C. tropicalis* and *C. krusei* (Bauters et al. 2002; Deorukhkar et al. 2014; Makanjuola et al. 2018). VVC caused by NAC species may be less severe than *C. albicans*, but present a higher risk of recurrence and lower susceptibility to azole antifungals (Makanjuola et al. 2018; McKloud et al. 2021).

Candida *Pathogenesis*

Candida exerts a dual-lifestyle within the vaginal microenvironment, shifting from commensal to pathogenic state (Jacobsen 2023). Although the exact pathogenesis is poorly defined, there are several key aspects which appear consistent among cases (Czechowicz et al. 2022). In asymptomatic women, it is believed that *Candida* species exist in yeast form, which is well tolerated by the host (Cauchie et al. 2017; Willems et al. 2020). In these cases, host immunity and local microbial populations form a delicate balance to attenuate *Candida* overgrowth and pathogenicity (Gaziano et al. 2023). For example, the yeast to hypha switch is inhibited in acidic environments (pH <4.5), which may be created by vaginal lactobacilli species (Buffo et al. 1984). However, hyphal formation may occur if this balance is disturbed through fluctuations in pH, microbiome composition or oestrogen level, promoting increased *Candida* virulence factors including secreted aspartyl proteases (SAPs) and candidalysin (Rosati et al. 2020; Willems et al. 2020). Furthermore, invasive hyphae are able to penetrate the epithelial surface as observed in tissue biopsies (Fig. 5.1b) (Swidsinski et al. 2019). Notably, *Candida* adhesion and biofilm formation have also been suggested as important factors in VVC development (Czechowicz et al. 2022; McKloud et al. 2021), which will be discussed in detail in the following section.

In addition to fungal adaptation, the host immune response plays a key role in VVC development and symptom onset. A seminal study with healthy volunteers receiving intravaginal *C. albicans* demonstrated that protection against VVC was non-inflammatory, while symptomatic VVC was associated with polymorphonuclear neutrophil (PMN) recruitment (Fidel et al. 2004). As a result, VVC is believed to be an immunopathology mediated by an aggravated PMN response (Rosati et al. 2020; Willems et al. 2020). In agreement with this, inflammatory mediators released by vaginal epithelial cells (alarmins, IL-1β) lead to PMN recruitment (Rosati et al. 2020; Yano et al. 2018), and neutrophils have been identified in vaginal lavage of VVC patients (Giraldo et al. 2012).

The Role of Fungal Biofilms in R/VVC

While it is widely accepted that biofilms are a significant contributing factor in other vaginal infections such as BV, the role for biofilms in R/VVC remains less defined. In particular, RVVC possesses the key clinical hallmarks that are associated with many biofilm-associated diseases, namely reoccurrence and chronicity of infection alongside treatment failure with the frontline antifungal agents. This is despite reported clinical sensitivities to these azole antifungals, suggesting non-intrinsic resistance traits facilitating this persistence.

Authors initially suggested that a function for biofilms in RVVC was “speculative at best”, addressing a lack of in vivo data to support the biofilm hypothesis (Sobel 2015). This editorial correctly highlighted the limitations with current VVC murine models and hypothetical conclusions drawn from studies using in vitro biofilm models. In addition, a study of tissue biopsies with culture confirmed VVC using 18S rRNA *Candida*-specific Fluorescence In Situ Hybridization (FISH) demonstrated evidence of hyphal invasion but not biofilm adherence, with invasion often polymicrobial in nature accompanied by other vaginal pathobionts such as *Gardnerella* and *Atopobium*, alongside *Lactobacillus* spp. (Swidsinski et al. 2019). This study does however have its own limitations including the anatomical location of the biopsy and potential disruption of biofilm during the procedure, as well as not aiming to quantify or visualise biofilm-specific factors such as the extracellular matrix. There is a body of evidence to support biofilm formation relating to the pathophysiology of RVVC using murine models (Harriott et al. 2010; Wu et al. 2020). However, the translational potential of these findings to humans is hindered due to a number of physiological limitations in these models including a lack of *C. albicans* as a normal inhabitant of the vaginal mycobiome, large quantities of organism needed to establish infection, pH elevated to neutral and the requirement for the exogenous supplementation of oestrogen to promote fungal colonisation. In addition to the biotic biofilms that form on the mucosa, *Candida* biofilms have been shown to colonise abiotic materials such as intrauterine devices (IUDs) (Auler et al. 2010; Cakiroglu et al. 2015). Microscopic analysis from a number of IUDs from Auler et al. (2010) demonstrated dense and sometimes polymicrobial biofilm formation on device surfaces, with clinical improvement demonstrated after removal of biofilm-positive devices (Cakiroglu et al. 2015).

Mechanistically, little is known about the overarching process of *Candida* biofilm formation on the vaginal mucosa. This likely follows fungal morphogenesis and initial adherence to vaginal epithelial cells, with successful tissue colonisation as a result of the production of extracellular enzymes such as phospolipases and the secreted aspartyl proteases family. In addition, the secretion of the cytolytic peptide toxin candidalysin (*ECE1*) has been shown to increase secretion of proinflammatory responses in vaginal epithelial cells and is required for neutrophil recruitment in a murine model; interestingly, candidalysin-deficient strains showed similar adherence profiles to the wild-type suggesting that it may not play a profound role in vaginal colonisation (Richardson et al. 2018). Through the aforementioned murine studies (Harriott et al. 2010; Wu et al. 2020), it has been shown that *C. albicans* can colonise and form biofilms on the murine vagina, with reduced colonisation observed using biofilm-defective mutants in key regulatory transcription factors (*BCR1* and *MED31*) as well as morphogenesis (*EFG1*). Furthermore, these biotic biofilms possess unique phenotypic traits related to biofilm formation such as elevated tolerance to antifungals and the formation of dormant subpopulations of persister cells (Wu et al. 2020).

Perhaps the most compelling evidence of the role for biofilms in RVVC was recently demonstrated by Pan et al., (2023). Using a combination of clinical and ex vivo biofilm analysis, the authors demonstrate that *C. albicans* and *N. glabratus* formed robust, azole tolerant biofilms on the vaginal epithelium of RVVC patients with limited biofilm evidence from an equivalent tissue sample of VVC patients (Pan et al. 2023). Biofilms from RVVC patients could be observed microscopically as filamentous aggregates and micro-colonies, whereas no visible *Candida* growth was observed within the VVC group. Through use of confocal laser scanning electron microscopy, secreted extracellular substances from *Candida* were also qualitatively identified through labelling with Concanavalin A, which stains both glycol- and mannoproteins within the biofilm matrix. Quantitatively, significant fungal bioburden was also enumerated from the RVVC cohort using colony-forming unit counting, with no detectable biomass from VVC patients (Pan et al. 2023). Further evidence has been suggested from ex vivo studies from McKloud et al. (2021), who, in addition to demonstrating that RVVC isolates could form biofilms in vitro, demonstrated biofilm characteristics directly from vaginal lavage of RVVC patients. It was shown that lavage fluid from RVVC patients contained visible hyphae and large aggregates and that biofilm-associated genes such as *HWP1*, *ALS3* and *SAP5* were highly expressed and up-regulated (McKloud et al. 2021).

Taken collectively, the cumulative evidence across in vitro, in vivo, ex vivo and human studies alongside treatment unresponsiveness despite clinical antifungal sensitivities supports the hypothesis of biofilm formation as a key virulence factor that drives this chronic infection. There remains a number of research gaps to fully understand the pathophysiology of RVVC, such as the role of the accompanying microbiome, including the dynamic interplay with *L. iners* (Papon and Van Dijck 2021). The role for both abiotic and hormonal contraceptives, alongside the discovery of new biofilm-active therapies, is outlined later in this chapter.

Clinical Manifestations and Diagnosis

Symptoms

From a clinical perspective, VVC commonly presents as vulvar itching, burning, pain and swelling, which is associated with a thick, white discharge (Fischer 2012; Jeanmonod et al. 2024; Yano et al. 2019). Although these symptoms appear to be hallmark features of VVC, the severity with which they present varies among different patients (Hong et al. 2014; Strydom et al. 2022). In addition, some patients report dyspareunia and dysuria, defined as pain before, during and post intercourse or urination (White and Vanthuyne 2006; Yano et al. 2019). Notably, VVC cases caused by NAC species have been shown to present milder symptoms when compared with *C. albicans* VVC (Ardizzoni et al. 2021; Dan et al. 2002). As such, NAC species are more commonly found among asymptomatic women compared to those with symptomatic infections (Dan et al. 2002; Makanjuola et al. 2018).

Diagnosis

As outlined previously, *Candida* may exist as a vaginal commensal in some women, and thus the presence of this organism alone is not diagnostic (Neal and Martens 2022). Furthermore, the symptoms of VVC are non-specific and overlap significantly with other vaginal conditions such as BV and trichomoniasis (Benedict et al. 2022a). Therefore, accurate diagnosis is performed by a combination of laboratory testing alongside clinical criteria evaluated by a physician (Fig. 5.2) (Benedict et al. 2022a; Ledger and Monif 2004; White and Vanthuyne 2006). Despite this, a significant number of VVC cases are self-diagnosed by patients themselves (Neal and Martens 2022), which is known to be of low accuracy (Rathod and Buffler 2014; Shen et al. 2021). Of note, one study demonstrated that of women who had self-diagnosed VVC and intended to purchase over-the counter (OTC) anti-fungal medication, only 33.7% had a VVC diagnosis confirmed (Ferris et al. 2002). The resulting implications of misdiagnosis include the ineffective treatment and resolution of the condition, in addition to the excessive and inappropriate use of OTC antifungals which may result in resistance and exacerbate later infections.

Laboratory Examination

It is recommended that all women with the signs and symptoms of VVC receive a wet mount microscopy (WMM) examination using 10% potassium hydroxide (Benedict et al. 2022a; Centers for Disease Control and Prevention 2021; Zhao et al. 2021). WMM is a rapid and easy method used to confirm clinical symptoms through identification of yeasts, hyphae and/or pseudohyphae in vaginal discharge. However, the diagnostic accuracy of WMM is dependent on operator skill (Zhao et al. 2021). It has been reported that this approach may have a high number of false-negative

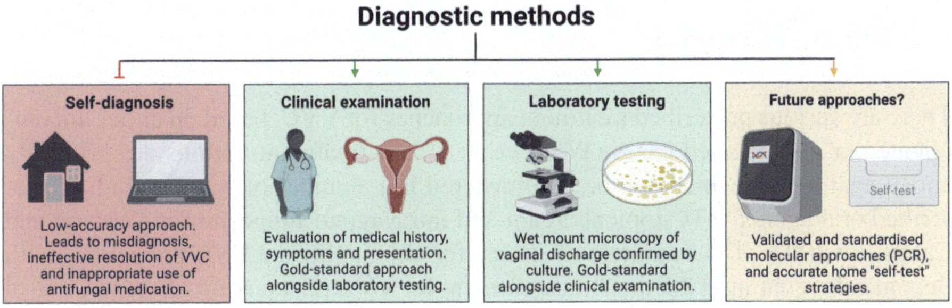

Fig. 5.2 Current and future diagnostic methods for R/VVC. Although known to be of low accuracy, a significant number of patients self-diagnose, leading to misdiagnosis and inappropriate use of OTC anti-fungals. Clinical and laboratory examination are the gold-standard diagnostic methods, while future approaches including validated molecular techniques and accurate self-test strategies may reduce processing times and provide greater scope of species identification. (Created in BioRender (https://BioRender.com/s00r929))

results, with only 50–80% of positive yeast cultures successfully identified by WMM (Abbott 1995; Chatwani et al. 2007). As a result, negative wet-mounts should be followed by culture on selective media (Sabouraud's agar) to limit false-negative results (Benedict et al. 2022a; Centers for Disease Control and Prevention 2021). However, culture is not always practical given extended processing times and associated costs (Chatwani et al. 2007), and alternative approaches have been suggested.

One example is standardised and validated molecular testing using polymerase chain reaction (PCR), which would reduce processing times, improve detection limits and allow for more accurate identification of the causative species (Nyirjesy et al. 2022). However, as noted by the Centers for Disease Control and Prevention (CDC), the vast majority of yeast PCR tests have yet to receive Food and Drug Administration (FDA) approval and users should be aware of the performance characteristics of different assays (Centers for Disease Control and Prevention 2021).

Another approach is the use of accurate self-testing devices which may reduce healthcare expenditure and reliance on self-identified symptoms (Nyirjesy et al. 2022). In line with this approach, one study demonstrated that a rapid yeast detection test was more sensitive than WMM, and results from patient-performed tests were identical to those performed by physicians (Chatwani et al. 2007). Given that the vaginal pH remains "normal" (i.e. <4.5) in VVC, additional approaches using self-testing of vaginal pH have been suggested to prevent misdiagnosis of VVC over BV and trichomoniasis, where vaginal pH is elevated (Shen et al. 2021). This may potentially reduce the inappropriate use of OTC anti-fungals; however, the clinical utility and cost-effectiveness of these approaches are currently unclear. As such, yeast culture remains the recommended reference-standard diagnostic test for R/VVC (Centers for Disease Control and Prevention 2021).

Treatment, Management and Prevention of VVC

Antifungal Therapies

There are various prescribed treatment approaches for VVC, based on either uncomplicated or complicated cases (Wang et al. 2024); treatment options are primarily anti-fungal agents which can be administered in a number of ways including prescribed oral agents, OTC topical cream and intravaginal suppositories (Watson and Calabretto 2007). The treatment of choice for uncomplicated VVC are azoles, with fluconazole remaining the first-line treatment for VVC (Centers for Disease Control and Prevention 2021; National Health Service 2023). Fluconazole demonstrates good bioavailability; the same concentration/dose (100–800 mg day^{-1}) can be administered via the oral and intravenous route (Powderly 2012).

A recent systemic review, published by Denison et al. (2020), revealed that there was no difference in azole effectiveness when administration route was compared (oral compared to intra-vaginal) to treat patients diagnosed with VVC (Denison et al. 2020). However, drug delivery routes play a pivotal role in side effects observed, with systemic drug delivery (such as oral administration) resulting in greater adverse side effects when compared against localised drug delivery routes (i.e. topical applications) (Satora et al. 2023). Furthermore, local administration of suppositories or creams avoids first-pass clearance and dilutionary effects (Campaña-Seoane et al. 2019; Swingle et al. 2023). In order to effectively treat VVC, different administration routes should be considered when developing and optimising novel treatment strategies.

In regard to treatment regimens, both the CDC and NHS recommend one, 150 mg oral dose of fluconazole for uncomplicated VVC (Centers for Disease Control and Prevention 2021; National Health Service 2023). First-line fluconazole treatment has been found to improve the quality of life in 96% of women; however, 63% of women have recurrent infections even after completing maintenance therapy (Galgiani et al. 2016; Pasko et al. 1990; Spampinato and Leonardi 2013). Product information of Diflucan® (a brand name of fluconazole) recommends maintenance doses of 50–400 mg/day and consideration of a double dose concentration of the initial treatment to achieve a steady-state equilibrium in plasma concentration by 48 h (Pfizer 2024; Sandaradura et al. 2021).

Challenges in Treatment

The treatment of VVC is achievable with the effective use of traditional therapeutic strategies (Wang et al. 2024). However, issues such as drug intolerance through suspected biofilm formation, the emergence of antimicrobial resistance and treatment failures can result in persisting and recurrent infections (Watson and Calabretto 2007). Commonly prescribed anti-fungal agents for vaginal and topical use are generally well tolerated (Farr et al. 2021). Safety has been a major advantage in the use of fluconazole to treat VVC, with side effects such as gastrointestinal upset and rashes being relatively uncommon (Powderly 2012; Sobel 2014). However, Azoles have been reported to cause slight localised burning in approximately 1–10% of cases (Mendling et al. 2004). Furthermore, the use of fluconazole may result in negative side effects such as hepatotoxicity, cytochrome P450 interactions and possible foetal harm when administered to pregnant women (Satora et al. 2023); for this reason, fluconazole is not recommended during pregnancy, in particular the first trimester (Patel et al. 2021; Zhu et al. 2020). The aforementioned side effects can often result in a reduction in patient compliance, and this may be misinterpreted as the *C. albicans* strain being resistant to therapy (das Neves et al. 2008; Farr et al. 2021).

The long-term, repeated use of fluconazole, and other azoles such as clotrimazole (which is used topically to treat VVC), increases the risk of the emergence of azole-resistant *Candida* species (Collins et al. 2020; Satora et al. 2023; Warrilow et al. 2014). A study in 2021 by Sandaradura et al. reported that current fluconazole therapeutic regimens are inadequate to treat organisms with a minimum inhibitory concentration ≥ 4 μg mL^{-1} (Sandaradura et al. 2021). The mean peak concentrations of fluconazole in plasma and vaginal secretions after a single 150 mg oral dose has been reported to be 2.82 μg mL^{-1} and 2.43 μg mL^{-1} respectively (Houang et al. 1990; Sobel et al. 2003). Repeated exposure to sub-minimum inhibitory concentration (MIC) has been shown to increase fungal tolerance to antifungals and can promote multidrug resistance (Arastehfar et al. 2020; Costa-de-Oliveira and Rodrigues 2020). Mechanisms of resistance include overexpression of efflux pumps, alteration of the target enzyme (for example, decreasing affinity of the binding site) and upregulation of the target enzyme (producing more enzyme so the antimicrobials' effect is negated) (Costa-de-Oliveira and Rodrigues 2020). Furthermore, *C. albicans biofilms* have been shown to be intrinsically resistant to fluconazole, with both pre-formed and sessile-to-biofilm studies demonstrating MICs >1024 μg mL^{-1} (Ramage et al. 2002).

Among NAC species, *N. glabratus* (formerly *C. glabrata*) is regarded as the second-leading cause of VVC (Chen et al. 2023; Parazzini et al. 2000; Richter et al. 2005). Notably, *N. glabratus* has the highest incidence rate of azole resistance, and demonstrates an intrinsic decrease in susceptibility toward the azole antifungals (Oxman et al. 2010; Pfaller et al. 2014; Whaley et al. 2017). Many molecular mechanisms which result in azole resistance have been reported in NAC species, and several mechanisms are related to the ergosterol biosynthesis pathway, specifically, overexpression or alteration of the drug target of *ERG11* (Berkow and Lockhart 2017). Another well-documented resistance mechanism of NAC species' is the over-expression of efflux pumps, such as ATP-binding cassette (ABC) transporters and major facilitator superfamily (MFS) transporters (Berkow and Lockhart 2017). The emergence of resistant *C. albicans* isolates and other NAC species can result in RVVC, and treatment often requires long-term antifungal prophylaxis with a maintenance regimen (Donders et al. 2022; Sobel and Sobel 2018). This can further exacerbate drug intolerance and safety concerns related with the long-term use of fluconazole, and therefore there is an urgent need for new anti-fungal therapies to be developed in order to effectively treat VVC.

Alternative and Emerging Treatment Strategies

Multiple novel treatment strategies for the treatment of VVC have been considered (Satora et al. 2023), and examples include novel antifungals (such as ibrexafungerp (Phillips et al. 2023)), probiotics (Oerlemans et al. 2020; Stabile et al. 2021), anti-infectives (Marrazzo et al. 2019), antiseptics (Mendling et al. 2016; Mirzaeei et al. 2021), immunotherapeutic vaccines (Edwards Jr et al. 2018) and antimicrobial peptides (Woodburn et al. 2019).

Novel Antifungals

In 2021, ibrexafungerp, a triterpenoid fungicidal agent, received FDA approval for oral use against VVC, thus becoming the first "non-azole" agent for VVC (Phillips et al. 2022). Dosage to effectively treat VVC is 300 mg via oral administration, twice daily (Schwebke et al. 2022). Ibrexafungerp has a similar mechanism of action to the echinocandins and exhibits fungicidal activity against *Candida* spp., by inhibiting (1 → 3)-β-D-glucan synthase (Ghannoum et al. 2020). This results in a reduction in (1 → 3)-β-D-glucan synthesis and eventual loss of cell wall integrity, resulting in fungal cell death (Pfaller et al. 2013). While Ibrexafungerp has a similar mechanism of action as the echinocandins, it is structurally distinct and acts differently with the target enzyme; it is the first-in-class triterpenoid antifungal or "fungerp" (Angulo et al. 2022; Ghannoum et al. 2020; Jallow and Govender 2021; Jiménez-Ortigosa et al. 2017). The use of ibrexafungerp has been reported to have an improved clinical cure rate and mycological eradication compared to fluconazole (He et al. 2024); it can also reach concentrations up to ninefold higher in vaginal tissues compared to plasma (indicating favourable vaginal tissue penetration) and it is not negatively impacted by the low pH environment of the vagina (Larkin et al. 2019; Schwebke et al. 2022; Sobel et al. 2021). In December 2022, Ibrexafungerp received further FDA approval for once-monthly dosing to decrease the incidence of RVVC (Phillips et al. 2023).

Alternative Anti-infectives, Antiseptics and Antimicrobial Peptides

Alternative antimicrobial agents have also been trialled as novel therapeutic options to treat VVC (David and Solomon 2023; Mendling et al. 2016; Mirzaeei et al. 2021; Swate and Weed 1974; Woodburn et al. 2019). Localised drug delivery results in fewer adverse side effects and can avoid first-pass clearance and dilutionary effects compared to systemic delivery (e.g. oral administration) (Campaña-Seoane et al. 2019; Palmeira-de-Oliveira et al. 2015; Satora et al. 2023; Swingle et al. 2023; Vanić et al. 2021). Current intravaginal dosage formulas include suppositories (tablets/capsules), creams, ointments, gels and antimicrobial-coated pessaries (Pandey et al. 2021).

In 2019, Marrazzo et al. reported that TOL-463, which is a boric acid based vaginal anti-infective enhanced with ethylenediaminetetraacetic acid (EDTA) was effective and safe at treating both BV and VVC (Marrazzo et al. 2019). During this randomised, single-blind, phase 2, controlled trial, 106 women were enrolled to test the efficacy of two TOL-463 formulations, a water-based vaginal gel and a polyethylene-glycol-based insert containing 250 mg and 500 mg of boric acid (per dose), respectively (Marrazzo et al. 2019). The selection criteria were 18–50 years of age and a diagnosis of BV and/or VVC; of the 106 enrollees, 53 had a clinical diagnosis of BV, 36 a clinical diagnosis of VVC, and 17 both BV and VVC (Marrazzo et al. 2019). The primary test of cure (TOC) was clinical cure at days 9–12, while safety was assessed at TOC and days 21–30 (Marrazzo et al. 2019). The clinical cure rate of VVC of the TOL-463 gel was 81% (95% CI, 57–93%) and for the TOL-463 insert 92% (95% CI, 67–99%). Both formulations also resulted in clinical cure of BV with the TOL-463 gel exhibiting a clinical cure rate of 50% compared to 59% of the TOL-463 insert (Marrazzo et al. 2019). Both products were well tolerated and safe with no secondary cases of VVC noted, the most common adverse event was vulvovaginal burning (9.6%) (Marrazzo et al. 2019).

Boric acid was first found to be effective in treating mycotic vaginitis in 1974 (Swate and Weed 1974), while intravaginal boric acid supplements are not FDA approved, they are currently recommended by the US Centres for Disease Control and Prevention for both recurrent BV and NAC-VVC (Control and Prevention 2008). The mechanism of action of boric acid is yet to be fully elucidated, but studies have shown that boric acid impairs hyphal morphogenesis and biofilm formation in *C. albicans* (Salama and Gerstein 2022; San Juan Galán et al. 2023). Future studies should look to provide key insights into the mechanism of action of boric acid in regard to VVC, the effect of boric acid on recurrent BV/VVC infections over time, and combining boric acid (and other biofilm disruptors) with traditional antifungals (such as azoles) to determine the effect on biofilm formation and VVC (Marrazzo et al. 2019).

Antimicrobial biocides are a distinct from chemotherapeutic antibiotics and are used in a wide range of applications such as disinfection, antisepsis and preservation (Maillard and Pascoe 2024). Although these biocides may be used for either application, disinfectant refers to non-living surfaces while antiseptic refers to living tissue (for example, the skin) (Maillard and Pascoe 2024). Typical classes of antiseptics that have been trialled against VVC include phenyl derivatives (e.g. chlorohexidine (Alvendal et al. 2020)), acids (e.g. salicylic acid and boric acid (Marrazzo et al. 2019; Rigo and Tasca 2020)), halogens (such as povidone-iodine (Younis et al. 2023)) and quaternary ammonium compounds (e.g. dequalinium chloride (DQC) (Eckel et al. 2024; Mendling et al. 2016)).

Dequalinium chloride (DQC) has shown antimicrobial activity against Gram-positive and Gram-negative bacteria, fungi and protozoa (Carter et al. 2024; Cox and D'Arcy 1961; Mendling et al. 2016); it has also shown promising antimicrobial activity against conditions such as BV and VVC (Della Casa et al. 2002; Eckel et al. 2024; Raba et al. 2024). DQC has multiple mechanisms of antimicrobial activity, such as increasing cell permeability which may result in membrane lysis; it can affect ribosomal protein synthesis leading to protein denaturation, and also targets mitochondria, leading to mitochondrial DNA depletion and inhibition of energy production (Bailly 2021; Brauss and Rieth 1974; Hugo and Frier 1969; Tisher et al. 2011). Having multiple mechanisms of action reduces the risk of resistance generation (Carter et al. 2024). In 2016, Mendling et al. reported that a number of different vaginal formulations of DCQ have been used to treat a total of 653 women with VVC in clinical trials and cure rates were reported to be between 70% and 90% (Mendling et al. 2016).

Antimicrobial peptides are small molecular peptides that play a vital role in the innate immune system of the host, conferring protection against a broad range of microorganisms including bacteria, fungi, parasites and viruses (Kang et al. 2017;

Lei et al. 2019; Mohammed et al. 2017; Wang et al. 2019; Zhang et al. 2021). In 2019, Woodburn et al. designed antimicrobial peptides (dAMPs) for RVVC treatment (Woodburn et al. 2019). Minimum inhibitory concentrations (MICs) of the designed antimicrobial peptides (dAMPs) were determined against 46 fluconazole-susceptible and -resistant *Candida* spp. (Woodburn et al. 2019). The dAMPs were then compared against fluconazole, trialled against *Candida* biofilms and topically applied to a rodent VVC infection model (Woodburn et al. 2019). In this study, the dAMPs demonstrated potent antimicrobial activity against *Candida*, reduced preformed biofilms and biofilm formation, and showed limited mammalian cytotoxicity with sustained topical activity in a rodent model (Woodburn et al. 2019). Furthermore, the dAMPs did not induce resistance when repeatedly exposed to *C. tropicalis* (Woodburn et al. 2019). Preclinical development is underway for one of the lead dAMPs (RP557; this was selected for further testing as it yielded the best *Candida* cell specificity over mammalian cells) as a potential VVC therapeutic option (Woodburn et al. 2019).

Probiotics

Lactobacillus spp., are the most abundant microorganisms in the vagina and are able to modulate the vaginal environment and confer protection against pathogen colonisation via a number of methods, including producing lactic acid (which lowers vaginal pH), antimicrobial production (such as production of hydrogen peroxide and bacteriocin-like substances), epithelial cell adhesion (competing against pathogens for attachment sites) and immune cell modulation (by regulating cytokine release to inhibit inflammation (Li et al. 2019; Liu et al. 2023; Miko and Barakonyi 2023)). Therefore, the use of *Lactobacilli* spp. as probiotics could maintain and recover the vaginal microbiome, and their ability to reduce *Candida* spp. has given rise to the concept of using probiotics to reduce VVC incidence and reoccurrence (Chee et al. 2020; Xie et al. 2017).

In 2021, in a random, placebo-controlled trial, Vahedpoor et al. investigated vaginal probiotics (including *L. acidophilus*, *L. plantarum*, *L. rhamnosus* and *L. gasseri*) and oral probiotics (*L. acidophilus*, *L. plantarum*, *L. fermentum* and *L. gasseri*) on 76 patients with VVC (age range: 18–48 years). These patients were administered fluconazole which was supplemented with a probiotic capsule or a placebo. Thirty-five days post intervention, the probiotic supplemented groups exhibited a reduction in VVC symptoms (including less vaginal discharge) when compared to the fluconazole and placebo group (Vahedpoor et al. 2021). Furthermore, the use of *Lactobacillus* spp., has been shown to reduce VVC recurrence (Chee et al. 2020). In 2019, Russo et al. demonstrated that oral coadministration of topical clotimazole and an oral capsule (containing *L. acidophilus* GLA-14, *L. rhamnosus* HN001 and bovine lactoferrin) significantly reduced RVVC, drastically reducing itching and vaginal discharge at 3 and 6 months (Russo et al. 2019). During the 6-month follow-up, those who had been administered the topical clotimazole and oral probiotic capsule demonstrated significantly lower recurrence (29.2%) compared to the placebo (topical clotimazole alone with oral capsule which contained maltodextrin) which had a recurrence rate of 100% (Russo et al. 2019). While studies into the efficacy of probiotics to treat VVC show promise, several limitations have been highlighted such as small sample size, absence of placebo groups, differences in dosage concentrations and treatment duration, the high heterogeneous variation of various probiotic strains and strain to strain variation in regard to a probiotic effect against *Candida* spp. (Falagas et al. 2006; Russo et al. 2019). These limitations demonstrate that further research is needed to fully elucidate the use of probiotics to treat R/VVC (Falagas et al. 2006; Russo et al. 2019).

Immunotherapeutic Vaccines

A number of vaccines are currently being developed against *Candida* spp., and more specifically for the treatment of RVVC (Edwards Jr et al. 2018; Levy et al. 1989; Pietrella et al. 2010; Schmidt et al. 2012). In 2018, Edward et al. evaluated a fungal immunotherapeutic vaccine (NDV-3A) by means of a Phase 2 randomised, double-blind, placebo-controlled trial (Edwards Jr et al. 2018). The vaccine contained the N-terminal portion of the agglutinin-like sequence 3 (Als3) protein of *C. albicans*, which is a virulence factor that is hyphal-specific and mediates adherence and invasion of human epithelial and vascular endothelial cells (Edwards Jr et al. 2018; Phan et al. 2007; Sheppard et al. 2004). In this study, the NDV-3A vaccine was given once, intramuscularly, to 188 women with RVVC ($n = 178$). Results showed that NDV-3A was safe and generated rapid and robust B- and T-cell immune responses (Edwards Jr et al. 2018). Post hoc analysis revealed a statistically significant increase in the percentage of symptom-free patients 12 months post vaccination (42% of vaccinated patients were symptom free compared to 22% of patients who were administered the placebo) (Edwards Jr et al. 2018). Interestingly, in 2019, another study combined bioinformatics with animal studies and demonstrated that the NDV-3A vaccine generated cross-reactive antibodies against *C. auris* clinical isolates and protected neutropenic mice from haematogenous *C. auris* infection (Singh et al. 2019).

Preventative Measures

In addition to therapy, there are effective preventative measures that can reduce the risk of VVC (Jeanmonod et al. 2024; Ramsay et al. 2009), such as keeping the vaginal area dry where possible (Jeanmonod et al. 2024; Ramsay et al. 2009). Other general recommendations which prevent the vaginal microbiome from becoming imbalanced include: not douching (Martino and Vermund 2002; Shaaban et al. 2013; van der Veer et al. 2019), avoiding the using local irritants (such as perfumed products e.g. perfumed soaps) (Otero et al. 1998; Sustr et al. 2020) and replacing soaps with vulval water-based moisturisers (Ramsay et al. 2009). Patients who are prescribed systemic antibiotics should also consider a prophylactic anti-fungal treatment to avoid opportunistic *Candida* spp., growth and proliferation (Shukla and Sobel 2019).

Conclusion

Vulvovaginal candidiasis (VVC) and recurrent VVC (RVVC) continue to pose significant health challenges for women worldwide, and their underlying pathophysiology remains to be fully understood. Previous work using in vitro or murine models had underpinned the biofilm-mediated hypothesis of R/VVC for many years, which carry inherent limitations. However, support for this hypothesis has been strengthened in recent years by evidence demonstrating biofilm formation in RVVC, suggesting a role for fungal biofilms in recurrent disease. Notably, the same study did not find comparable evidence in VVC (Pan et al. 2023). Collectively, these findings invite further research into the clinical implications of these biofilms in RVVC, particularly surrounding the consistency of biofilm formation among different patients, *Candida* strains and/or species, and the potential impact of biofilm formation on anti-fungal efficacy. Given the large association between *Candida* biofilms and anti-fungal tolerance, emerging approaches such as probiotic therapy and immunotherapeutic vaccines may offer promising avenues for improved management of RVVC moving forward.

References

Abbott J (1995) Clinical and microscopic diagnosis of vaginal yeast infection: a prospective analysis. Ann Emerg Med 25:587–591. https://doi.org/10.1016/s0196-0644(95)70168-0

Achkar JM, Fries BC (2010) Candida infections of the genitourinary tract. Clin Microbiol Rev 23:253–273. https://doi.org/10.1128/CMR.00076-09

Agnew KJ, Hillier SL (1995) The effect of treatment regimens for vaginitis and cervicitis on vaginal colonization by lactobacilli. Sex Transm Dis 22:269–273. https://doi.org/10.1097/00007435-199509000-00001

Alvendal C, Mohanty S, Bohm-Starke N, Brauner A (2020) Anti-biofilm activity of chlorhexidine digluconate against Candida albicans vaginal isolates. PLoS One 15:e0238428. https://doi.org/10.1371/journal.pone.0238428

Angulo DA, Alexander B, Rautemaa-Richardson R, Alastruey-Izquierdo A, Hoenigl M, Ibrahim AS, Ghannoum MA, King TR, Azie NE, Walsh TJ (2022) Ibrexafungerp, a novel triterpenoid antifungal in development for the treatment of mold infections. J. Fungi 8:1121. https://doi.org/10.3390/jof8111121

Antonio MA, Hawes SE, Hillier SL (1999) The identification of vaginal Lactobacillus species and the demographic and microbiologic characteristics of women colonized by these species. J Infect Dis 180:1950–1956. https://doi.org/10.1086/315109

Arastehfar A, Gabaldón T, Garcia-Rubio R, Jenks JD, Hoenigl M, Salzer HJF, Ilkit M, Lass-Flörl C, Perlin DS (2020) Drug-resistant fungi: an emerging challenge threatening our limited antifungal armamentarium. Antibiotics 9:877. https://doi.org/10.3390/antibiotics9120877

Ardizzoni A, Wheeler RT, Pericolini E (2021) It takes two to Tango: how a dysregulation of the innate immunity, coupled with Candida virulence, triggers VVC onset. Front Microbiol 12. https://doi.org/10.3389/fmicb.2021.692491

Auler ME, Morreira D, Rodrigues FFO, Abr Ao MS, Margarido PFR, Matsumoto FE, Silva EG, Silva BCM, Schneider RP, Paula CR (2010) Biofilm formation on intrauterine devices in patients with recurrent vulvovaginal candidiasis. Med Mycol 48:211–216. https://doi.org/10.3109/13693780902856626

Babula O, Lazdane G, Kroica J, Ledger WJ, Witkin SS (2003) Relation between recurrent vulvovaginal candidiasis, vaginal concentrations of mannose-binding lectin, and a mannose-binding lectin gene polymorphism in Latvian women. Clin Infect Dis 37:733–737. https://doi.org/10.1086/377234

Bailly C (2021) Medicinal applications and molecular targets of dequalinium chloride. Biochem Pharmacol 186:114467. https://doi.org/10.1016/j.bcp.2021.114467

Bauters TGM, Dhont MA, Temmerman MIL, Nelis HJ (2002) Prevalence of vulvovaginal candidiasis and susceptibility to fluconazole in women. Am J Obstet Gynecol 187:569–574. https://doi.org/10.1067/mob.2002.125897

Beigi RH, Meyn LA, Moore DM, Krohn MA, Hillier SL (2004) Vaginal yeast colonization in nonpregnant women: a longitudinal study. Obstet Gynecol 104:926–930. https://doi.org/10.1097/01.AOG.0000140687.51048.73

Benedict K, Lyman M, Jackson BR (2022a) Possible misdiagnosis, inappropriate empiric treatment, and opportunities for increased diagnostic testing for patients with vulvovaginal candidiasis—United States, 2018. PLoS One 17:e0267866. https://doi.org/10.1371/journal.pone.0267866

Benedict K, Singleton AL, Jackson BR, Molinari NAM (2022b) Survey of incidence, lifetime prevalence, and treatment of self-reported vulvovaginal candidiasis, United States, 2020. BMC Womens Health 22:147. https://doi.org/10.1186/s12905-022-01741-x

Berbudi A, Rahmadika N, Tjahjadi AI, Ruslami R (2020) Type 2 diabetes and its impact on the immune system. Curr Diabetes Rev 16:442–449. https://doi.org/10.2174/1573399815666191024085838

Berkow EL, Lockhart SR (2017) Fluconazole resistance in Candida species: a current perspective. Infect Drug Resist 10:237–245. https://doi.org/10.2147/IDR.S118892

Bongomin F, Gago S, Oladele RO, Denning DW (2017) Global and multi-national prevalence of fungal diseases-estimate precision. J Fungi Basel Switz 3:57. https://doi.org/10.3390/jof3040057

Borges S, Silva J, Teixeira P (2014) The role of lactobacilli and probiotics in maintaining vaginal health. Arch Gynecol Obstet 289:479–489. https://doi.org/10.1007/s00404-013-3064-9

Boris S, Barbés C (2000) Role played by lactobacilli in controlling the population of vaginal pathogens. Microbes Infect 2:543–546. https://doi.org/10.1016/s1286-4579(00)00313-0

Bradford LL, Ravel J, Bruno V (2013) Understanding vulvovaginal candidiasis through a community genomics approach. Curr Fungal Infect Rep 7:126–131. https://doi.org/10.1007/s12281-013-0135-0

Brauss FW, Rieth H (1974) Effectiveness of decamethylene aminoquinaldinium chloride against pathogenic microorganisms. MMW Munch Med Wochenschr 116:1519–1520

Buffo J, Herman MA, Soll DR (1984) A characterization of pH-regulated dimorphism in Candida albicans. Mycopathologia 85:21–30. https://doi.org/10.1007/BF00436698

Cakiroglu Y, Caliskan S, Doger E, Ozcan S, Caliskan E (2015) Does removal of CU-IUD in patients with biofilm forming candida really maintain regression of clinical symptoms? J Obstet Gynaecol 35:600–603. https://doi.org/10.3109/01443615.2014.986442

Campaña-Seoane M, Pérez-Gago A, Vázquez G, Conde N, González P, Martinez A, Martínez X, García Varela L, Herranz M, Aguiar P, Fernández-Ferreiro A, Laguna R, Otero-Espinar FJ (2019) Vaginal residence and pharmacokinetic preclinical study of topical vaginal mucoadhesive W/S emulsions containing ciprofloxacin. Int J Pharm 554:276–283. https://doi.org/10.1016/j.ijpharm.2018.11.022

Carter KA, Tuddenham S, Brotman RM (2024) Dequalinium chloride—an emerging option in the sparse landscape of bacterial vaginosis therapies. JAMA Netw Open 7:e248606. https://doi.org/10.1001/jamanetworkopen.2024.8606

Cauchie M, Desmet S, Lagrou K (2017) *Candida* and its dual lifestyle as a commensal and a pathogen. Res Microbiol 168:802–810. https://doi.org/10.1016/j.resmic.2017.02.005

Cavalheiro M, Teixeira MC (2018) Candida biofilms: threats, challenges, and promising strategies. Front Med 5. https://doi.org/10.3389/fmed.2018.00028

Centers for Disease Control and Prevention (2021) Vulvovaginal Candidiasis – STI Treatment Guidelines

Chaim W, Foxman B, Sobel JD (1997) Association of recurrent vaginal candidiasis and secretory ABO and Lewis phenotype. J Infect Dis 176:828–830. https://doi.org/10.1086/517314

Chatwani AJ, Mehta R, Hassan S, Rahimi S, Jeronis S, Dandolu V (2007) Rapid testing for vaginal yeast detection: a prospective study. Am J Obstet Gynecol 196:309.e1–309.e4. https://doi.org/10.1016/j.ajog.2006.11.025

Chatzivasileiou P, Vyzantiadis T-A (2019) Vaginal yeast colonisation: from a potential harmless condition to clinical implications and management approaches—a literature review. Mycoses 62:638–650. https://doi.org/10.1111/myc.12920

Chee WJY, Chew SY, Than LTL (2020) Vaginal microbiota and the potential of Lactobacillus derivatives in maintaining vaginal health. Microb Cell Factories 19:203. https://doi.org/10.1186/s12934-020-01464-4

Chen Z, Jin J, Chen H, Chen Y, Feng S (2023) The bacterial communities in vagina of different Candida species-associated vulvovaginal candidiasis. Microb Pathog 177:106037. https://doi.org/10.1016/j.micpath.2023.106037

Collins LM, Moore R, Sobel JD (2020) Prognosis and long-term outcome of women with idiopathic recurrent vulvovaginal candidiasis caused by Candida albicans. J Low Genit Tract Dis 24:48. https://doi.org/10.1097/LGT.0000000000000496

Control, C. for D., Prevention (2008) Sexually transmitted disease surveillance. http://www.cdc.gov/std/stats08/syphilis.htm

Costa-de-Oliveira S, Rodrigues AG (2020) Candida albicans antifungal resistance and tolerance in bloodstream infections: the triad yeast-host-antifungal. Microorganisms 8:154. https://doi.org/10.3390/microorganisms8020154

Cox WA, D'Arcy PF (1961) The activity of Polymethylene-Bis-4-Aminoquinaldinium salts against Pityrosporum Ovale and Candida Albicans. J Pharm Pharmacol 13:34–38. https://doi.org/10.1111/j.2042-7158.1961.tb11779.x

Czechowicz P, Nowicka J, Gościniak G (2022) Virulence factors of Candida spp. and host immune response important in the pathogenesis of vulvovaginal candidiasis. Int J Mol Sci 23:5895. https://doi.org/10.3390/ijms23115895

Dan M, Poch F, Levin D (2002) High rate of vaginal infections caused by non-C. albicans Candida species among asymptomatic women. Med Mycol 40:383–386. https://doi.org/10.1080/mmy.40.4.383.386

das Neves J, Pinto E, Teixeira B, Dias G, Rocha P, Cunha T, Santos B, Amaral MH, Bahia MF (2008) Local treatment of vulvovaginal candidosis. Drugs 68:1787–1802. https://doi.org/10.2165/00003495-200868130-00002

David H, Solomon AP (2023) Molecular association of Candida albicans and vulvovaginal candidiasis: focusing on a solution. Front Cell Infect Microbiol 13. https://doi.org/10.3389/fcimb.2023.1245808

Della Casa V, Noll H, Gonser S, Grob P, Graf F, Pohlig G (2002) Antimicrobial activity of dequalinium chloride against leading germs of vaginal infections. Arzneimittelforschung 52:699–705. https://doi.org/10.1055/s-0031-1299954

Denison HJ, Worswick J, Bond CM, Grimshaw JM, Mayhew A, Ramadoss SG, Robertson C, Schaafsma ME, Watson MC (2020) Oral versus intra-vaginal imidazole and triazole antifungal treatment of uncomplicated vulvovaginal candidiasis (thrush)

Dennerstein GJ, Ellis DH (2001) Oestrogen, glycogen and vaginal candidiasis. Aust N Z J Obstet Gynaecol 41:326–328. https://doi.org/10.1111/j.1479-828x.2001.tb01238.x

Denning DW, Kneale M, Sobel JD, Rautemaa-Richardson R (2018) Global burden of recurrent vulvovaginal candidiasis: a systematic review. Lancet Infect Dis 18:e339–e347. https://doi.org/10.1016/S1473-3099(18)30103-8

Deorukhkar SC, Saini S, Mathew S (2014) Non-albicans Candida infection: an emerging threat. Interdiscip Perspect Infect Dis 2014:615958. https://doi.org/10.1155/2014/615958

Dias LB, de Souza Carvalho Melhem M, Szeszs MW, Filho JM, Hahn RC (2011) Vulvovaginal candidiasis in Mato Grosso, Brazil: pregnancy status, causative species and drugs tests. Braz J Microbiol Publ Braz Soc Microbiol 42:1300–1307. https://doi.org/10.1590/S1517-83822011000400009

Donders G, Sziller IO, Paavonen J, Hay P, de Seta F, Bohbot JM, Kotarski J, Vives JA, Szabo B, Cepulienė R, Mendling W (2022) Management of recurrent vulvovaginal candidosis: narrative review of the literature and European expert panel opinion. Front Cell Infect Microbiol 12:934353. https://doi.org/10.3389/fcimb.2022.934353

Eckel F, Farr A, Deinsberger J, Kernmayer-Farr K, Foessleitner P (2024) Dequalinium chloride for the treatment of vulvovaginal infections: a systematic review and meta-analysis. J Low Genit Tract Dis 28:76. https://doi.org/10.1097/LGT.0000000000000790

Edwards JE Jr, Schwartz MM, Schmidt CS, Sobel JD, Nyirjesy P, Schodel F, Marchus E, Lizakowski M, DeMontigny EA, Hoeg J, Holmberg T, Cooke MT, Hoover K, Edwards L, Jacobs M, Sussman S, Augenbraun M, Drusano M, Yeaman MR, Ibrahim AS, Filler SG, Hennessey JP Jr (2018) A fungal immunotherapeutic vaccine (NDV-3A) for treatment of recurrent vulvovaginal candidiasis—a phase 2 randomized, double-blind, placebo-controlled trial. Clin Infect Dis 66:1928–1936. https://doi.org/10.1093/cid/ciy185

Falagas ME, Betsi GI, Athanasiou S (2006) Probiotics for prevention of recurrent vulvovaginal candidiasis: a review. J Antimicrob Chemother 58:266–272. https://doi.org/10.1093/jac/dkl246

Farhan MA, Moharram AM, Salah T, Shaaban OM (2019) Types of yeasts that cause vulvovaginal candidiasis in chronic users of corticosteroids. Med Mycol 57:681–687. https://doi.org/10.1093/mmy/myy117

Farr A, Effendy I, Frey Tirri B, Hof H, Mayser P, Petricevic L, Ruhnke M, Schaller M, Schaefer APA, Sustr V, Willinger B, Mendling W (2021) Guideline: Vulvovaginal candidosis (AWMF 015/072, level S2k). Mycoses 64:583–602. https://doi.org/10.1111/myc.13248

Ferrer J (2000) Vaginal candidosis: epidemiological and etiological factors. Int J Gynecol Obstet 71:21–27. https://doi.org/10.1016/S0020-7292(00)00350-7

Ferris HA, Kahn CR (2012) New mechanisms of glucocorticoid-induced insulin resistance: make no bones about it. J Clin Invest 122:3854–3857. https://doi.org/10.1172/JCI66180

Ferris DG, Nyirjesy P, Sobel JD, Soper D, Pavletic A, Litaker MS (2002) Over-the-counter antifungal drug misuse associated with patient-diagnosed vulvovaginal candidiasis. Obstet Gynecol 99:419–425. https://doi.org/10.1016/s0029-7844(01)01759-8

Fidel PL, Barousse M, Espinosa T, Ficarra M, Sturtevant J, Martin DH, Quayle AJ, Dunlap K (2004) An intravaginal live Candida challenge in humans leads to new hypotheses for the immunopathogenesis of vulvovaginal candidiasis. Infect Immun 72:2939–2946. https://doi.org/10.1128/IAI.72.5.2939-2946.2004

Fischer G (2012) Chronic vulvovaginal candidiasis: what we know and what we have yet to learn. Australas J Dermatol 53:247–254. https://doi.org/10.1111/j.1440-0960.2011.00860.x

Fischer G, Bradford J (2011) Vulvovaginal candidiasis in postmenopausal women: the role of hormone replacement therapy. J Low Genit Tract Dis 15:263–267. https://doi.org/10.1097/LGT.0b013e3182241f1a

Galgiani JN, Ampel NM, Blair JE, Catanzaro A, Geertsma F, Hoover SE, Johnson RH, Kusne S, Lisse J, MacDonald JD, Meyerson SL, Raksin PB, Siever J, Stevens DA, Sunenshine R, Theodore N (2016) 2016 Infectious Diseases Society of America (IDSA) clinical practice guideline for the treatment of coccidioidomycosis. Clin Infect Dis 63:e112–e146. https://doi.org/10.1093/cid/ciw360

Gaziano R, Sabbatini S, Monari C (2023) The interplay between Candida albicans, vaginal mucosa, host immunity and resident microbiota in health and disease: an overview and future perspectives. Microorganisms 11:1211. https://doi.org/10.3390/microorganisms11051211

Ghannoum M, Arendrup MC, Chaturvedi VP, Lockhart SR, McCormick TS, Chaturvedi S, Berkow EL, Juneja D, Tarai B, Azie N, Angulo D, Walsh TJ (2020) Ibrexafungerp: a novel oral triterpenoid antifungal in development for the treatment of Candida auris infections. Antibiotics 9:539. https://doi.org/10.3390/antibiotics9090539

Giraldo PC, de Carvalho JBJ, do Amaral RLG, da Silveira Gonçalves AK, Eleutério J Jr, Guimarães F (2012) Identification of immune cells by flow cytometry in vaginal lavages from women with vulvovaginitis and normal microflora. Am J Reprod Immunol 67:198–205. https://doi.org/10.1111/j.1600-0897.2011.01093.x

Gonçalves B, Ferreira C, Alves CT, Henriques M, Azeredo J, Silva S (2016) Vulvovaginal candidiasis: epidemiology, microbiology and risk factors. Crit Rev Microbiol 42:905–927. https://doi.org/10.3109/1040841X.2015.1091805

Gow NAR, van de Veerdonk FL, Brown AJP, Netea MG (2011) Candida albicans morphogenesis and host defence: discriminating invasion from colonization. Nat Rev Microbiol 10:112. https://doi.org/10.1038/nrmicro2711

Harriott MM, Lilly EA, Rodriguez TE, Fidel PL, Noverr MC (2010) Candida albicans forms biofilms on the vaginal mucosa. Microbiol Read Engl 156:3635–3644. https://doi.org/10.1099/mic.0.039354-0

He Y, Tang R, Deng J, Cai T, He P, Wu J, Cao Y (2022) Effects of oestrogen on vulvovaginal candidosis. Mycoses 65:4–12. https://doi.org/10.1111/myc.13385

He R, Lin F, Yu B, Huang L (2024) Efficacy and safety of ibrexafungerp in the treatment of vulvovaginal candidiasis: a meta-analysis of randomized controlled trials. Heliyon 10:e28776. https://doi.org/10.1016/j.heliyon.2024.e28776

Hillier SL, Lau RJ (1997) Vaginal microflora in postmenopausal women who have not received estrogen replacement therapy. Clin Infect Dis 25(Suppl 2):S123–S126. https://doi.org/10.1086/516221

Hong E, Dixit S, Fidel PL, Bradford J, Fischer G (2014) Vulvovaginal candidiasis as a chronic disease: diagnostic criteria and definition. J Low Genit Tract Dis 18:31–38. https://doi.org/10.1097/LGT.0b013e318287aced

Houang ET, Chappatte O, Byrne D, Macrae PV, Thorpe JE (1990) Fluconazole levels in plasma and vaginal secretions of patients after a 150-milligram single oral dose and rate of eradication of infection in vaginal candidiasis. Antimicrob Agents Chemother 34:909–910. https://doi.org/10.1128/aac.34.5.909

Hugo WB, Frier M (1969) Mode of action of the antibacterial compound dequalinium acetate. Appl Microbiol 17:118–127. https://doi.org/10.1128/am.17.1.118-127.1969

Irving G, Miller D, Robinson A, Reynolds S, Copas AJ (1998) Psychological factors associated with recurrent vaginal candidiasis: a preliminary study. Sex Transm Infect 74:334–338

Jabra-Rizk MA, Kong EF, Tsui C, Nguyen MH, Clancy CJ, Fidel PL, Noverr M (2016) Candida albicans pathogenesis: fitting within the host-microbe damage response framework. Infect Immun 84:2724–2739. https://doi.org/10.1128/iai.00469-16

Jacobsen ID (2023) The role of host and fungal factors in the commensal-to-pathogen transition of Candida albicans. Curr Clin Microbiol Rep 10:55–65. https://doi.org/10.1007/s40588-023-00190-w

Jallow S, Govender NP (2021) Ibrexafungerp: a first-in-class oral triterpenoid glucan synthase inhibitor. J Fungi 7:163. https://doi.org/10.3390/jof7030163

Jeanmonod R, Chippa V, Jeanmonod D (2024) Vaginal Candidiasis. In: StatPearls. StatPearls Publishing, Treasure Island (FL)

Jiménez-Ortigosa C, Perez WB, Angulo D, Borroto-Esoda K, Perlin DS (2017) De novo acquisition of resistance to SCY-078 in Candida glabrata involves FKS mutations that both overlap and are distinct from those conferring echinocandin resistance. Antimicrob Agents Chemother 61:e00833–e00817. https://doi.org/10.1128/AAC.00833-17

Johnston W, Ware A, Kuiters WF, Delaney C, Brown JL, Hagen S, Corcoran D, Cummings M, Ramage G, Kean R (2023) In vitro bacterial vaginosis biofilm community manipulation using endolysin therapy. Biofilms 5:100101. https://doi.org/10.1016/j.bioflm.2022.100101

Kang H-K, Kim C, Seo CH, Park Y (2017) The therapeutic applications of antimicrobial peptides (AMPs): a patent review. J Microbiol 55:1–12. https://doi.org/10.1007/s12275-017-6452-1

Kombade SP, Abhishek KS, Mittal P, Sharma C, Singh P, Nag VL (2021) Antifungal profile of vulvovaginal candidiasis in sexually active females from a tertiary care hospital of Western Rajasthan. J Fam Med Prim Care 10:398–402. https://doi.org/10.4103/jfmpc.jfmpc_1124_20

Larkin EL, Long L, Isham N, Borroto-Esoda K, Barat S, Angulo D, Wring S, Ghannoum M (2019) A novel 1,3-Beta-d-Glucan inhibitor, Ibrexafungerp (formerly SCY-078), shows potent activity in the lower pH environment of vulvovaginitis. Antimicrob Agents Chemother 63:e02611–e02618. https://doi.org/10.1128/AAC.02611-18

Ledger WJ, Monif GRG (2004) A growing concern: inability to diagnose vulvovaginal infections correctly. Obstet Gynecol 103:782–784. https://doi.org/10.1097/01.AOG.0000114988.61223.22

Lehtoranta L, Ala-Jaakkola R, Laitila A, Maukonen J (2022) Healthy vaginal microbiota and influence of probiotics across the female life span. Front Microbiol 13:819958. https://doi.org/10.3389/fmicb.2022.819958

Lei J, Sun L, Huang S, Zhu C, Li P, He J, Mackey V, Coy DH, He Q (2019) The antimicrobial peptides and their potential clinical applications. Am J Transl Res 11:3919–3931

Leli C, Mencacci A, Meucci M, Bietolini C, Vitali M, Farinelli S, D'Alò F, Bombaci JC, Perito S, Bistoni F (2013) Association of pregnancy and Candida vaginal colonization in women with or without symptoms of vulvovaginitis. Minerva Ginecol 65:303–309

Levy DA, Bohbot JM, Catalan F, Normier G, Pinel AM, Dussourd d'Hinterland L (1989) Phase II study of D.651, an oral vaccine designed to prevent recurrences of vulvovaginal candidiasis. Vaccine 7:337–340. https://doi.org/10.1016/0264-410X(89)90197-7

Li T, Liu Z, Zhang X, Chen X, Wang S (2019) Local probiotic Lactobacillus crispatus and Lactobacillus delbrueckii exhibit strong antifungal effects against vulvovaginal candidiasis in a rat model. Front Microbiol 10:1033. https://doi.org/10.3389/fmicb.2019.01033

Lietz A, Eckel F, Kiss H, Noe-Letschnig M, Farr A (2023) Quality of life in women with chronic recurrent vulvovaginal candidosis: a sub-analysis of the prospective multicentre phase IIb/III Prof-001 study. Mycoses 66:767–773. https://doi.org/10.1111/myc.13602

Liu P, Lu Y, Li R, Chen X (2023) Use of probiotic lactobacilli in the treatment of vaginal infections: in vitro and in vivo investigations. Front Cell Infect Microbiol 13:1153894. https://doi.org/10.3389/fcimb.2023.1153894

Luan T, Liu X, Mao P, Wang X, Rui C, Yan L, Wang Y, Fan C, Li P, Zeng X (2020) The role of 17β-estrogen in Candida albicans adhesion on human vaginal epithelial cells via FAK phosphorylation. Mycopathologia 185:425–438. https://doi.org/10.1007/s11046-020-00440-1

Machado D, Castro J, Palmeira-de-Oliveira A, Martinez-de-Oliveira J, Cerca N (2016) Bacterial vaginosis biofilms: challenges to current therapies and emerging solutions. Front Microbiol 6:1528. https://doi.org/10.3389/fmicb.2015.01528

Maillard J-Y, Pascoe M (2024) Disinfectants and antiseptics: mechanisms of action and resistance. Nat Rev Microbiol 22:4–17. https://doi.org/10.1038/s41579-023-00958-3

Makanjuola O, Bongomin F, Fayemiwo SA (2018) An update on the roles of non-albicans Candida species in vulvovaginitis. J Fungi 4:121. https://doi.org/10.3390/jof4040121

Marrazzo JM, Dombrowski JC, Wierzbicki MR, Perlowski C, Pontius A, Dithmer D, Schwebke J (2019) Safety and efficacy of a novel vaginal anti-infective, TOL-463, in the treatment of bacterial vaginosis and vulvovaginal candidiasis: a randomized, single-blind, phase 2, controlled trial. Clin Infect Dis 68:803–809. https://doi.org/10.1093/cid/ciy554

Martin Lopez JE (2015) Candidiasis (vulvovaginal). BMJ Clin Evid 2015:0815

Martino JL, Vermund SH (2002) Vaginal douching: evidence for risks or benefits to women's health. Epidemiol Rev 24:109. https://doi.org/10.1093/epirev/mxf004

McKloud E, Delaney C, Sherry L, Kean R, Williams S, Metcalfe R, Thomas R, Richardson R, Gerasimidis K, Nile CJ, Williams C, Ramage G (2021) Recurrent vulvovaginal candidiasis: a dynamic interkingdom biofilm disease of Candida and Lactobacillus. mSystems 6. https://doi.org/10.1128/msystems.00622-21

Mendling W, Krauss C, Fladung B (2004) A clinical multicenter study comparing efficacy and tolerability of topical combination therapy with clotrimazole (Canesten®, two formats) with oral single dose fluconazole (Diflucan®) in vulvovaginal mycoses. Mycoses 47:136–142. https://doi.org/10.1111/j.1439-0507.2004.00970.x

Mendling W, Weissenbacher ER, Gerber S, Prasauskas V, Grob P (2016) Use of locally delivered dequalinium chloride in the treatment of vaginal infections: a review. Arch Gynecol Obstet 293:469–484. https://doi.org/10.1007/s00404-015-3914-8

Miko E, Barakonyi A (2023) The role of hydrogen-peroxide (H2O2) produced by vaginal microbiota in female reproductive health. Antioxidants 12:1055. https://doi.org/10.3390/antiox12051055

Mirzaeei S, Zangeneh M, Veisi F, Parsa S, Hematti M (2021) Chlorhexidine, clotrimazole, metronidazole and combination therapy in the treatment of vaginal infections. J Med Life 14:250–256. https://doi.org/10.25122/jml-2019-0160

Mohammed I, Said DG, Dua HS (2017) Human antimicrobial peptides in ocular surface defense. Prog Retin Eye Res 61:1–22. https://doi.org/10.1016/j.preteyeres.2017.03.004

Mohammed L, Jha G, Malasevskaia I, Goud HK, Hassan A (2021) The interplay between sugar and yeast infections: do diabetics have a greater predisposition to develop oral and vulvovaginal candidiasis? Cureus 13:e13407. https://doi.org/10.7759/cureus.13407

National Health Service (2023) How and when to take fluconazole [WWW Document]. nhs.uk. URL https://www.nhs.uk/medicines/fluconazole/how-and-when-to-take-fluconazole/ (accessed 8.27.24)

Neal C, Martens M (2022) Clinical challenges in diagnosis and treatment of recurrent vulvovaginal candidiasis. SAGE Open Med 10:20503121221115201. https://doi.org/10.1177/20503121221115201

Nobile CJ, Johnson AD (2015) Candida albicans biofilms and human disease. Ann Rev Microbiol 69:71–92. https://doi.org/10.1146/annurev-micro-091014-104330

Noverr MC, Fidel PL (2019) Questions remain regarding the presence of fungal species biofilm in women with vulvovaginal candidiasis. Am J Obstet Gynecol 221:169. https://doi.org/10.1016/j.ajog.2019.04.017

Nyirjesy P, Brookhart C, Lazenby G, Schwebke J, Sobel JD (2022) Vulvovaginal candidiasis: a review of the evidence for the 2021 centers for disease control and prevention of sexually transmitted infections treatment guidelines. Clin Infect Dis 74:S162–S168. https://doi.org/10.1093/cid/ciab1057

Oerlemans EFM, Bellen G, Claes I, Henkens T, Allonsius CN, Wittouck S, van den Broek MFL, Wuyts S, Kiekens F, Donders GGG, Lebeer S (2020) Impact of a lactobacilli-containing gel on vulvovaginal candidosis and the vaginal microbiome. Sci Rep 10:7976. https://doi.org/10.1038/s41598-020-64705-x

Otero L, Palacio V, Carreno F, Mendez FJ, Vazquez F (1998) Vulvovaginal candidiasis in female sex workers. Int J STD AIDS 9:526–530. https://doi.org/10.1258/0956462981922764

Oxman DA, Chow JK, Frendl G, Hadley S, Hershkovitz S, Ireland P, McDermott LA, Tsai K, Marty FM, Kontoyiannis DP, Golan Y (2010) Candidaemia associated with decreased in vitro fluconazole susceptibility: is Candida speciation predictive of the susceptibility pattern? J Antimicrob Chemother 65:1460–1465. https://doi.org/10.1093/jac/dkq136

Palmeira-de-Oliveira R, Palmeira-de-Oliveira A, Martinez-de-Oliveira J (2015) New strategies for local treatment of vaginal infections. Adv. Drug Deliv. Rev. Vaginal Drug Delivery 92:105–122. https://doi.org/10.1016/j.addr.2015.06.008

Pan Y, Sun Y, Chen L, Cheng Y, Jin P, Zhang W, Zheng L, Liu J, Zhou T, Xu Z, Li C, Kostoulias X, Watson CJ, McGiffin D, Peleg AY, Qu Y (2023) Candida causes recurrent vulvovaginal candidiasis by forming morphologically disparate biofilms on the human vaginal epithelium. Biofilms 6:100162. https://doi.org/10.1016/j.bioflm.2023.100162

Pandey M, Choudhury H, Abdul-Aziz A, Bhattamisra SK, Gorain B, Carine T, Wee Toong T, Yi NJ, Win Yi L (2021) Promising drug delivery approaches to treat microbial infections in the vagina: a recent update. Polymers 13:26. https://doi.org/10.3390/polym13010026

Papon N, Van Dijck P (2021) A complex microbial interplay underlies recurrent vulvovaginal candidiasis pathobiology. mSystems 6:e01066. https://doi.org/10.1128/mSystems.01066-21

Parazzini F, Di Cintio E, Chiantera V, Guaschino S (2000) Determinants of different *Candida* species infections of the genital tract in women. Eur J Obstet Gynecol Reprod Biol 93:141–145. https://doi.org/10.1016/S0301-2115(00)00289-X

Pasko MT, Piscitelli SC, Van Slooten AD (1990) Fluconazole: a new triazole antifungal agent. DICP 24:860–867. https://doi.org/10.1177/106002809002400914

Patel MA, Aliporewala VM, Patel DA (2021) Common antifungal drugs in pregnancy: risks and precautions. J Obstet Gynecol India 71:577–582. https://doi.org/10.1007/s13224-021-01586-8

Pereira LC, Correia AF, da Silva ZDL, de Resende CN, Brandão F, Almeida RM, de Medeiros Nóbrega YK (2021) Vulvovaginal candidiasis and current perspectives: new risk factors and laboratory diagnosis by using MALDI TOF for identifying species in primary infection and recurrence. Eur J Clin Microbiol Infect Dis 40:1681–1693. https://doi.org/10.1007/s10096-021-04199-1

Pfaller MA, Messer SA, Motyl MR, Jones RN, Castanheira M (2013) In vitro activity of a new oral glucan synthase inhibitor (MK-3118) tested against Aspergillus spp. by CLSI and EUCAST broth microdilution methods. Antimicrob Agents Chemother 57:1065–1068. https://doi.org/10.1128/aac.01588-12

Pfaller MA, Jones RN, Castanheira M (2014) Regional data analysis of Candida non-albicans strains collected in United States medical sites over a 6-year period, 2006–2011. Mycoses 57:602–611. https://doi.org/10.1111/myc.12206

Pfizer (2024) Product Information – Difulcan

Phan QT, Myers CL, Fu Y, Sheppard DC, Yeaman MR, Welch WH, Ibrahim AS, Edwards JE, Filler SG (2007) Als3 is a Candida albicans invasin that binds to cadherins and induces endocytosis by host cells. PLoS Biol 5:e64. https://doi.org/10.1371/journal.pbio.0050064

Phillips NA, Bachmann G, Haefner H, Martens M, Stockdale C (2022) Topical treatment of recurrent vulvovaginal candidiasis: an expert consensus. Womens Health Rep 3:38–42. https://doi.org/10.1089/whr.2021.0065

Phillips NA, Rocktashel M, Merjanian L (2023) Ibrexafungerp for the treatment of vulvovaginal candidiasis: design, development and place in therapy. Drug Des Devel Ther 17:363–367. https://doi.org/10.2147/DDDT.S339349

Pietrella D, Rachini A, Torosantucci A, Chiani P, Brown AJP, Bistoni F, Costantino P, Mosci P, d'Enfert C, Rappuoli R, Cassone A, Vecchiarelli A (2010) A β-glucan-conjugate vaccine and anti-β-glucan antibodies are effective against murine vaginal candidiasis as assessed by a novel *in vivo* imaging technique. Vaccine 28:1717–1725. https://doi.org/10.1016/j.vaccine.2009.12.021

Powderly WG (2012) *Candida* in HIV infection. In: Volberding PA, Greene WC, Lange JMA, Gallant JE, Sewankambo N (eds) Sande's HIV/AIDS medicine. W.B. Saunders, London, pp 359–367. https://doi.org/10.1016/B978-1-4557-0695-2.00028-6

Raba G, Ďurkech A, Malík T, Bassfeld D, Grob P, Hurtado-Chong A, Fluomizin Study Group (2024) Efficacy of dequalinium chloride vs metronidazole for the treatment of bacterial vaginosis: a randomized clinical trial. JAMA Netw Open 7:e248661. https://doi.org/10.1001/jamanetworkopen.2024.8661

Ramage G, Bachmann S, Patterson TF, Wickes BL, López-Ribot JL (2002) Investigation of multidrug efflux pumps in relation to fluconazole resistance in Candida albicans biofilms. J Antimicrob Chemother 49:973–980. https://doi.org/10.1093/jac/dkf049

Ramage G, Saville SP, Thomas DP, López-Ribot JL (2005) Candida biofilms: an update. Eukaryot Cell 4:633–638. https://doi.org/10.1128/ec.4.4.633-638.2005

Ramsay S, Astill N, Shankland G, Winter A (2009) Practical management of recurrent vulvovaginal candidiasis. Trends Urol Gynaecol Sex Health 14:18–22. https://doi.org/10.1002/tre.127

Rathod S, Buffler P (2014) Highly-cited estimates of the cumulative incidence and recurrence of vulvovaginal candidiasis are inadequately documented. BMC Womens Health 14:43. https://doi.org/10.1186/1472-6874-14-43

Ravel J, Gajer P, Abdo Z, Schneider GM, Koenig SSK, McCulle SL, Karlebach S, Gorle R, Russell J, Tacket CO, Brotman RM, Davis CC, Ault K, Peralta L, Forney LJ (2011) Vaginal microbiome of reproductive-age women. Proc Natl Acad Sci USA 108(Suppl 1):4680–4687. https://doi.org/10.1073/pnas.1002611107

Richardson JP, Willems HME, Moyes DL, Shoaie S, Barker KS, Tan SL, Palmer GE, Hube B, Naglik JR, Peters BM (2018) Candidalysin drives epithelial signaling, neutrophil recruitment, and immunopathology at the vaginal mucosa. Infect Immun 86:e00645–17. https://doi.org/10.1128/IAI.00645-17

Richter SS, Galask RP, Messer SA, Hollis RJ, Diekema DJ, Pfaller MA (2005) Antifungal susceptibilities of Candida species causing vulvovaginitis and epidemiology of recurrent cases. J Clin Microbiol 43:2155–2162. https://doi.org/10.1128/jcm.43.5.2155-2162.2005

Rigo GV, Tasca T (2020) Vaginitis: review on drug resistance. Curr Drug Targets 21:1672–1686. https://doi.org/10.2174/1389450121666200804112340

Rosati D, Bruno M, Jaeger M, ten Oever J, Netea MG (2020) Recurrent vulvovaginal candidiasis: an immunological perspective. Microorganisms 8:144. https://doi.org/10.3390/microorganisms8020144

Russo R, Superti F, Karadja E, De Seta F (2019) Randomised clinical trial in women with recurrent vulvovaginal candidiasis: efficacy of probiotics and lactoferrin as maintenance treatment. Mycoses 62:328–335. https://doi.org/10.1111/myc.12883

Salama OE, Gerstein AC (2022) Differential response of Candida species morphologies and isolates to fluconazole and boric acid. Antimicrob Agents Chemother 66:e02406–e02421. https://doi.org/10.1128/aac.02406-21

Sampson T, George AP (2019) Evaluating the risk of Candida albicans associated with gestation amongst women in Port Harcourt, River State, Nigeria. Int J Pathog Res:1–7. https://doi.org/10.9734/ijpr/2019/v3i430098

San Juan Galán J, Poliquin V, Gerstein AC (2023) Insights and advances in recurrent vulvovaginal candidiasis. PLoS Pathog 19:e1011684. https://doi.org/10.1371/journal.ppat.1011684

Sandaradura I, Marriott DJE, Day RO, Norris RLG, Pang E, Stocker SL, Reuter SE (2021) Current fluconazole treatment regimens result in under-dosing of critically ill adults during early therapy. Eur J Clin Microbiol Infect Dis 40:1521–1528. https://doi.org/10.1007/s10096-021-04201-w

Satora M, Grunwald A, Zaremba B, Frankowska K, Żak K, Tarkowski R, Kułak K (2023) Treatment of vulvovaginal candidiasis—an overview of guidelines and the latest treatment methods. J Clin Med 12:5376. https://doi.org/10.3390/jcm12165376

Schmidt CS, White CJ, Ibrahim AS, Filler SG, Fu Y, Yeaman MR, Edwards JE, Hennessey JP (2012) NDV-3, a recombinant alum-adjuvanted vaccine for Candida and Staphylococcus aureus, is safe and immunogenic in healthy adults. Vaccine 30:7594–7600. https://doi.org/10.1016/j.vaccine.2012.10.038

Schwebke JR, Sobel R, Gersten JK, Sussman SA, Lederman SN, Jacobs MA, Chappell BT, Weinstein DL, Alfred H, Moffett J, Azie NE, Angulo DA, Harriott IA, Borroto-Esoda K, Ghannoum MA, Nyirjesy P, Sobel JD (2022) Ibrexafungerp versus placebo for vulvovaginal candidiasis treatment: a phase 3, randomized, controlled superiority trial (VANISH 303). Clin Infect Dis 74:1979. https://doi.org/10.1093/cid/ciab750

Shaaban OM, Youssef AEA, Khodry MM, Mostafa SA (2013) Vaginal douching by women with vulvovaginitis and relation to reproductive health hazards. BMC Womens Health 13:23. https://doi.org/10.1186/1472-6874-13-23

Shen C-J, Yang C-Y, Chen H-Y, Chen W-C, Chang T-C, Cheng C-M (2021) Clinical evaluation of a self-testing kit for vaginal infection diagnosis. J Healthc Eng 2021:4948954. https://doi.org/10.1155/2021/4948954

Sheppard DC, Yeaman MR, Welch WH, Phan QT, Fu Y, Ibrahim AS, Filler SG, Zhang M, Waring AJ, Edwards JE (2004) Functional and structural diversity in the Als protein family of Candida albicans*. J Biol Chem 279:30480–30489. https://doi.org/10.1074/jbc.M401929200

Sherry L, Kean R, McKloud E, O'Donnell LE, Metcalfe R, Jones BL, Ramage G (2017) Biofilms formed by isolates from recurrent vulvovaginal candidiasis patients are heterogeneous and insensitive to fluconazole. Antimicrob Agents Chemother 61:e01065. https://doi.org/10.1128/AAC.01065-17

Shukla A, Sobel JD (2019) Vulvovaginitis caused by Candida species following antibiotic exposure. Curr Infect Dis Rep 21:44. https://doi.org/10.1007/s11908-019-0700-y

Singh S, Uppuluri P, Mamouei Z, Alqarihi A, Elhassan H, French S, Lockhart SR, Chiller T, Edwards JE Jr, Ibrahim AS (2019) The NDV-3A vaccine protects mice from multidrug resistant Candida auris infection. PLoS Pathog 15:e1007460. https://doi.org/10.1371/journal.ppat.1007460

Sobel JD (2007) Vulvovaginal candidosis. Lancet 369:1961–1971. https://doi.org/10.1016/S0140-6736(07)60917-9

Sobel JD (2014) Factors involved in patient choice of oral or vaginal treatment for vulvovaginal candidiasis. Patient Prefer Adherence 8:31. https://doi.org/10.2147/PPA.S38984

Sobel JD (2015) Editorial commentary: vaginal biofilm: much ado about nothing, or a new therapeutic challenge? Clin Infect Dis 61:607–608. https://doi.org/10.1093/cid/civ358

Sobel JD (2016) Recurrent vulvovaginal candidiasis. Am J Obstet Gynecol 214:15–21. https://doi.org/10.1016/j.ajog.2015.06.067

Sobel JD, Sobel R (2018) Current treatment options for vulvovaginal candidiasis caused by azole-resistant Candida species. Expert Opin Pharmacother 19:971–977. https://doi.org/10.1080/14656566.2018.1476490

Sobel JD, Faro S, Force RW, Foxman B, Ledger WJ, Nyirjesy PR, Reed BD, Summers PR (1998) Vulvovaginal candidiasis: epidemiologic, diagnostic, and therapeutic considerations. Am J Obstet Gynecol 178:203–211. https://doi.org/10.1016/s0002-9378(98)80001-x

Sobel JD, Zervos M, Reed BD, Hooton T, Soper D, Nyirjesy P, Heine MW, Willems J, Panzer H (2003) Fluconazole susceptibility of vaginal isolates obtained from women with complicated Candida vaginitis: clinical implications. Antimicrob Agents Chemother 47:34–38. https://doi.org/10.1128/AAC.47.1.34-38.2003

Sobel JD, Borroto-Esoda K, Azie N, Angulo D (2021) In vitro pH activity of Ibrexafungerp against fluconazole-susceptible and -resistant Candida isolates from women with vulvovaginal candidiasis. Antimicrob Agents Chemother 65:e0056221. https://doi.org/10.1128/aac.00562-21

Sousa LGV, Pereira SA, Cerca N (2023) Fighting polymicrobial biofilms in bacterial vaginosis. Microb Biotechnol 16:1423–1437. https://doi.org/10.1111/1751-7915.14261

Spampinato C, Leonardi D (2013) *Candida* infections, causes, targets, and resistance mechanisms: traditional and alternative antifungal agents. Biomed Res Int 2013:1–13. https://doi.org/10.1155/2013/204237

Spinillo A, Capuzzo E, Nicola S, Baltaro F, Ferrari A, Monaco A (1995) The impact of oral contraception on vulvovaginal candidiasis. Contraception 51:293–297. https://doi.org/10.1016/0010-7824(95)00079-P

Stabile G, Gentile RM, Carlucci S, Restaino S, De Seta F (2021) A new therapy for uncomplicated vulvovaginal candidiasis and its impact on vaginal flora. Healthcare 9:1555. https://doi.org/10.3390/healthcare9111555

Strydom M, Walpola RL, McMillan S, Khan S, Ware RS, Tiralongo E (2022) Lived experience of medical management in recurrent vulvovaginal candidiasis: a qualitative study of an uncertain journey. BMC Womens Health 22:384. https://doi.org/10.1186/s12905-022-01973-x

Sustr V, Foessleitner P, Kiss H, Farr A (2020) Vulvovaginal Candidosis: current concepts, challenges and perspectives. J Fungi 6:267. https://doi.org/10.3390/jof6040267

Swate TE, Weed JC (1974) Boric acid treatment of vulvovaginal candidiasis. Obstet Gynecol 43:893

Swidsinski A, Guschin A, Tang Q, Dörffel Y, Verstraelen H, Tertychnyy A, Khayrullina G, Luo X, Sobel JD, Jiang X (2019) Vulvovaginal candidiasis: histologic lesions are primarily polymicrobial and invasive and do not contain biofilms. Am J Obstet Gynecol 220:91.e1–91.e8. https://doi.org/10.1016/j.ajog.2018.10.023

Swingle KL, Ricciardi AS, Peranteau WH, Mitchell MJ (2023) Delivery technologies for women's health applications. Nat Rev Bioeng 1:408–425. https://doi.org/10.1038/s44222-023-00040-w

Tarry W, Fisher M, Shen S, Mawhinney M (2005) Candida albicans: the estrogen target for vaginal colonization. J Surg Res 129:278–282. https://doi.org/10.1016/j.jss.2005.05.019

Tisher M, Pradel G, Ohlsen K, Holzgrabe U (2011) Quaternary ammonium salts and their antimicrobial potential: targets or nonspecific interactions? ChemMedChem 7:22. https://doi.org/10.1002/cmdc.201100404

Vahedpoor Z, Abastabar M, Sehat M, Talebian P, Felfelian Fini T, Dastanpour Z, Haghani I, Chelongarian R, Nazeri M (2021) Vaginal and oral use of probiotics as adjunctive therapy to fluconazole in patients with vulvovaginal candidiasis: a clinical trial on Iranian women. Curr Med Mycol 7:36–43. https://doi.org/10.18502/cmm.7.3.7803

van der Veer C, Bruisten SM, van Houdt R, Matser AA, Tachedjian G, van de Wijgert JHHM, de Vries HJC, van der Helm JJ (2019) Effects of an over-the-counter lactic-acid containing intra-vaginal douching product on the vaginal microbiota. BMC Microbiol 19:168. https://doi.org/10.1186/s12866-019-1545-0

Vanić Ž, Jøraholmen MW, Škalko-Basnet N (2021) Nanomedicines for the topical treatment of vulvovaginal infections: addressing the challenges of antimicrobial resistance. Adv Drug Deliv Rev 178:113855. https://doi.org/10.1016/j.addr.2021.113855

Wang J, Dou X, Song J, Lyu Y, Zhu X, Xu L, Li W, Shan A (2019) Antimicrobial peptides: promising alternatives in the post feeding antibiotic era. Med Res Rev 39:831–859. https://doi.org/10.1002/med.21542

Wang Y, Liu Z, Chen T (2024) Vaginal microbiota: potential targets for vulvovaginal candidiasis infection. Heliyon 10:e27239. https://doi.org/10.1016/j.heliyon.2024.e27239

Warrilow AGS, Hull CM, Parker JE, Garvey EP, Hoekstra WJ, Moore WR, Schotzinger RJ, Kelly DE, Kelly SL (2014) The clinical candidate VT-1161 is a highly potent inhibitor of Candida albicans CYP51 but fails to bind the human enzyme. Antimicrob Agents Chemother 58:7121–7127. https://doi.org/10.1128/aac.03707-14

Watson C, Calabretto H (2007) Comprehensive review of conventional and non-conventional methods of management of recurrent vulvovaginal candidiasis. Aust N Z J Obstet Gynaecol 47:262–272. https://doi.org/10.1111/j.1479-828X.2007.00736.x

Whaley SG, Berkow EL, Rybak JM, Nishimoto AT, Barker KS, Rogers PD (2017) Azole antifungal resistance in Candida albicans and emerging non-albicans Candida species. Front Microbiol 7. https://doi.org/10.3389/fmicb.2016.02173

White DJ, Vanthuyne A (2006) Vulvovaginal candidiasis. Sex Transm Infect 82:iv28–iv30. https://doi.org/10.1136/sti.2006.023168

Willems HME, Ahmed SS, Liu J, Xu Z, Peters BM (2020) Vulvovaginal candidiasis: a current understanding and burning questions. J Fungi 6:27. https://doi.org/10.3390/jof6010027

Woodburn KW, Clemens LE, Jaynes J, Joubert L-M, Botha A, Nazik H, Stevens DA (2019) Designed antimicrobial peptides for recurrent vulvovaginal candidiasis treatment. Antimicrob Agents Chemother 63. https://doi.org/10.1128/aac.02690-18

Workowski KA (2021) Sexually transmitted infections treatment guidelines, 2021. MMWR Recomm Rep 70:1. https://doi.org/10.15585/mmwr.rr7004a1

Wu X, Zhang S, Li H, Shen L, Dong C, Sun Y, Chen H, Xu B, Zhuang W, Deighton M, Qu Y (2020) Biofilm formation of Candida albicans facilitates fungal infiltration and persister cell formation in vaginal candidiasis. Front Microbiol 11. https://doi.org/10.3389/fmicb.2020.01117

Xie HY, Feng D, Wei DM, Mei L, Chen H, Wang X, Fang F (2017) Probiotics for vulvovaginal candidiasis in non-pregnant women. Cochrane Database Syst Rev 2017:CD010496. https://doi.org/10.1002/14651858.CD010496.pub2

Xu J, Schwartz K, Bartoces M, Monsur J, Severson RK, Sobel JD (2008) Effect of antibiotics on vulvovaginal candidiasis: a MetroNet study. J Am Board Fam Med JABFM 21:261–268. https://doi.org/10.3122/jabfm.2008.04.070169

Yano J, Peters BM, Noverr MC, Fidel PL (2018) Novel mechanism behind the immunopathogenesis of vulvovaginal candidiasis: "Neutrophil Anergy.". Infect Immun 86. https://doi.org/10.1128/iai.00684-17

Yano J, Sobel JD, Nyirjesy P, Sobel R, Williams VL, Yu Q, Noverr MC, Fidel PL (2019) Current patient perspectives of vulvovaginal candidiasis: incidence, symptoms, management and post-treatment outcomes. BMC Womens Health 19:48. https://doi.org/10.1186/s12905-019-0748-8

Younis SMK, Mahmoud MS, Elshorbagy AM (2023) Use of povidone iodine vaginal swabbing in treatment of recurrent vulvovaginal candidiasis. Int J Med Arts 5:3763–3768. https://doi.org/10.21608/ijma.2023.219025.1713

Zhang X, Essmann M, Burt ET, Larsen B (2000) Estrogen effects on Candida albicans: a potential virulence-regulating mechanism. J Infect Dis 181:1441–1446. https://doi.org/10.1086/315406

Zhang Q-Y, Yan Z-B, Meng Y-M, Hong X-Y, Shao G, Ma J-J, Cheng X-R, Liu J, Kang J, Fu C-Y (2021) Antimicrobial peptides: mechanism of action, activity and clinical potential. Mil Med Res 8:48. https://doi.org/10.1186/s40779-021-00343-2

Zhao Y, Yu Z, Yue X (2021) Evaluating the accuracy and diagnostic value of CFW and a new fluorescent reagents, fluorescent brightener 85, for the diagnosis of vulvovaginal candidiasis. J Clin Lab Anal 35:e23891. https://doi.org/10.1002/jcla.23891

Zhu Y, Bateman BT, Gray KJ, Hernandez-Diaz S, Mogun H, Straub L, Huybrechts KF (2020) Oral fluconazole use in the first trimester and risk of congenital malformations: population based cohort study. BMJ 369:m1494. https://doi.org/10.1136/bmj.m1494

Chapter 6
Prevalence and Influence of Fungal Biofilm in Wound Infections

Jontana Allkja, Bryn Short, Zainab Bilal, Jason L. Brown, Craig Williams, and Gordon Ramage

Abstract Wound infections disrupt the normal wound healing process leading to longer treatment, hospitalisation, patient discomfort, and in extreme cases severe outcomes like amputations and death, representing one of the major challenges in the healthcare system. Biofilms represent close to 78% of these infections, and while the role of bacterial pathogens and polymicrobial biofilms has been widely recognised in the field, fungi are generally ignored. This is also evident in the disproportionate focus on bacterial biofilm model development for the testing of wound therapies, with very few studies incorporating fungal species within them,

J. Allkja (✉)
Department of Chemistry, Open Innovation Hub for Antimicrobial Surfaces, Surface Science Research Centre, School of Physical Sciences, University of Liverpool, Liverpool, UK

Safeguarding Health through Infection Prevention (SHIP) Research Group, Research Centre for Health, School of Health and Life Sciences, Glasgow Caledonian University, Glasgow, UK

School of Medicine, Dentistry and Nursing, College of Medical, Veterinary and Life Sciences, University of Glasgow, Glasgow, UK
e-mail: Jontana.allkja@liverpool.ac.uk

B. Short · J. L. Brown
School of Medicine, Dentistry and Nursing, College of Medical, Veterinary and Life Sciences, University of Glasgow, Glasgow, UK

Z. Bilal
School of Medicine, Dentistry and Nursing, College of Medical, Veterinary and Life Sciences, University of Glasgow, Glasgow, UK

Department of Pure and Applied Chemistry, Centre for Molecular Nanometrology, Technology and Innovation Centre, University of Strathclyde, Glasgow, UK

C. Williams
Department of Microbiology, Lancaster Royal Infirmary, University of Lancaster, Lancaster, UK

G. Ramage
Safeguarding Health through Infection Prevention (SHIP) Research Group, Research Centre for Health, School of Health and Life Sciences, Glasgow Caledonian University, Glasgow, UK

G. Ramage, D. Andes (eds.), *Fungal Biofilms*, Springer Series on Biofilms 15,
https://doi.org/10.1007/978-3-031-93315-8_6

all primarily focused on *Candida albicans*. Hence, the role of fungi in these infections is poorly understood. In this chapter, we present a summary of studies evaluating fungal prevalence in various wound types, demonstrating that not only are fungi present in wounds at a larger rate than expected, but there is vast species diversity in these communities. We also provide an overview of the diagnostic methods used for fungal detection throughout these studies and their pitfalls. Finally, we discuss current developments in wound biofilm model development and the potential role of fungi in polymicrobial biofilms, highlighting the need for further research on fungi in wounds.

Introduction

Wound management constitutes a major burden on the healthcare system, with an estimated global cost of $281bn per annum associated to biofilm wound infections alone (Cámara et al. 2022). In the UK, there are over 3.8 million patients p. a. requiring wound management, leading to yearly costs of £8.3bn for the National Health Service (NHS) (Guest et al. 2020a, b). An aging population coupled with a rise in chronic conditions like obesity, diabetes and vascular health issues has increased the risk of developing chronic non-healing wounds susceptible to infections, leading to a rise in patients requiring wound care, resulting in higher costs (Nussbaum et al. 2018; Graves et al. 2022; Urwin et al. 2022). Apart from the economic cost, there is also a profound impact on the patient's quality of life, affected by extensive treatments, prolonged hospitalisation, and in extreme cases amputation and even death (Wolcott 2015; Guest et al. 2018; Piednoir et al. 2021). Thus, it is crucial to have a better understanding of wound infections, to be able to provide adequate support in the clinic.

Biofilm infections are a major concern as they constitute approximately 78% of wound infections (Malone et al. 2017). They can be defined as communities of microorganisms attached to biotic, abiotic surfaces or each other, encased in a self-produced extracellular matrix composed of extracellular polymeric substances (EPS) (Sauer et al. 2022). Biofilms are characterised by an increased tolerance to antimicrobial therapy, as many compounds cannot penetrate the EPS matrix and reach their intended target (Stewart 2015). They can also act as an optimal environment for the spread of antimicrobial resistance between different members of the community (Brauner et al. 2016). When taken into account, these facts explain the many difficulties in wound infection management due to treatment failure (Bowler 2018).

Research in this area has overwhelmingly focused on bacterial pathogens, especially *Staphylococcus aureus, Pseudomonas aeruginosa,* and *Acinetobacter baumannii,* as the main culprits in biofilm wound infections (Di Domenico et al. 2017; Schaal et al. 2020; Laliany et al. 2022). Moreover, recent reports have increasingly shown that these biofilms are polymicrobial in nature, with multiple species cohabiting in the same biofilm (Malone et al. 2017; Kalan and Brennan 2019; Uberoi et al. 2024). Despite many of these studies documenting fungal presence in these communities, they are often overlooked and as a result their impact and role in these infections are poorly understood (Short et al. 2023a, b). This chapter aims to provide an overview of fungal prevalence in various wound types, their potential role in propagating disease and the current state-of-the-art for wound biofilm models incorporating fungal pathogens.

Fungal Prevalence in Wounds

Wounds are a result of damage to the skin due to trauma or a consequence of a patient's comorbidity (Fig. 6.1). Generally, they are referred to as either acute or chronic wounds. Acute wounds are caused by trauma to the skin from cuts, burns, surgery or other injuries, e.g. combat-related wounds, and typically heal within 12 weeks (Morton and Phillips 2016). The wound repair process begins immediately, typically going through the stages of coagulation, inflammation, proliferation and remodelling, resulting in a closed wound (Wilkinson and Hardman 2020). Infection can disrupt this process, leading to major complications and in some cases

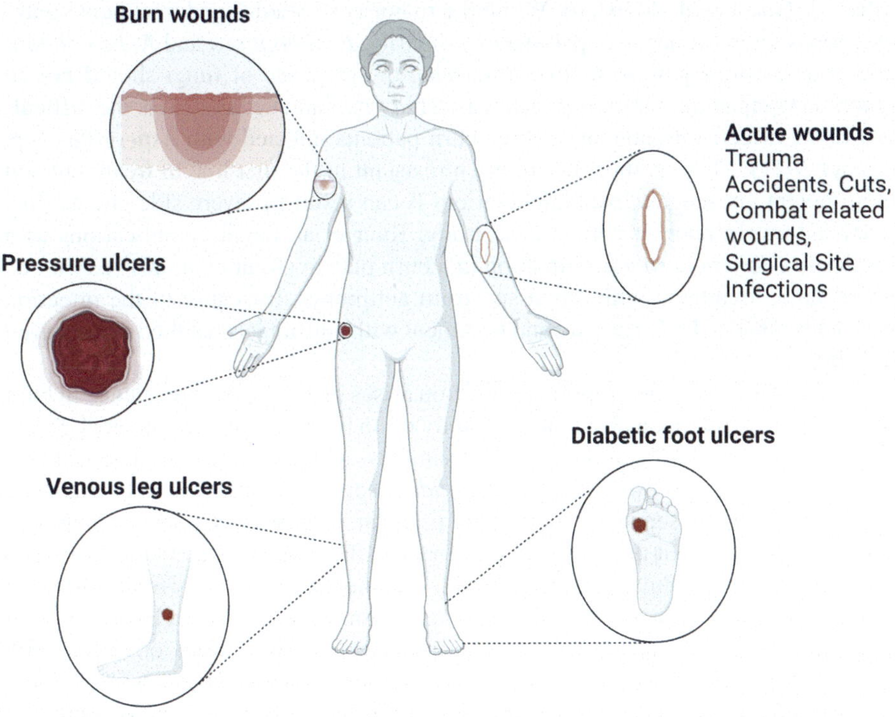

Fig. 6.1 Summary of wound types outlined in this chapter affected by fungal biofilm infections. Created with Biorender.com. Short. (2024) BioRender.com/q52r777

the formation of chronic non-healing wounds (Rodrigues et al. 2021; Sussman 2023). Chronic wounds can also form in patients with limited mobility and conditions like diabetes, often characterised by poor vascularisation and decreased sensitivity in lower limbs which lead to the formation of ulcers, e.g. diabetic foot ulcers (DFUs) (Deng et al. 2023).

Fungi in Acute Wounds

Table 6.1 provides a summary of reported fungal prevalence in what are commonly classified as acute wounds. Burn wounds are the most common acute wound in studies evaluating fungal prevalence found in our literature review. These wounds are very susceptible to infections due to deeper skin layers being exposed to the environment (Sussman 2023). Initially, as a result of the burn, the immune system produces a hyperinflammatory response, often associated with a cytokine storm, which can sometimes lead to systemic inflammatory response syndrome and in some cases sepsis. This is often followed by an influx of regulatory T cell and other immunosuppression mechanisms, which then leads to an inability to fight infections (Burgess et al. 2022). An estimated 60% of burn wounds in the UK are reported as infected (Guest et al. 2020a, b). While the majority of studies and treatment strategies focus on gram-negative pathogens primarily, *P. aeruginosa* and *A. baumannii*, and gram-positive pathogens like *S. aureus*, the presence of fungi should not be ignored (Wang et al. 2018). A patient case study by Pantet et al. details the difficulties in the effective treatment of three burn patients infected with *Aspergillus* spp. (Pantet 2021). These pathogens are often resistant to the first line of treatment and more potent antifungals like Amphotericin B can result in severe side effects, like renal failure. In another patient case study, Kaur et al. detail complications as a result of *Lichtheimia ramosa* infection in a burn patient (Kaur et al. 2014). After an initial failed treatment with broad-spectrum antibiotics, the cause of the infection was determined to be *L. ramosa*, and treatment with antifungals resulted in improved healing.

Considering these case reports, a question arises on the fungal prevalence in burn wounds in general. In an observational study of 36 burn patients, Okuno et al. found *Candida* spp. in 19.44% of the collected samples, with a majority of these samples being *Candida parapsilosis* (57.1% of *Candida* spp. detected) (Okuno et al. 2018). Interestingly, in subsequent experiments performed on these *C. parapsilosis* isolates, they noticed that they were able to form resilient biofilms that could successfully penetrate acellular dermal matrix fragments, a novel wound dressing technology, suggesting a potential impact on treatment success. In a larger study of 354 samples by Macedo et al., similar prevalence values were observed with *Candida tropicalis* being present in 10.3% of samples, followed by *C. parapasilosis* (4.2%) (Jefferson Lessa Soares de Macedo 2005). Similarly, Cen et al. found fungal presence in 13.4% out of 1942 patients tested in their burn ward over a 3-year period, most commonly reported as *C. albicans, C. tropicalis* and *Candida glabrata*

Table 6.1 Summary of fungal prevalence reports in acute wounds

Type of wound	Fungal species [percent prevalence]	Method of assessment	Reference
Blunt traumatic injury	*Candida* spp. [0.04%] Other fungi—*Ascomycota* (*Aspergillus, Curvularia, Fusarium*) and *Zygomycota* (*Mucor, Rhizopus*) [0.02%]	Culture Direct microscopy DNA sequencing	McGraw et al. (2019)
Burn wound	*Candida* spp. [19.44%], composed of: *C. parapsilosis* [57.1%], *C. albicans* [28.6%] and *C. glabrata* [14.3%].	Culture Biochemical assessment (VITEK2®)	Okuno et al. (2018)
Burn wound	*C. albicans* [7.9%], *Candida* spp. [1%]	Culture Direct microscopy	Sikwewa et al. (2023)
Burn wound	Patient case studies of *Aspergillus* spp. complications in burn wounds	Culture	Pantet (2021)
Burn wound	*C. tropicalis* [10.3%], *C. parapsilosis* [4.2%], *C. humiloca* [2.1%], *C. famata* [1.4%], *C. albicans* [1.1%]	Culture biochemical assessment (VITEK2®) Direct microscopy	Jefferson Lessa Soares de Macedo (2005)
Burn wound	Fungi [13.4%]: most common *C. albicans, C. tropicalis, C. glabrata*	Culture biochemical assessment (VITEK2®)	Hanghui Cen et al. (2015)
Burn wound	Fungi [39%] composed of: 48% filamentous moulds (*A. flavus, Fusarium* spp., *Lichtheimia corymbifera*), 37% yeasts (*C. tropicalis, C. parapsilosis*) and 15% combined moulds and yeast.	Culture Direct microscopy.	Jabeen et al. (2021)
Burn wound	*C. albicans* [12%], *A. flavus* [6%], non-albicans *Candida* spp. [8%]	Culture Direct microscopy	Sharma et al. (2016)
Burn wound	*Candida* spp. [22.2%]	Culture	Mousa and al-Bader (2001)
Burn wound	Patient case study of *Lichtheimia ramosa* in burn wound complication	Culture	Kaur et al. (2014)
Burn wound	Fungi [6.6%] composed of: *Aspergillus* spp. [82.9%], Mucor-like morphology [15.4%], yeast [36%] and mixed fungi [28.6%]	Culture Histology	Horvath et al. (2007)

(continued)

Table 6.1 (continued)

Type of wound	Fungal species [percent prevalence]	Method of assessment	Reference
Burn wound	Aspergillosis [1.2%], Mucormycosis [0.5%] and Fusariosis [0.2%]	Culture PCR	Schaal et al. (2015)
Burn wound	*C. tropicalis* [14%], *C. parapsilosis* [5.6%], *A. niger* [2.8%], *C. albicans* [1.4%], *C. glabrata* [1.4%], *Syncephalestrum* [1.4%] and *Fusarium solani* [1.4%]	Culture Direct microscopy Biochemical assessment (VITEK2®)	Sarabahii et al. (2012)
Burn wound	*Absidia corymbifera* [25.9%]	Culture	Christiaens et al. (2005)
Burn wound	Case studies of 6 patients with burn wound complication from *Rhizomucor* infection	Culture	Kyriopoulos et al. (2015)
Combat-related wound	*Monomicrobial infections*—Fungi [15.3%] composed of *Aspergillus* spp. [20%], *Candida* spp. [15%], *Mucorales* [20%] and other moulds [45%]. *Polymicrobial infections*—Fungi [39.2%] composed of: *Aspergillus* spp. [32.5%], *Candida* spp. [18.8%], *Mucorales* [23.8%], other moulds [57.5%] and other yeast [2.5%].	Culture Direct microscopy	Mende et al. (2019)
Combat-related wound	Mould contamination [60%] composed of *Aspergillus* spp. *Alternaria* spp., *Fusarium* spp., *Mucor* spp., *Penicillium* spp., *Saksenaea vasiformis*	Culture	Tribble et al. (2015)
Combat-related wound	Fungi [34%] composed of *Mucorales*, *Aspergillus* spp., *Fusarium* spp., and other filamentous fungi	Culture	Ganesan et al. (2019a, b)
Combat-related wound	Yeast [21.4%] (*Kloeckera spp* and *Rhodotorula* spp.) Moulds [43%] (*Aspergillus spp, Penicillium spp, Mucorales and Fusarium* spp.)	Culture	Wallum et al. (2015)
Combat-related wound	Case studies of 9 patients with confirmed fungal wound infections: *Aspergillus* spp. [66.7%], *Candida* spp. [22.2%], *Zygomycetes* [11.1%], *Alternaria* sp. [11.1%], *S. vasiformis* [11.1%], *Mucor* sp. [11.1%], *Scedosporium* sp. [11.1%], *Graphium* sp. [11.1%]	Culture	Radowsky et al. (2015)
Combat-related wound	Fungi [13%] composed of: *Mucorales* [35%], *Aspergillus* spp. [29%] and *Fusarium* spp. [21%].	Culture Histology PCR	Rodriguez et al. (2022)
Combat-related wound	Fungi [48%] composed of: *Mucorales* [35%], *Aspergillus* spp. [29%], *Fusarium* spp. [21%] and other moulds [33%]	Culture Histology	Tyler et al. (2015)

(continued)

Table 6.1 (continued)

Type of wound	Fungal species [percent prevalence]	Method of assessment	Reference
Deep lacerations due to sailing accident	Patient case study of complicated wound infection caused by *Saksenaea erythrospora*	Culture Direct microscopy DNA sequencing	Relloso et al. (2014)
Facial lacerations	*C. albicans* [1%]	Culture	Shin et al. (2023)
Surgical site infection	*C. albicans* [13.7%], non-albicans *Candida* spp. [3.3%]	Culture	Ballus et al. (2015)
Surgical site infection	Patient case study of SSI caused by *A. welwitschiae*	Culture Direct microscopy MALDI-TOF	Horiuchi et al. (2024)
Surgical site infection	Fungi [11.1%] composed of: *C. albicans* [43.4%], non-albicans *Candida* spp. [52.1%] and other fungi [4.5%]	Culture	Canadian et al. (2023)
Surgical site infection	Yeast [2%]: *Candida* spp. (*C. albicans, C. tropicalis* and *C. parapsilosis*)	Culture Direct microscopy biochemical assessment (API strips)	Demet and Yucel (2007)
Surgical site Infection	Patient case study of SSI infection due to *A. flavus*	Culture	Alquraish et al. (2023)
Surgical site infection	Fungi [10%] composed of *A. fumigatus, Paecilomyces* spp., *Scedosporium apiospermum*	Culture	Shields et al. (2013)
Surgical site infection (open heart surgery)	*C. glabrata* [2%], *C. tropicalis* [1.7%], *C. albicans* [1.7%], *C. kruzei* [1.1%], *C. parapsilosis* [1.1%]	Culture PCR	Gholamali Riahi (2023)
Trauma	*C. albicans* [1.86%]	Culture	Basavraj Nagoba et al. (2012)
Trauma	Case studies of 13 patients with confirmed fungal infections: *Apophysomyces trapeziformis* [100%]. Other fungi were isolated in the *Candida* [38.5%], *Fusarium* spp. [15%], *Aspergillus* spp. [8%], *Mucor circinelloides* [8%]	Culture Histology DNA sequencing	Neblett et al. (2012)
Trauma	Fungi [17.6%] composed of *Fusarium* spp. and *Curvularia* spp.	Culture	McNeil et al. (2018)

(continued)

Table 6.1 (continued)

Type of wound	Fungal species [percent prevalence]	Method of assessment	Reference
Trauma and surgical wounds	Incidence rate of 3.3 per 100,000 patients composed of *Lichtheimia ramosa* [26.5%], *Lichtheimia corymbifera* [16%], *Cunninghamella bertholletiae* [16%], *Rhizomucor pusillus* [10.5%], *M. circinelloides* [10.5%], *Rhizopus arrhizus* [10.5%], *Saksenaea vasiformis* [5%], unknown [5%]	Culture PCR	Guinea et al. (2017)
Wound following unexplained localised oedema and swelling	Case studies of 7 patients with complications in wound infections caused by *Mucorales* spp.	Culture	Gupta et al. (2018)

For each study: the wound type, fungal species and their respective percent prevalence, as well as the assessment method used are detailed

(Hanghui Cen 2015). Strikingly, in subsequent tests of these isolates, they found high rates (38–56%) of Ketoconazole resistance, as well as resistance to other antifungals like Amphotericin B, Clotrimazole, and other azoles, which limits treatment choices for these patients. Reports are not limited to yeast and *Candida* species only, Schaal et al. have reported on the incidence of filamentous fungal infections (1.7%) in 1849 burn patients, with a majority being attributed to *Aspergillus* spp. (Schaal et al. 2015). In their report, it became apparent that despite the low incidence, the presence of filamentous fungal species was associated with higher mortality, highlighting their importance as critical pathogens for monitoring in burn patients.

Another large subset of acute wounds is traumatic injuries, particularly combat-related wounds, which are more commonly exposed to dust and spores at the moment of injury (Ganesan et al. 2019a, b). This is likely a main factor why contrary to some of the burn wound studies detailed above, the most commonly found fungal species in combat wounds are moulds (Table 6.1). For example, in the study by Mende et al., they reported on fungal prevalence in single-species and multispecies infections of combat-related wounds (Mende et al. 2019). A total of 335 samples were assessed, 131 single-species infections and 204 polymicrobial ones. In the single-species group, 15.3% of samples were fungal positive, with the majority of these being attributed to unidentified moulds (45%), followed by *Aspergillus* spp. (20%) and *Mucorales* (20%). A larger portion of samples in the polymicrobial group were fungi positive (39.2%); of these, most were unidentified moulds (57.5%), followed by *Aspergillus* spp. (32.5%). Similarly, studies by Ganesan et al., and Warkentien et al., found fungal presence in 34% and 48% of patient samples, respectively, primarily colonised by *Mucorales, Aspergillus* spp., *Fusarium* spp., and other filamentous moulds (Tyler et al. 2015; Ganesan et al. 2019a, b). Likewise, Rodriguez et al. reported *Mucorales, Aspergillus* spp., and *Fusarium* spp., as the most commonly occurring fungi among 720 combat wounds examined over a 5-year period (Rodriguez et al. 2022). Furthermore, the presence of fungal infection corresponded with more severe complications and higher rates of amputation.

Surgical site infections (SSIs) are another common type of acute infection, representing a major burden on the healthcare system. In a recent study of the burden of SSIs in Australia, an estimated annual cost of A$323.5 million was calculated, due to extended hospital stays, longer treatment regiments, and devastatingly the majority of costs were associated to premature deaths of patients (Royle et al. 2023). As the name suggests, they occur following a surgical procedure, as vulnerable (often immunocompromised) patients are exposed to the myriad of pathogens circulating in the hospital environment (Piednoir et al. 2021). While bacteria are the more prevalent pathogens, fungal presence should not be ignored. In a surveillance study of 60 Canadian hospitals, between 2017 and 2021, 11.1% of SSI infections were attributed to fungal pathogens, with a majority being related to *Candida* yeasts (43.4% *C. albicans* and 52.1% other *Candida* spp.) (Canadian et al. 2023). Likewise, Shields et al. found fungal presence in 10% of SSI patient samples following lung transplants at the University of Pittsburgh medical centre (Shields et al. 2013). In contrast to the Canadian study, they did not find any yeast, only filamentous fungi (*A. fumigatus, Paecilomyces* spp., and *Scedosporium apiospermum*), likely due to their policy of prophylactic treatment with Voriconazole (Shields et al. 2013).

As detailed in Table 6.1 and above, many studies of wound samples have reported fungal presence, with a large diversity of species ranging from the most well-known yeasts like *C. albicans*, to relatively unknown species like *Saksenaea erythrospora* (Relloso et al. 2014); and prevalences varying between as low as 0.02% (McGraw et al. 2019) to as high as 60% (Horvath et al. 2007). As many of them highlight, despite their low prevalence compared to bacterial species, the fact that fungal presence was commonly associated with devastating outcomes suggests a need for better monitoring of these pathogens.

Fungi in Chronic Wounds

Table 6.2 provides a summary of fungal prevalence in chronic wounds. The overwhelming majority of studies focus on ulcers, open sores on skin often taking several weeks or months to heal (Nussbaum et al. 2018). They are commonly associated with poor circulation, ischemia, peripheral neuropathy, obesity, limited mobility and a poor immune system (Deng et al. 2023). These ulcers are particularly common in diabetic patients, presenting primarily as diabetic foot ulcers (DFUs) (Edmonds et al. 2021). Similar to acute wounds, ulcers are susceptible to infections by both bacterial and fungal pathogens, and several patient case studies of complications due to fungal pathogens have been reported. Examples include a chronic ulcer (non-DFU) resulting from *Sporothrix schenckii* (Jin et al. 2024), a DFU case study of a recurrent *Rhizomucor variabilis* infection (Monteiro et al. 2024), amputation due to *Fusarium oxysporum* in DFU infection (João et al. 2021), complications

Table 6.2 Summary of fungal prevalence reports in chronic wounds

Type of wound	Fungal species [percent prevalence]	Method of assessment	Reference
Buruli ulcer (secondary infection)	Fungi [62.5%], composed of: Yeast (*C. parapsilosis* and *C. duobushaemulonii*) [92%], Moulds (*Penicillium citrinum, Trichodema longibrachiatum*) [32%]	Culture PCR	Gyamfi et al. (2021)
Chronic leg ulcers (non-DFU)	Fungi [6%] composed of *C. albicans* and *C. parapsilosis*	Culture Direct microscopy	Majid Foroozan et al. (2011)
Chronic leg ulcers (non-DFU)	Fungi [7.4%] composed of: *C. albicans* [63.6%], *C. parapsilosis* [18.2%], *C. kruzei* [9.1%], *C. lipolytica* [9.1%]	Culture	Stefano Veraldi et al. (2008)
Chronic leg ulcers (non-DFU)	Case study of patient with complication due to *Sporothrix schenckii*	Culture Direct microscopy MALDI-TOF PCR	Jin et al. (2024)
Chronic trauma wound	Composition of wound in one patient—Fungi [32%] made up of: *C. albicans* [78%], *C. tropicalis* [20%]	Culture qPCR	Leake et al. (2009)
Chronic wound (non-DFU)	*C. tropicalis* [6%], Candida spp. [5%], *C. albicans* [3%], *Cladophiolophora* spp. [1%], *F. solani* [1%], *Geotrichium candidum* [1%], *Purpureocillium/Acremonium* [1%], *Zygosaccaromyces* spp. [1%]	Culture MALDI-TOF	Krumkamp et al. (2020)
Chronic wounds (including DFU)	Fungi [23%] composed of: *C. albicans* [46.09%], *C. parapsilosis* [45.19%], *C. lunata* [35.09%], *C. tropicalis* [17.79%], *T. mentagrophytes* [12.98%], *C. glabrata* [7.69%], *R. oryzae* [1.44%], *C. orthopsilosis* [28.37%], *F. monoliforme* [6.25%], *P. adscendens* [2.88%], *M. restricta* [20.67%], *P. rabiei* [13.46%], *F. equiseti* [7.21%], *C. herbarum* [21.1%], *F. sporotrichioides* [8.17%], *K. ohmeri* [0.48%], *P. betae* [0.96%], *A. terreus* [2.4%], *U. botrytis* [23.07%], *A. pullulans* [19.71%], *C. athensensis* [2.4%], *Myrothecium* spp. [10.09%], *A. niger* [2.4%], *C. silvicultrix* [0.96%], *Pyrenophora* spp. [8.65%], *B. eleusines* [7.69%], *M. sympodialis* [7.21%], *P. pyrolae* [3.36%], *E. album* [12.02%], *S. martiniae* [0.96%], *H. fuscoatra* [1.44%], *P. medicaginis* [10.1%], *C. oregonensis* [1.92%], *E. rostratum* [6.73%], *C. dubliensis* [21.6%], *C. viswanathii* [4.8%], *C. sojae* [1.92%], *P. janthinellum* [7.69%], *K. cladophila* [10.1%], *B. sorokiniana* [5.29%]	Culture PCR	Dowd et al. (2011)

(continued)

Table 6.2 (continued)

Type of wound	Fungal species [percent prevalence]	Method of assessment	Reference
Diabetic foot ulcer (DFU)	Fungi [48.38%] composed of: *C. tropicalis* [34.6%], *C. albicans* [29.3%], *C. krusei* [16.0%], *C. parapsilosis* [10.6%], *C. glabrata* [9.33%].	Culture DNA sequencing	Kumar et al. (2016)
DFU	*C. albicans* [14.3%]	Culture DNA sequencing	Hassan et al. (2019)
DFU	Case studies of four patients with *Trichophyton rubrum* infections	Culture Direct microscopy	Lee et al. (2014)
DFU	Yeast [20%] composed of: *C. albicans* [75%], *C. lusitaniae* [8%], *C. dubliniensis* [5%], *C. glabrata* [2%], *C. tropicalis* [2%], *C. famata* [2%], *C. parapsilosis* [2%]. Moulds [5%] composed of: *Penicillium* spp. [38%], *Aspergillus* spp. [25%], *Microsporum* spp. [25%], *T. mentagrophytes* [12%]	Culture Direct microscopy, biochemical analysis (Vitek2®) MALDI-TOF	Musyoki et al. (2022)
DFU	Fungi [21.4%] composed of: *C. albicans* [73.3%], *C. tropicalis* [53.3%], *C. glabrata* [66.7%], *T. mentagrophytes* [40%], *T. rubrum* [46.7%], *A. fumigates* [33.3%]	Culture	Sujatha et al. (2018)
DFU	Fungi [18%] composed of: *C. albicans* [89%] and *C. tropicalis* [11%]	Culture	Abilash et al. (2015)
DFU	Fungi [27.2%] composed of: *C. parapsilosis* [25.5%], *C. tropicalis* [22.7%], *T. asahii* [12.8%], *C. albicans* [10.6%], *Aspergillus* spp. [5.0%], (*C. guilliermondii,* Non-*albicans Candida* sp., *C. glabrata, Fusarium* sp., *C. sake, Zygosaccharomyces* sp., *Kodamaea ohmeri, C. globosa, C. krusei, Penicillium* sp., *C. lusitaniae, C. famata, C. melibiosica*) [<5%]	Culture Direct microscopy biochemical assessment (Vitek2®)	Chellan et al. (2010)
DFU	Fungi [6%]	Culture	Agrawal et al. (2017)
DFU	*C. Intermedia, T. asahii* [prevalence not assessed]	Culture PCR	Yas KhudhairAl-Ameri and Alrufae (2022)
DFU	Fungi [31.7%] composed of, *C. albicans* [33.33%], *C. tropicalis* [23.81%], *C. parapsilosis* [19.04%], *C. lustania* [4.76%], *C. glabrata* [4.76%], *Alternaria* spp. [4.76%], *Penicillium* spp. [4.76%], *Trichophyton spp* [4.76%]	Culture Direct microscopy	Kandregula et al. (2022)
DFU	*C. parapsilosis, C. albicans, C. glabrata, C. tropicalis and C. humicola, T. mentagrophytes, T. rubrum* [prevalence not assessed]	Culture Direct microscopy	Heald et al. (2001)

(continued)

Table 6.2 (continued)

Type of wound	Fungal species [percent prevalence]	Method of assessment	Reference
DFU	Patient case study of infection complication due to *Rhizomucor variabilis*	Culture Direct microscopy	Monteiro et al. (2024)
DFU	*C. albicans* [2%]	Culture Biochemical assessment (Vitek2®)	Chai et al. (2021)
DFU	*Biopolaris hawaiiensis* [5.5%], *T. schoenleinii* [3.7%], *A. niger* [3.0%], *T. rubrum* [3.0%], *F. oxysporum* [3.0%], *P. marneffei* [0.6%], *C. parapsilosis* [6.1%], *C. famata* [3.0%].	Culture	Gitau et al. (2011)
DFU	Patient case study of complication due to *F. oxysporum*	Culture Direct microscopy	João et al. (2021)
DFU	Fungi [17.38%] composed of: *C. parapsilosis* [30%], *C. albicans* [14.5%], *C. auris* [5.5%], *Trichosporon* spp. [6%], (*C. tropicalis, C. glabrata*) [<6%]	Culture Biochemical assessment (Vitek2®)	Arun et al. (2019)
DFU	Fungi [17.5%] composed of: *C. albicans* [42.85%], *C. tropicalis* [21.42%], *C. glabrata* [14.28%], *T. mentagrophytes* [7.14%], *T. rubrum* [7.14%], *A. fumigatus* [7.14%]	Culture Direct microscopy biochemical analysis (Vitek2®)	Kalshetti et al. (2017)
DFU	28.6% fungi, of which 19% Apergillus sp., 57% Candida spp., 14% Fusarium spp., 10% Trichophyton spp.	Culture Direct microscopy	Mehra et al. (2017)
DFU	Fungi [65%] composed of: *C. albicans* [49%], *C. tropicalis* [23%], *C. parapsilosis* [18%], *C. guillermondi* [5 %], *C. krusei* [5%], (*T. cutaneum and T. capitatum*) [<5%]. Moulds [34%] composed of: *A. flavus* [60%], *A. fumigatus* [40 %], *F. solani* [9%], *P. marneffei* [9%], *B. ranarum* [4%]	Culture Direct microscopy	Nair et al. (2006)
DFU	Patient case report of complication due to *Scedosporium apiospermum*	Culture Direct microscopy	Vijaya et al. (2013)
DFU	Fungi before antibiotics [12.1%], fungi after antibiotic therapy [20%]	Culture	Wu et al. (2018)
DFU	*C. albicans* [9.1%], *C. tropicalis* [4.1%], *C. parapsilosis* [0.83%], *C. glabrata* [0.83%], *C. krusei* [0.83%], *Candida* spp. [3.3%], *T. mentagrophytes* [2.5%], *Rhodotorula* spp., *Acremonium* spp. [0.84%], *Scopulariopsis* spp. [0.84%], *A. fumigatus* [0.84%]	Culture Direct microscopy	Fata et al. (2011)

(continued)

Table 6.2 (continued)

Type of wound	Fungal species [percent prevalence]	Method of assessment	Reference
DFU	Composition of wound in one patient—Fungi [55%] made up of: *C. albicans* [88%] and *C. tropicalis* [11%]	Culture qPCR	Leake et al. (2009)
DFU	Fungi [80%] composed of: *Cladosporium herbarum* [41%], *C. albicans* [22%], *C. parapsilosis* [15%], *A. cibarius* [12%], *E. nigrum* [9%], *Penicillium* spp. [9%], *L. chartarum* [7%], P. bialowiezense [6%], *G. zeae* [6%], *Hypocreales* sp. [4%], Unclassified *Ascomycota* [16%], Family *Nectriaceae* [16%], Order *Capnodiales* [4%], T. asahii [10%], *Trichosporon* spp. [4%], *R. diobovatum* [5%], Unclassified fungi [32%]	ITS sequencing	Kalan et al. (2016)
DFU	Patient case study of complication due to *C. haemulonii*	MALDI-TOF	Coles et al. (2020)
DFU	Fungi [19%] composed of: *Yeast [31.58%]* (*Candida* spp. [83.33%], *Trichosporon spp* [16.66%]); *Moulds [68.42%]* (*Aspergillus* spp. [30.77%], *Fusarium* spp. [23.08%], *Trichophyton* spp. [23.08%], *Penicillium* spp. [15.38%], *Acremonium* [7.69%])	Culture Direct microscopy biochemical assessment (Vitek2®)	Raza and Anurshetru (2017)
DFU	*Candida* spp. [7.7%]	Culture	Oates et al. (2012)
DFU	*C. parapsilosis* [2.94%], *C. albicans* [0.98%]	Culture MALDI-TOF	Goncalves et al. (2024)
DFU	Fungi [9%] composed of: *C. tropicalis* [29%], *C. albicans* [14%], *C. guililermondii* [7%], *Aspergillus flavus* [21%], *A. niger* [14%], *Fusarium* sp. [14%]	Culture	Bansal et al. (2008)
DFU	Case studies 13 patients with confirmed fungal infection composed of: *C. albicans* [31%], *C. glabrata* [23%], *C. parapsilosis* [8%], *C. krusei* [8%], *C. lipolytica* [8%], *Fusarium* spp. [15%], combined *T. asahii* and *F. solani* [8%]	Culture Direct microscopy, MALDI-TOF	Öztürk et al. (2019)
Surgical non-healing wound	Composition of wound in one patient—Fungi [25%] made up of: *C. parapsilosis* [90%], *C. tropicalis* [8%]	Culture qPCR	Leake et al. (2009)

For each study: the wound type, fungal species and their respective percent prevalence, as well as the assessment method used are detailed

due to *Candida haemulonii* (Coles et al. 2020) and the case studies of four patients with *Trichophyton rubrum* DFU infections (Weon Ju Lee et al. 2014). The occurrence of these fungal infections is often tied to the use of antibiotics as prophylaxis or first line of treatment. Predictably, when assessing fungal prevalence in DFUs, Wu et al. showed an increase in fungi from 12.1% before antibiotic treatment to 20% post-antibiotic treatment (Wu et al. 2018). Moreover, in a study of secondary infections in Buruli ulcers (ulcers formed following *Mycobacterium ulcerans* skin infection) by Gyamfi et al., 62.5% of patients were colonised by fungi, primarily yeast (92%) (Gyamfi et al. 2021).

It is important to remember that these infections are most commonly polymicrobial (Malone et al. 2017). In a study by Leake et al., the wound composition of three patients with different types of chronic wounds was evaluated (Leake 2009). In the chronic trauma wound, fungi made up 32% of the total microbial burden, with *C. albicans* comprising 78% of the fungal DNA and *C. tropicalis* 20%. In the DFU patient, fungal burden was higher (55%), with similar distributions of *C. albicans* (88%) and *C. tropicalis* (11%). In contrast, in the case of a surgical non-healing wound, only 25% of the microbial burden was due to fungi, with *C. parapsilosis* being the most abundant species at 90%.

Several studies have evaluated the microbiome and mycobiome of chronic wounds (mainly DFU) in large patient cohorts. In a large study of 915 samples collected from different chronic wound types (decubitus ulcers, DFUs, non-healing surgical wounds, and venous leg ulcers) by Dowd et al., 23% of samples were positive for fungal species (Dowd et al. 2011). Among them, *C. albicans* (46%), *C. parapsilosis* (45%) and *Curvularia lunata* (35%) were the most prevalent species. Moreover, they also noticed associations between the more dominant/prevalent species and wound types. For example, in DFUs, *C. paparasilosis* and *C. albicans* were the most prevalent, whereas in chronic non-DFU wounds *C. lunata* and *Ulocladium botryti* were more common. Smaller studies by Chellan et al. and Musyoki et al. reported similar rates of 27.2% and 25% fungal prevalence in DFUs, respectively (Chellan et al. 2010; Victor Moses Musyoki 2022). Similarly to Dowd et al., in the study by Chellan et al., *C. parapsilosis* was the most prevalent species (25.5%), followed closely by *C. tropicalis* (22.7%) (Chellan et al. 2010). In the Musyoki et al. study, *C. albicans* was by far the most prevalent species (75%), while *C. parapsilosis* prevalence was very low (2%) (Victor Moses Musyoki 2022). All three studies found the presence of filamentous fungi of various genera such as *Aspergillus, Penicillium* and *Fusarium,* though at lower prevalence compared to yeasts (Chellan et al. 2010; Dowd et al. 2011; Victor Moses Musyoki 2022).

Among the population studies summarised in Table 6.2, the study by Kalan et al. stands out as fungi were detected at a much larger rate compared to all other studies; 80% of their DFU samples used ITS sequencing (Kalan et al. 2016). The most prevalent species was *Cladosporium herbarum* (41%), a common allergen frequently found in the environment, typically regarded as non-pathogenic (Bush 2012). Among pathogenic fungi, *C. albicans* (22%) and *C. parapsilosis* (15%) were the most prevalent. Interestingly, their study offered a longitudinal view of the patients journey and how changes in their mycobiome correlated to certain clinical outcomes. Higher rates of pathogenic taxa were associated with more negative outcomes, such as amputation and necrosis when compared to allergen taxa. This study emphasises the potential importance of fungal species in propagating chronic wounds and the need for further studies to better understand these mechanisms and how to combat them.

Methods for Fungal Detection

In the previous section, we outlined several studies assessing fungal prevalence in various wounds (Tables 6.1 and 6.2). A notable detail is the methodology used to detect fungi. Most studies, regardless of location, cohort size or wound type, used routine culture as the main and occasionally only method. Microbiological culture remains the gold standard for diagnostics to this day (Mendonça et al. 2022). Samples are plated on selective media optimal for fungal growth, often with antibiotics to prevent bacterial growth or other kinds of reagents for species identification. For example, Chromogenic agars contain substrates which when metabolised by fungal cells release chromophores resulting in coloured colonies, specific to each species within a group or genus (Bayona et al. 2022). Culture plates are incubated at optimal temperatures for fungal growth, 25 °C and 30 °C, for up to 14 days (up to 1 month in rare cases), as many fungal species are slow growing. This is a cheap and easy-to-use methods, well established in diagnostics labs which does not require any specialised equipment. However, it is not a very sensitive technique, meaning that fungal presence at lower concentrations might not always be detected, or can be inhibited by bacteria. Additionally, the long incubation periods make it susceptible to contamination from fungal spores in the environment (Mendonça et al. 2022).

Culture is often coupled with direct microscopy evaluation and/or biochemical assays. Different techniques can be applied for direct visualisation of fungal species, depending on the sample type. For skin sample processing, potassium hydroxide wet mounting and methylene blue staining are among the most common techniques (Knoll et al. 2023). These can be combined with dyes such as Calcofluor white, which stains polysaccharides in chitin on the cell wall to better visualise fungal morphological traits (Borman et al. 2022). These methods offer a faster alternative to traditional culture, limiting potential sources of contamination as no long-term incubation is required. However, sample preparation can be difficult at times and some fungal features may be lost during processing, e.g. if tissue needs to be digested before visualising. Moreover, many different species appear very similar under the microscope, making genus and species identification difficult (Terrero-Salcedo and Powers-Fletcher 2020; Fang et al. 2023). On the other hand, biochemical assays often rely on commercial platforms such as VITEK2® (Loïez et al. 2006). These platforms use an automated multi-well system, where each well contains nutrients and reagents for specific biochemical tests. As the sample reacts to each test, the resulting profiles are compared to a database which then allows identification, the same as manual analytical profile index (API) strip tests (Barry et al. 2003; Figueiredo et al. 2014). The system can also be used to simultaneously perform antifungal testing, which helps in selecting appropriate treatments (Melhem et al. 2014). While the technique is relatively inexpensive, apart from the initial equipment cost, it is also limited to the availability of information within the database. Hence, many species that might not have been characterised previously would not be detected.

More sensitive methods relying on mass spectrometry like the Matrix-assisted laser desorption ionisation-time of flight (MALDI-TOF) system were also used in a few studies. This method allows for the identification of various fungal species based on their signature peaks, and this has become widely used in clinical diagnostics (Terrero-Salcedo and Powers-Fletcher 2020). However, equipment costs are high, and while detection is quick, sample preparation often requires colony isolation through traditional culture techniques prior to use (Weiss et al. 2021). Additionally, similarly to biochemical assessment systems, this technique is reliant on the quality and availability of libraries within their database.

Lastly, molecular techniques like real-time PCR and DNA sequencing are the least used methods in the studies summarised in Tables 6.1 and 6.2. This is likely due to their high costs, for both equipment and reagents. A major advantage of these methods is their high sensitivity and the detection of unculturable samples. This is strikingly evident in the study by Kalan et al., where fungal detection by ITS sequencing was 80%, compared to 5% detection by culture (Kalan et al. 2016). They are faster than traditional culture methods as no incubation is necessary and commercial diagnostic platforms are becoming more readily available (Wickes et al. 2018). However, they are also reliant on the quality of available databases and data analysis can be very complex requiring expert knowledge of bioinformatics analysis platforms (Jenks et al. 2023). This was also evident in the Kalan study, where 32% of ITS sequences could not be identified, highlighting the need for better databases, particularly for fungal species (Kalan et al. 2016). These techniques are also limited however, as they can detect DNA from dead fungi that may or may not have had a role in these infections.

Fungi in Biofilm Wound Models

As detailed in previous sections, biofilms play a key role in wound infections. Despite a lack of study into the role of biofilms in acute wounds, there exists some evidence of their presence. A recent systematic review reported that of all manuscripts examined (n = 13), 69% of studies reported the presence of biofilms after 14 days (Patenall et al. 2023). The same systematic review reported that 38% of studies showed evidence of biofilms after only 14 h. However, the role of biofilms in wounds is better known in chronic wounds. The chronicity of wounds due to infection is a well-documented phenomenon and is often driven by the presence of biofilms (Percival et al. 2015). These wounds are typically recalcitrant to therapeutics and lead to poorer patient outcomes. Previous research reports that biofilms can be identified in approximately 78% of chronic wounds (Malone et al. 2017), highlighting the need to study and understand not only the host–biofilm interactions but the interactions between cells in these complex communities.

One of the most widely utilised methods of studying the impact that pathogens have on wounds is through the use of biofilm models (Fig. 6.2). Previous studies have utilised biofilm models of pathogens commonly isolated from wound

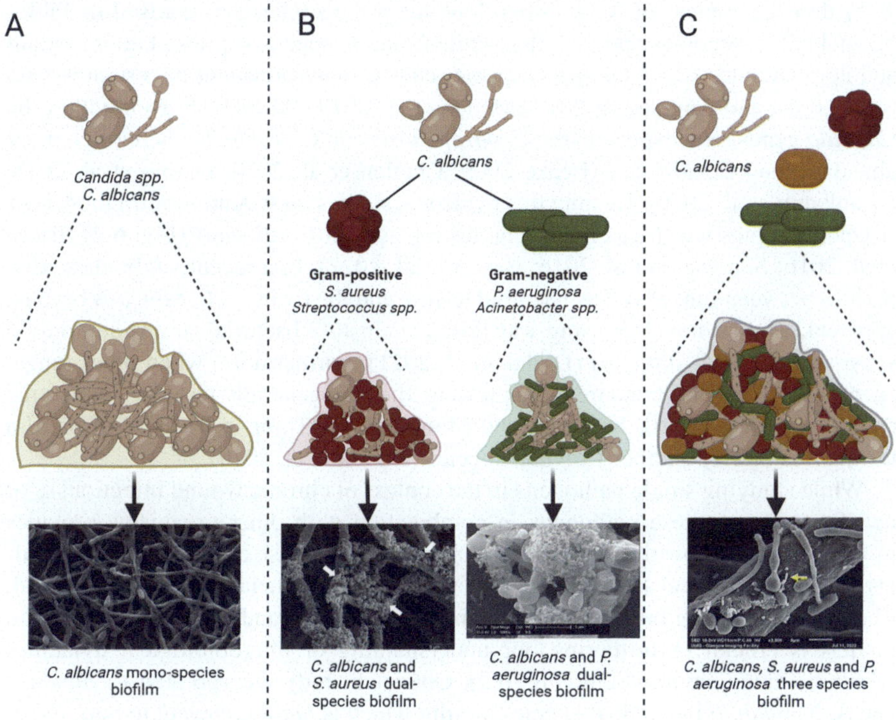

Fig. 6.2 Overview of the most common scenarios for fungal species incorporation in wound biofilm models. A. Single species biofilms primarily of *C. albicans* and a few other *Candida* spp. SEM image of *C. albicans* mono-species biofilm repurposed from Haque et al. (2016). B. Dual-species biofilms primarily *C. albicans* and either a key gram-positive or gram-negative pathogen. SEM image of *C. albicans* and *S. aureus* dual-species biofilm repurposed from Kean et al. (2017). SEM image of *C. albicans* and *P. aeruginosa* dual-species biofilm repurposed from Alam et al. (2020) (Alam et al. 2023). C. Polymicrobial biofilms usually containing the trio: *C. albicans*, *P. aeruginosa* and *S. aureus*. SEM image of *C. albicans, S. aureus* and *P. aeruginosa* three species biofilm repurposed from Baz et al. (2023). (Figure created in BioRender. Short (2025) https://BioRender.com/j36y751)

infections such as *P. aeruginosa* or *S. aureus* (Trøstrup et al. 2017; Brandenburg et al. 2019; Roy et al. 2020; Paleczny et al. 2021). Many other studies have interrogated wound pathogens such as these and the fungal opportunistic pathogen, *C. albicans* (Gil et al. 2022). These organisms are commonly identified in a myriad of wound types making them highly relevant to incorporate into chronic wound biofilm models (Oh et al. 2014; Wolcott et al. 2016; Verbanic et al. 2020). Although these organisms are important features of the microbiome of many wounds, they are not the sole inhabitants of these sites. Due to varying oxygen levels depending on depth and location, wounds are home to a mixed community of aerobic and anaerobic microorganisms, many of which are common skin commensals linked to opportunistic infections (Oh et al. 2014; Smith et al. 2016; Verbanic et al. 2020).

Within the context of fungi, when looking at the studies summarised in Tables 6.1 and 6.2, it becomes apparent that while *Candida* species represent an important aetiology, moulds like *Aspergillus* species can be more prevalent in certain wound types (Raza and Anurshetru 2017; Mende et al. 2019). Moreover, even within the *Candida* genus, other species like *C. parapsilosis* and *C. tropicalis* were often more abundant than *C. albicans* (Leake 2009, Chellan et al. 2010, Okuno et al. 2018, Krumkamp et al. 2020). Despite this evidence, studies on wound biofilm models or wound therapies will focus mainly on testing against *C. albicans* (Fig. 6.2) (Bjorn et al. 2016; Sancineto et al. 2016; Besser et al. 2022). In a recent study, these disparities between fungal versus bacterial testing and the focus on *C. albicans* became apparent, as their novel dressing was tested against 48 bacterial strains and only 6 yeast strains, all *C. albicans* (Dydak et al. 2021). Furthermore, when other fungal species are used for wound treatment testing, they are generally tested as planktonic cells (Jones et al. 2010; Krishnen and Muniandy 2023), or in the case of biofilm testing it is largely another *Candida* species (Karnik et al. 2019).

While studying single pathogens in the context of chronic wound infections is of benefit, studying these pathogens in combination with other organisms provides more insight into how they may behave in vivo. As emphasised in many of the studies in Tables 6.1 and 6.2, wound infections are overwhelmingly polymicrobial. Therefore, designing biofilm studies using relevant dual- and multi-species biofilm models is crucial in furthering our understanding of microbiological dynamics within a wound environment (Fig. 6.2). One such study incorporated *C. albicans* and *A. baumannii* into a dual-species biofilm which is highly relevant to burn infections (Ušjak et al. 2023). *C. albicans* potentially acts as a biofilm substrate for *A. baumannii,* a phenomenon reported amongst other interkingdom interactions (Kean et al. 2017), indicating that it may promote wound colonisation by *A. baumannii* (Ušjak et al. 2023).

An innovative model in the in vitro study of wound biofilms is the Lubbock chronic wound biofilm model. This technique utilises a pipette tip which acts as the biofilm substrate and contains the organisms of choice which is subsequently submerged in a mixture of growth media, plasma and lysed blood (Sun et al. 2008). This is an effective model as coagulase-positive organisms such as *S. aureus,* form aggregates, more closely resembling in vivo biofilms (Bjarnsholt et al. 2013), rather than the traditional mushroom-like structure typically associated with biofilm formation. With this in mind, more recent biofilm models have aimed to reproduce this structure in a multi-species environment (Townsend et al. 2016, 2017).

These novel biofilm models often utilise a semi-solid substrate which aids organisms in forming an aggregate phenotype. One such model combined *S. aureus* and *P. aeruginosa* with *C. albicans* to capture an underappreciated community of wound inhabitants, fungi (Townsend et al. 2017). This biofilm model displayed high levels of tolerance to antimicrobials and reported that this recalcitrance to antimicrobials was driven by the presence of *C. albicans*, highlighting the importance of testing for the presence of fungi in wounds before administering antimicrobial therapies. *C. albicans* drives this recalcitrance by providing a protective mesh supported by dense extracellular matrix components (Kean et al. 2017; Townsend et al. 2017).

The authors also showed that using antibiotics or antifungals as monotherapies was not sufficient to reduce overall biofilm viability and tended to target a single organism within the biofilm. This, however, created a niche for another member of the biofilm to fill, which supports some of the findings discussed in previous sections, where fungal prevalence increased post-antibiotic treatment (Kaur et al. 2014; Wu et al. 2018). Only with a combination of antibiotics and antifungals was a significant reduction in biofilm viability reported (Townsend et al. 2017). Similar findings have also been reported by Dalton and colleagues who reported that multi-species bacterial biofilms impaired wound healing and displayed increased tolerance to antimicrobials compared to single-species biofilms (Dalton et al. 2011). Some anaerobic bacteria such as *Prevotella bivia* have also been shown to increase the virulence of *S. aureus* (Mikamo et al. 1998).

Despite the success of studies using dual- and tri-species biofilm models, to better capture the complexity of the wound environment, more complex models are required. The use of multi-species biofilm models is limited throughout the literature. This is partly attributed to the difficulty in growing fastidious organisms such as anaerobic bacteria in a complex environment which aims to recreate host conditions in vitro. One such model developed by Brown and colleagues implemented an 11-species model comprising aerobes, anaerobes and fungi (*C. albicans*) within a 3-dimensional growth structure to replicate the wound bed and dressings (Brown et al. 2022). This biofilm model displayed high levels of compositional stability over 9 days and a degree of tolerability towards conventional antiseptics (hydrogen peroxide, chlorhexidine and povidone-iodine). Whilst all antiseptics exhibited similar levels of activity against this multi-species interkingdom biofilm, only hydrogen peroxide and povidone-iodine caused a downregulation in biofilm immunogenicity compared to untreated biofilms. This highlights the importance of wound therapies accounting for downstream effects that biofilm treatments may have on host tissue.

A study published in 2021 sought to develop a novel, layered, wound biofilm model substrate (Chen et al. 2021). This model was generated by stacking semi-solid agar to mimic dermis and subcutaneous fat layers. This model was capable of supporting growth of *P. aeruginosa* and *S. aureus* over multiple days and promoted these bacteria to form biofilm aggregates as seen in vivo. As part of this model, authors reported mild anti-biofilm effects of antimicrobial wound dressings. Despite poor anti-biofilm activity, this agrees with clinical findings and emphasises the importance of selecting appropriate growth substrates for biofilm studies. Whilst this study makes significant progress regarding in vitro wound biofilm models, no fungi were tested.

To reiterate, a large number of studies investigating the role of pathogenic biofilms in wounds solely focus on bacterial biofilms, making the biofilm models discussed herein part of the minority for including fungal elements. Despite this, evidence supporting the importance of fungi within wounds is steadily growing. Continuing to incorporate fungi such as *C. albicans* or other clinically relevant fungi into in vitro biofilm models will help further our knowledge of how fungi and interkingdom biofilms impact wound healing dynamics which can, in turn, improve treatment regimens and patient outcomes.

The Role of Fungi in Biofilm Wound Infections

The presence of fungi in wounds is an undeniable fact when looking at the studies summarised in Tables 6.1 and 6.2. However, as we have repeatedly seen throughout this chapter, particularly in the in vitro study of biofilm wound infections, fungi are generally omitted. As a result, little is known about their role with regard to wound healing, and within the polymicrobial communities in the wound environment. The exception is *C. albicans*. As we pointed out in the previous section, when fungi are included in in vitro biofilm models, it is overwhelmingly *C. albicans*. As a result, we can look to these studies to understand how *C. albicans* might be affecting the growth and behaviour of key bacterial pathogens and extrapolate potential roles of fungi within the wound environment.

C. albicans *and* S. aureus

The relationship between *C. albicans* and *S. aureus* is reasonably well understood, with the two being isolated from a variety of clinical samples including those taken from dentures, the urinary tract, DFUs and burn wounds (Gupta et al. 2005; Peng 2022). The major factor which drives the synergy between the two pathogens in biofilms is the scaffold-like structure provided by *C. albicans* hyphae. Dual-species biofilms composed of both *C. albicans* and *S. aureus* had reportedly greater biomass when compared to mono-species biofilms, with confocal laser scanning microscopy confirming clustering of *S. aureus* cells on the hyphal structures (Kean et al. 2017). Moreover, coating of *S. aureus* cells in the *C. albicans* extracellular matrix appears to significantly increase their resistance to vancomycin in polymicrobial biofilms (Harriott and Noverr 2009). A recent study found that this also increased toxin production due to upregulation of the bacterial *agr* by *C. albicans* (Paul et al. 2024). The *agr* locus is responsible for *S. aureus* α-toxin production; this upregulation in co-cultures of *S. aureus* and *C. albicans* had lethal effects in murine models. This beneficial effect of co-habitation is reciprocated, as *S. aureus* secretes factors increasing *C. albicans* biofilm formation (Eichelberger and Cassat 2021). A main factor contributing to this is the suggestion that hyphae development is triggered by the presence of peptidoglycan in *S. aureus*. This synergistic effect could be contributing to biofilm persistence within the wound and the increased tolerance to treatment.

C. albicans *and* P. aeruginosa

The interaction between *C. albicans* and *P. aeruginosa* is much more complex, with reports of the bacteria both promoting and inhibiting the growth and antifungal resistance of *C. albicans* simultaneously (Short et al. 2023a, b). *P. aeruginosa* dominates the microbiome of infected burn wounds due to its ability to thrive in burn wound secretions, and has been reported to outcompete *C. albicans* in a rodent burn wound model by colonising the hyphae and leading to fungal cell death (Gonzalez et al. 2016, Brandenburg et al. 2021, Kamath 2024). Additionally, there have also been reports of *P. aeruginosa* inducing fungal cell death through the upregulation of *C. albicans* stress pathways (Fourie et al. 2021). Similarly, isolation of bacterial species from samples obtained from burn wound patients revealed a statistically significant reduction of *Candida* species isolated from samples where *P. aeruginosa* was also present (Gupta et al. 2005). Recently, Hattab and colleagues found that *P. aeruginosa* sensitises *C. albicans* to fluconazole treatment in a polymicrobial model, resulting in a 3-log reduction of *C. albicans* growth (Hattab et al. 2022). Similarly, transcriptional analysis of a dual-species biofilm showed altered regulation of *C. albicans* genes associated with oxidative stress, which could possibly be induced by the secretion of phenazines (redox-active secondary metabolite) by *P. aeruginosa* (Alam et al. 2023). While not beneficial to *C. albicans*, in the case of *P. aeruginosa* an upregulation in the expression of iron uptake proteins in dual-species biofilms has been reported (Fourie et al. 2016). This enhanced iron uptake resulted in increased growth of the bacteria. Thus, the presence of *C. albicans* in the wound environment could be contributing to the observed dominance of *P. aeruginosa* in these infections.

C. albicans *and* A. baumannii

A. baumannii is a common coloniser of infected wounds; as a multi-drug resistant pathogen, it poses a significant hurdle in the treatment of burns and in particular, combat wounds (Ruamsap et al. 2023). Very little research has been conducted on the combined effect of *C. albicans* and *A. baumannii* in chronic wounds; however, a few mechanisms explaining their interaction in biofilms have been suggested. It is known that the *C. albicans* Hyr1p protein (cell surface protein) is able to bind the *A. baumannii* outer membrane protein A (OmpA) and anchor it to the hyphae, increasing its adherence in multi-species biofilms (Uppuluri et al. 2018). The relationship between *C. albicans* and *A. baumannii* is complex. Like *P. aeuruginosa,* the interaction is multifaceted with both synergistic and antagonistic effects observed. For example, production of farnesol (a quorum sensing molecule) by *C. albicans* has been observed to disrupt growth of *A. baumanni* by compromising the stability of the plasma membrane and decreasing virulence; however, this was not observed in multidrug-resistant *A. baumannii* isolates (Kostoulias et al. 2015).

Similarly, Peleg and colleagues presented research showing inhibition of *C. albicans* hyphal development in a *C. elegans* model co-infected with *A. baumannii* (Peleg et al. 2008). Although the mechanism underlying the inhibition was not resolved, it was suggested that an *A. baumannii* quorum sensing molecule may be involved, as filtrate from *A. baumannii* grown to stationary phase had an inhibitory effect on *C. albicans* but exponential phase filtrate did not (Peleg et al. 2008). Overall, it is clear that these relationships are complex and important, and further understanding of the mechanisms that drive them is crucial in informing decision-making when dealing with infections.

C. albicans *and Anaerobes*

Anaerobic bacteria have been found to colonise more than 50% of chronic wounds, with the most prevalent species being *Bacteroides*, *Peptostreptococcus*, *Clostridium* and *Fusobacterium* (Stephens et al. 2003). Similarly to fungi, their presence is often ignored as they are fastidious organisms, difficult to grow in vitro and very hard to detect with traditional routine culture methods (Malik Asif et al. 2016). As expected, anaerobes colonise the deep layers of chronic wounds, where oxygen concentrations are low (Oh et al. 2014). Research conducted on the growth of anaerobes in *C. albicans* biofilms found that oxygen concentrations were significantly lower in the deep layers of biofilms (< 50 μM) compared to the upper layers (> 300 μM) (Fox et al. 2014). This was observed in both *C. albicans* mono-species biofilms and dual-species biofilms with *C. perfringens* or *B. fragilis*, leading to the hypothesis that *C. albicans* biofilms can create hypoxic microenvironments conducive to the growth of anaerobic bacteria (Fox et al. 2014). Similar reports have been published with regard to oral biofilms, which are also susceptible to colonisation by both *C. albicans* and a variety of anaerobic species (Stichternoth and Ernst 2009; Bartnicka et al. 2019; Delaney et al. 2023).

Conclusions

In conclusion, fungi are present in different kinds of acute and chronic wounds. *Candida* species remain an important pathogenic group, though data presented in this chapter suggests a necessary shift from the current focus on *C. albicans* and more research on other species like *C. parapsilosis* and *C. tropicalis*. Moreover, the presence of other genera especially various moulds should not be ignored. The current diagnostic methods and the need for more complete fungal libraries and databases could be partly responsible for the lack of recognition of fungal pathogens. As a result, little is known about their role within the wound environment. Available studies suggest that they could be taking over post broad-spectrum antibiotic treatments or be supporting biofilm development for several key pathogens such as

S. aureus and *P. aeruginosa*, potentially propagating persistence in chronic wounds. In vitro biofilm models are a crucial tool to expand our understanding of these mechanisms, but as we discussed, very few studies incorporate fungi in their models. Moreover, this is almost exclusively limited to the use of *C. albicans* as the sole fungal species, which is why our understanding of bacterial–fungal interactions is primarily an understanding of *C. albicans*-bacterial interactions. Thus, moving forward, it is crucial to expand research into fungal pathogens, their relationship with bacterial pathogens and their role in wound infections. This will help the development of new treatment strategies and ultimately improve patient outcomes.

References

Abilash S, Kannan NS, Rajan KV, Pramodhini M, Ramanathan M (2015) Clinical study on the prevalance of fungal infections in diabetic foot ulcers. Int J Curr Res Rev 7(23):8–8

Agrawal KS, Sarda AV, Shrotriya R, Bachhav M, Puri V, Nataraj G (2017) Acetic acid dressings: finding the Holy Grail for infected wound management. Ind J Plast Surg 50(3):273–280

Alam F, Blackburn SA, Davis J, Massar K, Correia J, Tsai H-J, Blair JMA, Hall RA (2023) Pseudomonas aeruginosa increases the susceptibility of Candida albicans to amphotericin B in dual-species biofilms. J Antimicrob Chemother 78(9):2228

Alquraish FA, AlQattan AS, Al-Shammari T (2023) Surgical site invasive aspergillosis in immunocompetent patient secondary to Kramericeae herb: case report. Int J Surg Case Rep 105:108026

Arun CS, Raju P, Lakshmanan V, Kumar A, Bal A, Kumar H (2019) Emergence of fluconazole-resistant Candida infections in diabetic foot ulcers: implications for public health. Ind J Commun Med 44(Suppl 1):S74–S76

Ballus J, Lopez-Delgado JC, Sabater-Riera J, Perez-Fernandez XL, Betbese AJ, Roncal JA (2015) Surgical site infection in critically ill patients with secondary and tertiary peritonitis: epidemiology, microbiology and influence in outcomes. BMC Infect Dis 15:304

Bansal E, Garg A, Bhatia S, Attri AK, Chander J (2008) Spectrum of microbial flora in diabetic foot ulcers. Indian J Pathol Microbiol 51(2):204–208

Barry J, Brown A, Ensor V, Lakhani U, Petts D, Warren C, Winstanley T (2003) Comparative evaluation of the VITEK 2 advanced expert system (AES) in five UK hospitals. J Antimicrob Chemother 51(5):1191

Bartnicka D, Karkowska-Kuleta J, Zawrotniak M, Satała D, Michalik K, Zielinska G, Bochenska O, Kozik A, Ciaston I, Koziel J, Dutton LC, Nobbs AH, Potempa B, Baster Z, Rajfur Z, Potempa J, Rapala-Kozik M, Bartnicka D, Karkowska-Kuleta J, Zawrotniak M, Satała D, Michalik K, Zielinska G, Bochenska O, Kozik A, Ciaston I, Koziel J, Dutton LC, Nobbs AH, Potempa B, Baster Z, Rajfur Z, Potempa J, Rapala-Kozik M (2019) Adhesive protein-mediated cross-talk between Candida albicans and Porphyromonas gingivalis in dual species biofilm protects the anaerobic bacterium in unfavorable oxic environment. Scientific Report 9(1):9

Basavraj Nagoba RG, Wadher B, Rao A, Selkar S (2012) Simple and effective approach for the treatment of traumatic wounds in non-diabetic patients: a prospective open study. Int Wound J 10:585

Bayona JVM, García CS, Palop NT, Martín AV, Padrón CG, Rodríguez JC, Pemán J, Cardona CG (2022) Novel chromogenic medium CHROMagar™ Candida plus for detection of Candida auris and other Candida species from surveillance and environmental samples: A multicenter study. J Fungi 8(3)

Baz A, Bakri A, Butcher M, Short B, Ghimire B, Gaur N, Jenkins T, Short RD, Riggio M, Williams C, Ramage G, Brown JL (2023) Staphylococcus aureus strains exhibit heterogenous tolerance to direct cold atmospheric plasma therapy. Biofilms 5:100123–100123

Besser M, Schaeler L, Plattfaut I, Brill FHH, Kampe A, Geffken M, Smeets R, Debus ES, Stuermer EK (2022) Pulsed low-intensity laser treatment stimulates wound healing without enhancing biofilm development in vitro. J Photochem Photobiol B 233:112504

Bjarnsholt T, Alhede M, Alhede M, Eickhardt-Sørensen SR, Moser C, Kühl M, Jensen PØ, Høiby N (2013) The in vivo biofilm. Trends Microbiol 21(9):466

Bjorn C, Mahlapuu M, Mattsby-Baltzer I, Hakansson J (2016) Anti-infective efficacy of the lactoferrin-derived antimicrobial peptide HLR1r. Peptides 81:21–28

Borman AM, Mohammed S, Palmer MD, Childs N, Johnson EM (2022) The importance of appropriate processing and direct microscopic examination for the timely diagnosis and management of invasive infections caused by filamentous fungi. Med Mycol 60(12)

Bowler PG (2018) Antibiotic resistance and biofilm tolerance: a combined threat in the treatment of chronic infections. J Wound Care 27(5):273–277

Brandenburg KS, Weaver AJ, Karna SLR, You T, Chen P, Stryk SV, Qian L, Pineda U, Abercrombie JJ, Leung KP, Brandenburg KS, Weaver AJ, Karna SLR, You T, Chen P, Stryk SV, Qian L, Pineda U, Abercrombie JJ, Leung KP (2019) Formation of Pseudomonas aeruginosa biofilms in full-thickness scald burn wounds in rats. Scientific Report 9(1):9

Brandenburg KS, Weaver AJ, Karna SLR, Leung KP (2021) The impact of simultaneous inoculation of Pseudomonas aeruginosa, Staphylococcus aureus, and Candida albicans on rodent burn wounds. Burns 47(8):1818

Brauner A, Fridman O, Gefen O, Balaban NQ (2016) Distinguishing between resistance, tolerance and persistence to antibiotic treatment. Nat Rev Microbiol 14(5):320–330

Brown JL, Townsend E, Short RD, Williams C, Woodall C, Nile CJ, Ramage G (2022) Assessing the inflammatory response to in vitro polymicrobial wound biofilms in a skin epidermis model. NPJ Biofilm Microb 8(1):19–19

Burgess M, Valdera F, Varon D, Kankuri E, Nuutila K (2022) The immune and regenerative response to burn injury. Cells 11(19):3073

Bush A (2012) In: Wilmott RW, Boat TF, Bush A et al (eds) Severe asthma. Kendig & Chernick's disorders of the respiratory tract in children, 8th edn. Saunders, Philadelphia, W.B, pp 736–743

Cámara M, Green W, MacPhee CE, Rakowska PD, Raval R, Richardson MC, Slater-Jefferies J, Steventon K, Webb JS (2022) Economic significance of biofilms: a multidisciplinary and cross-sectoral challenge. npj Biofilm Microb 8(1):42

Canadian, Nosocomial, Infection, Surveillance and Program (2023) Device and surgical procedure-related infections in Canadian acute care hospitals, 2017–2021. Can Commun Dis Rep 49(5):221–234

Chai W, Wang Y, Zheng H, Yue S, Liu Y, Wu Y, Li X (2021) The profile of microbiological pathogens in diabetic foot ulcers. Front Med 8:656467–656467

Chellan G, Shivaprakash S, Karimassery Ramaiyar S, Varma AK, Varma N, Thekkeparambil Sukumaran M, Rohinivilasam Vasukutty J, Bal A, Kumar H (2010) Spectrum and prevalence of fungi infecting deep tissues of lower-limb wounds in patients with type 2 diabetes. J Clin Microbiol 48(6):2097–2102

Chen X, Lorenzen J, Xu Y, Jonikaite M, Thaarup IC, Bjarnsholt T, Kirketerp-Møller K, Thomsen TR (2021) A novel chronic wound biofilm model sustaining coexistence of Pseudomonas aeruginosa and Staphylococcus aureus suitable for testing of antibiofilm effect of antimicrobial solutions and wound dressings. Wound Repair Regen 29(5):820

Christiaens G, Hayette MP, Jacquemin D, Melin P, Mutsers J, De Mol P (2005) An outbreak of Absidia corymbifera infection associated with bandage contamination in a burns unit. J Hosp Infect 61(1):88

Coles M, Cox K, Chao A (2020) Candida haemulonii: an emerging opportunistic pathogen in the United States? IDCases 21:e00900

Dalton T, Dowd SE, Wolcott RD, Sun Y, Watters C, Griswold JA, Rumbaugh KP (2011) An in vivo polymicrobial biofilm wound infection model to study interspecies interactions. PLoS One 6(11):e27317 de Figueiredo DSY, de Almeida JN Jr, Motta AL, Silva DMC e, Szeszs MW, Del Negro GMB (2014) Evaluation of VITEK 2 for discriminating Trichosporon species: misidentification of Trichosporon non–T. Asahii. Diagn Microbiol Infect Dis 80(1):59

Delaney C, Alapati S, Alshehri M, Kubalova D, Veena CLR, Abusrewil S, Short B, Bradshaw D, Brown JL (2023) Investigating the role of Candida albicans as a universal substrate for oral bacteria using a transcriptomic approach: implications for interkingdom biofilm control? APMIS 131(11):601–612

Demet Kaya CAA, Yucel M (2007) Fungal agents as a cause of surgical wound Infections_ an overview of host factors. A Compendium of Clinical Research and Practice, Wounds

Deng H, Li B, Shen Q, Zhang C, Kuang L, Chen R, Wang S, Ma Z, Li G (2023) Mechanisms of diabetic foot ulceration: a review. J Diabetes 15(4):299–312

Di Domenico EG, Farulla I, Prignano G, Gallo MT, Vespaziani M, Cavallo I, Sperduti I, Pontone M, Bordignon V, Cilli L, De Santis A, Di Salvo F, Pimpinelli F, Lesnoni La Parola I, Toma L, Ensoli F (2017) Biofilm is a major virulence determinant in bacterial colonization of chronic skin ulcers independently from the multidrug resistant phenotype. Int J Mol Sci 18(5)

Dowd SE, Delton Hanson J, Rees E, Wolcott RD, Zischau AM, Sun Y, White J, Smith DM, Kennedy J, Jones CE (2011) Survey of fungi and yeast in polymicrobial infections in chronic wounds. J Wound Care 20(1):40–47

Dydak K, Junka A, Dydak A, Brozyna M, Paleczny J, Fijalkowski K, Kubielas G, Aniolek O, Bartoszewicz M (2021) In vitro efficacy of bacterial cellulose dressings chemisorbed with antiseptics against biofilm formed by pathogens isolated from chronic wounds. Int J Mol Sci 22(8)

Edmonds M, Manu C, Vas P (2021) The current burden of diabetic foot disease. J Clin Orthop Trauma 17:88–93

Eichelberger KR, Cassat JE (2021) Frontiers | metabolic adaptations during Staphylococcus aureus and Candida albicans co-Infection. Front Immunol 12

Alam F, Catlow D, Alessandro Di Maio, Blair JMA, Hall RA (2020) Candida albicans enhances meropenem tolerance of Pseudomonas aeruginosa in a dual-species biofilm, Journal of Antimicrobial Chemotherapy, Vol 75(4): 925–935. https://doi.org/10.1093/jac/dkz514

Fang W, Wu J, Cheng M, Zhu X, Du M, Chen C, Liao W, Zhi K, Pan W, Fang W, Wu J, Cheng M, Zhu X, Du M, Chen C, Liao W, Zhi K, Pan W (2023) Diagnosis of invasive fungal infections: challenges and recent developments. J Biomed Sci 30(1):30

Fata S, Saeid MMH, Feyzi R, Najafzadeh MJ, Afzal AM, Ghasemi M, Mohammadian M, Naseri A, Meshkat M, Fata AAM (2011) Mycotic infections in diabetic foot ulcers in Emam Reza hospital. Mashhad, 2006–2008

Fourie R, Ells R, Swart CW, Sebolai OM, Albertyn J, Pohl CH (2016) Frontiers | Candida albicans and Pseudomonas aeruginosa interaction, with focus on the role of eicosanoids. Front Physiol 7

Fourie R, Cason ED, Albertyn J, Pohl CH (2021) Transcriptional response of Candida albicans to Pseudomonas aeruginosa in a polymicrobial biofilm. G3: Genes|Genomes|Genetics 11(4)

Fox EP, Cowley ES, Nobile CJ, Hartooni N, Newman DK, Johnson AD (2014) Anaerobic bacteria grow within Candida albicans biofilms and induce biofilm formation in suspension cultures. Curr Biol 24(20):2411

Ganesan A, Shaikh F, Bradley W, Blyth DM, Bennett D, Petfield JL, Carson ML, Wells JM, Tribble DR, G. Infectious Disease Clinical Research Program Trauma Infectious Disease Outcomes Study (2019a) Classification of trauma-associated invasive fungal infections to support wound treatment decisions. Emerg Infect Dis 25(9):1639–1647

Ganesan A, Wells J, Shaikh F, Peterson P, Bradley W, Carson ML, Petfield JL, Klassen-Fischer M, Akers KS, Downing K, Bialek R, Tribble DR, Wickes BL (2019b) Molecular detection of filamentous fungi in formalin-fixed paraffin-embedded specimens in invasive fungal wound infections is feasible with high specificity. J Clin Microbiol 58(1)

Gholamali Riahi EK, Hosseini SS (2023) Fungal and bacterial co-infection in the superficial and deep sternal wound after open cardiac surgery. Iran J Microbiol

Gil J, Solis M, Higa A, Davis SC, Gil J, Solis M, Higa A, Davis SC (2022) Candida albicans Infections: a novel porcine wound model to evaluate treatment efficacy. BMC Microbiol 22(1):22

Gitau AM, Ng'ang'a ZW, Sigilai W, Bii C, Mwangi M (2011) Fungal infections among diabetic foot ulcer-patients attending diabetic clinic in Kenyatta National Hospital, Kenya. East Afr Med J 88:9

Goncalves J, Guimaraes AR, Ferreira HU, Ribeiro S, Moreno T, Borges-Canha M, Meira I, Menino J, Silva F, Pedro J, Neves N, Simao RS, Santos L, Queiros J, de Grupo Pe DC (2024) Microbiological characterization of neuropathic diabetic foot infection: a retrospective study at a Portuguese tertiary hospital. BMC Infect Dis 24(1):791

Gonzalez MR, Fleuchot B, Lauciello L, Jafari P, Applegate LA, Raffoul W, Que Y-A, Perron K (2016) Effect of human burn wound exudate on Pseudomonas aeruginosa virulence. mSphere 1(2)

Graves N, Phillips CJ, Harding K (2022) A narrative review of the epidemiology and economics of chronic wounds. Br J Dermatol 187(2):141–148

Guest JF, Fuller GW, Vowden P (2018) Diabetic foot ulcer management in clinical practice in the UK: costs and outcomes. Int Wound J 15(1):43–52

Guest JF, Fuller GW, Edwards J (2020a) Cohort study evaluating management of burns in the community in clinical practice in the UK: costs and outcomes. BMJ Open 10(4):e035345

Guest JF, Fuller GW, Vowden P (2020b) Cohort study evaluating the burden of wounds to the UK's National Health Service in 2017/2018: update from 2012/2013. BMJ Open 10(12):e045253

Guinea J, Escribano P, Vena A, Munoz P, Martinez-Jimenez MDC, Padilla B, Bouza E (2017) Increasing incidence of mucormycosis in a large Spanish hospital from 2007 to 2015: epidemiology and microbiological characterization of the isolates. PLoS One 12(6):e0179136

Gupta N, Haque A, Mukhopadhyay G, Narayan RP, Prasad R (2005) Interactions between bacteria and Candida in the burn wound. Burns 31(3):375–378

Gupta AK, Parwal C, Mangal M, Gambhir SS, Nanda BS, Sarangi K (2018) Treating mucormycosis using a multimodality approach: a case series. J Wound Care 27:735

Gyamfi E, Narh CA, Quaye C, Abbass A, Dzudzor B, Mosi L (2021) Microbiology of secondary infections in Buruli ulcer lesions; implications for therapeutic interventions. BMC Microbiol 21(1):4

Hanghui Cen ZW, Wang F, Han C (2015) Pathogen distribution and drug resistance in a burn ward: a three-year retrospective analysis of a single center in China. Int J Clin Exp Med 8(10)

Haque F, Alfatah M, Ganesan K, Bhattacharyya MS (2016) Inhibitory effect of Sophorolipid on Candida albicans biofilm formation and hyphal growth. Sci Rep 6:23575

Harriott MM, Noverr MC (2009) Candida albicans and Staphylococcus aureus form Polymicrobial biofilms: effects on antimicrobial resistance. Antimicrob Agents Chemother 53(9):3914

Hassan MA, Tamer TM, Rageh AA, Abou-Zeid AM, Abd El-Zaher EHF, Kenawy E-R (2019) Insight into multidrug-resistant microorganisms from microbial infected diabetic foot ulcers. Diab Metab Syndr 13(2):1261–1270

Hattab S, Dagher A-M, Wheeler RT (2022) Pseudomonas synergizes with fluconazole against Candida during treatment of polymicrobial infection. Infect Immun 90(4):e0062621

Heald AH, O'Halloran DJ, Richards K, Webb F, Jenkins S, Hollis S, Denning DW, Young RJ (2001) Fungal infection of the diabetic foot: two distinct syndromes. Diab Med 18(7):567–572

Horiuchi H, Watanabe A, Yaguchi T, Ban S, Otsuka T, Miyazaki H, Furuya R (2024) Superficial abdominal surgical site infection caused by Aspergillus welwitschiae: a case report. BMC Infect Dis 24(1):990

Horvath EE, Murray CK, Vaughan GM, Chung KK, Hospenthal DR, Wade CE, Holcomb JB, Wolf SE, Mason AD Jr, Cancio LC (2007) Fungal wound infection (not colonization) is independently associated with mortality in burn patients. Ann Surg 245(6):978–985

Jabeen K, Khan M, Umar S, Shaheen N, Farooqi J (2021) Spectrum of fungal pathogens in burn wound specimens: data from a tertiary Care Hospital Laboratory in Pakistan. J Burn Care Res 42(2):241–244

Jefferson Lessa Soares de Macedo JBS (2005) Bacterial and fungal colonization of burn wounds. Mem Inst Oswaldo Cruz

Jenks JD, White PL, Kidd SE, Goshia T, Fraley SI, Hoenigl M, GR. T. III (2023) An update on current and novel molecular diagnostics for the diagnosis of invasive fungal infections. Expert Rev Mol Diagn 23(12):1135

Jin W, Liu Y, Ning Q, Wu S, Su S, Zheng D, Ma S, Zou J, Yang M, Hu D, Ding H (2024) A case of chronic wounds caused by Sporothrix schenckii infection was rapidly detected by metagenomic next generation sequencing. Heliyon 10(2):e24420

João AL, Lencastre A, Dutra E, Pessoa T, Costa E, Formiga A, Neves J (2021) Fusarium spp.-an emerging pathogen in chronic diabetic ulcer: case report and review of the literature. Int J Low Extrem Wounds 20(1):67–72

Jones ML, Ganopolsky JG, Labbe A, Prakash S (2010) A novel nitric oxide producing probiotic patch and its antimicrobial efficacy: preparation and in vitro analysis. Appl Microbiol Biotechnol 87(2):509–516

Kalan LR, Brennan MB (2019) The role of the microbiome in nonhealing diabetic wounds. Ann N Y Acad Sci 1435(1):79–92

Kalan L, Loesche M, Hodkinson BP, Heilmann K, Ruthel G, Gardner SE, Grice EA (2016) Redefining the chronic-wound microbiome: fungal communities are prevalent, dynamic, and associated with delayed healing. mBio 7(5)

Kalshetti VT, Wadile R, Bothikar S, Ambade V, Bhate V (2017) Study of fungal infections in diabetic foot ulcer. Ind J Microbiol Res 4(1):87–89

Kamath PPS, Valdes J, Gil J, Solis M, Higa A, Davis SC (2024) In vitro analysis of interactions between Pseudomonas aeruginosa and Candida albicans treated with silver sulfadiazine in wound infections—PubMed. JAC-Antimicrob Resist 6(3)

Kandregula S, Behura A, Behera CR, Pattnaik D, Mishra A, Panda B, Mohanty S (2022) A clinical significance of fungal infections in diabetic foot ulcers. Cureus 14(7):e26872–e26872

Karnik T, Dempsey SG, Jerram MJ, Nagarajan A, Rajam R, May BCH, Miller CH (2019) Ionic silver functionalized ovine forestomach matrix—a non-cytotoxic antimicrobial biomaterial for tissue regeneration applications. Biomater Res 23:6

Kaur R, Bala K, Ahuja RB, Srivastav P, Bansal U (2014) Primary cutaneous mucormycosis in a patient with burn wounds due to Lichtheimia ramosa. Mycopathologia 178(3–4):291–295

Kean R, Rajendran R, Haggarty J, Townsend EM, Short B, Burgess KE, Lang S, Millington O, Mackay WG, Williams C, Ramage G (2017) Frontiers | Candida albicans Mycofilms support Staphylococcus aureus colonization and enhances Miconazole resistance in dual-species interactions. Front Microbiol 8

Knoll MA, Steixner S, Lass-Flörl C (2023) How to use direct microscopy for diagnosing fungal infections. Clin Microbiol Infect 29(8):1031

Kostoulias X, Murray GL, Cerqueira GM, Kong JB, Bantun F, Mylonakis E, Khoo CA, Peleg AY (2015) Impact of a cross-kingdom signaling molecule of Candida albicans on Acinetobacter baumannii physiology. Antimicrob Agents Chemother 60(1):161

Krishnen R, Muniandy S (2023) Fungicidal effectiveness of super-oxidized solution in vitro: potential implications for wound care. Wounds 35(8):E243–E247

Krumkamp R, Oppong K, Hogan B, Strauss R, Frickmann H, Wiafe-Akenten C, Boahen KG, Rickerts V, McCormick Smith I, Gross U, Schulze M, Jaeger A, Loderstadt U, Sarpong N, Owusu-Dabo E, May J, Dekker D (2020) Spectrum of antibiotic resistant bacteria and fungi isolated from chronically infected wounds in a rural district hospital in Ghana. PLoS One 15(8):e0237263

Kumar D, Banerjee T, Chakravarty J, Singh SK, Dwivedi A, Tilak R (2016) Identification, antifungal resistance profile, in vitro biofilm formation and ultrastructural characteristics of Candida species isolated from diabetic foot patients in northern India. Indian J Med Microbiol 34(3):308–314

Kyriopoulos EJ, Kyriakopoulos A, Karonidis A, Gravvanis A, Gamatsi I, Tsironis C, Tsoutsos D (2015) Burn injuries and soft tissue traumas complicated by mucormycosis infection: a report of six cases and review of the literature. Ann Burn Fire Disaster 28(4):280–287

Laliany G, Jamehdar SA, Makhdoumi A (2022) In vitro virulence activity of Pseudomonas aeruginosa, enhanced by either Acinetobacter baumannii or Enterococcus faecium through the polymicrobial interactions. Arch Microbiol 204(12):709

Leake JL, S. E. D, Wolcott RD, Zischkau AM (2009) Identification of yeast in chronic wounds using new pathogen detection technologies. J Wound Care 18:103

Lee WJ, Park KH, Jang YH, Lee SJ, Kim DW (2014) Isolation of Trichophyton rubrum from diabetic foot ulcers of 4 patients. Korean J Med Mycol 19(1):9–12

Loïez C, Wallet F, Sendid B, Courcol RJ (2006) Evaluation of VITEK 2 colorimetric cards versus fluorimetric cards for identification of yeasts. Diagn Microbiol Infect Dis 56(4):455

Majid Foroozan NC-A, Granel-Brocard F, Barbaud A, Schmutz JL (2011) Prevalence analysis of fungi in chronic lower extremity ulcers. Wounds 23(3)

Malik Asif H, Irani Udeshika R, Flavia H (2016) The importance of anaerobic bacteria in nonhealing wounds. Wound Practice Res 24(4)

Malone M, Bjarnsholt T, McBain AJ, James GA, Stoodley P, Leaper D, Tachi M, Schultz G, Swanson T, Wolcott RD (2017) The prevalence of biofilms in chronic wounds: a systematic review and meta-analysis of published data. J Wound Care 26(1):20–25

McGraw C, Carrick M, Ekengren F, Berg G, Lieser M, Orlando A, Madayag R, Tanner Ii A, Kelly M, Banton K, Bar-Or D (2019) Severe fungal infections following blunt traumatic injuries: a 5-year multicenter descriptive study. Injury 50(12):2234–2239

McNeil JC, Vallejo JG, Hulten KG, Kaplan SL (2018) Osteoarticular infections following open or penetrating trauma in children in the post-community-acquired methicillin-resistant Staphylococcus aureus era: the impact of Enterobacter cloacae. Pediatr Infect Dis J 37(12):1204–1210

Mehra BK, Singh AK, Gupta D, Narang R, Patil R (2017) A Clinicomicrobiological study on incidence of Mycotic infections in diabetic foot ulcers. Int J Scientific Study 4(12):50–54

Melhem M, Bertoletti A, Lucca H, Silva R, Meneghin F, Szeszs M (2014) Use of the VITEK 2 system to identify and test the antifungal susceptibility of clinically relevant yeast species. Braz J Microbiol 44(4):1257

Mende K, Stewart L, Shaikh F, Bradley W, Lu D, Krauss MR, Greenberg L, Yu Q, Blyth DM, Whitman TJ, Petfield JL, Tribble DR (2019) Microbiology of combat-related extremity wounds: trauma infectious disease outcomes study. Diagn Microbiol Infect Dis 94(2):173–179

Mendonça A, Santos H, Franco-Duarte R, Sampaio P (2022) Fungal infections diagnosis—past, present and future. Res Microbiol 173(3):103915

Mikamo H, Kawazoe K, Izumi K, Watanabe K, Ueno K, Tamaya T (1998) Studies on the pathogenicity of anaerobes, especially Prevotella bivia, in a rat Pyometra model. Infect Dis Obstet Gynecol 6(2):61

Monteiro AY, Min AW, See JS, Pan JY (2024) Cutaneous mucormycosis infection owing to Rhizomucor variabilis presenting as recurrent lower limb ulceration and cellulitis in a diabetic patient. Dermatol Online J 30(2)

Morton LM, Phillips TJ (2016) Wound healing and treating wounds: differential diagnosis and evaluation of chronic wounds. J Am Acad Dermatol 74(4):589–605. quiz 605–586

Mousa HA, Al-Bader SM (2001) Yeast infection of burns. Mycoses 44(5):147–149

Musyoki VM, Mutai W, Ngugi N, Otieno F, Masika MM (2022) Speciation and antifungal susceptibility of Candida isolates from diabetic foot ulcer patients in a tertiary hospital in Kenya. Pan Afr Med J 41:34–34

Nair S, Peter S, Sasidharan A, Sistla S, Unni AKK (2006) Incidence of mycotic infections in diabetic foot tissue. J Cult Collect 5(1):85–89

Neblett Fanfair R, Benedict K, Bos J, Bennett SD, Lo YC, Adebanjo T, Etienne K, Deak E, Derado G, Shieh WJ, Drew C, Zaki S, Sugerman D, Gade L, Thompson EH, Sutton DA, Engelthaler DM, Schupp JM, Brandt ME, Harris JR, Lockhart SR, Turabelidze G, Park BJ (2012)

Necrotizing cutaneous mucormycosis after a tornado in Joplin, Missouri, in 2011. N Engl J Med 367(23):2214–2225

Nussbaum SR, Carter MJ, Fife CE, DaVanzo J, Haught R, Nusgart M, Cartwright D (2018) An economic evaluation of the impact, cost, and Medicare policy implications of chronic nonhealing wounds. Value Health 21(1):27–32

Oates A, Bowling FL, Boulton AJ, McBain AJ (2012) Molecular and culture-based assessment of the microbial diversity of diabetic chronic foot wounds and contralateral skin sites. J Clin Microbiol 50(7):2263–2271

Oh J, Byrd AL, Deming C, Conlan S, Kong HH, Segre JA, Oh J, Byrd AL, Deming C, Conlan S, Kong HH, Segre JA (2014) Biogeography and individuality shape function in the human skin metagenome. Nature 514:7520

Okuno E, Jarros IC, Bonfim-Mendonca PS, Vicente de Rezende G, Negri M, Svidzinski TE (2018) Candida parapsilosis isolates from burn wounds can penetrate an acellular dermal matrix. Microb Pathog 118:330–335

Öztürk AM, Taşbakan M, Metin DY, Yener C, Uysal S, Yıldırım Şımşır I, Ertam İ, Pullukçu H, Arda B, Çetinkalp Ş (2019) A neglected causative agent in diabetic foot infection: a retrospective evaluation of 13 patients with fungal etiology. Turkish J Med Sci 49(1):81–86

Paleczny J, Junka A, Brożyna M, Dydak K, Oleksy-Wawrzyniak M, Ciecholewska-Juśko D, Dziedzic E, Bartoszewicz M, Paleczny J, Junka A, Brożyna M, Dydak K, Oleksy-Wawrzyniak M, Ciecholewska-Juśko D, Dziedzic E, Bartoszewicz M (2021) The high impact of Staphylococcus aureus biofilm culture medium on in vitro outcomes of antimicrobial activity of wound antiseptics and antibiotic. Pathogens 10:1385

Pantet O, Laurent A, Abdel-Sayed P, Hirt-Burri N, Michetti M, Raffoul W, Monod M, Applegate LA, de Buys Roessingh AS (2021) Effective management of topical nosocomial Aspergillus spp. infections in three severely burned patients. Ann Burns Fire Disast 34:235

Patenall BL, Ridgley JD, Jenkins ATA, Young AE (2023) Evidence of bacterial biofilms within acute wounds: a systematic review. J Wound Care 32(5):273

Paul S, Todd OA, Eichelberger KR, Tkaczyk C, Sellman BR, Noverr MC, Cassat JE, Fidel PL, Peters BM, Paul S, Todd OA, Eichelberger KR, Tkaczyk C, Sellman BR, Noverr MC, Cassat JE, Fidel PL, Peters BM (2024) A fungal metabolic regulator underlies infectious synergism during Candida albicans-Staphylococcus aureus intra-abdominal co-infection. Nat Commun 15(1)

Peleg AY, Tampakakis E, Fuchs BB, Eliopoulos GM, Robert J, Moellering C, Mylonakis E, Peleg AY, Tampakakis E, Fuchs BB, Eliopoulos GM, Moellering RC, Mylonakis E (2008) Prokaryote–eukaryote interactions identified by using Caenorhabditis elegans. Proc Natl Acad Sci 105(38):14585

Peng K, L. M., Himawan A, Domínguez-Robles J, Vora LK, Duncan R, Dai X, Zhang C, Zhao L, Li L, Larrañeta E, Donnelly RF (2022) Amphotericin B- and levofloxacin-loaded chitosan films for potential use in antimicrobial wound dressings: analytical method development and its application—PubMed. Pharmaceutics 14(11)

Percival SL, McCarty SM, Benjamin L (2015) Biofilms and wounds: an overview of the evidence. Adv Wound Care 4(7):431

Piednoir E, Robert-Yap J, Baillet P, Lermite E, Christou N (2021) The socioeconomic impact of surgical site infections. Front Public Health 9:712461

Radowsky JS, Brown TS, Lisboa FA, Rodriguez CJ, Forsberg JA, Elster EA (2015) Serum inflammatory cytokine markers of invasive fungal Infection in previously immunocompetent Battle casualties. Surg Infect 16(5):526–532

Raza M, Anurshetru BS (2017) Clinical study of coexistence of fungal infections in diabetic foot ulcers and its management. Int Surg J 4(12):3943–3950

Relloso S, Romano V, Landaburu MF, Herrera F, Smayevsky J, Vecino C, Mujica MT (2014) Saksenaea erythrospora infection following a serious sailing accident. J Med Microbiol 63(Pt 2):317–321

Rodrigues CF, Kaushik KS, Light C (2021) Biofilms in wounds: new advances in therapy and in healing management. Biomedicines 9(2)

Rodriguez RCJ, Ganesan A, Shaikh F, Carson ML, Bradley W, Warkentien TE, Tribble DR (2022) Combat-related invasive fungal wound infections. Mil Med 187(Suppl 2):34–41

Roy S, Santra S, Das A, Dixith S, Sinha M, Ghatak S, Ghosh N, Banerjee P, Khanna S, Mathew-Steiner S, Ghatak PD, Blackstone BN, Powell HM, Bergdall VK, Wozniak DJ, Sen CK (2020) Staphylococcus aureus biofilm Infection compromises wound healing by causing deficiencies in granulation tissue collagen. Ann Surg 271(6):1174

Royle R, Gillespie BM, Chaboyer W, Byrnes J, Nghiem S (2023) The burden of surgical site infections in Australia: A cost-of-illness study. J Infect Public Health 16(5):792–798

Ruamsap N, Thomas CS, Imerbsin R, Reed MC, Gonwong S, Lurchachaiwong W, Islam D, Wojnarski M, Vesely BA, Lugo-Roman LA, Waters NC, Zurawski DV, Demons ST (2023) Chronic wound Infection model of Acinetobacter baumannii in outbred mice. Mil Med 188(7–8):e1708

Sancineto L, Piccioni M, De Marco S, Pagiotti R, Nascimento V, Braga AL, Santi C, Pietrella D (2016) Diphenyl diselenide derivatives inhibit microbial biofilm formation involved in wound infection. BMC Microbiol 16(1):220

Sarabahi S, Tiwari VK, Arora S, Capoor MR, Pandey A (2012) Changing pattern of fungal infection in burn patients. Burns 38(4):520–528

Sauer K, Stoodley P, Goeres DM, Hall-Stoodley L, Burmølle M, Stewart PS, Bjarnsholt T (2022) The biofilm life cycle: expanding the conceptual model of biofilm formation. Nat Rev Microbiol 20(10):608–620

Schaal JV, Leclerc T, Soler C, Donat N, Cirrode A, Jault P, Bargues L (2015) Epidemiology of filamentous fungal infections in burned patients: A French retrospective study. Burns 41(4):853–863

Schaal JV, Donat N, Ragot C, Soler C, Leclerc T (2020) Carbapenem-resistant Acinetobacter baumannii in military burn Centre. Burns 46(3):747–748

Sharma S, Bajaj D, Sharma P (2016) Fungal infection in thermal burns: a prospective study in a tertiary care centre. J Clin Diagn Res 10(9):PC05–PC07

Shields RK, Clancy CJ, Minces LR, Shigemura N, Kwak EJ, Silveira FP, Abdel-Massih RC, Bhama JK, Bermudez CA, Pilewski JM, Crespo M, Toyoda Y, Nguyen MH (2013) Epidemiology and outcomes of deep surgical site infections following lung transplantation. Am J Transplant 13(8):2137–2145

Shin SH, Woo SS, Lee JH, Kim HJ, Kim SH, Kim JH, Suh IS (2023) Analysis and management of pathogens isolated from patients with complicated facial lacerations and abrasions. Int Wound J 20(1):85–91

Short B, Bakri A, Baz A, Williams C, Brown J, Ramage G (2023a) There is more to wounds than bacteria: fungal biofilms in chronic wounds. Curr Clin Microbiol Rep 10(1):9–16

Short B, Bakri A, Baz A, Williams C, Brown J, Ramage G, Short B, Bakri A, Baz A, Williams C, Brown J, Ramage G (2023b) There is more to wounds than bacteria: fungal biofilms in chronic wounds. Curr Clinic Microb Report 10(1):10

Sikwewa K, Simusika P, Mumbula M, Mwenya DM, Mandona C, Mulundu G (2023) The occurrence of fungi from burn wound patients and antifungal susceptibility patterns: a cross-sectional study in Lusaka, Zambia. Afr Health Sci 23(3):506

Smith K, Collier A, Townsend EM, O'Donnell LE, Bal AM, Butcher J, Mackay WG, Ramage G, Williams C, Smith K, Collier A, Townsend EM, O'Donnell LE, Bal AM, Butcher J, Mackay WG, Ramage G, Williams C (2016) One step closer to understanding the role of bacteria in diabetic foot ulcers: characterising the microbiome of ulcers. BMC Microbiol 16(1):16

Stefano Veraldi AMT, Lunardon L, Persico MC, Francia C (2008) Mycologic evaluations in chronic leg ulcers. Wounds 20(9):250

Stephens P, Wall IB, Wilson MJ, Hill KE, Davies CE, Hill CM, Harding KG, Thomas DW (2003) Anaerobic cocci populating the deep tissues of chronic wounds impair cellular wound healing responses in vitro. Br J Dermatol 148(3):456

Stewart PS (2015) Antimicrobial tolerance in biofilms. Microbiol Spectr 3(3)

Stichternoth C, Ernst JF (2009) Hypoxic adaptation by Efg1 regulates biofilm formation by Candida albicans. Appl Environ Microbiol 75(11):3663

Sujatha R, Chauhan P, Sameer D (2018) A study of fungal infections of diabetic foot in tertiary care hospital of UP. Journal of Medical Science And clinical Research

Sun Y, Dowd SE, Smith E, Rhoads DD, Wolcott RD (2008) In vitro multispecies Lubbock chronic wound biofilm model. Wound Repair Regen 16(6):805–813

Sussman G (2023) An update on wound management. Aust Prescr 46(2):29–35

Terrero-Salcedo D, Powers-Fletcher MV (2020) Updates in laboratory diagnostics for invasive fungal infections. J Clin Microbiol 58(6)

Townsend EM, Sherry L, Rajendran R, Hansom D, Butcher J, Mackay WG, Williams C, Ramage G (2016) Development and characterisation of a novel three-dimensional inter-kingdom wound biofilm model. Biofouling 32(10):1259–1270

Townsend EM, Sherry L, Kean R, Hansom D, Mackay WG, Williams C, Butcher J, Ramage G (2017) Implications of antimicrobial combinations in complex wound biofilms containing fungi. Antimicrob Agents Chemother 61(9)

Tribble DR, Rodriguez CJ, Weintrob AC, Shaikh F, Aggarwal D, Carson ML, Murray CK, Masuoka P, G. Infectious Disease Clinical Research Program Trauma Infectious Disease Outcomes Study (2015) Environmental factors related to fungal wound contamination after combat trauma in Afghanistan, 2009-2011. Emerg Infect Dis 21(10):1759–1769

Trøstrup H, Lerche CJ, Christophersen LJ, Thomsen K, Jensen PØ, Hougen HP, Høiby N, Moser C (2017) Chronic Pseudomonas aeruginosa biofilm infection impairs murine S100A8/A9 and neutrophil effector cytokines—implications for delayed wound closure? Pathog Disease 75(7)

Tyler E, Warkentien FS, Weintrob AC, Rodriguez CJ, Murray CK, Lloyd BA, Ganesan DAA, Carson ML, Tribble DR, on behalf of the Infectious Disease Clinical Research Program Trauma Infectious Disease Outcomes Study Group (2015) Impact of mucorales and other invasive molds on clinical outcomes of Polymicrobial traumatic wound infections. J Clin Microbiol 53:2262

Uberoi A, McCready-Vangi A, Grice EA (2024) The wound microbiota: microbial mechanisms of impaired wound healing and infection. Nat Rev Microbiol 22:507

Uppuluri P, Lin L, Alqarihi A, Luo G, Youssef EG, Alkhazraji S, Yount NY, Ibrahim BA, Bolaris MA, Edwards JE Jr, Swidergall M, Filler SG, Yeaman MR, Ibrahim AS (2018) The Hyr1 protein from the fungus Candida albicans is a cross kingdom immunotherapeutic target for Acinetobacter bacterial infection. PLoS Pathog 14(5):e1007056

Urwin S, Dumville JC, Sutton M, Cullum N (2022) Health service costs of treating venous leg ulcers in the UK: evidence from a cross-sectional survey based in the north west of England. BMJ Open 12(1):e056790

Ušjak D, Novović K, Ivković B, Tomić B, Đorđević V, Milenković MT (2023) Targeting outer membrane protein A (OmpA)—inhibitory effect of 2'-hydroxychalcone derivatives on Acinetobacter baumannii and Candida albicans dual-species biofilm formation. Biofouling 39(3):316

Verbanic S, Shen Y, Lee J, Deacon JM, Chen IA, Verbanic S, Shen Y, Lee J, Deacon JM, Chen IA (2020) Microbial predictors of healing and short-term effect of debridement on the microbiome of chronic wounds. npj Biofilms Microb 6(1):6

Victor Moses Musyoki WM, Ngugi N, Otieno F, Masika MM (2022) Speciation and antifungal susceptibility of Candida isolates from diabetic foot ulcer patients in a tertiary hospital in Kenya. Pan Afr Med J 41:34

Vijaya D, Nagaratnamma T, Sathish J (2013) Diabetic foot ulcer due to scedosporium apiospermum. J Clin Diagn Res 7(11):2579–2580

Wallum TE, Yun HC, Rini EA, Carter K, Guymon CH, Akers KS, Tyner SD, White CE, Murray CK (2015) Pathogens present in acute mangled extremities from Afghanistan and subsequent pathogen recovery. Mil Med 180(1):97–103

Wang Y, Beekman J, Hew J, Jackson S, Issler-Fisher AC, Parungao R, Lajevardi SS, Li Z, Maitz PKM (2018) Burn injury: challenges and advances in burn wound healing, infection, pain and scarring. Adv Drug Deliv Rev 123:3–17

Weiss ZF, Leon A, Koo S, Freeman Weiss Z, Leon A, Koo S (2021) The Evolving Landscape of Fungal Diagnostics, Current and Emerging Microbiological Approaches. J Fungi 7:127

Weon Ju Lee KHP, Jang YH, Lee S-J, Kim DW (2014) Isolation of Trichophyton rubrum from diabetic foot ulcers of 4 patients. Korean J Med Mycol 19:9

Wickes BL, Wiederhold NP (2018) Molecular diagnostics in medical mycology. Nat Commun 9:1–9

Wilkinson HN, Hardman MJ (2020) Wound healing: cellular mechanisms and pathological outcomes. Open Biol 10(9):200223

Wolcott R (2015) Economic aspects of biofilm-based wound care in diabetic foot ulcers. J Wound Care 24(5):189–194

Wolcott RD, Hanson JD, Rees EJ, Koenig LD, Phillips CD, Wolcott RA, Cox SB, White JS (2016) Analysis of the chronic wound microbiota of 2,963 patients by 16S rDNA pyrosequencing. Wound Repair Regen 24(1):163

Wu M, Pan H, Leng W, Lei X, Chen L, Liang Z (2018) Distribution of microbes and drug susceptibility in patients with diabetic foot infections in Southwest China. J Diabetes Res 2018:9817308–9817308

Yas KhudhairAl-Ameri A, Alrufae MM (2022) Phenotypic and genotypic identification of yeast species isolated from diabetic foot patients in Al-Najaf Province, Iraq. Arch Razi Inst 77(2):727–731

Chapter 7
Candidozyma auris: Implications for Infection Prevention and Control in the Management of This Resilient Biofilm Former

Gordon Ramage, Ryan Kean, William G. MacKay, Alessandra Vaccaro, Jose L. Lopez-Ribot, and Mireille H. Howe

Abstract *Candidozyma auris* (formerly *Candida auris*) is now recognized by the World Health Organization (WHO) as a critical priority pathogen. Not only does it exhibit multidrug resistant properties, but also has the intrinsic capacity to form adherent biofilm communities in the hospital setting that impact and negatively influence clinical management. Its biological properties are primed to maintain its population within harsh environmental niches, which has supported its successful emergence on the global stage. Clinically, these behaviors and its antifungal tolerant phenotype have created an infection prevention and control problem. This chapter review will explain its clinical importance, its importance as a transmissible pathogen, and its key biological features that support biofilm growth. We will also consider why antifungal agents fail, and offer insights into how to optimally prevent and/or control these biofilms within healthcare facilities.

G. Ramage (✉) · R. Kean
Safeguarding Health through Infection Prevention (SHIP) Research Group, Research Centre for Health, School of Health and Life Sciences, Glasgow Caledonian University, Glasgow, UK
e-mail: gordon.ramage@gcu.ac.uk; ryan.kean@gcu.ac.uk

W. G. MacKay
Centre for Infection and Microbiology, School of Health and Life Sciences, University of the West of Scotland, High Blantyre, UK
e-mail: W.Mackay@uws.ac.uk

A. Vaccaro · J. L. Lopez-Ribot
Department of Molecular Microbiology & Immunology and South Texas Center for Emerging Infectious Diseases, The University of Texas San Antonio, San Antonio, TX, USA
e-mail: alessandra.vaccaro@utsa.edu; jose.lopezribot@utsa.edu

M. H. Howe
Antimicrobial Resistance and Healthcare Associated Infection (ARHAI) Scotland, NHS National Services Scotland, Glasgow, UK
e-mail: mireille.howe@nhs.scot

G. Ramage, D. Andes (eds.), *Fungal Biofilms*, Springer Series on Biofilms 15,
https://doi.org/10.1007/978-3-031-93315-8_7

Introduction

Candidozyma auris (formerly *Candida auris*) is an emerging fungal pathogen with an increased ability to cause hospital outbreaks and elevated mortality (Egger et al. 2022). In 2019 the Center for Disease Control (CDC) ranked this nosocomial pathogen as an "urgent threat" due to its healthcare outbreak pervasiveness (CDC 2019). Its ability to form biofilms on medical equipment and surfaces (e.g., catheters, hospital beds, etc.) and multidrug resistance profiles has made this fungal pathogen a challenging prospect for infection prevention and control (Chowdhary et al. 2023; Du et al. 2020; Kean et al. 2020; Vazquez-Munoz et al. 2020). In 2022 the WHO published a fungal priority list, where it highlighted *C. auris* as a critical priority pathogen, emphasizing the need for better surveillance and treatment options against this difficult to treat pathogen (Fisher and Denning 2023; Parums 2022). Its simultaneous and unprecedented emergence across the globe in six different clades (at the time of writing) is mysterious, though informed speculation has suggested a link to climate change and misuse of agricultural fungicides (Casadevall et al. 2021). As of December 2023, it has been reported in 61 countries in 6 different continents (Jones et al. 2024), and this is likely to continue to rise.

C. auris possesses the 'super-bug' like traits typically associated with common hospital acquired infections, such as MRSA, in that it can often be multidrug resistance (MDR) and survive and persist in the nosocomial environment. Crude mortality rates are variable, but have been reported to be as high as 66% (de Jong and Hagen 2019). The principal factor that makes this organism so enigmatic is its intrinsic resistance to conventional front-line antifungal agents, as well as tolerance to antiseptics and disinfectants. Its resilience and adaptability within a variety of clinical environments has afforded it with the opportunity to emerge and cause alarm among medical professions worldwide. This chapter will specifically focus on exploring the biofilm-related factors driving its tolerance to clinical management in the broadest sense that we have uncovered since its discovery in 2009.

Transmission Considerations

C. auris biofilm formation supports its resilience and persistence in the clinical environment. It has been readily found on fingertips, palms and nares, and in rooms before and after disinfection (Sansom et al. 2024). The first outbreak in a European hospital was reported in the UK that involved 50 patients (Schelenz et al. 2016). Here, environmental sampling revealed persistent presence of *C. auris* in between bed spaces in a cardio thoracic unit in London (UK). Strict infection control procedures were implemented, which included chlorhexidine skin decontamination and environmental cleaning with chlorine-based disinfectants and hydrogen peroxide vapor. Indeed, recommended interventions for this pathogen include screening, standard infection control precautions (SICPs), inpatient transfer, outbreak management and treatment (Kenters et al. 2019). Despite the emphasis on decontamination of equipment and the environment in the available guidance, issues can still occur. In a Neurosciences unit in Oxford (UK), it was identified that an outbreak was linked to reusable axillary temperature probes, despite a bundle of infection-control interventions being implemented (Eyre et al. 2018). These data further support the notion that biofilm growth is an important factor in its success as a hospital pathogen. Key risk factors for predisposing to this organism include prolonged stay in the ICU, mechanical ventilation, presence of a central-venous catheter or urinary catheter (Briano et al. 2022; Rudramurthy et al. 2017; Rossow et al. 2021), all of which are associated with the surface attached biofilm mode of growth. Indeed, it has been shown that a range of contaminated hospital equipment can support the transmission of *C. auris*, including oximeters, thermometers, temperature probes, blood pressure cuffs, and even cloth lanyards (de Almeida et al. 2021; Nobrega de Almeida et al. 2021; Patterson et al. 2021). Beyond the hospital environment, it was recently shown that *C. auris* could be detected in a wastewater facility in Utah where no known clinical cases were located prior to the documented transfer of colonized patients from bordering Nevada (Chavez et al. 2024). Further, simulated survival studies of planktonic and biofilm *C. auris* over 30 days in river water or seawater demonstrated persistent colonization of high-density polyethylene (HDPE) and glass that would readily transfer to wet sand (Akinbobola et al. 2024). There is also evidence of long-term survival and persistence (>3 weeks) on wet wood (Dire et al. 2023). These studies exemplify the transmission risk from hospital environments to the wider environment, with attached biofilm communities playing a key role. There is further suggestion that *C. auris* could be transmitted by air dispersion, based on data from an open-cubicle ward setting (Didik et al. 2023). Taken together, this yeast's universal ability to survive in a harsh environment and readily transmit throughout the hospital environment by direct and indirect means makes this an important infection, prevention and control (IPC) issue. Figure 7.1 illustrates the variety of substrates and environments where *C. auris* has been isolated.

Frequency of community transmission is difficult to pinpoint due to *C. auris's* ability to colonize and persist at multiple bodily sites in immunocompetent hosts without causing visible symptoms (Bradley 2019), effectively discarding them from high-invasive fungal infections (IFI) risk groups. It is important to note that *C. auris* can colonize the skin without causing any symptoms (Bradley 2019), increasing risk of contact transmission among patients in healthcare facilities, polymicrobial interactions due to wound colonization, and reciprocal transmission between patients and fomites. It is imperative to perform comprehensive screening studies by obtaining swabs from multiple body sites across immunocompetent and immunocompromised hosts at different settings to contribute to containment and mitigation efforts against *C. auris*. This is especially important given that in the US alone skin colonization increased by 200% between 2020 and 2021, which was picked up by enhanced screening protocols (Lyman et al. 2023). Worryingly, transmission and spread may disproportionately impact lower socioeconomic groups (Smith et al. 2024). Many researchers have stressed the importance of a collective contribution between clinical, laboratory, public health, and government entities globally to

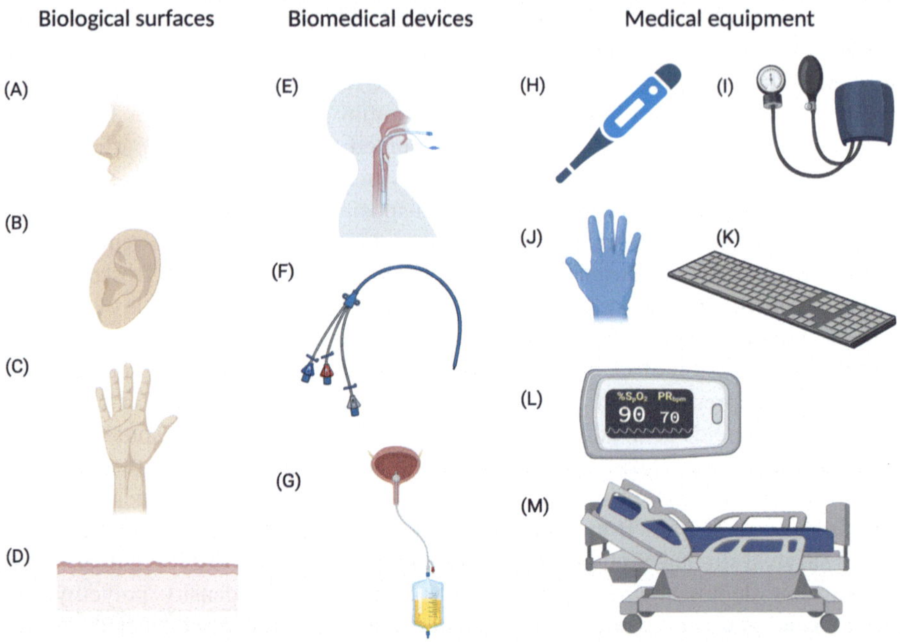

Fig. 7.1 Overview of the different sites for *Candidozyma auris* biofilms are located on biological, biomedical or equipment. Biofilms form on biological substrates (not limited to) (**a**) anterior nares, (**b**) ear, (**c**) hands, (**d**) skin, and biotic surfaces (not limited to) (**e**) endotracheal tube, (**f**) central line, (**g**) urinary catheter. It can also be isolated from (**h**) temperature probes, (**i**) blood pressure cuff, (**j**) gloves, (**k**) keyboards, (**l**) pulsometer, and (**m**) hospital bed). (Figures produced on BioRender under license)

reduce nosocomial spread by implementing commercial methods for screening and surveillance at the localized level (Kordalewska and Perlin 2019). This includes accurate surveying and reporting of positive symptomatic and asymptomatic cases, development of de novo disinfection protocols with decreased possibility of inducing acquired resistance, and community outreach for education in prevention strategies at the individual level. However, for this to happen, these collective groups must have a basic understanding of *C. auris* biology and its role as a biofilm pathogen. The global implications of the spread of this yeast are yet to be fully realized, but it is reasonable to argue that climate change will have an important role in driving transmission (Baker et al. 2022; Casadevall et al. 2021).

Microbiological Characteristics

C. auris produces ovoid-shaped cells that can be between 2.0–3.0 and/or 2.5–5.0 μm in diameter. Cells can appear in pairs, groups or singles. It has a white to gray appearance on malt agar and does not typically form hyphae or pseudohyphae

(Satoh et al. 2009; Du et al. 2020). Although the budding yeast morphology predominates in *C. auris*, a study found that pseudohyphae are formed upon HSP90 depletion (Kim et al. 2019). Indeed, the transcription factor Ume6 has been shown to have a role in *C. auris* morphogenesis, where hyperactivation of Ume6 induced filamentation and aggregation (Louvet et al. 2024). Pseudohyphal growth is characterized by a wide cell diameter compared to yeast cells, elongated cell morphology due to delayed or incomplete cell division (mother-daughter cell attachments remain), and lack of cytoplasmic bridging and parallel divisions (Sudbery et al. 2004). Pseudohyphae growth can also be induced in hypersaline conditions (e.g., yeast extract) (Du et al. 2020), via genotoxic stress (Ruiz et al. 2020), and adverse environmental conditions (e.g. passage through a mammalian host) (Yue et al. 2018). *C. auris* grows well at 42 °C, and its carbon assimilation profile differs from other closely related *Candida* species (Satoh et al. 2009). Moreover, *C. auris* is unable to produce chlamydospores (Larkin et al. 2017; Rossato and Colombo 2018), which are described as thick-walled, hypha-specific cells produced by many fungi capable of filamentation. When grown on fetal bovine serum, it fails to produce germ tubes compared to *C. albicans,* and its ability to adhere to inert surfaces is significantly less than *C. albicans* (Larkin et al. 2017). These are both fundamental properties that make *C. albicans* such an excellent biofilm former, though *C. auris* does have its own architecturally reduced capacity to develop surface attached communities. However, *C. auris* is unlikely to operate in isolation and is often found within complex skin microbiomes and has been shown co-isolated with different microbial species that may impact biofilm formation (Tharp et al. 2023; Proctor et al. 2023; Janeczko and Skrzypek 2025).

Comparative Molecular Characteristics

Comparative genomics determined that *C. auris* inherited a multitude of genes orthologous to those of *C. albicans* and *C. lusitaniae* (Fanning and Mitchell 2012). Some of these orthologs found in *C. auris* are associated with virulence factors, including the secretion of destructive enzymes (e.g. lipases, phospholipases, proteinases and aspartyl proteases). Moreover, adhesins and major facilitator superfamily (MFS) families (efflux pumps and transporters) involved in drug resistance and nutrient acquisition, metabolic regulation in stress conditions, and biofilm formation have also been identified (Muñoz et al. 2018; Ahmad and Alfouzan 2021). These characteristics enhance *C. auris* virulence and pathogenicity, though not to the same extent as *C. albicans*, as demonstrated in a series of in vivo studies in mice and human neutrophils (Bruno et al. 2020; Johnson et al. 2018).

C. auris differs from *C. albicans* and other dimorphic fungi in that it largely exists as a yeast (Bravo Ruiz and Lorenz 2021). In response to atypical or unfavorable growth conditions, certain *C. auris* isolates may form pseudohyphae-like and/or even hyphae-resembling structures (Ahmad and Alfouzan 2021). After passage through a mammalian host, *C. auris* exhibited a unique phenotypic switch characterized by heritable and nonheritable changes between filamentous-form, yeast cells and filamentous-competent cells. Interestingly, some orthologs of *C. albicans* needed for attachment and filamentation like *ALS4* and *HGC1* were differentially expressed in the filamentous morphology of *C. auris* (Yue et al. 2018). Other key filamentation-associated transcriptional factors like *EFG1* and *WOR1*, which are upregulated during *C. albicans* white-to-opaque transitions, appear to be downregulated in *C. auris* filamentous cells. In *C. auris* yeast cells, much like *C. albicans,* factors like *NRG1* and *CUP9* were differentially expressed (these factors typically suppress filamentation) (Yue et al. 2018). Moreover, genotoxic stress induced by treatment with hydroxyurea, methyl methanesulfonate, and 5-fluorocytosine triggered pseudohyphal morphogenesis in *C. auris* Δ*tup1* strains (a transcription factor belonging to the family of negative hyphal growth regulators in *C. auris*). Altogether, these findings indicate that, although *C. auris* shares many evolutionarily conserved alleles with other resistant *Candida* spp., it has diverged in some capacity due to stress and environmental selection pressure. Genome-wide association studies are practical tools that continue to provide insight into *C. auris* copy number variants, conserved and expanded gene families and how these may impact virulence, host-pathogen interactions, and persistence (Ruiz et al. 2020).

Aggregating Phenotype

C. auris has also been found to include strains characterized by clumping, or aggregation, and non-aggregation, phenotypes that seem to have a relationship with biofilm formation and virulence (Borman et al. 2016; Sherry et al. 2017). Borman and colleagues compared pathogenicity of *C. auris* against other pathogenic *Candida* spp. using a *Galleria mellonella* model, where they found that a non-aggregating phenotype of *C. auris* exhibited pathogenicity virtually comparable to that of *C. albicans* and *C. tropicalis* (Borman et al. 2016). The larvae were inoculated with several virulent *Candida* strains, incubated at 37 °C, and dissected 18 h post-infection to reveal no *C. auris* hyphal or pseudohyphal formation compared to *C. albicans.* However, they did find that non-aggregate strains of *C. auris* resulted in similar larval killing rates to those of *C. albicans* when incubated at 37 °C, while aggregate strains displayed lower virulence (Borman et al. 2016). It is important to note that aggregate strains were difficult to eradicate in vitro, and regardless of larval inoculation with single yeast cells obtained from aggregates, the hemolymph-formed larvae exhibited aggregation post-infection. This suggests that certain *C. auris* strains can maintain aggregation both in vivo and in vitro in physiologically unfavorable conditions but this does not necessarily correlate to increased virulence (Borman et al. 2016). Despite the finding that aggregative strains are less virulent than non-aggregative strains, the former may still be associated with increased fungal burden, tissue colonization and biofilm formation (Brown et al. 2020). Indeed, in animal studies it was shown that *C. auris* aggregates were ubiquitous in kidney histopathology samples, which may denote aggregate morphology as an immune evasion mechanism (Ben-Ami et al. 2017). Together, these findings support the notion that aggregation is an important factor in survival and resilience in the human host and external environment, which relates to the biofilm growth modality (Singh et al. 2019; Hernando-Ortiz et al. 2021). Indeed, recent work suggested that while aggregation is strain and temperature dependent, it is linked with extracellular matrix production and surface amyloids (Malavia-Jones et al. 2023), which is a critical feature of biofilm formation (Nett and Andes 2020).

Biofilm Formation

Biofilms are complex and intricate structures that are self-organized, three-dimensional assemblies, characterized by cell-cell and cell-surface interactions within an extracellular polymer (Ramage et al. 2025). Interestingly, biofilm biogenesis relies on nutrient availability, sensory mechanisms, negative and positive transcriptional regulation, and divergent gene expression profiles that can induce metabolic and adaptive changes optimal for nutrient acquisition and survival (Fanning and Mitchell 2012). *C. auris* has the ability to persist as biofilms much like other *Candida* spp. and other medically relevant fungal species (Fanning and Mitchell 2012). The first formal description and screening attempt of *C. auris* biofilm formation was described by Oh et al. (2011), though these authors were unable to explicitly show a biofilm phenotype (Oh et al. 2011). Sherry et al. (2017) were the first to formally demonstrate a highly antifungal resistant biofilm phenotype, but with sensitivity to chlorhexidine (Sherry et al. 2017). Welsh and colleagues simultaneously modeled the capacity of this yeast to survive on plastic healthcare surfaces and showed that *C. auris* could survive for 14 days under controlled conditions of humidity and temperature, indicating its capacity for environmental resilience (Welsh et al. 2017). Together, these reports created a platform from which to study this yeast in its biofilm-forming capacity.

Biofilms are characterized by an amalgamation of sessile cells that can form tightly interwoven, heterogeneous communities of multi-morphic cells, embedded in an extracellular matrix (ECM) via quorum-sensing mechanisms, which allow them to adhere to biotic and abiotic surfaces (Fanning and Mitchell 2012; Wall et al. 2019). Figure 7.2 illustrates the basic developmental profile of the key phases of biofilm development, which like *C. albicans* takes place over 4 major steps: adherence, proliferation, maturation and dispersal (Wall et al. 2019; Gulati et al. 2018). In broad and basic terms, biofilm formation starts with an adhesion phase where individual planktonic yeast cells adhere to a biotic or abiotic substrate, initiating surface attachment. In the subsequent proliferation phase, cells undergo rapid replication and expansion, leading to the formation of microcolonies (small clusters of cells), which is considered to have increased transcriptional and metabolic activity. As the biofilm biomass increases through cellular division and co-aggregation, the micro-environment within the biofilm becomes increasingly heterogeneous, promoting niche differentiation and microbial interactions. In the maturation phase, there is a

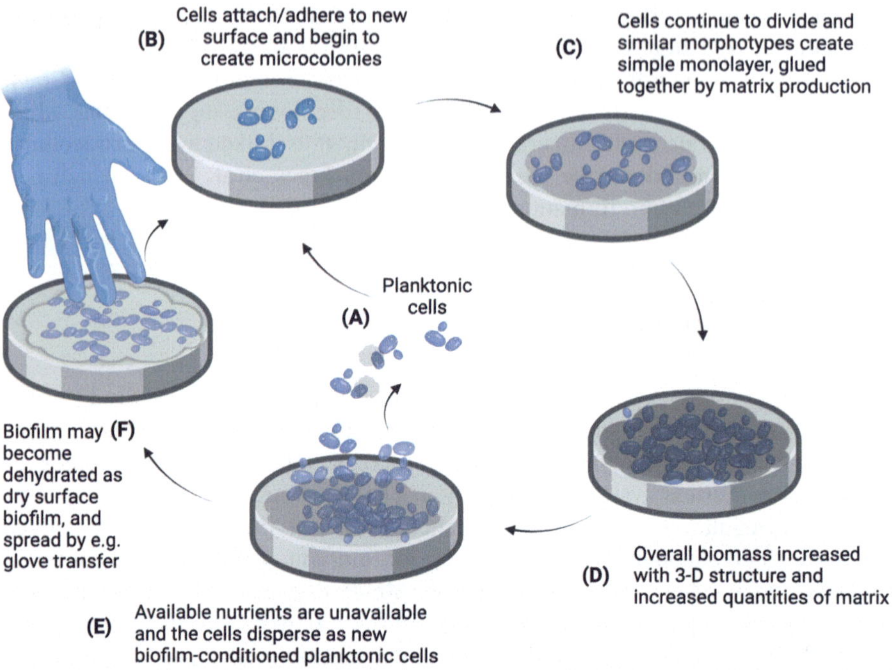

Fig. 7.2 Biofilm developmental phases for *Candidozyma auris* biofilms. Biofilm formation is initiated from either (**a**) dispersed planktonic cells from a liquid or aerosolized environment or from transmission from a medical device or fomite. Biofilm formation is generally comprised into four main phases, including the (**b**) initial colonization (30–90 mins), (**c**) proliferation (8–12 h), (**d**) maturation phase (24–48 h) and (**e**) dispersal phase (>24–48 h). As biofilm mature, ECM is produced and covers the cellular mats (denoted by gray clouds). Dispersed biofilm primed cells or aggregates can then recolonize new surfaces. Often the biofilm will dehydrate as (**f**) dry-surface biofilms but will be readily transmitted through contact. (Figures produced on BioRender under license)

notable escalation in complexity, and the full expansion of a three-dimensional biofilm structure (though this is dependent on the occupied space). Architectural features, such as spatially arranged channels and voids, and fluid exchange channels, emerge within the biofilm to facilitate nutrient exchange, waste removal, and intercellular communication. Dispersion completes the cycle, typically occurring in response to unfavorable environmental conditions (i.e. waste accumulation, nutrient depletion, osmotic stressors and pH fluctuations). It is generally considered as an active process in which sessile, matrix-encased biofilm cells actively release from the biofilm (Uppuluri et al. 2018).

Specifically, *C. auris* biofilms are less architecturally complex than *C. albicans* but are still highly organized structurally. Despite being substantially scant and weaker than those of *C. albicans,* they have a persistent ability to adhere to multiple abiotic surfaces (Short et al. 2019). ECM present in most *Candida* spp. biofilms contributes to antimycotic sequestration, and thus, biofilm-associated drug resistance (Taff et al. 2013). Phenotypically, *C. auris* biofilms differ in that they are not comprised of true hyphae, but relatively clumpy aggregates or single yeast cell layers with occasional pseudohyphal formation (Wang et al. 2018; Du et al. 2020; Ahmad and Alfouzan 2021; Rossato and Colombo 2018; Larkin et al. 2017). Biofilm spatial architecture largely depends on whether the strain is aggregate or non-aggregate. Vila et al. showcased fluorescent confocal images of *C. auris* isolates (high-biofilm former [dense and homogenous multilayer] and low biofilm former [dispersed biofilm with scattered cell aggregates]) (Vila et al. 2020). Interestingly, both formed biofilms were associated with large secretions of ECM, suggesting that aggregation may not be a single feature that promotes biofilm formation, despite studies stating otherwise (Singh et al. 2019). Despite these nuanced differences in aggregation and strain heterogeneity, the capacity to adhere to a surface, its ability to produce sufficient quantities of extracellular matrix 'glue' and its ability to disperse and recolonize surfaces are all critically important in its life cycle, and are features governed by complex molecular machinery.

Molecular Basis of Biofilm Formation

Sticking to a surface, or to another cell, is an important starting point for all biofilms, and also for skin colonization. In *Candida* spp., members of the agglutinin-like sequence (ALS) family of adhesins play a fundamental role in adherence (Sundstrom 2002). Several virulence genes that encode glycophosphatidylinositol (GPI)-linked cell wall adhesins, integrins and ALS become upregulated during biofilm formation and maturation in *C. auris.* Interestingly, seven of those upregulated GPI-anchored proteins (*IFF4, PLB3, PGA52, PGA26, CSA1, HYR3, PLB3,* and *PGA7)* are evolutionarily conserved between *C. auris* and its phylogenetic relatives *C. haelumonii, C. pseudohaelumonii,* and *C. duobushaemuloni* (Muñoz et al. 2018; Ahmad and Alfouzan 2021; Chybowska et al. 2020). In *C. auris* the adhesins Als4 and Als5 has been shown to be important for biofilm formation (Bing et al. 2023; Kean et al. 2018a), as has the Surface Colonization Factor (Scf1) protein, which is required for attachment to inert and mammalian cells and biofilm formation (Santana et al. 2023). These two proteins appear to have a functional redundancy according to recent work by Wang et al. (2024), so that individually they have an abrogated function, but in combination the cells could co-adhere and aggregate, supporting biofilm formation (Wang et al. 2024). Recent studies in a Ume6-hyperactivated strain (*UME6*^{HA}) background further confirm the importance of Scf1, by revealing that Ume6 controls filamentation via Hgc1 (hypha-specific G1 cyclin-related protein) and aggregation via Als4498 (agglutin-like sequence adhesin) and Scf1 (Louvet et al. 2024). This indicated that Ume6 is a universal regulator that is important for biofilm formation. Recent work also suggests higher levels of biofilm control in studies with the *C. auris* Hog1 mitogen-activated protein kinase (MAPK), which is essential for skin colonization an intradermal persistence, but also shown to facilitate biofilm formation (Shivarathri et al. 2020). It has been reported that mannan, with highly enriched in β-1,2-linkages, was important for binding to IgG (12–20 greater than for *C. albicans*) (Yan et al. 2020). Together, these data imply a pathogen that is highly regulated with the capacity to persist and adhere to inert and biological environments.

A key feature in conferring biofilm-associated tolerance is the secretion of extracellular vesicles (EVs), which is essential for ECM production. The most abundant polysaccharides in the biofilm matrix are mannans (both α-1,2 branched and α-1,6 mannans) and glucans (both β-1,6 and β-1,3 glucans), which together comprise the mannan–glucan complexes (MGCxs) (Dominguez et al. 2018; Dominguez et al. 2019). Zarnowski and colleagues (2018) reported that over 30% of the protein cargo was biofilm specific, and that many conserved payloads were up to 100-fold more abundant in biofilms than planktonic cells. Intriguingly, using a proteomics, lipidomics, and transcriptomics approach, it was reported that *C. auris* released EVs with cargoes that potentiated the adhesion of this yeast to an epithelial cell monolayer, and that these were significantly different from those of *C. albicans* (Zamith-Miranda et al. 2021). These EVs are also critical for supporting extracellular matrix production (ECM). ECM is a prerequisite for biofilm mechanics (Taff et al. 2012), and has shown to be present in *C. auris* (Dominguez et al. 2019). The studies by Kean and colleagues (2018) were the first to highlight the possibility of the importance of the matrix through identifying differential expression of matric modifying enzymes through transcriptomic analyses (Kean et al. 2018a). It was initially shown that matrix was able to sequester approximately 70% of triazole, suggesting it soaked it up like a sponge, and conferred azole tolerance (Dominguez et al. 2019). This was characterized as being rich in MGCxs, a conserved feature of *Candida* spp. biofilms. It has been shown that *C. auris,* as well as three other key species (*C. tropicalis, C. parapsilosis* and *C. glabrata*) contain a group of 36 orthologous proteins to *C. albicans* within the biofilm EV proteome (Zarnowski et al. 2022). In *C. auris* these EVs appear to have an immunomodulatory property, which suggests that in addition to biofilm formation these may play a more direct role in pathogenesis (Zamith-Miranda et al. 2021; Zarnowski et al. 2022).

As the biofilm matures with complex architecture, stressful environmental conditions may influence how this structure may maintain itself. Fluctuations in nutrients and waste products, changes in pH and oxygen tension all have the possibility to induce transcriptional changes. While in *C. albicans* this has been extensively described (Wall et al. 2019), to date there are no studies that have tackled the subject directly. Mature *Candida* spp. biofilms are often associated with the notion of quorum-sensing (QS) molecules, which has been identified as farnesol and tyrosol (Kovacs and Majoros 2020). Elevated levels of these small QS molecules are reported to be associated with morphological form and lead to altered biofilm phenotypes (Ramage et al. 2002; Alem et al. 2006). Tyrosol has been identified from *C. auris* cultures (Semreen et al. 2019), and in transcriptional studies of these organisms in response to these QS molecules then it has been shown that genes involved in biofilm, glycolysis, ergosterol biosynthesis, fatty acid oxidation and iron metabolism were primarily affected in treated cells (Jakab et al. 2024). It remains unclear as yet whether QS is crucial in *C. auris* dispersal, but we can speculate that metabolites like these and others will be critical in orchestrating biofilm dispersal, though given the variety of niches it occupies then it is highly likely. How and why dispersal would be important in harsher desiccated environments remains to be seen, as it is logical that these yeast cells would aim to maintain a tightly bound protected consortia to protect the cells within dry-surface biofilms. Adding to this is the concept of persister cells, though not well understood in *Candida* spp., is a form of fungal biofilm tolerance through the formation of dormant, phenotypic variants of the wild-type population (Wuyts et al. 2018). These low frequency dormant cells can tolerate very high concentrations of fungicidal drugs, predominantly amphotericin B, and have been recently identified in *C. auris* (Rossoni et al. 2020). Their purpose is linked to the ability to survive reactive oxygen species (ROS) accumulation induced by fungicidal drugs, or maybe a trait linked to generalized survival in harsh environments.

Assessing Resistance and Tolerance

Resistance tends to suggest an inheritable and genetically maintained phenotype, whereas tolerance is a responsive and reversible phenotype through the induction and expression of various evolved defenses. There are a variety of reviews that describe *C. auris* resistance (Rhodes and Fisher 2019; Kean and Ramage 2019), though generally speaking fluconazole resistance, while not universal per se, is considered high and therefore this antifungal is not considered for therapeutic management (Jones et al. 2024). Moreover, dual resistance is frequent (approx. 40%) (Lockhart et al. 2017), and even pan-fungal resistance (to a lower extent) to the main antifungal classes (Lyman et al. 2023). Therefore, the clinical landscape is challenging against the simplest of cellular forms to kill or inhibit, i.e. planktonic (free-floating) cells. When we add to this that the skin, mucosa, medical devices, surgical wounds and near patient environment is filled with *C. auris* biofilms, then it is unsurprising that control of this biofilm pathogen is difficult. For *C. auris*, official clinical cut-offs have not been established, but the CDC suggested tentative breakpoints based on common pathogenic *Candida* spp. and expert opinion for all major antifungals in drug-resistant strains (Ahmad and Alfouzan 2021). The MIC breakpoints are summarized in Table 7.1. Given that >90% of *C. auris* strains are intrinsically resistant to fluconazole, which has the same mechanism of action as other triazoles, MICs have not been established for second generation triazoles as susceptibility profiles vary across isolates (CDC 2020).

The first step in combating antifungal resistance and tolerance is utilizing methods to screen for and identify resistant strains and associated determinants. Surveillance and epidemiological studies have been essential in characterizing microorganisms that are unresponsive to currently available therapies, and entities like CLSI and EUCAST have established standardized protocols for in vitro susceptibility testing (Fisher et al. 2022; Du et al. 2020; Ahmad and Alfouzan 2021;

Table 7.1 *Candidozyma auris* tentative breakpoints (expert opinion)

Antifungal (class)	CLSI breakpoint
Fluconazole (AZL)	^{3}32 ug/mL
Amphotericin B (POL)	^{3}2 ug/mL
Caspofungin (ECH)	^{3}2 ug/mL
Micafungin (ECH)	^{3}4 ug/mL
Anidulafungin (ECH)	^{3}4 ug/mL

Source: Centers for Disease Control and Prevention (https://www.cdc.gov/candida-auris/hcp/laboratories/antifungal-susceptibility-testing.html); European Committee on Antimicrobial Susceptibility Testing (EUCAST)

Chowdhary et al. 2023). Methods include broth microdilution, disk diffusion, strip-based E-test, and azole agar screening. Though these reference methods are considered the gold standard in medical mycology for establishing clinical cut-offs, many breakpoints in understudied species are tentative to nonexistent, and these parameters need to be evaluated in all clinical isolates being tested, where species-specific variability could posit reproducibility and sensitivity issues (Fisher et al. 2022; Arastehfar et al. 2019; Ahmad and Alfouzan 2021; Du et al. 2020). Moreover, and in the context of this chapter, these standardized methods are limited to planktonic cell susceptibility, and many pathogenic fungi are known to form biofilms (Pierce et al. 2008). Methods to grow and assess fungal biofilms are available and are an important resource to be able to assess existing and new antifungal agents (Brown et al. 2022; Pierce et al. 2008; Van Dijck et al. 2018).

More recently, novel automated methods like the MALDI Biotyper antibiotic susceptibility test rapid assay (MBT ASTRA) based on MALDI-TOF mass spectrometry and even isothermal microcalorimetry (IMC) show promise as future susceptibility testing methods (Vatanshenassan et al. 2018; Di Luca et al. 2019). Nevertheless, MALDI-TOF MS requires highly skilled personnel for operation, and equipment costs may be too demanding for smaller facilities (Arastehfar et al. 2019; Kordalewska and Perlin 2019). IMC may be a cost-effective alternative, as it has shown reliability when tested on biofilm and planktonic conditions in *C. auris* and compared against other available metabolic signature testing methods like VITEK2 (Di Luca et al. 2019). Limitations still linger, so further validation studies are required to determine reliance of use in surveillance studies.

Sherry and colleagues initially showed that *C. auris* biofilms were tolerant to azoles, polyenes and echinocandins, but susceptible to the antiseptic chlorhexidine (Sherry et al. 2017). Mechanistically, this was supported by data from one of the first transcriptome experiments, which demonstrated it was a dynamically tolerant pathogen by virtue of its up-regulation of efflux pumps (both ABC and major facilitator superfamily [MFS] transporters) and associated factors (glucan-modifying enzymes) that were driving antifungal recalcitrance (Kean et al. 2018a). This study also showed a phase-dependent increase in tolerance across all classes. Some of the same gene classes have been also identified independently in screening of sensitive and resistant clinical isolates (Zhou et al. 2021). Therefore, as it stands the conventional treatment options are limited for biofilms and likely to be further depleted as more pan-fungal strains emerge, though new antifungal agents are on the horizon (Hoenigl et al. 2021; Neoh et al. 2023).

Novel Strategies to Combat Biofilm Tolerance

While not strictly specific for *C. auris* biofilms, there are new compound classes, such as the first-in-class oral triterpenoid Ibrexafungerp for use in recurrent vulvovaginal candidiasis (Wiederhold 2022; Wiederhold et al. 2021; Hoenigl et al. 2021; Larkin et al. 2017; Marcos-Zambrano et al. 2017). Clinical data of a large cohort indicates an excellent clinical cure ratio and eradication of fungi in a biofilm-associated disease (He et al. 2024). Ibrexafungerp also exhibits antibiofilm activity against *C. auris* by targeting the beta-glucan synthase pathway, alongside the newly released novel echinocandin rezafungin. Clinically this has been shown to be non-inferior compared to caspofungin for all-cause mortality (Locke et al. 2024; Thompson et al. 2024), though data is limited for biofilm activity (Chandra and Ghannoum 2018). Finally, Fosmanogepix, is a new antifungal that targets Gwt1, an essential enzyme in the glycosylphosphatidylinositol (GPI) anchor biosynthesis pathway (Hoenigl et al. 2021; Pfizer 2022; Shaw and Ibrahim 2020), which has a wide spectrum of activity that includes *C. auris*, though there is no information on its antibiofilm activity.

High-throughput screening (HTS) of chemical libraries is an attractive method for the identification of compounds with antifungal activity, which can be performed using in vivo and in vitro models coupled with robotics and software (Szymański et al. 2012). For instance, protocols have been developed for HTS against *Candida* spp. biofilms using 96-well and 384-well microtiter plates (Pierce et al. 2008; Ajetunmobi et al. 2023). The protocols are simple, reproducible, cost-effective, rapid and easy to implement across small and large laboratory settings. They are typically coupled with an XTT 2,3-Bis-(2-Methoxy-4-Nitro-5-Sulfophenyl)-2H-Tetrazolium-5-Carboxanilide reduction colorimetric assay, whereby water-soluble XTT is converted by actively respiring cells into a water-soluble, orange-colored formazan product (Pierce et al. 2008; Singh et al. 2011; Ajetunmobi et al. 2023). Results can be read spectrophotometrically using a microtiter plate reader, whereby metabolically active sessile cell activity has a linear relationship to cell density and, subsequently, signals detected from absorbance readings. HTS has also been used to identify compounds with repositioning potential. These compounds will generally have a high safety profile because they are already FDA-approved, as well as characterized mechanisms of action and ADMET properties, making drug repurposing an attractive solution to high attrition rates and time-consuming research and development facing the pharmaceutical industry (Pushpakom et al. 2019).

Alternative strategies such as screening non-exhaustive natural repertoire of bioactive compounds, phenolic compounds, and even synthesizing silver nanoparticles with natural antioxidant and anti-virulence properties have shown great promise and continue to be adopted globally in the war against multidrug resistance

(Frías-De-León et al. 2020; Vazquez-Munoz et al. 2020; AlJindan and AlEraky 2022). Combination therapy approaches have also proven successful, whereby classic antifungals have been administered in tandem for synergistic and/or additive effects. For example, the monoterpene carvacrol (CAR) was used in combination with AMB, fluconazole (FLU), nystatin (NYS) and caspofungin (CAS) in an in vitro screen against *C. auris* virulence factors using buccal epithelial cells. The authors found that CAR had anti-adherence and anti-proteinase activity against *C. auris* at sub-clinical concentrations, as there was no pseudohyphal morphogenesis or penetration into epithelial cells. Furthermore, CAR-AMB and CAR-NYS combinations resulted in synergy across 28% of tested strains, of which 12% and 28% out of 25 total strains were resistant to both antifungals respectively. Additivity was also reported for 52% and 24% of strains in the CAR-FLU and CAR-CAS combinations, out of which 48% were FLU resistant and none were CAS resistant (Shaban et al. 2020). We have also shown the possibility of using chitosan as a potential augmentative strategy to support existing antifungals (Arias et al. 2020). Other innovative approaches include the experimental use of turbinmicin which targets the EV delivery pathway used for ECM production (Zhao et al. 2021). Other creative strategies include probiotics, antimicrobial peptides, heat shock inhibition, and photodynamic therapy, among others, as well as innovations targeting catheter-related candidemia, such as novel surface coatings to prevent biofilm formation or controlled release of antifungals to a wound site (Butcher et al. 2022). Recently, a novel keto-alkyl-pyridinium antifungal agents was shown to effectively function in both a vascular catheter and skin colonization model (Beattie et al. 2023). There is even potential to use truncated staphylococcal peptides in combination with antifungal agents (amphotericin B and caspofungin), which has been shown to be effective in a *C. auris* biofilm catheter model (Kumari et al. 2024), or other bacterial derived antifungal agents (Kovacs et al. 2021). The Andes group recently reported a novel non-toxic antifungal agent (PQA-Az-13), containing a combination of indazole, pyrrolidine, and arylpiperazine scaffolds (substituted with a trifluoromethyl moiety), that in liposomal formulations demonstrated excellent antibiofilm activity in vitro, and an in ex vivo skin model (Jaromin et al. 2024). While these approaches are experimental, together they have potential to improve the management of *C. auris* biofilm-associated infections.

Infection Prevention and Control

It goes without saying that prevention is better than the cure and given the importance of skin and the surrounding clinical environment that this organism inhabits, then without debate this is the most critical element in the clinical management of *C. auris*. The hierarchy of controls provides a structured approach to infection prevention and control (IPC), with elimination of the hazard being the most effective strategy at the top of the hierarchy, and the use of protective personal equipment (PPE) as the last level of control (ARHAI Scotland 2022; CDC 2024b). There are a range of simple and yet efficacious IPC measures available, such as effective hand hygiene (Aldejohann et al. 2022; Biswal et al. 2017), but a combination of various levels of control is recommended by IPC experts to minimize the risk of failure in any one method and to create a more robust defense against the spread of infection (Jones et al. 2024; Aldejohann et al. 2022; CDC 2024a). The CDC recommends among other things a combination of transmission-based precautions (TBPs) and enhanced barrier precautions, such as PPE (long sleeve gown and gloves) and single-use items for the duration of stay; ideally in a single room occupancy. Due to the persistence of *C. auris* on surfaces in healthcare settings, other precautions such as decontamination of equipment and the environment are essential.

Decontamination

It is noteworthy that the majority of disinfection studies focus on suspension testing using the ASTM E2197-11 methodology, which does not take into consideration complex substrate topographies and the 3-dimensional architecture and associated biological material of the *C. auris* biofilm. Similarly, EN standards 13,624 (quantitative suspension test) and EN 16615 (activity on non-porous surfaces with mechanical action employing wipes) have been used to assess the efficacy of alcohol and quartenary ammonium disinfectants against *C. auris*. Both disinfectants demonstrated efficacy at a 1-minute contact (Muller et al. 2020). However, these methods do not include biofilm testing and may over-estimate efficacy. In the USA, the CDC recommends using an Environmental Protection Agency (EPA)-registered hospital-grade disinfectant effective against *C. auris* outline on list P (disinfectant products for claims against *C. auris*), which have been reported to kill *C. auris* (CDC 2024a). This list contains 42 different formulations, including chemicals such as sodium hypochlorite, isopropyl alcohol, ethanol, quaternary ammonium, hydrogen peroxide, peracetic acid, sodium dichloro-s-triazinetrione, citric acid and DDAC. These appear as single agent or combinational formulations, and with different recommended exposure contact times on a variety of surfaces. However, the evidence-base to support their antibiofilm efficacy is limited (Kean et al. 2018b; Ledwoch and Maillard 2018; Short et al. 2019; Zatorska et al. 2021).

Interpretation of experimental data is increasingly difficult due to the non-standardized approaches researchers take to conduct studies of these nature. Researchers must also be mindful to consider the real-world application of the intervention being tested. The earliest studies quickly showed that clinically relevant concentrations of sodium hypochlorite and peracetic acid on relevant substrates have reduced efficacy against adherent *C. auris* cells (Kean et al. 2018c). In an attempt to standardize, the ASTM2967-15 method was used on stainless steel coupons, where a dry *C. auris* biofilm model was used to demonstrate the recovery of viable *C. auris* following treatment with various biocides (sodium hypochlorite and peracetic acid). Interestingly, while there was a general success in reducing the viable bioburden of cells, over half of the disinfectants tested failed to prevent its transferability, and 75% did not delay biofilm regrowth (Ledwoch and Maillard 2018). Our own studies showed that *C. auris* cells (preferentially the aggregative phenotype) were able to survive for up to 14 days as desiccated dry biofilms (Short et al. 2019), whereas other studies using an enzymatic based detection methodology suggest up to 28 days (Welsh et al. 2017), which is an important consideration as it has been isolated from bedding of patients up to 7 days (Biswal et al. 2017). These studies highlighted that hypochlorite based disinfectants (NaOCl) and dichloroisocyanurate are preferentially used for controlling this yeast (Ku et al. 2018; Moore et al. 2017), though there is evidence ethanol and certain quaternary ammonium compounds were able to achieve 5$\log_{10}$ reductions in biofilm viability (Zatorska et al. 2021). This is at odds with competing studies and is not recommended as there is concerns of enhanced tolerance with long-term exposure (Cadnum et al. 2017; Dire et al. 2023). Moreover, inferior antibiofilm efficacy and resistance to some quaternary ammonium compounds may explain the failure in disinfecting the axillary temperature probes in the Oxford outbreak (Eyre et al. 2018). A range of studies summarized by Jones and colleagues show that despite its preferential study and use, these hypochlorite compounds often failed to obtain sterility, so peracetic acid and hydrogen peroxide vapor have been explored (Abdolrasouli et al. 2017; Kean et al. 2018c; Jones et al. 2024). Hydrogen peroxide was shown to offer promising results for environmental decontamination (Abdolrasouli et al. 2017), though the level of evidence remains limited (Cadnum et al. 2017). Other disinfection strategies including glutaraldehyde and phenol, as well as ultraviolet-C (UVC) light, have been considered but there is as yet no solid body of evidence to support their implementation (Jones et al. 2024).

One theory that links environmental condition and extended periods of survival of *C. auris* within nosocomial settings is through dry-surface biofilm (DSB) formation. DSB's are defined as microbial communities that are embedded within a matrix of extracellular substances, that can colonize a wide range of substrates, subjected to low water availability in a dry state (Ledwoch and Maillard 2018) Importantly, these are more complex than planktonic cells that are simply desiccated on to a surface. This phenomenon has been shown to impact biocide efficacy, withstand high temperatures such as autoclaving and increase transmissibility across a number of notorious nosocomial pathogens (Almatroudi et al. 2018), and they have been identified from a variety of different high touch surfaces in hospitals including keyboard, sanitizing bottles and patient folders (Ledwoch et al. 2018). Less is known about the role for DSB's in fungal pathogens, although initial studies have suggested that *C. auris* is able to form robust and tolerant DSBs. Ledwoch and Maillard (2018) developed the initial *C. auris* DSB model, simulating growth and biological soiling on stainless steel. Morphologically, *C. auris* formed sparse, mono-layered communities that lacked any distinguishable ECM. Across a variety of biocide products including sodium dichloroisocyanurate, peracetic acid, benzalkonium chloride and sodium hypochlorite, antifungal efficacy was demonstrated (removal range: 1–7 $\log_{10}$). Importantly, this study concurrently investigated mechanical action of removal through wiping, although post-wiping transferability and regrowth were observed across all tested products, highlighting potential issues with current disinfection guidelines. Expanding on this work, Ware et al. (2023), further assessed the efficacy of sodium hypochlorite against these communities. Using clinical isolates from outbreak associated clades they were able to show that DSBs could tolerate clinically relevant concentrations of sodium hypochlorite (2–4 log10-reductions) in a phase-dependent manner (Ware et al. 2023). In addition, they also provided the first mechanistic insights into DSB formation from *C. auris*, identifying iron acquisition and scavenging, alongside ABC transporter up-regulation as key processes in the DSB life cycle, suggesting a key role in acquiring small molecules for survival in stress inducing environments.

Collectively, the fundamental role for DSBs in the nosocomial environment is not concrete, with evidence now identifying links between the environment, surface adhesion and inefficacy of biocides. Moving forward, it is important to consider the role for DSBs in disinfection protocols and testing standards/guidelines should be developed considering this tolerant phenotype.

Patient Decolonization

Patient decolonization of *C. auris* is currently not recommended as a routine IPC measure due to the lack of firm evidence to support this (Kenters et al. 2019; UKHSA 2025). However, it could be considered on a case-by-case basis when other control measures have failed or in high-risk procedures and patients (UKHSA 2025). Various studies demonstrate good anti-*C. auris* decolonization activity (Wu et al. 2024; Chesnay et al. 2021), though biofilm studies remain limited. There are in vitro studies that have shown the applicability of using chlorhexidine digluconate to effectively reduce and kill *C. auris* biofilms in the context of an antiseptic application (Sherry et al. 2017). Indeed, this is a useful antiseptic given its universal use in healthcare. Moreover, in studies of interkingdom *Staphylococcus aureus* and *C. auris* biofilms, it was shown that chlorhexidine is effective (Gulmez et al. 2022). However, we need to be cognizant of taking these limited data at face value and should instead look to a range of alternative model systems to develop a consensus of evidence. For example, mature *C. auris* biofilms grown in a three-dimensional cellulose model system demonstrated chlorhexidine and hydrogen peroxide tolerance, whereas they were marginally more susceptible to povidone iodine (Kean et al. 2018b), as has been shown elsewhere (Chesnay et al. 2021). Chlorhexidine has also been shown to have a variable and formulation dependent efficacy, with resuspension of the active in 70% isopropyl alcohol yielding more significant killing at a short contact time (Moore et al. 2017). Supporting clinical studies reported detectable *C. auris* despite daily chlorhexidine digluconate washing (Biswal et al. 2017). Together, these data indicate promise for its use, but with being mindful that it will not perform with 100% efficacy. To this end, a range of other studies assessed the relative efficacy of this and other routinely used commercially available biocides against *C. auris* (Abdolrasouli et al. 2017; Cadnum et al. 2017; Moore et al. 2017; Ku et al. 2018). Moreover, some newer antiseptics have now been incorporated into chitosan based bandages for the effective use of octinidene against *C. auris* biofilms (Pandian et al. 2021). We are now in an active period of investigation to tackle these important questions of which are the most effective strategies and antiseptic formulations to control *C. auris* on the human body. Established agents, such as boric acid, triclosan or clotrimazole may also offer potential (Gavilanes-Martinez et al. 2021; Ghannoum et al. 2023), and agents such as polyhexamethylenebiguanide, though active against *Candida* species, remain to be tested against *C. auris* (Montemezzo et al. 2021; Ntow-Boahene et al. 2023). As it has become established as an endemic pathogen in certain regions of the globe (Chowdhary et al. 2023), and threatens to do so more widely (Guitard et al. 2024), then consideration of decolonization protocols are likely to be required and developing the evidence base is of the utmost importance.

Conclusions and Future Perspectives

Biofilms are an important aspect of how *C. auris* persists and contributes to human disease. We have highlighted how its biological properties have enhanced its survival and tolerance to antimicrobial agents and the clinical environment in which it resides. Surface attached biofilm cells on both biological and inert surfaces pose potential risks to the patient directly or to transmission within the healthcare facility or beyond. Therefore, standard and transmission-based IPC measures are essential to manage this globally important pathogen that can be supported by advanced surveillance and epidemiological monitoring. There are significant efforts underway to find methods to enhance existing antifungal strategies, in addition to novel therapeutic control molecules in development that can support clinical management, particularly with consideration for the biofilm lifestyle. However, patients and hospitals will benefit best from understanding the optimal ways to use our existing antifungal, antiseptic and disinfectant resources and being cognizant of the role that *C. auris* biofilms play.

Compliance and Ethical Standards The authors confirm that they have no conflicts and have complied with all ethical standards.

References

Abdolrasouli A, Armstrong-James D, Ryan L, Schelenz S (2017) In vitro efficacy of disinfectants utilised for skin decolonisation and environmental decontamination during a hospital outbreak with Candida auris. Mycoses 60(11):758–763. https://doi.org/10.1111/myc.12699

Ahmad S, Alfouzan W (2021) Candida auris: epidemiology, diagnosis, pathogenesis, antifungal susceptibility, and infection control measures to combat the spread of infections in healthcare facilities. Microorganisms 9(4). https://doi.org/10.3390/microorganisms9040807

Ajetunmobi OH, Wall G, Bonifacio BV, Montelongo-Jauregui D, Lopez-Ribot JL (2023) A 384-well microtiter plate model for Candida biofilm formation and its application to high-throughput screening. Methods Mol Biol 2658:53–64. https://doi.org/10.1007/978-1-0716-3155-3_5

Akinbobola A, Kean R, Quilliam RS (2024) Plastic pollution as a novel reservoir for the environmental survival of the drug resistant fungal pathogen Candida auris. Mar Pollut Bull 198:115841. https://doi.org/10.1016/j.marpolbul.2023.115841

Aldejohann AM, Wiese-Posselt M, Gastmeier P, Kurzai O (2022) Expert recommendations for prevention and management of Candida auris transmission. Mycoses 65(6):590–598. https://doi.org/10.1111/myc.13445

Alem MA, Oteef MD, Flowers TH, Douglas LJ (2006) Production of tyrosol by Candida albicans biofilms and its role in quorum sensing and biofilm development. Eukaryot Cell 5(10):1770–1779. https://doi.org/10.1128/EC.00219-06

AlJindan R, AlEraky DM (2022) Silver nanoparticles: a promising antifungal agent against the growth and biofilm formation of the emergent Candida auris. J Fungi (Basel) 8(7). https://doi.org/10.3390/jof8070744

Almatroudi A, Tahir S, Hu H, Chowdhury D, Gosbell IB, Jensen SO, Whiteley GS, Deva AK, Glasbey T, Vickery K (2018) Staphylococcus aureus dry-surface biofilms are more resistant to heat treatment than traditional hydrated biofilms. J Hosp Infect 98(2):161–167. https://doi.org/10.1016/j.jhin.2017.09.007

Arastehfar A, Daneshnia F, Kord M, Roudbary M, Zarrinfar H, Fang W, Hashemi SJ, Najafzadeh MJ, Khodavaisy S, Pan W, Liao W, Badali H, Rezaie S, Zomorodian K, Hagen F, Boekhout T (2019) Comparison of 21-Plex PCR and API 20C AUX, MALDI-TOF MS, and rDNA sequencing for a wide range of clinically isolated yeast species: improved identification by combining 21-Plex PCR and API 20C AUX as an alternative strategy for developing countries. Front Cell Infect Microbiol 9. https://doi.org/10.3389/fcimb.2019.00021

ARHAI Scotland (2022) Appendix 17—risk assessment principles for health and care settings: hierarchy of controls. NHS National Service Scotland

Arias LS, Butcher MC, Short B, McKloud E, Delaney C, Kean R, Monteiro DR, Williams C, Ramage G, Brown JL (2020) Chitosan Ameliorates Candida auris Virulence in a Galleria mellonella infection model. Antimicrob Agents Chemother 64(8). https://doi.org/10.1128/AAC.00476-20

Baker RE, Mahmud AS, Miller IF, Rajeev M, Rasambainarivo F, Rice BL, Takahashi S, Tatem AJ, Wagner CE, Wang LF, Wesolowski A, Metcalf CJE (2022) Infectious disease in an era of global change. Nat Rev Microbiol 20(4):193–205. https://doi.org/10.1038/s41579-021-00639-z

Beattie SR, Esan T, Zarnowski R, Eix E, Nett JE, Andes DR, Hagen T, Krysan DJ (2023) Novel Keto-alkyl-pyridinium antifungal molecules active in models of in vivo Candida albicans vascular catheter infection and ex vivo Candida auris skin colonization. Antimicrob Agents Chemother 67(5):e0008123. https://doi.org/10.1128/aac.00081-23

Ben-Ami R, Berman J, Novikov A, Bash E, Shachor-Meyouhas Y, Zakin S, Maor Y, Tarabia J, Schechner V, Adler A, Finn T (2017) Multidrug-resistant Candida haemulonii and C. auris, Tel Aviv, Israel. Emerg Infect Dis 23(1):195–203. https://doi.org/10.3201/eid2302.161486

Bing J, Guan Z, Zheng T, Zhang Z, Fan S, Ennis CL, Nobile CJ, Huang G (2023) Clinical isolates of Candida auris with enhanced adherence and biofilm formation due to genomic amplification of ALS4. PLoS Pathog 19(3):e1011239. https://doi.org/10.1371/journal.ppat.1011239

Biswal M, Rudramurthy SM, Jain N, Shamanth AS, Sharma D, Jain K, Yaddanapudi LN, Chakrabarti A (2017) Controlling a possible outbreak of Candida auris infection: lessons learnt from multiple interventions. J Hosp Infect 97(4):363–370. https://doi.org/10.1016/j.jhin.2017.09.009

Borman AM, Szekely A, Johnson EM (2016) Comparative Pathogenicity of United Kingdom Isolates of the Emerging Pathogen Candida auris and Other Key Pathogenic Candida Species. mSphere 1(4). https://doi.org/10.1128/mSphere.00189-16

Bradley SF (2019) Candida auris infection. JAMA 322(15):1526–1526. https://doi.org/10.1001/jama.2019.13857

Bravo Ruiz G, Lorenz A (2021) What do we know about the biology of the emerging fungal pathogen of humans Candida auris? Microbiol Res 242:126621. https://doi.org/10.1016/j.micres.2020.126621

Briano F, Magnasco L, Sepulcri C, Dettori S, Dentone C, Mikulska M, Ball L, Vena A, Robba C, Patroniti N, Brunetti I, Gratarola A, D'Angelo R, Di Pilato V, Coppo E, Marchese A, Pelosi P, Giacobbe DR, Bassetti M (2022) Candida auris Candidemia in critically ill, colonized patients: cumulative incidence and risk factors. Infect Dis Ther 11(3):1149–1160. https://doi.org/10.1007/s40121-022-00625-9

Brown JL, Delaney C, Short B, Butcher MC, McKloud E, Williams C, Kean R, Ramage G (2020) Candida auris phenotypic heterogeneity determines pathogenicity in vitro. mSphere 5:3. https://doi.org/10.1128/mSphere.00371-20

Brown JL, Short B, Ware A, Sherry L, Kean R, Ramage G (2022) Cell viability assays for Candida auris. Methods Mol Biol 2517:129–153. https://doi.org/10.1007/978-1-0716-2417-3_10

Bruno M, Kersten S, Bain JM, Jaeger M, Rosati D, Kruppa MD, Lowman DW, Rice PJ, Graves B, Ma Z, Jiao YN, Chowdhary A, Renieris G, van de Veerdonk FL, Kullberg BJ, Giamarellos-Bourboulis EJ, Hoischen A, Gow NAR, Brown AJP, Meis JF, Williams DL, Netea MG (2020) Transcriptional and functional insights into the host immune response against the emerging fungal pathogen Candida auris. Nat Microbiol 5(12):1516–1531. https://doi.org/10.1038/s41564-020-0780-3

Butcher MC, Brown JL, Hansom D, Wilson-Van Os R, Delury C, Laycock PA, Ramage G (2022) Antifungal-loaded calcium sulfate beads as a potential therapeutic in combating Candida auris. Antimicrob Agents Chemother 66(1):e0171321. https://doi.org/10.1128/AAC.01713-21

Cadnum JL, Shaikh AA, Piedrahita CT, Sankar T, Jencson AL, Larkin EL, Ghannoum MA, Donskey CJ (2017) Effectiveness of disinfectants against Candida auris and other Candida species. Infect Control Hosp Epidemiol 38(10):1240–1243. https://doi.org/10.1017/ice.2017.162

Casadevall A, Kontoyiannis DP, Robert V (2021) Environmental Candida auris and the global warming emergence hypothesis. mBio 12(2):10. https://doi.org/10.1128/mbio.00360-21

CDC (2019) Antibiotic resistance threats in the United States 2019. https://www.cdc.gov/drugresistance/pdf/threats-report/2019-ar-threats-report-508.pdf. Accessed 18 Sept 2022

CDC (2020) Antifungal Susceptibility Testing and Interpretation

CDC (2024a) Infection Control Guidance: Candida auris. https://www.cdc.gov/candida-auris/hcp/infection-control/index.html

CDC (2024b) National Institute for Occupational Safety and Health (NIOSH). About Hierarchy of Controls. CDC

Chandra J, Ghannoum MA (2018) CD101, a novel Echinocandin, possesses potent Antibiofilm activity against early and mature Candida albicans biofilms. Antimicrob Agents Chemother 62(2). https://doi.org/10.1128/AAC.01750-17

Chavez J, Crank K, Barber C, Gerrity D, Iverson T, Mongillo J, Weil A, Rider L, Lacross N, Oakeson K, Rossi A (2024) Early introductions of Candida auris detected by wastewater surveillance, Utah, USA, 2022–2023. Emerg Infect Dis 30(10):2107–2117. https://doi.org/10.3201/eid3010.240173

Chesnay A, Bailly E, Desoubeaux G (2021) Demonstration of the yeasticidal efficacy of povidone-iodine-based commercial antiseptic solutions against Candida auris. J Mycol Med 31(4):101173. https://doi.org/10.1016/j.mycmed.2021.101173

Chowdhary A, Jain K, Chauhan N (2023) Candida auris genetics and emergence. Ann Rev Microbiol 77:583. https://doi.org/10.1146/annurev-micro-032521-015858

Chybowska AD, Childers DS, Farrer RA (2020) Nine things genomics can tell us about Candida auris. Front Genet 11. https://doi.org/10.3389/fgene.2020.00351

de Almeida JN Jr, Francisco EC, Hagen F, Brandao IB, Pereira FM, Presta Dias PH, de Miranda Costa MM, de Souza Jordao RT, de Groot T, Colombo AL (2021) Emergence of Candida auris in Brazil in a COVID-19 intensive care unit. J Fungi (Basel) 7(3). https://doi.org/10.3390/jof7030220

de Jong AW, Hagen F (2019) Attack, defend and persist: how the fungal pathogen Candida auris was able to emerge globally in healthcare environments. Mycopathologia 184:353. https://doi.org/10.1007/s11046-019-00351-w

Di Luca M, Koliszak A, Karbysheva S, Chowdhary A, Meis JF, Trampuz A (2019) Thermogenic characterization and antifungal susceptibility of Candida auris by microcalorimetry. J Fungi (Basel) 5(4). https://doi.org/10.3390/jof5040103

Didik T, Yau AP, Cheung HL, Lee SY, Chan NH, Wah YT, Luk HK, Choi GK, Cheng NH, Tse H, Li Y, Wong SCY, Lung DC (2023) Long-range air dispersion of Candida auris in a cardiothoracic unit outbreak in Hong Kong. J Hosp Infect 142:105–114. https://doi.org/10.1016/j.jhin.2023.09.019

Dire O, Ahmad A, Duze S, Patel M (2023) Survival of Candida auris on environmental surface materials and low-level resistance to disinfectant. J Hosp Infect 137:17–23. https://doi.org/10.1016/j.jhin.2023.04.007

Dominguez E, Zarnowski R, Sanchez H, Covelli AS, Westler WM, Azadi P, Nett J, Mitchell AP, Andes DR (2018) Conservation and divergence in the Candida species biofilm matrix Mannan-Glucan complex structure, function, and genetic control. mBio 9(2). https://doi.org/10.1128/mBio.00451-18

Dominguez EG, Zarnowski R, Choy HL, Zhao M, Sanchez H, Nett JE, Andes DR (2019) Conserved role for biofilm matrix polysaccharides in Candida auris drug resistance. mSphere 4(1). https://doi.org/10.1128/mSphereDirect.00680-18

Du H, Bing J, Hu T, Ennis CL, Nobile CJ, Huang G (2020) Candida auris: epidemiology, biology, antifungal resistance, and virulence. PLoS Pathog 16(10):e1008921. https://doi.org/10.1371/journal.ppat.1008921

Egger NB, Kainz K, Schulze A, Bauer MA, Madeo F, Carmona-Gutierrez D (2022) The rise of Candida auris: from unique traits to co-infection potential. Microb Cell 9(8):141–144. https://doi.org/10.15698/mic2022.08.782

Eyre DW, Sheppard AE, Madder H, Moir I, Moroney R, Quan TP, Griffiths D, George S, Butcher L, Morgan M, Newnham R, Sunderland M, Clarke T, Foster D, Hoffman P, Borman AM, Johnson EM, Moore G, Brown CS, Walker AS, Peto TEA, Crook DW, Jeffery KJM (2018) A Candida auris outbreak and its control in an intensive care setting. N Engl J Med 379(14):1322–1331. https://doi.org/10.1056/NEJMoa1714373

Fanning S, Mitchell AP (2012) Fungal biofilms. PLoS Pathog 8(4):e1002585. https://doi.org/10.1371/journal.ppat.1002585

Fisher MC, Denning DW (2023) The WHO fungal priority pathogens list as a game-changer. Nat Rev Microbiol 21(4):211–212. https://doi.org/10.1038/s41579-023-00861-x

Fisher MC, Alastruey-Izquierdo A, Berman J, Bicanic T, Bignell EM, Bowyer P, Bromley M, Brüggemann R, Garber G, Cornely OA, Gurr SJ, Harrison TS, Kuijper E, Rhodes J, Sheppard DC, Warris A, White PL, Xu J, Zwaan B, Verweij PE (2022) Tackling the emerging threat of antifungal resistance to human health. Nat Rev Microbiol 20(9):557–571. https://doi.org/10.1038/s41579-022-00720-1

Frías-De-León MG, Hernández-Castro R, Vite-Garín T, Arenas R, Bonifaz A, Castañón-Olivares L, Acosta-Altamirano G, Martínez-Herrera E (2020) Antifungal resistance in Candida auris: molecular determinants. Antibiotics (Basel) 9(9). https://doi.org/10.3390/antibiotics9090568

Gavilanes-Martinez MA, Coral-Garzon A, Caceres DH, Garcia AM (2021) Antifungal activity of boric acid, triclosan and zinc oxide against different clinically relevant Candida species. Mycoses 64(9):1045–1052. https://doi.org/10.1111/myc.13302

Ghannoum M, Herrada J, McCormick TS, Long L (2023) A novel transdermal application for clearing skin colonization by Candida auris. Antimicrob Agents Chemother 95(5). https://doi.org/10.1128/AAC.02303-20

Guitard J, Bellanger AP, Dorin J, Cassaing S, Capitaine A, Gabriel F, Nicolas M, Coron N, Penn P, Moniot M, Quinio D, Ranque S, Sasso M, Lepape P, Dannaoui E, Brun S, Lacroix C, Cornu M, Debourgogne A, Durieux MF, Laurent G, Bru V, Bourgeois N, Brunet K, Chouaki T, Huguenin A, Hasseine L, Maubon D, Gangneux JP, Desbois-Nogard N, Houze S, Dalle F, Bougnoux ME, Alanio A, Costa D, Botterel F, Hennequin C (2024) Current knowledge and practice of Candida auris screening in France: a nationwide survey from the French Society of Medical Mycology (SFMM). J Mycol Med 34(3):101490. https://doi.org/10.1016/j.mycmed.2024.101490

Gulati M, Lohse MB, Ennis CL, Gonzalez RE, Perry AM, Bapat P, Arevalo AV, Rodriguez DL, Nobile CJ (2018) In vitro culturing and screening of Candida albicans biofilms. Curr Protoc Microbiol 50(1):e60. https://doi.org/10.1002/cpmc.60

Gulmez D, Brown JL, Butcher MC, Delaney C, Kean R, Ramage G, Short B (2022) Investigating dual-species Candida auris and staphylococcal biofilm antiseptic challenge. Antibiotics (Basel) 11(7). https://doi.org/10.3390/antibiotics11070931

He R, Lin F, Yu B, Huang L (2024) Efficacy and safety of ibrexafungerp in the treatment of vulvovaginal candidiasis: a meta-analysis of randomized controlled trials. Heliyon 10(8):e28776. https://doi.org/10.1016/j.heliyon.2024.e28776

Hernando-Ortiz A, Mateo E, Perez-Rodriguez A, de Groot PWJ, Quindos G, Eraso E (2021) Virulence of Candida auris from different clinical origins in Caenorhabditis elegans and Galleria mellonella host models. Virulence 12(1):1063–1075. https://doi.org/10.1080/21505594.2021.1908765

Hoenigl M, Sprute R, Egger M, Arastehfar A, Cornely OA, Krause R, Lass-Flörl C, Prattes J, Spec A, Thompson GR 3rd, Wiederhold N, Jenks JD (2021) The antifungal pipeline: Fosmanogepix, Ibrexafungerp, Olorofim, Opelconazole, and Rezafungin. Drugs 81(15):1703–1729. https://doi.org/10.1007/s40265-021-01611-0

Jakab A, Kovacs F, Balla N, Nagy-Koteles C, Ragyak A, Nagy F, Borman AM, Majoros L, Kovacs R (2024) Comparative transcriptional analysis of Candida auris biofilms following farnesol and tyrosol treatment. Microbiol Spectr 12(4):e0227823. https://doi.org/10.1128/spectrum.02278-23

Janeczko M, Skrzypek T (2025) Relationships between Candida auris and the rest of the Candida world—analysis of dual-species biofilms and infections. Pathogens 8:40

Jaromin A, Zarnowski R, Markowski A, Zagorska A, Johnson CJ, Etezadi H, Kihara S, Mota-Santiago P, Nett JE, Boyd BJ, Andes DR (2024) Liposomal formulation of a new antifungal hybrid compound provides protection against Candida auris in the ex vivo skin colonization model. Antimicrob Agents Chemother 68(1):e0095523. https://doi.org/10.1128/aac.00955-23

Johnson CJ, Davis JM, Huttenlocher A, Kernien JF, Nett JE (2018) Emerging fungal pathogen Candida auris evades neutrophil attack. mBio 9(4). https://doi.org/10.1128/mBio.01403-18

Jones CR, Neill C, Borman AM, Budd EL, Cummins M, Fry C, Guy RL, Jeffery K, Johnson EM, Manuel R, Mirfenderesky M, Moore G, Patel B, Schelenz S, Staniforth K, Taori SK, Brown CS (2024) The laboratory investigation, management, and infection prevention and control of Candida auris: a narrative review to inform the 2024 national guidance update in England. J Med Microbiol 73(5). https://doi.org/10.1099/jmm.0.001820

Kean R, Ramage G (2019) Combined antifungal resistance and biofilm tolerance: the global threat of Candida auris. mSphere 4(4). https://doi.org/10.1128/mSphere.00458-19

Kean R, Delaney C, Sherry L, Borman A, Johnson EM, Richardson MD, Rautemaa-Richardson R, Williams C, Ramage G (2018a) Transcriptome assembly and profiling of Candida auris reveals novel insights into biofilm-mediated resistance. mSphere 3(4). https://doi.org/10.1128/mSphere.00334-18

Kean R, McKloud E, Townsend EM, Sherry L, Delaney C, Jones BL, Williams C, Ramage G (2018b) The comparative efficacy of antiseptics against Candida auris biofilms. Int J Antimicrob Agents 52(5):673–677. https://doi.org/10.1016/j.ijantimicag.2018.05.007

Kean R, Sherry L, Townsend E, McKloud E, Short B, Akinbobola A, Mackay WG, Williams C, Jones BL, Ramage G (2018c) Surface disinfection challenges for Candida auris: an in-vitro study. J Hosp Infect 98(4):433–436. https://doi.org/10.1016/j.jhin.2017.11.015

Kean R, Brown J, Gulmez D, Ware A, Ramage G (2020) Candida auris: a decade of understanding of an enigmatic pathogenic yeast. J Fungi (Basel) 6(1). https://doi.org/10.3390/jof6010030

Kenters N, Kiernan M, Chowdhary A, Denning DW, Peman J, Saris K, Schelenz S, Tartari E, Widmer A, Meis JF, Voss A (2019) Control of Candida auris in healthcare institutions: outcome of an International Society for Antimicrobial Chemotherapy expert meeting. Int J Antimicrob Agents 54(4):400–406. https://doi.org/10.1016/j.ijantimicag.2019.08.013

Kim SH, Iyer KR, Pardeshi L, Muñoz JF, Robbins N, Cuomo CA, Wong KH, Cowen LE (2019) Genetic analysis of Candida auris implicates Hsp90 in morphogenesis and azole tolerance and Cdr1 in azole resistance. mBio 10(1). https://doi.org/10.1128/mBio.02529-18

Kordalewska M, Perlin DS (2019) Molecular diagnostics in the times of surveillance for Candida auris. Journal of Fungi 5(3):77

Kovacs R, Majoros L (2020) Fungal quorum-sensing molecules: a review of their antifungal effect against Candida biofilms. J Fungi (Basel) 6(3). https://doi.org/10.3390/jof6030099

Kovacs R, Nagy F, Toth Z, Forgacs L, Toth L, Varadi G, Toth GK, Vadaszi K, Borman AM, Majoros L, Galgoczy L (2021) The Neosartorya fischeri antifungal protein 2 (NFAP2): a new potential weapon against multidrug-resistant Candida auris biofilms. Int J Mol Sci 22(2). https://doi.org/10.3390/ijms22020771

Ku TSN, Walraven CJ, Lee SA (2018) Candida auris: disinfectants and implications for infection control. Front Microbiol 9:726. https://doi.org/10.3389/fmicb.2018.00726

Kumari A, Sharma A, Kumari L, Pawar SV, Singh R (2024) Antibiofilm activity of truncated Staphylococcus aureus phenol soluble modulin alpha2 (SaDelta1Delta2PSMalpha2) against Candida auris in vitro and in an animal model of catheter-associated infection. Microb Pathog 196:106943. https://doi.org/10.1016/j.micpath.2024.106943

Larkin E, Hager C, Chandra J, Mukherjee PK, Retuerto M, Salem I, Long L, Isham N, Kovanda L, Borroto-Esoda K, Wring S, Angulo D, Ghannoum M (2017) The emerging pathogen Candida auris: growth phenotype, virulence factors, activity of antifungals, and effect of SCY-078, a novel Glucan synthesis inhibitor, on growth morphology and biofilm formation. Antimicrob Agents Chemother 61(5). https://doi.org/10.1128/AAC.02396-16

Ledwoch K, Maillard JY (2018) Candida auris dry surface biofilm (DSB) for disinfectant efficacy testing. Materials (Basel) 12(1). https://doi.org/10.3390/ma12010018

Ledwoch K, Dancer SJ, Otter JA, Kerr K, Roposte D, Rushton L, Weiser R, Mahenthiralingam E, Muir DD, Maillard JY (2018) Beware biofilm! Dry biofilms containing bacterial pathogens on multiple healthcare surfaces; a multi-Centre study. J Hosp Infect 100(3):e47–e56. https://doi.org/10.1016/j.jhin.2018.06.028

Locke JB, Pillar CM, Castanheira M, Carvalhaes CG, Andes D, Aram JA, Andrzejewski C, Bartizal K, Das AF, Sandison T, Thompson GR 3rd, Pappas PG (2024) Outcomes by Candida spp. in the ReSTORE Phase 3 trial of rezafungin versus caspofungin for candidemia and/or invasive candidiasis. Antimicrob Agents Chemother 68(5):e0158423. https://doi.org/10.1128/aac.01584-23

Lockhart SR, Etienne KA, Vallabhaneni S, Farooqi J, Chowdhary A, Govender NP, Colombo AL, Calvo B, Cuomo CA, Desjardins CA, Berkow EL, Castanheira M, Magobo RE, Jabeen K, Asghar RJ, Meis JF, Jackson B, Chiller T, Litvintseva AP (2017) Simultaneous emergence of multidrug-resistant Candida auris on 3 continents confirmed by whole-genome sequencing and epidemiological analyses. Clin Infect Dis 64(2):134–140. https://doi.org/10.1093/cid/ciw691

Louvet M, Li J, Brandalise D, Bachmann D, Sala de Oyanguren F, Labes D, Jacquier N, Genoud C, Mucciolo A, Coste AT, Sanglard D, Lamoth F (2024) Ume6-dependent pathways of morphogenesis and biofilm formation in Candida auris. Microbiol Spectr 12:e0153124. https://doi.org/10.1128/spectrum.01531-24

Lyman M, Forsberg K, Sexton DJ, Chow NA, Lockhart SR, Jackson BR, Chiller T (2023) Worsening spread of Candida auris in the United States, 2019 to 2021. Ann Intern Med 176(4):489–495. https://doi.org/10.7326/M22-3469

Malavia-Jones D, Farrer RA, Stappers MHT, Edmondson MB, Borman AM, Johnson EM, Lipke PN, Gow NAR (2023) Strain and temperature dependent aggregation of Candida auris is attenuated by inhibition of surface amyloid proteins. Cell Surf 10:100110. https://doi.org/10.1016/j.tcsw.2023.100110

Marcos-Zambrano LJ, Gomez-Perosanz M, Escribano P, Bouza E, Guinea J (2017) The novel oral glucan synthase inhibitor SCY-078 shows in vitro activity against sessile and planktonic Candida spp. J Antimicrob Chemother 72(7):1969–1976. https://doi.org/10.1093/jac/dkx010

Montemezzo M, Ferrari MD, Kerstner E, Santos VD, Victorazzi Lain V, Wollheim C, Frozza C, Roesch-Ely M, Baldo G, Brandalise RN (2021) PHMB-loaded PDMS and its antimicrobial properties for biomedical applications. J Biomater Appl 36(2):252–263. https://doi.org/10.1177/08853282211011921

Moore G, Schelenz S, Borman AM, Johnson EM, Brown CS (2017) Yeasticidal activity of chemical disinfectants and antiseptics against Candida auris. J Hosp Infect 97(4):371–375. https://doi.org/10.1016/j.jhin.2017.08.019

Muller P, Tan CK, Issleib U, Passvogel L, Eilts B, Steinhauer K (2020) Investigation of the susceptibility of Candida auris and Candida albicans to chemical disinfectants using European standards EN 13624 and EN 16615. J Hosp Infect 105(4):648–656. https://doi.org/10.1016/j.jhin.2020.05.026

Muñoz JF, Gade L, Chow NA, Loparev VN, Juieng P, Berkow EL, Farrer RA, Litvintseva AP, Cuomo CA (2018) Genomic insights into multidrug-resistance, mating and virulence in Candida auris and related emerging species. Nat Commun 9(1):5346. https://doi.org/10.1038/s41467-018-07779-6

Neoh CF, Jeong W, Kong DC, Slavin MA (2023) The antifungal pipeline for invasive fungal diseases: what does the future hold? Expert Rev Anti-Infect Ther 21(6):577–594. https://doi.org/10.1080/14787210.2023.2203383

Nett JE, Andes DR (2020) Contributions of the biofilm matrix to Candida pathogenesis. J Fungi (Basel) 6(1). https://doi.org/10.3390/jof6010021

Nobrega de Almeida J Jr, Brandao IB, Francisco EC, de Almeida SLR, de Oliveira Dias P, Pereira FM, Santos Ferreira F, de Andrade TS, de Miranda Costa MM, de Souza Jordao RT, Meis JF, Colombo AL, Candida auris Brazilian Study G (2021) Axillary digital thermometers uplifted a multidrug-susceptible Candida auris outbreak among COVID-19 patients in Brazil. Mycoses 64(9):1062–1072. https://doi.org/10.1111/myc.13320

Ntow-Boahene W, Papandronicou I, Miculob J, Good L (2023) Fungal cell barriers and organelles are disrupted by polyhexamethylene biguanide (PHMB). Sci Rep 13(1):2790. https://doi.org/10.1038/s41598-023-29756-w

Oh BJ, Shin JH, Kim MN, Sung H, Lee K, Joo MY, Shin MG, Suh SP, Ryang DW (2011) Biofilm formation and genotyping of Candida haemulonii, Candida pseudohaemulonii, and a proposed new species (Candida auris) isolates from Korea. Med Mycol 49(1):98–102. https://doi.org/10.3109/13693786.2010.493563

Pandian M, Kumar VA, Jayakumar R (2021) Antiseptic chitosan bandage for preventing topical skin infections. Int J Biol Macromol 193(Pt B):1653–1658. https://doi.org/10.1016/j.ijbiomac.2021.11.002

Parums DV (2022) Editorial: the World Health Organization (WHO) fungal priority pathogens list in response to emerging fungal pathogens during the COVID-19 pandemic. Med Sci Monit 28:e939088. https://doi.org/10.12659/MSM.939088

Patterson CA, Wyncoll D, Patel A, Ceesay Y, Newsholme W, Chand M, Mitchell H, Tan M, Edgeworth JD (2021) Cloth lanyards as a source of intermittent transmission of Candida auris on an ICU. Crit Care Med 49(4):697–701. https://doi.org/10.1097/CCM.0000000000004843

Pfizer (2022) A Study to Learn About the Study Medicine (Called Fosmanogepix/PF-07842805) in People With Candidemia and/or Invasive Candidiasis—ClinicalTrials.Gov https://clinicaltrials.gov/ct2/show/NCT05421858

Pierce CG, Uppuluri P, Tristan AR, Wormley FL Jr, Mowat E, Ramage G, Lopez-Ribot JL (2008) A simple and reproducible 96-well plate-based method for the formation of fungal biofilms and its application to antifungal susceptibility testing. Nat Protoc 3(9):1494–1500. https://doi.org/10.1038/nport.2008.141

Proctor DM, Drummond RA, Lionakis MS, Segre JA (2023) One population, multiple lifestyles: commensalism and pathogenesis in the human mycobiome. Cell Host Microbe 31(4):539–553. https://doi.org/10.1016/j.chom.2023.02.010

Pushpakom S, Iorio F, Eyers PA, Escott KJ, Hopper S, Wells A, Doig A, Guilliams T, Latimer J, McNamee C, Norris A, Sanseau P, Cavalla D, Pirmohamed M (2019) Drug repurposing: progress, challenges and recommendations. Nat Rev Drug Discov 18(1):41–58. https://doi.org/10.1038/nrd.2018.168

Ramage G, Saville SP, Wickes BL, Lopez-Ribot JL (2002) Inhibition of Candida albicans biofilm formation by farnesol, a quorum-sensing molecule. Appl Environ Microbiol 68(11):5459–5463. https://doi.org/10.1128/AEM.68.11.5459-5463.2002

Ramage G, Kean R, Rautemaa-Richardson R, Williams C, Lopez-Ribot JL (2025) Fungal biofilms in human health and disease. Nat Rev Microbiol. https://doi.org/10.1038/s41579-025-01147-0

Rhodes J, Fisher MC (2019) Global epidemiology of emerging Candida auris. Curr Opin Microbiol 52:84–89. https://doi.org/10.1016/j.mib.2019.05.008

Rossato L, Colombo AL (2018) Candida auris: what have we learned about its mechanisms of pathogenicity? Front Microbiol 9. https://doi.org/10.3389/fmicb.2018.03081

Rossoni RD, de Barros PP, Mendonca IDC, Medina RP, Silva DHS, Fuchs BB, Junqueira JC, Mylonakis E (2020) The postbiotic activity of lactobacillus paracasei 28.4 against Candida auris. Front Cell Infect Microbiol 10:397. https://doi.org/10.3389/fcimb.2020.00397

Rossow J, Ostrowsky B, Adams E, Greenko J, McDonald R, Vallabhaneni S, Forsberg K, Perez S, Lucas T, Alroy KA, Jacobs Slifka K, Walters M, Jackson BR, Quinn M, Chaturvedi S, Blog D, New York Candida auris Investigation W (2021) Factors associated with Candida auris colonization and transmission in skilled nursing facilities with ventilator units, New York, 2016–2018. Clin Infect Dis 72(11):e753–e760. https://doi.org/10.1093/cid/ciaa1462

Rudramurthy SM, Chakrabarti A, Paul RA, Sood P, Kaur H, Capoor MR, Kindo AJ, Marak RSK, Arora A, Sardana R, Das S, Chhina D, Patel A, Xess I, Tarai B, Singh P, Ghosh A (2017) Candida auris candidaemia in Indian ICUs: analysis of risk factors. J Antimicrob Chemother 72(6):1794–1801. https://doi.org/10.1093/jac/dkx034

Ruiz GB, Ross ZK, Gow NAR, Lorenz A (2020) Pseudohyphal growth of the emerging pathogen Candida auris is triggered by genotoxic stress through the S phase checkpoint. mSphere 5(2). https://doi.org/10.1128/msphere.00151-20

Sansom SE, Gussin GM, Schoeny M, Singh RD, Adil H, Bell P, Benson EC, Bittencourt CE, Black S, Guzman M DMV, Froilan MC, Fukuda C, Barsegyan K, Gough E, Lyman M, Makhija J, Marron S, Mikhail L, Noble-Wang J, Pacilli M, Pedroza R, Saavedra R, Sexton DJ, Shimabukuro J, Thotapalli L, Zahn M, Huang SS, Hayden MK (2024) Rapid environmental contamination with Candida auris and multidrug-resistant bacterial pathogens near colonized patients. Clin Infect Dis 78(5):1276–1284. https://doi.org/10.1093/cid/ciad752

Santana DJ, Anku JAE, Zhao G, Zarnowski R, Johnson CJ, Hautau H, Visser ND, Ibrahim AS, Andes D, Nett JE, Singh S, O'Meara TR (2023) A Candida auris-specific adhesin, Scf1, governs surface association, colonization, and virulence. Science 381(6665):1461–1467. https://doi.org/10.1126/science.adf8972

Satoh K, Makimura K, Hasumi Y, Nishiyama Y, Uchida K, Yamaguchi H (2009) Candida auris sp. nov., a novel ascomycetous yeast isolated from the external ear canal of an inpatient in a Japanese hospital. Microbiol Immunol 53:41. https://doi.org/10.1111/j.1348-0421.2008.00083.x

Schelenz S, Hagen F, Rhodes JL, Abdolrasouli A, Chowdhary A, Hall A, Ryan L, Shackleton J, Trimlett R, Meis JF, Armstrong-James D, Fisher MC (2016) First hospital outbreak of the globally emerging Candida auris in a European hospital. Antimicrob Resist Infect Control 5:35. https://doi.org/10.1186/s13756-016-0132-5

Semreen MH, Soliman SSM, Saeed BQ, Alqarihi A, Uppuluri P, Ibrahim AS (2019) Metabolic profiling of Candida auris, a newly-emerging multi-drug resistant Candida species, by GC-MS. Molecules 24(3). https://doi.org/10.3390/molecules24030399

Shaban S, Patel M, Ahmad A (2020) Improved efficacy of antifungal drugs in combination with monoterpene phenols against Candida auris. Sci Rep 10(1):1162. https://doi.org/10.1038/s41598-020-58203-3

Shaw KJ, Ibrahim AS (2020) Fosmanogepix: a review of the first-in-class broad Spectrum agent for the treatment of invasive fungal infections. J Fungi 6(4):239. https://doi.org/10.3390/jof6040239

Sherry L, Ramage G, Kean R, Borman A, Johnson EM, Richardson MD, Rautemaa-Richardson R (2017) Biofilm-forming capability of highly virulent, multidrug-resistant Candida auris. Emerg Infect Dis 23(2):328–331. https://doi.org/10.3201/eid2302.161320

Shivarathri R, Jenull S, Stoiber A, Chauhan M, Mazumdar R, Singh A, Nogueira F, Kuchler K, Chowdhary A, Chauhan N (2020) The two-component response regulator Ssk1 and the mitogen-activated protein kinase Hog1 control antifungal drug resistance and Cell Wall architecture of Candida auris. mSphere 5:5. https://doi.org/10.1128/mSphere.00973-20

Short B, Brown J, Delaney C, Sherry L, Williams C, Ramage G, Kean R (2019) Candida auris exhibits resilient biofilm characteristics in vitro: implications for environmental persistence. J Hosp Infect 103:92. https://doi.org/10.1016/j.jhin.2019.06.006

Singh U, Akhtar S, Mishra A, Sarkar D (2011) A novel screening method based on menadione mediated rapid reduction of tetrazolium salt for testing of anti-mycobacterial agents. J Microbiol Methods 84(2):202–207. https://doi.org/10.1016/j.mimet.2010.11.013

Singh R, Kaur M, Chakrabarti A, Shankarnarayan SA, Rudramurthy SM (2019) Biofilm formation by Candida auris isolated from colonising sites and candidemia cases. Mycoses 62(8):706–709. https://doi.org/10.1111/myc.12947

Smith LL, Falvey J, Grace B, Vaeth E, Rubin J, Perlmutter R, Blythe D, Hawkins D, Mbuthia M, Roghmann MC, Rock C, Leekha S (2024) C. auris and neighborhood socioeconomic vulnerability in the state of Maryland from 2019 to 2022. Infect Control Hosp Epidemiol 1–7:1183. https://doi.org/10.1017/ice.2024.98

Sudbery P, Gow N, Berman J, Sudbery P, Gow N, Berman J (2004) The distinct morphogenic states of Candida albicans. Trends Microbiol 12:317–324. https://doi.org/10.1016/j.tim.2004.05.008

Sundstrom P (2002) Adhesion in Candida spp. Cell Microbiol 4(8):461–469. https://doi.org/10.1046/j.1462-5822.2002.00206.x

Szymański P, Markowicz M, Mikiciuk-Olasik E (2012) Adaptation of high-throughput screening in drug discovery-toxicological screening tests. Int J Mol Sci 13(1):427–452. https://doi.org/10.3390/ijms13010427

Taff HT, Nett JE, Zarnowski R, Ross KM, Sanchez H, Cain MT, Hamaker J, Mitchell AP, Andes DR (2012) A Candida biofilm-induced pathway for matrix glucan delivery: implications for drug resistance. PLoS Pathog 8(8):e1002848. https://doi.org/10.1371/journal.ppat.1002848

Taff HT, Mitchell KF, Edward JA, Andes DR (2013) Mechanisms of Candida biofilm drug resistance. Future Microbiol 8(10):1325–1337. https://doi.org/10.2217/fmb.13.101

Tharp B, Zheng R, Bryak G, Litvintseva AP, Hayden MK, Chowdhary A, Thangamani S (2023) Role of microbiota in the skin colonization of Candida auris. mSphere 8(1):e0062322. https://doi.org/10.1128/msphere.00623-22

Thompson GR 3rd, Soriano A, Honore PM, Bassetti M, Cornely OA, Kollef M, Kullberg BJ, Pullman J, Hites M, Fortun J, Horcajada JP, Kotanidou A, Das AF, Sandison T, Aram JA, Vazquez JA, Pappas PG (2024) Efficacy and safety of rezafungin and caspofungin in candidaemia and invasive candidiasis: pooled data from two prospective randomised controlled trials. Lancet Infect Dis 24(3):319–328. https://doi.org/10.1016/S1473-3099(23)00551-0

UKHSA (2025) Candidozyma auris (formerly Candida auris): guidance for acute healthcare settings. https://www.gov.uk/government/publications/candida-auris-laboratory-investigation-management-and-infection-prevention-and-control

Uppuluri P, Acosta Zaldivar M, Anderson MZ, Dunn MJ, Berman J, Lopez Ribot JL, Kohler JR (2018) Candida albicans dispersed cells are developmentally distinct from biofilm and planktonic cells. mBio 9:4. https://doi.org/10.1128/mBio.01338-18

Van Dijck P, Sjollema J, Cammue BP, Lagrou K, Berman J, d'Enfert C, Andes DR, Arendrup MC, Brakhage AA, Calderone R, Canton E, Coenye T, Cos P, Cowen LE, Edgerton M, Espinel-Ingroff A, Filler SG, Ghannoum M, Gow NAR, Haas H, Jabra-Rizk MA, Johnson EM, Lockhart SR, Lopez-Ribot JL, Maertens J, Munro CA, Nett JE, Nobile CJ, Pfaller MA, Ramage G, Sanglard D, Sanguinetti M, Spriet I, Verweij PE, Warris A, Wauters J, Yeaman MR, Zaat SAJ, Thevissen K (2018) Methodologies for in vitro and in vivo evaluation of efficacy of antifungal and antibiofilm agents and surface coatings against fungal biofilms. Microb Cell 5(7):300–326. https://doi.org/10.15698/mic2018.07.638

Vatanshenassan M, Boekhout T, Lass-Flörl C, Lackner M, Schubert S, Kostrzewa M, Sparbier K (2018) Proof of concept for MBT ASTRA, a rapid matrix-assisted laser desorption ionization-time of flight mass spectrometry (MALDI-TOF MS)-based method to detect caspofungin resistance in Candida albicans and Candida glabrata. J Clin Microbiol 56(9). https://doi.org/10.1128/jcm.00420-18

Vazquez-Munoz R, Lopez FD, Lopez-Ribot JL (2020) Silver Nanoantibiotics display strong antifungal activity against the emergent multidrug-resistant yeast Candida auris under both planktonic and biofilm growing conditions. Front Microbiol 11. https://doi.org/10.3389/fmicb.2020.01673

Vila T, Montelongo-Jauregui D, Ahmed H, Puthran T, Sultan AS, Jabra-Rizk MA (2020) Comparative evaluations of the pathogenesis of Candida auris phenotypes and Candida albicans using clinically relevant murine models of infections. mSphere 5(4). https://doi.org/10.1128/msphere.00760-20

Wall G, Montelongo-Jauregui D, Vidal Bonifacio B, Lopez-Ribot JL, Uppuluri P (2019) Candida albicans biofilm growth and dispersal: contributions to pathogenesis. Curr Opin Microbiol 52:1–6. https://doi.org/10.1016/j.mib.2019.04.001

Wang X, Bing J, Zheng Q, Zhang F, Liu J, Yue H, Tao L, Wang Y, Wang H, Huang G (2018) The first isolate of Candida auris in China: clinical and biological aspects. Emerging Microbes & Infections 7:1. https://doi.org/10.1038/s41426-018-0095-0

Wang TW, Sofras D, Montelongo-Jauregui D, Paiva TO, Carolus H, Dufrene YF, Alfaifi AA, McCracken C, Bruno VM, Van Dijck P, Jabra-Rizk MA (2024) Functional redundancy in Candida auris cell surface Adhesins crucial for cell-cell interaction and aggregation. Res Sq. https://doi.org/10.21203/rs.3.rs-4077218/v1

Ware A, Johnson W, Delaney C, Butcher MC, Price L, Ramage G, Butcher J, Kean R (2023) Dry surface biofilm formation by Candida auris facilitates persistence and tolerance to sodium hypochlorite. bioRxiv. https://doi.org/10.1101/2023.10.02.560537

Welsh RM, Bentz ML, Shams A, Houston H, Lyons A, Rose LJ, Litvintseva AP (2017) Survival, persistence, and isolation of the emerging multidrug-resistant pathogenic yeast Candida auris on a plastic health care surface. J Clin Microbiol 55(10):2996–3005. https://doi.org/10.1128/JCM.00921-17

Wiederhold NP (2022) Pharmacodynamics, mechanisms of action and resistance, and spectrum of activity of new antifungal agents. J Fungi 8(8):857. https://doi.org/10.3390/jof8080857

Wiederhold NP, Najvar LK, Olivo M, Morris KN, Patterson HP, Catano G, Patterson TF (2021) Ibrexafungerp demonstrates in vitro activity against fluconazole-resistant Candida auris and in vivo efficacy with delayed initiation of therapy in an experimental model of invasive candidiasis. Antimicrob Agents Chemother 65(6):e02694–e02620. https://doi.org/10.1128/AAC.02694-20

Wu WG, Luk KS, Hung MF, Tsang WY, Lee KP, Lam BH, Cheng KL, Cheung WS, Tang HL, To WK (2024) Antifungal efficacy of natural antiseptic products against Candida auris. Med Mycol 62(7). https://doi.org/10.1093/mmy/myae060

Wuyts J, Van Dijck P, Holtappels M (2018) Fungal persister cells: the basis for recalcitrant infections? PLoS Pathog 14(10):e1007301. https://doi.org/10.1371/journal.ppat.1007301

Yan L, Xia K, Yu Y, Miliakos A, Chaturvedi S, Zhang F, Chen S, Chaturvedi V, Linhardt RJ (2020) Unique cell surface Mannan of yeast pathogen Candida auris with selective binding to IgG. ACS Infect Dis 6(5):1018–1031. https://doi.org/10.1021/acsinfecdis.9b00450

Yue H, Bing J, Zheng Q, Zhang Y, Hu T, Du H, Wang H, Huang G (2018) Filamentation in Candida auris, an emerging fungal pathogen of humans: passage through the mammalian body induces a heritable phenotypic switch. Emerg Microbes Infect 7(1):188. https://doi.org/10.1038/s41426-018-0187-x

Zamith-Miranda D, Heyman HM, Couvillion SP, Cordero RJB, Rodrigues ML, Nimrichter L, Casadevall A, Amatuzzi RF, Alves LR, Nakayasu ES, Nosanchuk JD (2021) Comparative Molecular and Immunoregulatory Analysis of Extracellular Vesicles from Candida albicans and Candida auris. mSystems 6(4):e0082221. https://doi.org/10.1128/mSystems.00822-21

Zarnowski R, Sanchez H, Jaromin A, Zarnowska UJ, Nett JE, Mitchell AP, Andes D (2022) A common vesicle proteome drives fungal biofilm development. Proc Natl Acad Sci USA 119(38):e2211424119. https://doi.org/10.1073/pnas.2211424119

Zatorska B, Moser D, Diab-Elschahawi M, Ebner J, Lusignani LS, Presterl E (2021) The effectiveness of surface disinfectants and a micellic H(2)O(2) based water disinfectant on Candida auris. J Mycol Med 31(4):101178. https://doi.org/10.1016/j.mycmed.2021.101178

Zhao M, Zhang F, Zarnowski R, Barns K, Jones R, Fossen J, Sanchez H, Rajski SR, Audhya A, Bugni TS, Andes DR (2021) Turbinmicin inhibits Candida biofilm growth by disrupting fungal vesicle-mediated trafficking. J Clin Invest 131(5). https://doi.org/10.1172/JCI145123

Zhou W, Li X, Lin Y, Yan W, Jiang S, Huang X, Yang X, Qiao D, Li N (2021) A comparative transcriptome between anti-drug sensitive and resistant Candida auris in China. Front Microbiol 12:708009. https://doi.org/10.3389/fmicb.2021.708009

Chapter 8
Composition and Role of the Extracellular Matrix in Fungal Biofilm Drug Resistance

Min-Ju Kim, Robert Zarnowski, and David Andes

Abstract This chapter reviews the role of the extracellular matrix (ECM) in fungal biofilm resistance, particularly in *Candida* species, *Cryptococcus neoformans*, and *Aspergillus fumigatus*. ECM serves as a protective barrier that contributes to biofilm resistance to both host immune defenses and antifungal drug treatment. In *Candida* biofilms, the ECM is involved in drug sequestration and limits the efficacy of antifungal agents across multiple layers. Previous studies of matrix composition in *Candida* revealed critical components such as mannan-glucan complexes that protect against drug penetration. Similarly, ECMs in *C. neoformans* and *A. fumigatus* contribute to biofilm integrity and drug resistance with specific matrix elements like glucuronoxylomannan in *C. neoformans* and galactosaminogalactan in *A. fumigatus*. The chapter highlights matrix-mediated drug resistance mechanisms including the sequestration of antifungal drugs and modulation of host immune responses. It suggests that targeting ECM components could offer new avenues for biofilm-related infection therapies.

Introduction

Biofilms exhibit characteristics distinct from planktonic growth. Two phenotypes that are unique to biofilms yet common across microbes are the presence of a community encasing extracellular polymeric matrix and resistance to both host immune and antimicrobial drug clearance. The extracellular matrix has been shown to provide protection for the microbial biofilm population from external threats including the host immune system and antimicrobial drugs. The present chapter chronicles the role of the extracellular matrix role in fungal biofilm persistence, including phenotypic, biochemical, and genetic data that delineate this pathogenesis phenomenon.

M.-J. Kim · R. Zarnowski · D. Andes (✉)
University of Wisconsin, Madison, WI, USA
e-mail: dra@medicine.wisc.edu

G. Ramage, D. Andes (eds.), *Fungal Biofilms*, Springer Series on Biofilms 15,
https://doi.org/10.1007/978-3-031-93315-8_8

Candida Biofilm Antifungal Drug Resistance

Numerous studies have examined the mechanisms behind *Candida* biofilm resistance to antifungal drugs, identifying two main mechanistic categories (Fig. 8.1) (Taff et al. 2013). The first involves mechanisms shared by both planktonic and biofilm cells, such as the upregulation of drug efflux pumps, Hsp90 (heat shock protein), and the calcineurin pathway, which primarily affect resistance to triazole drugs (Ramage et al. 2002a; Uppuluri et al. 2008; Robbins et al. 2011). However, these mechanisms account for less than half of the overall drug resistance phenotype, leaving the majority unexplained by planktonic resistance mechanisms alone. Two additional drug resistance mechanisms are specific to the biofilm state, namely quorum sensing and protection provided by the extracellular matrix (Ramage et al.

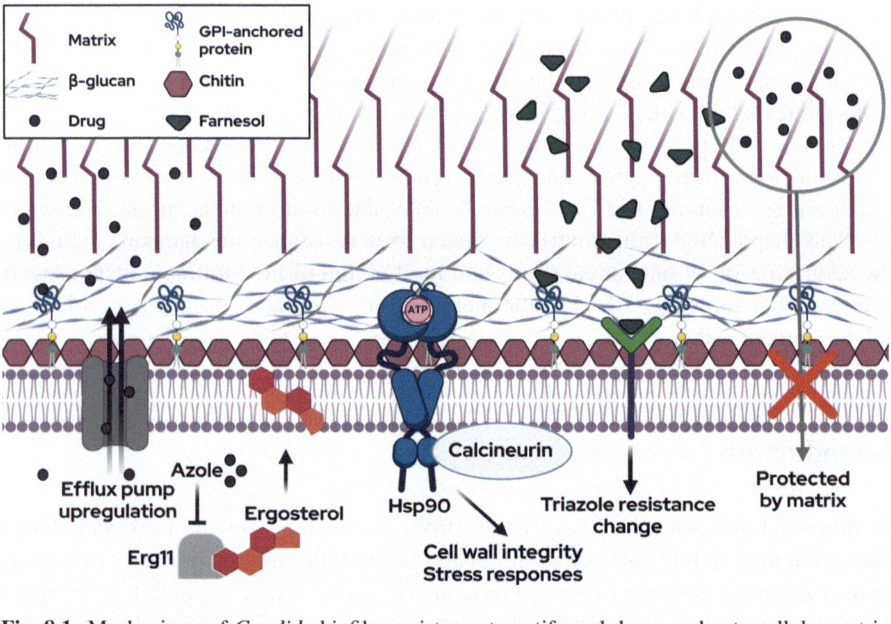

Fig. 8.1 Mechanisms of *Candida* biofilm resistance to antifungal drugs and extracellular matrix composition. Ergosterol is the key sterol in fungal cell membranes. Azole drugs inhibit ergosterol synthesis by targeting *ERG11* gene. *Upregulation of drug efflux pumps* is a key mechanism of antifungal drug resistance, leading to reduced drug accumulation and increased resistance to antifungal drugs. *Hsp90* (Heat shock protein 90), a chaperone protein, is essential for *Candida albicans* capacity to cause infection and develop resistance to antifungal treatments. Hsp90 plays a crucial role in stabilizing calcineurin, a key protein involved in fungal stress responses and cell wall integrity. By maintaining calcineurin's stability, Hsp90 helps the fungus endure the effects of antifungal agents, thereby contributing to the development of drug resistance. *Quorum sensing* mechanism impacts virulence factors such as biofilm formation and hyphal development, with molecules like farnesol and tyrosol. *Extracellular matrix (ECM)*, especially the mannan-glucan complex, can sequester antifungal drugs, hindering their ability to reach fungal cells. As a physical barrier, the ECM limits the penetration of antifungal agents into the biofilm. Additionally, the ECM shields fungal cells from the host immune system and various environmental stressors

2002b). Quorum sensing contributes to only modest triazole resistance changes (Perumal et al. 2007), whereas the extracellular matrix's protective properties appear to work across drug classes (Mitchell et al. 2016). Early evidence for the extracellular matrix's role showed that polyene antifungal amphotericin B was less effective when conditions optimized matrix production, likely due to increased matrix presence under higher flow conditions (Al-Fattani and Douglas 2004, 2006; Vediyappan et al. 2010). More recent studies suggest that the extracellular matrix sequesters antifungal drugs, preventing them from reaching their target sites. For instance, radiolabeled fluconazole remains primarily within the matrix of mature *C. albicans* biofilms (Nett et al. 2010a), and spectroscopy analysis with amphotericin B finds a similar matrix association (Vediyappan et al. 2010). While the specific interaction between antifungal drugs and the matrix remains undefined, the range of physicochemical properties of sequestered compounds suggests a noncovalent, complex interaction.

Candida Species Extracellular Matrix Composition, Production, and Function

Candida albicans is a major fungal pathogen known for frequently forming biofilms on implanted medical devices. Initial studies of the *C. albicans* extracellular matrix focused on identifying its composition, revealing a complex mixture of nucleic acids, proteins, lipids, and carbohydrates (Thomas et al. 2006; Silva et al. 2009; Martins et al. 2010; Zarnowski et al. 2014, 2016). More detailed investigations have analyzed the proteome and glycome. Granular assessment of the glycome identified a novel mannan-glucan complex containing α-1,2 branched α-1,6 mannan, β-1,6 glucan, and β-1,3 glucan that is distinct from related structures in the cell wall (Zarnowski et al. 2014). Candidate genetic approaches subsequently then identified several components of pathways in *C. albicans* essential for producing these matrix components, with key roles observed for genes with putative roles in synthesizing and modifying glucans and mannans (Nett et al. 2010a, 2011; Taff et al. 2012; Mitchell et al. 2015). Studies on deletion mutants further demonstrated that these genes and matrix components, especially the polysaccharide components, contribute to triazole resistance, notably through matrix sequestration of the antifungal drug, fluconazole. For example, *C. albicans* mutants deficient in β-1,3 glucan synthesis (via *FKS1* deletion) displayed fluconazole susceptibility at concentrations as low as 4 μg/ml (near the planktonic MIC), while wild-type biofilms resisted concentrations as high as 2 mg/ml (Nett et al. 2007, 2010a).

Further research has suggested potential therapeutic targets within these matrix components. Genetic deletion of mannan production and modification genes, pharmacologic inhibition of the mannan-glucan complex (e.g., with tunicamycin), and enzymatic degradation of matrix polysaccharides (e.g., with mannosidase or glucanase) reduced biofilm burden and enhanced antifungal efficacy (Mitchell et al.

2015). These studies also revealed the protective role of the matrix across multiple antifungal drug classes—including polyenes, echinocandins, flucytosine, triazoles (Nett et al. 2010b), and even antibacterial drugs (Harriott and Noverr 2009, 2010; Kong et al. 2016). Specifically, in mixed *Candida*–bacterial biofilms, the *Candida* extracellular matrix shields *Staphylococcus aureus* and *Escherichia coli* from glycopeptide and quinolone antibiotics (Harriott and Noverr 2009, 2010; Kong et al. 2016).

Investigations have expanded to assess the matrix's role across various *Candida* species. Analyses of matrices from *C. glabrata*, *C. parapsilosis*, *C. tropicalis*, and the emerging *C. auris* identified similar macromolecular compositions and conserved mannan-glucan complexes, suggesting a shared biofilm defense mechanism among *Candida* species (Silva et al. 2009; Dominguez et al. 2018, 2019). Proteomic and nucleic acid studies confirmed composition conservation, while genetic and pharmacologic approaches further supported a similar matrix-sequestration mechanism in biofilm protection across these species (Mitchell et al. 2013).

Studies on the composition and function of the biofilm extracellular matrix highlight its critical role in antifungal drug resistance, but the mechanisms of matrix assembly and delivery remained unclear. A breakthrough came from experiments combining *C. albicans* pathway mutants defective in glucan and mannan production. Individually, these mutants failed to produce a robust matrix and were susceptible to fluconazole. However, mixing mutants from the two pathways restored biofilm matrix production and drug resistance, suggesting extracellular assembly of the mannan-glucan complex components (Mitchell et al. 2015). This pointed to a cooperative process where components from different cells and pathways contribute to the functional matrix.

Further insights into matrix delivery and assembly emerged from studies on the lipid components of the matrix. Analyses revealed phospholipids and sphingolipids, likely derived from either fungal plasma membranes or extracellular vesicles (Zarnowski et al. 2014; Brown et al. 2015). High-magnification SEM imaging confirmed vesicles on biofilm cell surfaces, and time course quantitative analysis of vesicle production correlated with matrix accumulation during biofilm maturation (Zarnowski et al. 2018). Characterization of vesicle cargo showed overlap with matrix constituents, including nucleic acids, proteins, and mannan-glucan polysaccharides, suggesting vesicles are delivery vehicles for matrix assembly. These findings underscore the role of vesicle-mediated transport in constructing the biofilm matrix and promoting drug resistance.

Research into biofilm vesicle production in *C. albicans* discovered the role of the ESCRT pathway, a mechanism known to regulate exosome production in other systems (Abels and Breakefield 2016). Screening mutants of ESCRT pathway homologs revealed 16 with significant defects in biofilm vesicle production (Zarnowski et al. 2018, 2021). These mutants also exhibited deficiencies in biofilm matrix formation and antifungal drug resistance, suggesting a role for vesicles in delivering functional extracellular matrix components. To assess this further, researchers developed a vesicle 'add-back' assay in which vesicles from the wild-type strain are added exogenously to mutant biofilms. Restoration of both matrix production and drug resistance phenotypes with vesicle addition is supportive of the idea that functional vesicle cargo plays a critical role in biofilm formation and antifungal resistance.

Subsequent studies used machine learning to analyze vesicle cargo proteomic data from ESCRT mutants, narrowing nearly 1000 vesicle proteins to about 60 candidates potentially associated with drug resistance (Zarnowski et al. 2021). Screening deletion mutants identified approximately 20 proteins involved in biofilm matrix formation and drug protection, most of which were linked to mannan or glucan modification. Wild-type vesicle addition again restored biofilm matrix and resistance phenotypes in these mutants, whereas vesicles from cargo-deficient mutants did not, reinforcing the specificity of vesicle cargo function. Further studies are needed to elucidate the enzymatic and structural roles of these proteins in extracellular matrix maturation.

Extracellular Matrix of *Cryptococcus neoformans*

Cryptococcus neoformans, a fungus capable of causing meningoencephalitis in humans, can form biofilms on polystyrene plates and medical devices such as intracranial shunts used to treat elevated intracranial pressure (Martinez and Casadevall 2015). During biofilm development, the extracellular matrix gradually accumulates, providing structure and added complexity to clusters of adherent cells (Fig. 8.2) (Martinez et al. 2010; Martinez and Casadevall 2007).

This basidiomycete grows in a yeast form encased by a polysaccharide capsule composed of glucuronoxylomannan (GXM), which is essential for its virulence. GXM is released into the surroundings, and this capsule structure is necessary for biofilm formation (Martinez and Casadevall 2005). Early studies of the *C. neoformans* biofilm matrix used monoclonal antibodies to identify GXM, confirming its presence around biofilm cells. GC/MS analysis further revealed monosaccharides consistent with GXM, along with additional monosaccharides—glucose, ribose, and fucose—indicating the matrix may contain other types of polysaccharides (Martinez and Casadevall 2007).

It is theorized the *Cryptococcus* matrix likely shields biofilm cells from environmental and infection-related stressors (Martinez and Casadevall 2006a, 2007; Alvarez et al. 2008). Compared to planktonic cells, *C. neoformans* biofilms demonstrate greater resistance to temperature fluctuations, pH changes, and UV light, supporting the idea that biofilm growth offers survival advantages in nature (Martinez and Casadevall 2007). Similar to biofilms formed by other pathogenic fungi, *C. neoformans* biofilms exhibit higher resistance to antifungal treatments than planktonic cells (Martinez and Casadevall 2006a) and also show increased resistance to antimicrobial agents produced by the innate immune system (Martinez and Casadevall 2006b). Notably, a study involving both *C. neoformans* and *C. gattii* revealed that biofilm-like microcolonies are associated with resistance to phagocytosis and escape from macrophages. These microcolonies were surrounded by a

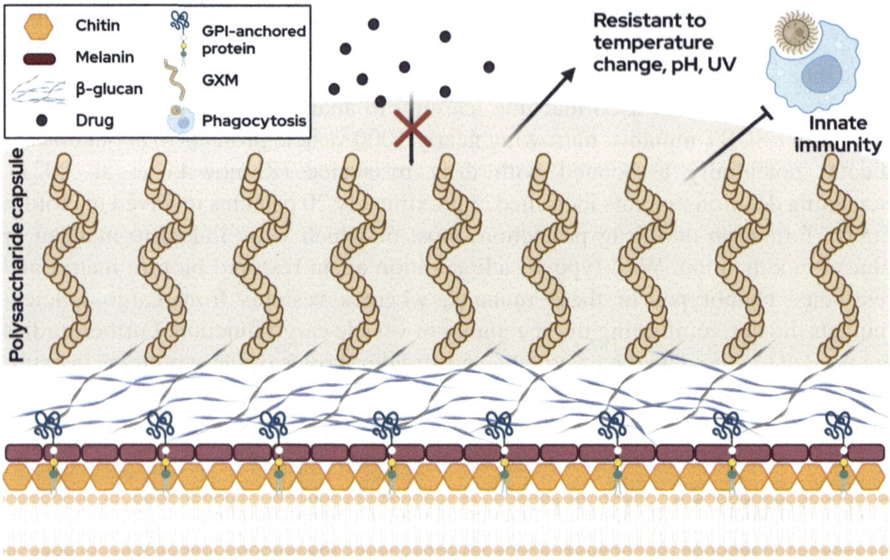

Fig. 8.2 Extracellular matrix composition and functions in *Cryptococcus neoformans*. In *C. neoformans*, the extracellular matrix (ECM) is composed of a polysaccharide capsule with glucuronoxylomannan (GXM) and other monosaccharides such as glucose, ribose, and fucose. GXM is a key component of the *C. neoformans* capsule, a sugar-based outer layer that shields the yeast from attacks by host immune cells. It plays an essential role in immunomodulation, influencing the host's immune response. GXM helps the fungus escape immune detection by interfering with phagocytosis, inhibiting leukocyte migration, and regulating cytokine production. Additionally, the ECM protects fungal cells from environmental stresses and contributes to biofilm formation, thus enhancing the fungal persistence and resistance to antifungal treatments

polysaccharide matrix containing antibodies, suggesting that the fungus may leverage antibody-mediated agglutination to its advantage (Alvarez et al. 2008). However, the role of matrix and specific matrix components in drug resistance remains to be elucidated.

Extracellular Matrix Composition and Function in *Aspergillus fumigatus* Biofilms

Aspergillus fumigatus is a prevalent airborne filamentous fungal pathogen and can cause pulmonary infections, including acute invasive aspergillosis (IA) and chronic pulmonary aspergillosis. *A. fumigatus* produces biofilms in both acute and chronic pulmonary infections, though the structure and composition of these biofilms may differ between these conditions (Loussert et al. 2010). Moreover, *A. fumigatus* can colonize the airways of patients with cystic fibrosis (CF) including biofilm formation (Nelson et al. 1979; Al Shakirchi et al. 2021).

A. fumigatus biofilms offer protection against antifungal treatments and host immune responses similar to yeast biofilms. The composition of *A. fumigatus* biofilms differs from that of yeast biofilms. For example, *Candida* biofilms are characterized by a dense network of mixed morphological forms, including yeast cells, hyphae, and pseudohyphae (Lohse et al. 2018). In contrast, *A. fumigatus* biofilms primarily consist of interconnected, branched multinucleate vegetative hyphae (Morelli et al. 2021; Kowalski et al. 2019).

One feature of biofilms in all fungi is the existence of extracellular matrix. In *Aspergillus* spp., the extracellular matrix has been shown to promote surface adherence and increasing resistance to antifungal agents, and host defenses (Fig. 8.3). The initial investigations into the content of the *Aspergillus* biofilm matrix cultured under aerial-static conditions in glucose yeast extract (GYE) medium identified galactomannan, α-1,3 glucan, proteins, polyols, and melanin (Beauvais et al. 2007).

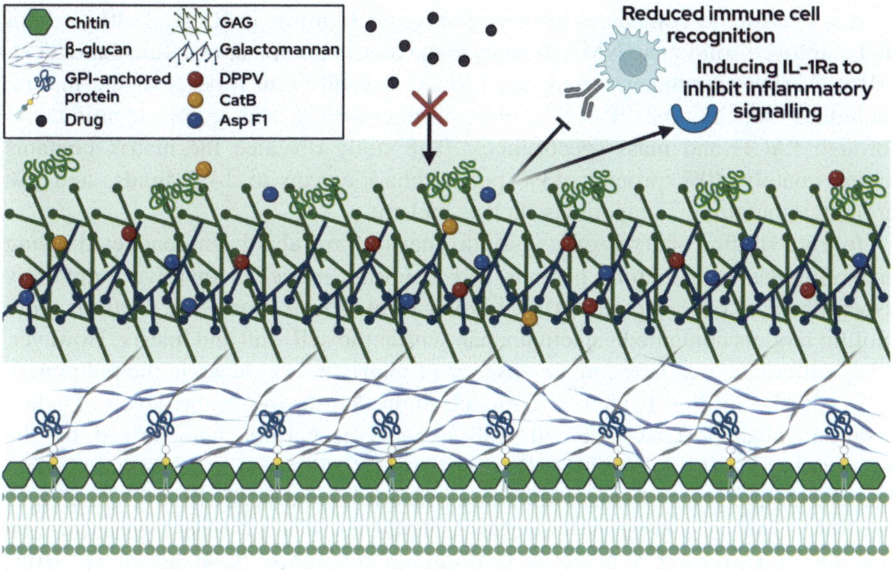

Fig. 8.3 Extracellular matrix composition and functions in *Aspergillus fumigatus*. In *A. fumigatus*, the extracellular matrix (ECM) is composed of galactosaminogalactan (GAG), galactomannan, hydrophobins, major antigen proteins including dipeptidyl-peptidase V (DPPV), catalase B (CatB), and ribotoxin (Asp F1). GAG is crucial for the development of *A. fumigatus* biofilms, which are resistant to antifungal treatments and play a role in chronic infections. It acts as an adhesin, facilitating the attachment of *A. fumigatus* to host tissues, such as pulmonary epithelial cells, endothelial cells, and fibronectin. GAG interacts with the host immune system, possibly influencing immune responses and helping the pathogen evade host defenses. It alters host immune cell recognition by masking β-glucan exposure on the surface of the fungus during hyphal growth. Also, GAG stimulates the production of interleukin-1 receptor antagonist (IL-1Ra), a natural anti-inflammatory cytokine that blocks the pro-inflammatory effects of IL-1 signaling. In *A. fumigatus* biofilms, the ECM serves as a binding layer, holding the hyphae (fungal filaments) together. It offers structural support and protection to the biofilm, boosting its resilience against environmental stressors and antifungal treatments

Many studies demonstrated the presence of galactomannan in both the cell wall and the matrix, while α-1,3 glucan was primarily located extracellularly near the surface of hyphae (Beauvais et al. 2007; Gravelat et al. 2013; Sheppard and Howell 2016). The protein content, which comprised 2% of the total matrix, included three major secreted antigens: dipeptidyl peptidase V (DPPV), catalase B (*catB*), and ribotoxin (ASPF1). A group of hydrophobin proteins play a role in intercellular adhesion during aerial growth. Extracellular DNA (eDNA) is considered important as a component of ECM of both fungal and bacterial biofilms (Martins et al. 2010; Whitchurch et al. 2002). eDNA matching genomic DNA sequence was discovered in the *A. fumigatus* biofilm matrix (Rajendran et al. 2013). Interestingly, host DNA may also contribute to the *Aspergillus* matrix (Shopova et al. 2013). Adding exogenous DNA enhanced the matrix's structural integrity and carbohydrate production (Shopova et al. 2013). This finding is particularly relevant to the cystic fibrosis (CF) lung, where human DNA concentrations are relatively high. This offers insight into the possible role of host factors in biofilm matrix development during infections.

Reichhardt and colleagues grew *A. fumigatus* biofilms in RPMI 1640 medium and employed solid-state NMR to analyze the matrix composition (Reichhardt et al. 2015). Various approaches were used to explore different aspects of the matrix, including ^{13}C, ^{15}N, and ^{31}P-NMR, electron microscopy, and protein identification through PAGE and mass spectrometry. The study revealed the matrix contains approximately 40% protein, 43% polysaccharides, up to 14% lipids, and 3% aromatic-containing components such as melanin.

In vivo studies of *Aspergillus* biofilm matrix have also been conducted, using aspergillomas from human patients and a murine model of invasive pulmonary aspergillosis (Loussert et al. 2010). Similar to in vitro *A. fumigatus* biofilms, both biofilm models contained galactomannan within the cell wall and matrix. However, a key difference was noted in the absence of matrix α-1,3 glucan in the pulmonary aspergillosis biofilms. Both in vivo models, however, showed higher levels of galactosaminogalactan (GAG), a cell wall exopolysaccharide characterized by the Sheppard group (Lee et al. 2014; Bamford et al. 2015). GAG (galactosaminogalactan) is a unique glycan found within the extracellular matrix (ECM) of *A. fumigatus* biofilms. GAG is a heteropolysaccharide made up of α-1,4-linked galactose and partially deacetylated N-acetyl galactosamine (GalNAc) (Fontaine et al. 2011; Briard et al. 2016). Its biosynthesis is regulated by a cluster of five genes located on chromosome 3. The *uge3* gene encodes a glucose 4-epimerase that is responsible for producing the nucleotide monosaccharides UDP-galactopyranose and UDP-GalNAc (Gravelat et al. 2013). The *gtb3* gene encodes a transmembrane glycosyltransferase that polymerizes and exports these sugars (Briard et al. 2020). The absence of either *uge3* or *gtb3* completely disrupts GAG synthesis (Gravelat et al. 2013; Briard et al. 2020). Additionally, two glycoside hydrolases, *ega3* and *sph3*, target different regions within the GAG polymer, with *sph3* being necessary for GAG production (Bamford et al. 2015, 2019). Another critical gene, *agd3*, encodes a polysaccharide deacetylase that deacetylates GalNAc residues, making the GAG polycationic (Bamford et al. 2015). Mutants lacking *agd3* produce fully acetylated, non-adherent GAG that cannot support biofilm formation (Lee et al. 2016; Bamford et al. 2020). GAG in the biofilm matrix has been shown to play a critical role in adherence and modulates the host immune response by masking β-1,3 glucan and inducing the interleukin-1 receptor antagonist (IL-1Ra), which inhibits proinflammatory IL-1 signaling (Gravelat et al. 2013; Bamford et al. 2015,2019; Briard et al. 2020; Gresnigt et al. 2014).

Similar to the *Candida albicans* biofilm matrix, the *Aspergillus* biofilm matrix is believed to contribute to antifungal resistance, potentially by slowing drug diffusion or through a specific sequestration mechanism (Beauvais et al. 2007; Rajendran et al. 2013; Bugli et al. 2013). One component identified as contributing to *A. fumigatus* biofilm resistance is eDNA (Rajendran et al. 2013; Shopova et al. 2013). An in vitro study by Rajendran et al. showed that eDNA, which accumulates during biofilm maturation, could be degraded by DNase, increasing susceptibility to amphotericin B and caspofungin (Rajendran et al. 2013). Autolysis, evidenced by increased chitinase activity, is speculated to be responsible for releasing eDNA into the matrix. Another study found that treating *A. fumigatus* biofilms with alginate lyase, which degrades uronic acid-containing carbohydrates, enhanced susceptibility to amphotericin B (Bugli et al. 2013). In *C. albicans*, eDNA plays a role in maintaining and stabilizing mature biofilms and is felt to play some role in resistance to antifungal treatments as well (Martins et al. 2012). Beyond, eDNA, the specific contributions of individual ECM components to antifungal resistance in *A. fumigatus* biofilms has not been defined. Specifically, the roles of other polysaccharides in the ECM of *A. fumigatus* biofilms are less understood.

Summary

The polymeric community encasing organism-derived extracellular matrix is a distinct feature of all biofilms. Yet the specific composition and function varies among pathogens. Matrix roles range from biofilm initiation and structure to microbe protection from clearance by the host immune system and antimicrobials. The importance of the matrix for biofilm formation and persistence would seem a promising target for preventive and therapeutic drug development.

References

Abels ER, Breakefield XO (2016) Introduction to extracellular vesicles: biogenesis, RNA cargo selection, content, release, and uptake. Cell Mol Neurobiol 36:301–312

Al Shakirchi M, Sorjonen K, Klingspor L, Bergman P, Hjelte L, de Monestrol I (2021) The effects of aspergillus fumigatus colonization on lung function in patients with cystic fibrosis. J Fungi (Basel) 7(11):944

Al-Fattani MA, Douglas LJ (2004) Penetration of Candida biofilms by antifungal agents. Antimicrob Agents Chemother 48:3291–3297

Al-Fattani MA, Douglas LJ (2006) Biofilm matrix of Candida albicans and Candida tropicalis: chemical composition and role in drug resistance. J Med Microbiol 55:999–1008

Alvarez M, Saylor C, Casadevall A (2008) Antibody action after phagocytosis promotes Cryptococcus neoformans and Cryptococcus gattii macrophage exocytosis with biofilm-like microcolony formation. Cell Microbiol 10:1622–1633

Bamford NC, Snarr BD, Gravelat FN, Little DJ, Lee MJ, Zacharias CA, Chabot JC, Geller AM, Baptista SD, Baker P, Robinson H, Howell PL, Sheppard DC (2015) Sph3 is a glycoside hydrolase required for the biosynthesis of Galactosaminogalactan in aspergillus fumigatus. J Biol Chem 290:27438–27450

Bamford NC, Le Mauff F, Subramanian AS, Yip P, Millan C, Zhang Y, Zacharias C, Forman A, Nitz M, Codee JDC, Uson I, Sheppard DC, Howell PL (2019) Ega3 from the fungal pathogen aspergillus fumigatus is an endo-alpha-1,4-galactosaminidase that disrupts microbial biofilms. J Biol Chem 294:13833–13849

Bamford NC, Le Mauff F, Van Loon JC, Ostapska H, Snarr BD, Zhang Y, Kitova EN, Klassen JS, Codee JDC, Sheppard DC, Howell PL (2020) Structural and biochemical characterization of the exopolysaccharide deacetylase Agd3 required for aspergillus fumigatus biofilm formation. Nat Commun 11:2450

Beauvais A, Schmidt C, Guadagnini S, Roux P, Perret E, Henry C, Paris S, Mallet A, Prevost MC, Latge JP (2007) An extracellular matrix glues together the aerial-grown hyphae of aspergillus fumigatus. Cell Microbiol 9:1588–1600

Briard B, Muszkieta L, Latge JP, Fontaine T (2016) Galactosaminogalactan of aspergillus fumigatus, a bioactive fungal polymer. Mycologia 108:572–580

Briard B, Fontaine T, Samir P, Place DE, Muszkieta L, Malireddi RKS, Karki R, Christgen S, Bomme P, Vogel P, Beau R, Mellado E, Ibrahim-Granet O, Henrissat B, Kalathur RC, Robinson C, Latge JP, Kanneganti TD (2020) Galactosaminogalactan activates the inflammasome to provide host protection. Nature 588:688–692

Brown L, Wolf JM, Prados-Rosales R, Casadevall A (2015) Through the wall: extracellular vesicles in gram-positive bacteria, mycobacteria and fungi. Nat Rev Microbiol 13:620–630

Bugli F, Posteraro B, Papi M, Torelli R, Maiorana A, Paroni Sterbini F, Posteraro P, Sanguinetti M, De Spirito M (2013) In vitro interaction between alginate lyase and amphotericin B against aspergillus fumigatus biofilm determined by different methods. Antimicrob Agents Chemother 57:1275–1282

Dominguez E, Zarnowski R, Sanchez H, Covelli AS, Westler WM, Azadi P, Nett J, Mitchell AP, Andes DR (2018) Conservation and divergence in the Candida species biofilm matrix Mannan-Glucan complex structure, function, and genetic Control. mBio 9(2):e00451-18

Dominguez EG, Zarnowski R, Choy HL, Zhao M, Sanchez H, Nett JE, Andes DR (2019) Conserved role for biofilm matrix polysaccharides in Candida auris drug resistance. mSphere 4(1):e00680-18

Fontaine T, Delangle A, Simenel C, Coddeville B, van Vliet SJ, van Kooyk Y, Bozza S, Moretti S, Schwarz F, Trichot C, Aebi M, Delepierre M, Elbim C, Romani L, Latge JP (2011) Galactosaminogalactan, a new immunosuppressive polysaccharide of aspergillus fumigatus. PLoS Pathog 7:e1002372

Gravelat FN, Beauvais A, Liu H, Lee MJ, Snarr BD, Chen D, Xu W, Kravtsov I, Hoareau CM, Vanier G, Urb M, Campoli P, Al Abdallah Q, Lehoux M, Chabot JC, Ouimet MC, Baptista SD, Fritz JH, Nierman WC, Latge JP, Mitchell AP, Filler SG, Fontaine T, Sheppard DC (2013) Aspergillus galactosaminogalactan mediates adherence to host constituents and conceals hyphal beta-glucan from the immune system. PLoS Pathog 9:e1003575

Gresnigt MS, Bozza S, Becker KL, Joosten LA, Abdollahi-Roodsaz S, van der Berg WB, Dinarello CA, Netea MG, Fontaine T, De Luca A, Moretti S, Romani L, Latge JP, van de Veerdonk FL (2014) A polysaccharide virulence factor from aspergillus fumigatus elicits anti-inflammatory effects through induction of Interleukin-1 receptor antagonist. PLoS Pathog 10:e1003936

Harriott MM, Noverr MC (2009) Candida albicans and Staphylococcus aureus form polymicrobial biofilms: effects on antimicrobial resistance. Antimicrob Agents Chemother 53:3914–3922

Harriott MM, Noverr MC (2010) Ability of Candida albicans mutants to induce Staphylococcus aureus vancomycin resistance during polymicrobial biofilm formation. Antimicrob Agents Chemother 54:3746–3755

Kong EF, Tsui C, Kucharikova S, Andes D, Van Dijck P, Jabra-Rizk MA (2016) Commensal Protection of Staphylococcus aureus against Antimicrobials by Candida albicans. Biofilm Matrix mBio 7:e01365-16

Kowalski CH, Kerkaert JD, Liu KW, Bond MC, Hartmann R, Nadell CD, Stajich JE, Cramer RA (2019) Fungal biofilm morphology impacts hypoxia fitness and disease progression. Nat Microbiol 4:2430–2441

Lee MJ, Gravelat FN, Cerone RP, Baptista SD, Campoli PV, Choe SI, Kravtsov I, Vinogradov E, Creuzenet C, Liu H, Berghuis AM, Latge JP, Filler SG, Fontaine T, Sheppard DC (2014) Overlapping and distinct roles of aspergillus fumigatus UDP-glucose 4-epimerases in galactose metabolism and the synthesis of galactose-containing cell wall polysaccharides. J Biol Chem 289:1243–1256

Lee MJ, Geller AM, Bamford NC, Liu H, Gravelat FN, Snarr BD, Le Mauff F, Chabot J, Ralph B, Ostapska H, Lehoux M, Cerone RP, Baptista SD, Vinogradov E, Stajich JE, Filler SG, Howell PL, Sheppard DC (2016) Deacetylation of Fungal Exopolysaccharide Mediates Adhesion and Biofilm Formation. mBio 7:e00252–e00216

Lohse MB, Gulati M, Johnson AD, Nobile CJ (2018) Development and regulation of single- and multi-species Candida albicans biofilms. Nat Rev Microbiol 16:19–31

Loussert C, Schmitt C, Prevost MC, Balloy V, Fadel E, Philippe B, Kauffmann-Lacroix C, Latge JP, Beauvais A (2010) In vivo biofilm composition of aspergillus fumigatus. Cell Microbiol 12:405–410

Martinez LR, Casadevall A (2005) Specific antibody can prevent fungal biofilm formation and this effect correlates with protective efficacy. Infect Immun 73:6350–6362

Martinez LR, Casadevall A (2006a) Susceptibility of Cryptococcus neoformans biofilms to antifungal agents in vitro. Antimicrob Agents Chemother 50:1021–1033

Martinez LR, Casadevall A (2006b) Cryptococcus neoformans cells in biofilms are less susceptible than planktonic cells to antimicrobial molecules produced by the innate immune system. Infect Immun 74:6118–6123

Martinez LR, Casadevall A (2007) Cryptococcus neoformans biofilm formation depends on surface support and carbon source and reduces fungal cell susceptibility to heat, cold, and UV light. Appl Environ Microbiol 73:4592–4601

Martinez LR, Casadevall A (2015) Biofilm formation by Cryptococcus neoformans. Microbiol Spectr 3

Martinez LR, Mihu MR, Han G, Frases S, Cordero RJ, Casadevall A, Friedman AJ, Friedman JM, Nosanchuk JD (2010) The use of chitosan to damage Cryptococcus neoformans biofilms. Biomaterials 31:669–679

Martins M, Uppuluri P, Thomas DP, Cleary IA, Henriques M, Lopez-Ribot JL, Oliveira R (2010) Presence of extracellular DNA in the Candida albicans biofilm matrix and its contribution to biofilms. Mycopathologia 169:323–331

Martins M, Henriques M, Lopez-Ribot JL, Oliveira R (2012) Addition of DNase improves the in vitro activity of antifungal drugs against Candida albicans biofilms. Mycoses 55:80–85

Mitchell KF, Taff HT, Cuevas MA, Reinicke EL, Sanchez H, Andes DR (2013) Role of matrix beta-1,3 glucan in antifungal resistance of non-albicans Candida biofilms. Antimicrob Agents Chemother 57:1918–1920

Mitchell KF, Zarnowski R, Sanchez H, Edward JA, Reinicke EL, Nett JE, Mitchell AP, Andes DR (2015) Community participation in biofilm matrix assembly and function. Proc Natl Acad Sci USA 112:4092–4097

Mitchell KF, Zarnowski R, Andes DR (2016) Fungal super glue: the biofilm matrix and its composition, assembly, and functions. PLoS Pathog 12:e1005828

Morelli KA, Kerkaert JD, Cramer RA (2021) Aspergillus fumigatus biofilms: toward understanding how growth as a multicellular network increases antifungal resistance and disease progression. PLoS Pathog 17:e1009794
Nelson LA, Callerame ML, Schwartz RH (1979) Aspergillosis and atopy in cystic fibrosis. Am Rev Respir Dis 120:863–873
Nett J, Lincoln L, Marchillo K, Massey R, Holoyda K, Hoff B, Van Handel M, Andes D (2007) Putative role of beta-1,3 glucans in Candida albicans biofilm resistance. Antimicrob Agents Chemother 51:510–520
Nett JE, Sanchez H, Cain MT, Andes DR (2010a) Genetic basis of Candida biofilm resistance due to drug-sequestering matrix glucan. J Infect Dis 202:171–175
Nett JE, Crawford K, Marchillo K, Andes DR (2010b) Role of Fks1p and matrix glucan in Candida albicans biofilm resistance to an echinocandin, pyrimidine, and polyene. Antimicrob Agents Chemother 54:3505–3508
Nett JE, Sanchez H, Cain MT, Ross KM, Andes DR (2011) Interface of Candida albicans biofilm matrix-associated drug resistance and cell wall integrity regulation. Eukaryot Cell 10:1660–1669
Perumal P, Mekala S, Chaffin WL (2007) Role for cell density in antifungal drug resistance in Candida albicans biofilms. Antimicrob Agents Chemother 51:2454–2463
Rajendran R, Williams C, Lappin DF, Millington O, Martins M, Ramage G (2013) Extracellular DNA release acts as an antifungal resistance mechanism in mature aspergillus fumigatus biofilms. Eukaryot Cell 12:420–429
Ramage G, Bachmann S, Patterson TF, Wickes BL, Lopez-Ribot JL (2002a) Investigation of multidrug efflux pumps in relation to fluconazole resistance in Candida albicans biofilms. J Antimicrob Chemother 49:973–980
Ramage G, Saville SP, Wickes BL, Lopez-Ribot JL (2002b) Inhibition of Candida albicans biofilm formation by farnesol, a quorum-sensing molecule. Appl Environ Microbiol 68:5459–5463
Reichhardt C, Ferreira JA, Joubert LM, Clemons KV, Stevens DA, Cegelski L (2015) Analysis of the aspergillus fumigatus biofilm extracellular matrix by solid-state nuclear magnetic resonance spectroscopy. Eukaryot Cell 14:1064–1072
Robbins N, Uppuluri P, Nett J, Rajendran R, Ramage G, Lopez-Ribot JL, Andes D, Cowen LE (2011) Hsp90 governs dispersion and drug resistance of fungal biofilms. PLoS Pathog 7:e1002257
Sheppard DC, Howell PL (2016) Biofilm exopolysaccharides of pathogenic fungi: lessons from bacteria. J Biol Chem 291:12529–12537
Shopova I, Bruns S, Thywissen A, Kniemeyer O, Brakhage AA, Hillmann F (2013) Extrinsic extracellular DNA leads to biofilm formation and colocalizes with matrix polysaccharides in the human pathogenic fungus aspergillus fumigatus. Front Microbiol 4:141
Silva S, Henriques M, Martins A, Oliveira R, Williams D, Azeredo J (2009) Biofilms of non-Candida albicans Candida species: quantification, structure and matrix composition. Med Mycol 47:681–689
Taff HT, Nett JE, Zarnowski R, Ross KM, Sanchez H, Cain MT, Hamaker J, Mitchell AP, Andes DR (2012) A Candida biofilm-induced pathway for matrix glucan delivery: implications for drug resistance. PLoS Pathog 8:e1002848
Taff HT, Mitchell KF, Edward JA, Andes DR (2013) Mechanisms of Candida biofilm drug resistance. Future Microbiol 8:1325–1337
Thomas DP, Bachmann SP, Lopez-Ribot JL (2006) Proteomics for the analysis of the Candida albicans biofilm lifestyle. Proteomics 6:5795–5804
Uppuluri P, Nett J, Heitman J, Andes D (2008) Synergistic effect of calcineurin inhibitors and fluconazole against Candida albicans biofilms. Antimicrob Agents Chemother 52:1127–1132
Vediyappan G, Rossignol T, d'Enfert C (2010) Interaction of Candida albicans biofilms with antifungals: transcriptional response and binding of antifungals to beta-glucans. Antimicrob Agents Chemother 54:2096–2111

Whitchurch CB, Tolker-Nielsen T, Ragas PC, Mattick JS (2002) Extracellular DNA required for bacterial biofilm formation. Science 295:1487
Zarnowski R, Westler WM, Lacmbouh GA, Marita JM, Bothe JR, Bernhardt J, Lounes-Hadj Sahraoui A, Fontaine J, Sanchez H, Hatfield RD, Ntambi JM, Nett JE, Mitchell AP, Andes DR (2014) Novel entries in a fungal biofilm matrix encyclopedia. MBio 5:e01333–e01314
Zarnowski R, Sanchez H, Andes DR (2016) Large-scale production and isolation of Candida biofilm extracellular matrix. Nat Protoc 11:2320–2327
Zarnowski R, Sanchez H, Covelli AS, Dominguez E, Jaromin A, Bernhardt J, Mitchell KF, Heiss C, Azadi P, Mitchell A, Andes DR (2018) Candida albicans biofilm-induced vesicles confer drug resistance through matrix biogenesis. PLoS Biol 16:e2006872
Zarnowski R, Noll A, Chevrette MG, Sanchez H, Jones R, Anhalt H, Fossen J, Jaromin A, Currie C, Nett JE, Mitchell A, Andes DR (2021) Coordination of fungal biofilm development by extracellular vesicle cargo. Nat Commun 12:6235

Chapter 9
Fungal Biofilm Dispersal

Sunna Nabeela and Priya Uppuluri

Abstract Biofilm dispersal is a critical stage in the biofilm life cycle, enabling fungal cells to spread, colonize new environments, and establish persistent infections. Dispersed cells exhibit distinct phenotypic and transcriptional characteristics, often displaying enhanced virulence, increased adhesion, and immune evasion compared to their planktonic counterparts. Recent studies have revealed that biofilm dispersal in fungi is influenced by multiple factors, including epigenetic modifications, extracellular vesicle-mediated regulation, and metabolic adaptation under nutrient deprivation. Understanding fungal biofilm dispersal is essential for devising effective strategies to prevent and treat fungal infections, especially in the context of medical devices, where biofilms are notoriously resilient and dispersing cells can directly enter the bloodstream, increasing the risk of systemic infection.

Introduction

Over the past two decades, the field of fungal biofilm research has seen remarkable advancements. The introduction of cutting-edge molecular technologies and animal models has revolutionized our understanding of the biofilm lifestyle. These breakthroughs have not only deepened our knowledge but also opened new avenues for exploring how biofilms function, evolve, and interact with their environments, leading to innovative applications in medicine, industry, and environmental management.

S. Nabeela
Division of Infectious Diseases, The Lundquist Institute for Biomedical Innovation, Harbor-University of California at Los Angeles (UCLA) Medical Center, Torrance, CA, USA

P. Uppuluri (✉)
Division of Infectious Diseases, The Lundquist Institute for Biomedical Innovation, Harbor-University of California at Los Angeles (UCLA) Medical Center, Torrance, CA, USA

David Geffen School of Medicine at UCLA, Los Angeles, CA, USA
e-mail: puppuluri@lundquist.org

G. Ramage, D. Andes (eds.), *Fungal Biofilms*, Springer Series on Biofilms 15,
https://doi.org/10.1007/978-3-031-93315-8_9

The intense focus on biofilms is largely driven by their profound clinical implications. Biofilms formed on indwelling medical devices, such as catheters, are notoriously difficult to eradicate (Donlan 2001). Compounding this issue is the fact that biofilms serve as reservoirs, continuously shedding daughter cells as they grow (Uppuluri et al. 2010a). In recent years, the study of how biofilms transition from their surface-bound state to free-floating, planktonic cells has garnered significant attention. In the context of medical devices, this is particularly problematic, as cells released from drug-resistant biofilms can enter the bloodstream and disseminate throughout the body, leading to widespread infections such as disseminated candidiasis (Uppuluri et al. 2010a; Guembe et al. 2014). Notably, these dispersal cells exhibit virulence factors comparable to those of their parent biofilms and are primed for attachment to new surfaces and tissue invasion (Uppuluri et al. 2010a, 2018).

Emerging evidence suggests that the dispersal of cells from biofilms plays a crucial role in the pathogenic success of microbes (Rumbaugh and Sauer 2020; Uppuluri and Lopez-Ribot 2016). In this chapter, we provide a comprehensive overview of current knowledge on biofilm dispersal in medically relevant fungal species. We explore the molecular and nutritional triggers that prompt the release of cells from these sessile fungal communities and examine the unique physiological and genetic traits that characterize dispersal cells. Finally, we discuss how these insights can be applied to the clinical management of biofilm-associated infections.

Biofilm Life Cycle The transition from a free-living state to a biofilm is driven by environmental cues and orchestrated through complex regulatory networks that modulate gene expression, resulting in the spatial and temporal organization of microbial communities. Regardless of the pathogenic species, the biofilm formation process follows a series of well-defined developmental stages (Gulati and Nobile 2016; Ramage et al. 2005, 2009) and shares key characteristics, including the secretion of exopolymeric substances (EPS), heightened resistance to host immune responses, and significant tolerance to antimicrobial treatments.

Biofilms typically progress through three distinct phases: initial adherence mediated by surface adhesins (Nobile et al. 2008), maturation involving robust cell proliferation and EPS production (Guzmán-Soto et al. 2021), and dispersal, where biofilm-associated cells transition back to a planktonic state (Uppuluri et al. 2010a). The release of cells from biofilms is a critical process that ensures microbial survival, colonization of new niches, and persistence in host environments. In fungal pathogens, dispersal mechanisms vary significantly among species, with some releasing yeast-like cells (e.g., *C. albicans*) (Uppuluri and Lopez-Ribot 2016), while others, such as *Aspergillus fumigatus*, disperse conidia (Subroto et al. 2022).

Biofilm Dispersal

Fungal biofilm dispersal has long been an overlooked area of research, with significant advancements only recently emerging, primarily centered on *C. albicans*. As a result, *C. albicans* has become the principal model for unraveling the underlying mechanisms of biofilm dispersal in pathogenic fungi, offering valuable insights that may be applicable to a broader range of fungal species.

Research utilizing in vitro flow biofilm models has provided key insights into the dynamics of fungal biofilm dispersal (Uppuluri et al. 2009, 2011). Dispersal is not a random process but is highly regulated and peaks during the active growth phase (mid-log phase), continuing at a steady rate once the biofilm reaches maturity (Uppuluri et al. 2010a). This suggests that dispersal is an integral part of the biofilm life cycle rather than a mere byproduct of aging or structural instability.

Environmental conditions play a crucial role in modulating dispersal rates. The composition of the surrounding medium significantly influences the extent of dispersal; for example, exposure to glucose has been shown to enhance dispersal in mature *C. albicans* biofilms, while nutrient deprivation—such as growth in alternative carbon sources like maltose or galactose, or in phosphate-buffered saline (PBS)—markedly reduces dispersal (Uppuluri et al. 2010a). This indicates that biofilm cells actively respond to nutrient availability, potentially regulating dispersal as a means of colonizing more favorable environments.

pH also serves as a key environmental cue, with acidic conditions promoting dispersal and alkaline environments suppressing it (Uppuluri et al. 2010a). This pH-dependent regulation suggests a mechanistic link to fungal morphogenesis, particularly the hyphae-to-yeast transition. Studies indicate that dispersal primarily involves yeast-form cells being shed from the uppermost layers of hyphal biofilms in response to environmental triggers. This process is likely governed by intricate signaling pathways that integrate nutrient sensing, pH response, and morphogenetic transitions, enabling fungi to optimize dispersal timing for successful dissemination and colonization of new niches.

Given that biofilm-associated infections are notoriously difficult to treat due to their resistance to antifungal drugs and immune clearance, understanding the molecular regulation of dispersal could have clinical implications. Targeting dispersal mechanisms may offer a novel therapeutic strategy, potentially facilitating biofilm breakdown and enhancing the efficacy of antifungal treatments.

Biofilm Dispersal in *A. fumigatus* *A. fumigatus* disperses conidia (asexual spores) from its mycelial biofilm. González-Ramírez et al. demonstrated that *A. fumigatus* biofilm development is asynchronous, particularly during maturation, where newly formed conidia germinate and initiate fresh mycelial growth (González-Ramírez et al. 2016). Structural adaptations, such as hyphal modifications, likely optimize dispersal efficiency and promote successful colonization of new environments.

Biofilm dispersal in *A. fumigatus* is not merely a passive process but is regulated by environmental factors, including nutrient availability, oxygen levels, and pH fluctuations (González-Ramírez et al. 2016). In pulmonary *Aspergillus* infections, conidia released from biofilms within the lungs can disseminate, contributing to disease progression in conditions such as invasive aspergillosis and chronic pulmonary aspergillosis (González-Ramírez et al. 2016; Janssens et al. 2024).

Beyond clinical settings, *Aspergillus* biofilm dispersal has significant industrial and environmental implications, including contamination in bioreactors and food-processing environments (Ramage et al. 2011). The ability to disperse conidia enables *A. fumigatus* to adapt to changing environments, promoting genetic diversity and persistence.

Biofilm Dispersal in *Scedosporium/Lomentospora* Species *Scedosporium* and *Lomentospora* species are fungal pathogens that primarily affect immunocompromised individuals, such as organ transplant recipients, patients with chronic granulomatous disease, hematological malignancies, and those with cystic fibrosis (Cortez Karoll et al. 2008; Ramirez-Garcia et al. 2018). These filamentous fungi form durable biofilms on both abiotic surfaces (e.g., polystyrene, glass, and catheters) and biotic surfaces (e.g., lung epithelial cells), making them highly resistant to all major classes of antifungal drugs, including azoles, echinocandins, and polyenes (Mello et al. 2016, 2022). The resistance of mature biofilms is attributed to their complex extracellular matrix, which includes polysaccharides, glycoproteins, and eDNA, along with the activity of efflux pumps and robust antioxidant defenses (Mello et al. 2022). Studies on biofilm dispersal revealed that the early biofilm development phase releases more dispersed cells compared to the maturation phase, with conidia being the predominant form (Mello et al. 2023). Unlike *C. albicans*, their release is not affected by elevated glucose levels. Notably, mucin, found in the lungs of cystic fibrosis patients, stimulates biofilm cell dispersion, while nutrient-poor environments like phosphate-buffered saline hinder dispersal (Mello et al. 2023). These findings were consistent with those in *Pseudomonas aeruginosa*, where mucin induces biofilm cell dispersion in cystic fibrosis lungs (Co et al. 2018), perhaps by disruption of the biofilm matrix by glycosylated mucins (Co et al. 2018), or by inhibition of cell aggregation, keeping cells planktonic (Caldara et al. 2012). These results suggest that mucin-induced dispersal can enhance microbial dissemination in cystic fibrosis patients.

Biofilm Dispersal in *Cryptococcus* Species *C. neoformans* and *C. gattii* form biofilms composed of encapsulated yeast cells embedded in a polysaccharide-rich extracellular matrix (Ravi et al. 2009; Martinez and Casadevall 2015). Unlike filamentous fungi, *Cryptococcus* species lack hyphae, relying instead on capsule-mediated adherence and biofilm formation (Ramage et al. 2011; Martinez and Casadevall 2015). The capsule not only aids in immune evasion but also contributes to biofilm stability and integrity (Aslanyan et al. 2017). Dispersal in *C. neoformans* biofilms is primarily driven by the production of yeasts and capsule-deficient variants (Aslanyan et al. 2017).

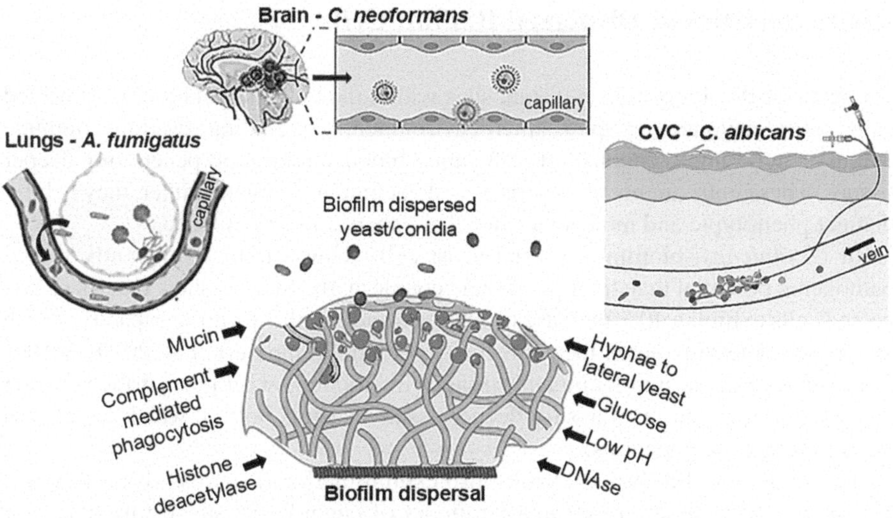

Fig. 9.1 Biofilm dispersal

Shotgun proteomic studies have revealed that *C. neoformans* biofilm cells exhibit upregulated expression of enzymes involved in proteolysis and oxidative stress resistance, suggesting a survival advantage within the host (Santi et al. 2014). The mode of opsonization prior to phagocytosis also influences dispersal: complement-mediated phagocytosis results in individually dispersed fungal cells, whereas antibody-mediated opsonization promotes aggregation and biofilm-like microcolony formation (Alvarez et al. 2008). These microcolonies resemble biofilm subunits, allowing *C. neoformans* to establish new biofilms in host tissues, evade immune clearance, and persist in latent infections (Aslanyan et al. 2017).

The biofilm-like dispersal of *C. neoformans* is particularly relevant in cryptococcal meningitis, where the fungus forms large, encapsulated cryptococcomas in the brain (Aslanyan et al. 2017). These structures, rich in capsular polysaccharides, provide a protective niche, making them highly resistant to antifungal treatment. The ability of *C. neoformans* to form cryptococcomas in the central nervous system underscores its neurotropic nature and highlights the clinical challenges associated with treating cryptococcal infections (Fig. 9.1).

In conclusion, fungal biofilm dispersal is a complex, species-specific process governed by environmental and molecular cues. In *A. fumigatus*, conidial dispersal facilitates environmental adaptation and pathogenesis, while in *C. neoformans*, biofilm-like microcolony formation enhances persistence within host tissues. Understanding the regulatory mechanisms underlying biofilm dispersal has significant clinical implications, as targeting dispersal pathways could improve antifungal therapy outcomes.

Characteristics of Dispersed Biofilm Cells

For successful colonization at distant sites within the host, fungal biofilm-dispersed cells must navigate complex microenvironments, including evading immune defenses, adhering to endothelial cells lining blood vessels, and penetrating deeper tissues. These cells are not simply passive detachment products; rather, they exhibit distinct phenotypic and molecular traits that enhance their virulence.

In *C. albicans*, biofilm-dispersed yeast cells demonstrate significantly higher pathogenic potential than their planktonic counterparts. Studies show that these dispersed cells exhibit a 40% increase in adhesion and biofilm-forming capacity, which are critical determinants of their infectious potential (Uppuluri et al. 2010a, 2018). This enhanced adhesion facilitates stronger interactions with host surfaces, while increased filamentation allows for deeper tissue invasion, both of which are critical steps in fungal pathogenesis.

Strikingly, biofilm-dispersed cells also exhibit markedly increased lethality in a hematogenously disseminated murine model of candidiasis, suggesting that their heightened virulence is not transient but persists over multiple generations (Uppuluri et al. 2010a, 2018). This observation raises important questions about the underlying molecular mechanisms that sustain this virulent phenotype. One compelling hypothesis is that heritable epigenetic modifications drive the long-term retention of enhanced adhesion, filamentation, and virulence, potentially priming dispersed cells for successful host invasion and persistence. This aligns with growing evidence that *C. albicans* utilizes epigenetic and transcriptional plasticity to rapidly adapt to changing environments, reinforcing the pivotal role of biofilm dispersal in fungal pathogenesis (Iracane et al. 2021; Nobile et al. 2012; Nobile and Johnson 2015). Indeed, Ribiero et al. demonstrated that *C. albicans* isolates persisting in the bloodstream for 4 days or longer, despite antifungal treatment, exhibited enhanced adherence, filamentation, and biofilm formation in vitro, along with increased expression of associated virulence-related genes (Ribeiro et al. 2024). Notably, cells dispersed from second-generation biofilms formed by these persistent isolates displayed even more pronounced pathogenic characteristics, including improved immune evasion. This suggests that dispersed cells not only inherit adaptively acquired virulent traits but may also reinforce these traits through successive generations, likely via stable epigenetic modifications or transcriptional memory mechanisms. Additionally, the persistence of this hypervirulent phenotype in successive biofilm-derived populations raises the possibility of an adaptive prediction model, wherein *C. albicans* anticipates and primes its dispersed progeny for survival in hostile host environments. This model is supported by RNA sequencing studies showing that dispersed cells exhibit pre-activated stress response pathways, increased metabolic plasticity, and upregulation of genes involved in adhesion (*ALS5*, *ALS6*, and *ECM33*), antifungal resistance (*MDR1*, *QDR1*, and *ERG*), and host interaction (*SAP* family genes) (Uppuluri et al. 2018). These findings underscore the pivotal role of biofilm dispersal not merely as a mechanism for colonization but as a sophisticated survival strategy that enhances the long-term persistence and pathogenic potential of *C. albicans*. In murine models of disseminated candidiasis, these pre-adapted dispersed cells cause notably more severe infections compared to planktonic yeast cells (Uppuluri et al. 2010a). Their upregulated virulence factors not only facilitate host colonization but also contribute to persistence and resilience against antifungal treatments, making them particularly challenging to eradicate. These findings underscore the critical role of biofilm dispersal in fungal disease progression.

Molecular Mechanism of Biofilm Dispersal

The hyphae-to-yeast transition is a critical driver of biofilm dispersal in *C. albicans*. Research by the Uppuluri group has revealed that dispersed yeast cells originate from the lateral septal regions of biofilm hyphae, a phenomenon known as the 'lateral yeast' transition (Uppuluri et al. 2010a). However, the molecular mechanisms governing this process remain largely unexplored. Gaining a deeper understanding of the regulatory pathways involved in this hyphae-to-lateral yeast transition is essential for unraveling the broader mechanisms of biofilm dispersal in biofilms formed by fungi that filament. Such insights could open new avenues for developing targeted interventions aimed at disrupting biofilm dissemination, offering potential strategies to combat biofilm-associated infections more effectively.

Role of *PES1* in Biofilm Dispersal Kohler and colleagues, including Shen et al. (2008) (Shen et al. 2008), demonstrated that the expression of the pescadillo homolog *PES1* is both necessary and sufficient to drive lateral yeast formation in *C. albicans*. In yeast cells, *PES1* is essential, whereas it is dispensable in hyphal forms. Overexpression of *PES1* leads to an increase in lateral yeast formation, while reduced expression—achieved through a doxycycline-regulated allele—significantly decreases lateral yeast formation (Uppuluri et al. 2010a, 2012). Additionally, reports showed *PES1* being implicated in the generation of biofilm dispersal cells, further underscoring its critical role in biofilm dynamics and dispersal (Uppuluri et al. 2012). These findings position *PES1* as a key regulatory factor in the transition from hyphae to yeast, offering potential therapeutic targets for inhibiting biofilm-related infections. Indeed, *PES1* expression of *C. albicans* is required for full virulence in mice models of disseminated candidiasis, because curtailing lateral yeast cell release from tissue-invading hyphae by depleting *PES1* in vivo, significantly reduces dissemination and virulence in mice (Uppuluri et al. 2012; 2018). Furthermore, in a jugular vein-catheterized mouse model, dispersed yeast cells regulated by *PES1* were found as the link between biofilm-infected catheters and the foci of infection in a deep organ (Uppuluri et al. 2018). Specifically, compared to infections caused by the wildtype or *PES1*-overexpressing strains, repression of *PES1* in vivo in *C. albicans*-infected central venous catheter (CVC), significantly curtailed biofilm dispersal and subsequent dissemination to distal primary organs such as kidneys.

Recently, Krysan and colleagues (2024) demonstrated that heterologous expression of *PES1* is sufficient to induce lateral yeast formation in vivo (Wakade et al. 2024). However, during infections using wild-type *C. albicans* in the murine subdermal ear tissue, *PES1* expression was found to be extremely low or undetectable. This anatomical site shares many characteristics with the stromal tissue underlying the colonized epithelium of the mouth and gastrointestinal tract, suggesting that environmental factors may significantly influence *PES1* expression during infection (Wakade et al. 2024). This outcome aligns with the limited lateral yeast production observed in hyphal cells during the early stages of infection (4–24 h). Given its role in virulence, it is likely that *PES1* effects become more significant only after the initial establishment of infection. Indeed, *PES1* plays a critical role post-establishment of infection, as evidenced by its time-dependent effects on fungal burden observed at days 3 and 5 (Ravi et al. 2009). Furthermore, unpublished data from the same research group indicate that central venous catheter (CVC) infections require at least 3 days to establish before cells begin to disperse from the infected catheters into the bloodstream and other organs.

While the precise mechanism by which *PES1* regulates the hyphae-to-yeast transition remains unclear, studies have shown that farnesol, a quorum-sensing molecule, can induce the formation of lateral yeast cells and promote the dispersal of yeast cells from hyphal biofilms at high frequencies (Sebaa et al. 2019; Ramage et al. 2002). Farnesol inhibits hyphal morphogenesis by downregulating the Ras1-cAMP signaling pathway (Lindsay et al. 2012). These findings suggest that the emergence of lateral yeast cells may be linked to the suppression of the Ras1-cAMP pathway, highlighting its importance as a key regulatory axis in biofilm dispersal.

Role of Regulation of Ace2 and Morphogenesis (RAM) Pathway in Biofilm Dispersal The RAM pathway is a crucial regulatory network involved in various aspects of *C. albicans* pathobiology. The RAM pathway's two most well-studied components, the NDR/Lats kinase Cbk1 and its putative substrate, the transcription factor Ace2, have a wide range of phenotypes and functions (Saputo et al. 2012). During hyphal morphogenesis, *ACE2* RNA and protein levels are initially low and then increase to peak at approximately 5 h hyphal postinduction. Studies have shown that a significant set of genes are upregulated in the absence of Ace2 (or Cbk1 that phosphorylates Ace2), suggesting its role as a repressor of gene expression during hyphal growth (Wakade et al. 2020). Notably, both *HWP1*, a gene expressed exclusively during hyphal growth, and *PES1,* associated with the hypha-to-yeast transition, are upregulated under these conditions followed by increased lateral yeast growth. This expression profile of both hyphal and yeast-related programs suggests that the Cbk1-Ace2 axis plays an important role in regulating the balance between morphogenic transcriptional programs during hyphal formation in *C. albicans*. Specifically, Ace2 appears to play a significant role in the subapical compartments, maintaining the transcriptional program of hyphae while suppressing the yeast program (Wakade et al. 2020). This function of Ace2 is yet to be explored in the biofilm setting, where its downregulation is expected to reduce biofilm dispersal.

Other Regulators of Biofilm Dispersal Recent genome-wide studies have uncovered a transcriptional network governing *C. albicans* biofilm development, driven by six master regulators—Bcr1, Tec1, Efg1, Ndt80, Rob1, and Brg1 (Nobile et al. 2012). These regulators form an interwoven circuit that controls over 1000 downstream genes. Notably, Sfu1, Crz2, and Nrg1 are targeted by all six, playing key roles in biofilm processes: Sfu1 represses iron uptake (Chen et al. 2011), Crz2 aids initial cell adherence (Finkel et al. 2012), and Nrg1 regulates biofilm dispersal (Uppuluri et al. 2010b). Nrg1 facilitates dispersal by repressing genes involved in hyphal maintenance, effectively promoting a switch from the hyphal form back to the yeast form (Uppuluri et al. 2010b). Indeed, it was also demonstrated that the essential regulator of dispersal, and the yeast-specific gene *PES1*, acts downstream of *NRG1* (Uppuluri et al. 2018). Whether Nrg1 directly regulates *PES1* has not yet been investigated. Thus, Nrg1 functions as a key regulator in shifting the biofilm's lifecycle toward dispersal, making it an essential factor in the propagation of biofilm-associated infections.

Mamouei et al. employed genetic, bioinformatic, and biochemical approaches to identify NDU1, a *Candida* gene encoding a mitochondrial protein that is evolutionarily divergent from other eukaryotic orthologs (Mamouei et al. 2021). Notably, loss of NDU1 resulted in a striking reduction in *C. albicans* cell wall thickness, primarily due to a significant decrease in both chitin and cell surface mannan content (Mamouei et al. 2021). This structural deficiency rendered *ndu1Δ* mutants hypersusceptible to cell wall-perturbing agents, highlighting the critical role of *NDU1* in maintaining cell wall integrity. Intriguingly, deletion of *NDU1* also led to enhanced biofilm dispersal and increased detachment of biofilm cells (Mamouei et al. 2021). This premature detachment was attributed to the essential role of *NDU1* in supporting *C. albicans* respiration under conditions of nutrient deprivation, such as those encountered in the deeper layers of the biofilm, where oxygen and nutrient availability are severely limited (Fox et al. 2014). In the absence of *NDU1*, *C. albicans* cells likely experience metabolic collapse, leading to a loss of viability in the innermost, nutrient-starved layers of the biofilm. Additionally, defects in cell wall composition may contribute to a weakened adhesion phenotype, further accelerating biofilm detachment.

Recent studies have shown that chromatin and chromatin-modifying enzymes play crucial roles in regulating the expression of morphogenesis-related genes in *C. albicans*. For instance, the Hda1 histone deacetylase (Lu et al. 2012) and NuA4 histone acetyltransferase modulate histone acetylation at hypha-specific gene promoters (Lu et al. 2008). Mutations in chromatin modifiers like Set1, Rtt109, and Set3 result in abnormal morphogenesis phenotypes (Lopes da Rosa et al. 2010; Raman et al. 2006). Hnisz et al. identified the Set3 complex as a key regulator of biofilm dispersal, binding to the coding regions of master biofilm regulators (*BRG1, TEC1, EFG1, NDT80,* and *ROB1*) and *NRG1* (Hnisz et al. 2012). The Set3 complex regulates the transcription of *NRG1*, a key factor in biofilm dispersal. Given that Nrg1 represses filamentation and Set3 complex mutants are hyperfilamentous, it's likely that Set3 influences dispersal by controlling *NRG1* levels (Nobile et al. 2014). Since cells dispersed from *C. albicans* biofilms are typically in the yeast form, increasing filamentous cells while reducing yeast cells may decrease biofilm dispersal (Nobile et al. 2014). Thus, manipulating *SET3* or *NRG1* could be a strategy to limit biofilm dispersal.

Extensive studies have demonstrated that *C. albicans* releases extracellular vesicles during biofilm growth, contributing to matrix formation, drug tolerance, immunomodulation, and cell dissemination (Massey et al. 2023; Kulig et al. 2022; Zarnowski et al. 2018). This process is genetically regulated by the endosomal sorting complexes required for transport (ESCRT) pathway. Zarnowski et al. identified 17 *Candida* ESCRT cargo genes, including *CHT3* and *MP65*, whose deletion led to increased biofilm dispersion across multiple *Candida* species (Zarnowski et al. 2021, 2022). The restoration of wild-type phenotypes upon gene complementation confirmed their critical role in biofilm stability and dispersal. These findings underscore vesicle-mediated regulation as a key determinant of biofilm integrity and dispersal, presenting novel targets for therapeutic intervention in biofilm-associated infections.

Implications for Clinical Management

The primary goal of research on biofilm dispersal is to develop new therapeutic and preventive strategies for biofilm-related infections. Fungal biofilms, like their bacterial counterparts, exhibit extreme resilience to environmental threats and antimicrobial agents. In bacterial biofilms, the most effective approach involves dispersing cells into their planktonic state, where they become more susceptible to antibiotics (Rumbaugh and Sauer 2020, b). Various dispersal agents, such as Dispersin B and DNase I, have demonstrated efficacy in enhancing antibiotic sensitivity (Rumbaugh and Sauer 2020; Kaplan et al. 2003; McDougald et al. 2011).

However, targeting fungal biofilm dispersal remains a significant challenge due to the structural complexity of fungal biofilms and the absence of a well-defined dispersal mechanism. Unlike bacterial biofilms, fungal biofilms consist of dense hyphal networks embedded in an extracellular polymeric substance (EPS) matrix, making them highly resistant to dissociation, even under external treatments (Uppuluri and Lopez-Ribot 2016). While no fungal equivalent of "Dispersin" has been identified, enzymatic approaches show promise. DNase treatment disrupts *C. albicans* biofilms by targeting the biofilm matrix rather than directly affecting cell viability (Martins et al. 2010), while β-1,3-glucanase weakens and disperses *Candida* biofilms by degrading the structural β(1 → 3)-glucan component (Tan et al. 2017). Notably, β-1,3-glucanase enhances the efficacy of amphotericin B against mature biofilms, providing a dual strategy that both disrupts the biofilm structure and improves antifungal penetration (Tan et al. 2017).

A novel anti-biofilm approach involves turbinmicin, an antifungal agent that inhibits vesicle-mediated trafficking, a critical pathway in biofilm matrix assembly (Zhao et al. 2021). By disrupting extracellular vesicle (EV) delivery during biofilm growth, turbinmicin impairs matrix formation, increases biofilm detachment, and enhances susceptibility to conventional antifungal drugs (Zhao et al. 2021). This unique mechanism underscores the potential of targeting fungal vesicle-mediated biofilm assembly as a therapeutic strategy, positioning turbinmicin as a promising anti-biofilm agent.

Given that most currently approved antifungals, except for echinocandins, neither inhibit biofilms nor prevent dispersal, and there is a critical need to discover new molecules capable of addressing both challenges. One promising avenue involves fungal Hsp90 inhibitors, which, in combination with fluconazole, have been shown to completely eradicate biofilms from species such as *C. albicans* and *A. fumigatus* (Robbins et al. 2011). Additionally, these inhibitors alone rendered more than 90% of dispersed biofilm cells inviable, offering a potentially groundbreaking therapeutic approach. Another FDA-approved inhibitor repurposed for anti-biofilm and anti-dispersal strategies is niclosamide (NCL). NCL is an anthelmintic, indicated in the treatment of tapeworm infections in humans (Chen et al. 2018). It has impressive anticancer activity (NCI-lead phase I and II clinical trials against prostate and colon cancer) and has shown remarkable promise for its broad-spectrum antiviral effect, including on Severe Acute Respiratory Syndrome Coronavirus 2 (SARS-CoV-2) (Chen et al. 2018). NCL nanoparticles have demonstrated the ability to penetrate mature fungal (*C. albicans*) biofilms, exhibiting anti-biofilm activity at low microgram concentrations and effectively detaching biofilm cells from their substrate (Sutar et al. 2022). This enhanced efficacy is attributed to NCL's capacity to induce reactive oxygen species (ROS) activity in *Candida*, significantly reducing the fungus's oxygen consumption rate (Sutar et al. 2022). Additionally, NCL nanoparticles have been shown to effectively diminish mucosal biofilms in vivo, such as those associated with oropharyngeal and vulvovaginal candidiasis (Sutar et al. 2022).

Several new antifungal classes in the late stages of clinical development demonstrate promising activities. These include fosmanogepix (a Gwt1 enzyme inhibitor), ibrexafungerp (a first-in-class triterpenoid), olorofim (a dihyroorotate dehydrogenase enzyme inhibitor), and rezafungin (an echinocandin designed for weekly dosing) (Hoenigl et al. 2021). While these antifungal drugs demonstrate impressive efficacy against invasive fungal infections in both in vitro and in vivo studies (Hoenigl et al. 2021; Chu et al. 2021; Larkin et al. 2017), their potential for anti-biofilm activity remains largely unassessed in experimental settings. Considering that at least 60–80% of all disseminated infections originate from a biofilm setting (Sharma et al. 2023; Schulze et al. 2021), research into the effects of these novel drugs for biofilm disruption may open new avenues for treating conditions where biofilms play a critical role, ultimately improving patient outcomes in fungal infections.

Concluding Comments

Biofilm communities on indwelling medical devices, such as CVCs and urinary catheters, act as reservoirs for future infections, making biofilm dispersal a clinically significant process. Dispersed cells retain enhanced virulence and adaptive traits, underscoring the need to decipher the molecular drivers of this phenomenon (Uppuluri et al. 2010a, 2018). Emerging evidence highlights the role of epigenetic modifications, molecular signaling, and extracellular vesicle-mediated regulation in controlling biofilm stability and detachment (Iracane et al. 2021; Nobile et al. 2012; Zarnowski et al. 2021). Additionally, research is expanding to identify dispersal processes in non-albicans *Candida* species, which are frequently implicated in device-related infections (Rodrigues et al. 2017; De la Cruz-Claure et al. 2019).

Despite these advances, targeting fungal biofilm dispersal remains an unresolved challenge due to the complex regulatory networks that govern this process. However, promising antifungal agents such as ibrexafungerp and fosmanogepix, as well as strategies that impair extracellular vesicle delivery and biofilm matrix assembly, offer potential therapeutic avenues. Future efforts should focus on translating these findings into clinically viable treatments that can disrupt biofilm integrity and enhance the efficacy of conventional antifungal drugs. By further elucidating biofilm dispersal mechanisms and their clinical relevance, novel therapeutic strategies can be developed to combat biofilm-related infections more effectively.

References

Alvarez M, Saylor C, Casadevall A (2008) Antibody action after phagocytosis promotes Cryptococcus neoformans and Cryptococcus gattii macrophage exocytosis with biofilm-like microcolony formation. Cell Microbiol 10:1622–1633. https://doi.org/10.1111/j.1462-5822.2008.01152.x

Aslanyan L et al (2017) The crucial role of biofilms in Cryptococcus neoformans survival within macrophages and colonization of the central nervous system. J Fungi (Basel, Switzerland) 3:10. https://doi.org/10.3390/jof3010010

Caldara M et al (2012) Mucin biopolymers prevent bacterial aggregation by retaining cells in the free-swimming state. Curr Biol 22:2325–2330. https://doi.org/10.1016/j.cub.2012.10.028

Chen C, Pande K, French SD, Tuch BB, Noble SM (2011) An iron homeostasis regulatory circuit with reciprocal roles in Candida albicans commensalism and pathogenesis. Cell Host Microbe 10:118–135. https://doi.org/10.1016/j.chom.2011.07.005

Chen W, Mook RA Jr, Premont RT, Wang J (2018) Niclosamide: beyond an antihelminthic drug. Cell Signal 41:89–96. https://doi.org/10.1016/j.cellsig.2017.04.001

Chu S et al (2021) A second-generation Fungerp analog, SCY-247, shows potent in vivo activity in a murine model of Hematogenously disseminated Candida albicans. Antimicrob Agents Chemother 65:e01989-20. https://doi.org/10.1128/aac.01989-20

Co JY et al (2018) Mucins trigger dispersal of Pseudomonas aeruginosa biofilms. npj Biofilms Microbi 4:23. https://doi.org/10.1038/s41522-018-0067-0

Cortez Karoll J et al (2008) Infections caused by Scedosporium spp. Clin Microbiol Rev 21:157–197. https://doi.org/10.1128/cmr.00039-07

De la Cruz-Claure ML et al (2019) Inhibition-disruption of Candida glabrata biofilms: symmetrical Selenoesters as potential anti-biofilm agents. Microorganisms 7:664. https://doi.org/10.3390/microorganisms7120664

Donlan RM (2001) Biofilms and device-associated infections. Emerg Infect Dis J 7:277. https://doi.org/10.3201/eid0702.700277

Finkel JS et al (2012) Portrait of Candida albicans adherence regulators. PLoS Pathog 8:e1002525. https://doi.org/10.1371/journal.ppat.1002525

Fox EP et al (2014) Anaerobic bacteria grow within Candida albicans biofilms and induce biofilm formation in suspension cultures. Curr Biol 24:2411–2416. https://doi.org/10.1016/j.cub.2014.08.057

González-Ramírez AI, Ramírez-Granillo A, Medina-Canales MG, Rodríguez-Tovar AV, Martínez-Rivera MA (2016) Analysis and description of the stages of Aspergillus fumigatus biofilm formation using scanning electron microscopy. BMC Microbiol 16:243. https://doi.org/10.1186/s12866-016-0859-4

Guembe M et al (2014) Is biofilm production a predictor of catheter-related candidemia? Med Mycol 52:407–410. https://doi.org/10.1093/mmy/myt031

Gulati M, Nobile CJ (2016) Candida albicans biofilms: development, regulation, and molecular mechanisms. Microbes Infect 18:310–321. https://doi.org/10.1016/j.micinf.2016.01.002

Guzmán-Soto I et al (2021) Mimicking biofilm formation and development: recent progress in in vitro and in vivo biofilm models. iScience 24:102443. https://doi.org/10.1016/j.isci.2021.102443

Hnisz D et al (2012) A histone deacetylase adjusts transcription kinetics at coding sequences during Candida albicans morphogenesis. PLoS Genet 8:e1003118. https://doi.org/10.1371/journal.pgen.1003118

Hoenigl M et al (2021) The antifungal pipeline: Fosmanogepix, Ibrexafungerp, Olorofim, Opelconazole, and Rezafungin. Drugs 81:1703–1729. https://doi.org/10.1007/s40265-021-01611-0

Iracane E, Vega-Estévez S, Buscaino A (2021) On and Off: epigenetic regulation of C. albicans morphological switches. Pathogens (Basel, Switzerland) 10:1463. https://doi.org/10.3390/pathogens10111463

Janssens I, Lambrecht BN, Van Braeckel E (2024) Aspergillus and the lung. Semin Respir Crit Care Med 45:3–20. https://doi.org/10.1055/s-0043-1777259

Kaplan JB, Ragunath C, Ramasubbu N, Fine DH (2003) Detachment of Actinobacillus actinomycetemcomitans biofilm cells by an endogenous beta-hexosaminidase activity. J Bacteriol 185:4693–4698. https://doi.org/10.1128/jb.185.16.4693-4698.2003

Kulig K et al (2022) Insight into the properties and Immunoregulatory effect of extracellular vesicles produced by Candida glabrata, Candida parapsilosis, and Candida tropicalis biofilms. Front Cell Infect Microbiol 12:879237. https://doi.org/10.3389/fcimb.2022.879237

Larkin E et al (2017) The emerging pathogen Candida auris: growth phenotype, virulence factors, activity of antifungals, and effect of SCY-078, a novel Glucan synthesis inhibitor, on growth morphology and biofilm formation. Antimicrob Agents Chemother 61:e02396-16. https://doi.org/10.1128/aac.02396-16

Lindsay AK, Deveau A, Piispanen AE, Hogan DA (2012) Farnesol and cyclic AMP signaling effects on the hypha-to-yeast transition in Candida albicans. Eukaryot Cell 11:1219–1225. https://doi.org/10.1128/EC.00144-12

Lopes da Rosa J, Boyartchuk VL, Zhu LJ, Kaufman PD (2010) Histone acetyltransferase Rtt109 is required for Candida albicans pathogenesis. Proc Natl Acad Sci USA 107:1594–1599. https://doi.org/10.1073/pnas.0912427107

Lu Y et al (2008) Efg1-mediated recruitment of NuA4 to promoters is required for hypha-specific Swi/Snf binding and activation in Candida albicans. Mol Biol Cell 19:4260–4272. https://doi.org/10.1091/mbc.e08-02-0173

Lu Y, Su C, Liu H (2012) A GATA transcription factor recruits Hda1 in response to reduced Tor1 signaling to establish a hyphal chromatin state in Candida albicans. PLoS Pathog 8:e1002663. https://doi.org/10.1371/journal.ppat.1002663

Mamouei Z et al (2021) An evolutionarily diverged mitochondrial protein controls biofilm growth and virulence in Candida albicans. PLoS Biol 19:e3000957. https://doi.org/10.1371/journal.pbio.3000957

Martinez LR, Casadevall A (2015) Biofilm formation by Cryptococcus neoformans. Microbiol Spectr 3. https://doi.org/10.1128/microbiolspec.MB-0006-2014

Martins M et al (2010) Presence of extracellular DNA in the Candida albicans biofilm matrix and its contribution to biofilms. Mycopathologia 169:323–331. https://doi.org/10.1007/s11046-009-9264-y

Massey J, Zarnowski R, Andes D (2023) Role of the extracellular matrix in Candida biofilm antifungal resistance. FEMS Microbiol Rev 47:fuad059. https://doi.org/10.1093/femsre/fuad059

McDougald D, Rice SA, Barraud N, Steinberg PD, Kjelleberg S (2011) Should we stay or should we go: mechanisms and ecological consequences for biofilm dispersal. Nat Rev Microbiol 10:39–50. https://doi.org/10.1038/nrmicro2695

Mello TP et al (2016) Assessment of biofilm formation by Scedosporium apiospermum, S. aurantiacum, S. minutisporum and Lomentospora prolificans. Biofouling 32:737–749. https://doi.org/10.1080/08927014.2016.1192610

Mello TP, Oliveira SSC, Branquinha MH, Santos ALS (2022) Decoding the antifungal resistance mechanisms in biofilms of emerging, ubiquitous and multidrug-resistant species belonging to the Scedosporium/Lomentospora genera. Med Mycol 60:myac036. https://doi.org/10.1093/mmy/myac036

Mello TP, Barcellos IC, Branquinha MH, Santos ALS (2023) Cell dispersion during biofilm formation by Scedosporium apiospermum, Scedosporium aurantiacum, Scedosporium minutisporum and Lomentospora prolificans. Curr Res Microb Sci 4:100191. https://doi.org/10.1016/j.crmicr.2023.100191

Nobile CJ, Johnson AD (2015) Candida albicans biofilms and human disease. Ann Rev Microbiol 69:71–92. https://doi.org/10.1146/annurev-micro-091014-104330

Nobile CJ et al (2008) Complementary adhesin function in C. albicans biofilm formation. Curr Biol 18:1017–1024. https://doi.org/10.1016/j.cub.2008.06.034

Nobile CJ et al (2012) A recently evolved transcriptional network controls biofilm development in Candida albicans. Cell 148:126–138. https://doi.org/10.1016/j.cell.2011.10.048

Nobile CJ et al (2014) A histone deacetylase complex mediates biofilm dispersal and drug resistance in Candida albicans. MBio 5:e01201–e01214. https://doi.org/10.1128/mBio.01201-14

Ramage G, Saville SP, Wickes BL, López-Ribot JL (2002) Inhibition of Candida albicans biofilm formation by farnesol, a quorum-sensing molecule. Appl Environ Microbiol 68:5459–5463. https://doi.org/10.1128/aem.68.11.5459-5463.2002

Ramage G, Saville SP, Thomas DP, Lopez-Ribot JL (2005) Candida biofilms: an update. Eukaryot Cell 4:633–638. https://doi.org/10.1128/EC.4.4.633-638.2005

Ramage G, Mowat E, Jones B, Williams C, Lopez-Ribot J (2009) Our current understanding of fungal biofilms. Crit Rev Microbiol 35:340–355. https://doi.org/10.3109/10408410903241436

Ramage G, Rajendran R, Gutierrez-Correa M, Jones B, Williams C (2011) Aspergillus biofilms: clinical and industrial significance. FEMS Microbiol Lett 324:89–97. https://doi.org/10.1111/j.1574-6968.2011.02381.x

Raman SB et al (2006) Candida albicans SET1 encodes a histone 3 lysine 4 methyltransferase that contributes to the pathogenesis of invasive candidiasis. Mol Microbiol 60:697–709. https://doi.org/10.1111/j.1365-2958.2006.05121.x

Ramirez-Garcia A et al (2018) Scedosporium and Lomentospora: an updated overview of underrated opportunists. Med Mycol 56:S102–S125. https://doi.org/10.1093/mmy/myx113

Ravi S, Pierce C, Witt C, Wormley FL Jr (2009) Biofilm formation by Cryptococcus neoformans under distinct environmental conditions. Mycopathologia 167:307–314. https://doi.org/10.1007/s11046-008-9180-6

Ribeiro F, d. C. et al (2024) Candida albicans recovered from persistent Candidemia exhibits enhanced virulence traits. J Infect Dis jiae631:e803. https://doi.org/10.1093/infdis/jiae631

Robbins N et al (2011) Hsp90 governs dispersion and drug resistance of fungal biofilms. PLoS Pathog 7:e1002257. https://doi.org/10.1371/journal.ppat.1002257

Rodrigues CF, Rodrigues ME, Silva S, Henriques M (2017) Candida glabrata biofilms: how far have we come? J Fungi (Basel, Switzerland) 3:3. https://doi.org/10.3390/jof3010011

Rumbaugh KP, Sauer K (2020) Biofilm dispersion. Nat Rev Microbiol 18:571–586. https://doi.org/10.1038/s41579-020-0385-0

Santi L et al (2014) Proteomic profile of Cryptococcus neoformans biofilm reveals changes in metabolic processes. J Proteome Res 13:1545–1559. https://doi.org/10.1021/pr401075f

Saputo S, Chabrier-Rosello Y, Luca FC, Kumar A, Krysan DJ (2012) The RAM network in pathogenic fungi. Eukaryot Cell 11:708–717. https://doi.org/10.1128/ec.00044-12

Schulze A, Mitterer F, Pombo JP, Schild S (2021) Biofilms by bacterial human pathogens: clinical relevance – development, composition and regulation – therapeutical strategies. Microb Cell 8:28–56. https://doi.org/10.15698/mic2021.02.741

Sebaa S, Boucherit-Otmani Z, Courtois P (2019) Effects of tyrosol and farnesol on Candida albicans biofilm. Mol Med Rep 19:3201–3209. https://doi.org/10.3892/mmr.2019.9981

Sharma S et al (2023) Microbial biofilm: a review on formation, infection, antibiotic resistance, control measures, and innovative treatment. Microorganisms 11:1614. https://doi.org/10.3390/microorganisms11061614

Shen J, Cowen LE, Griffin AM, Chan L, Kohler JR (2008) The Candida albicans pescadillo homolog is required for normal hypha-to-yeast morphogenesis and yeast proliferation. Proc Natl Acad Sci USA 105:20918–20923. https://doi.org/10.1073/pnas.0809147105

Subroto E, van Neer J, Valdes I, de Cock H (2022) Growth of Aspergillus fumigatus in Biofilms in Comparison to Candida albicans. J Fungi (Basel, Switzerland) 8:48. https://doi.org/10.3390/jof8010048

Sutar Y et al (2022) Niclosamide-loaded nanoparticles disrupt Candida biofilms and protect mice from mucosal candidiasis. PLoS Biol 20:e3001762. https://doi.org/10.1371/journal.pbio.3001762

Tan Y, Leonhard M, Ma S, Moser D, Schneider-Stickler B (2017) Dispersal of single and mixed non-albicans Candida species biofilms by β-1,3-glucanase in vitro. Microb Pathog 113:342–347. https://doi.org/10.1016/j.micpath.2017.10.057

Uppuluri P, Lopez-Ribot JL (2016) Go forth and colonize: dispersal from clinically important microbial biofilms. PLoS Pathog 12:e1005397. https://doi.org/10.1371/journal.ppat.1005397

Uppuluri P, Chaturvedi AK, Lopez-Ribot JL (2009) Design of a simple model of Candida albicans biofilms formed under conditions of flow: development, architecture, and drug resistance. Mycopathologia 168:101–109. https://doi.org/10.1007/s11046-009-9205-9

Uppuluri P et al (2010a) Dispersion as an important step in the Candida albicans biofilm developmental cycle. PLoS Pathog 6:e1000828. https://doi.org/10.1371/journal.ppat.1000828

Uppuluri P et al (2010b) The transcriptional regulator Nrg1p controls Candida albicans biofilm formation and dispersion. Eukaryot Cell 9:1531–1537. https://doi.org/10.1128/ec.00111-10

Uppuluri P, Srinivasan A, Ramasubramanian A, Lopez-Ribot JL (2011) Effects of fluconazole, amphotericin B, and caspofungin on Candida albicans biofilms under conditions of flow and on biofilm dispersion. Antimicrob Agents Chemother 55:3591–3593. https://doi.org/10.1128/AAC.01701-10

Uppuluri P et al (2012) Physiologic expression of the Candida albicans pescadillo homolog is required for virulence in a murine model of hematogenously disseminated candidiasis. Eukaryot Cell 11:1552–1556. https://doi.org/10.1128/EC.00171-12

Uppuluri P et al (2018) Candida albicans dispersed cells are developmentally distinct from biofilm and planktonic cells. MBio 9:e01338-18. https://doi.org/10.1128/mBio.01338-18

Wakade RS, Ristow LC, Stamnes MA, Kumar A, Krysan DJ (2020) The Ndr/LATS kinase Cbk1 regulates a specific subset of Ace2 functions and suppresses the hypha-to-yeast transition in Candida albicans. mBio 11:e01900-20. https://doi.org/10.1128/mBio.01900-20

Wakade RS, Wellington M, Krysan DJ (2024) Temporal dynamics of Candida albicans morphogenesis and gene expression reveals distinctions between in vitro and in vivo filamentation. mSphere **9**:e0011024. https://doi.org/10.1128/msphere.00110-24
Zarnowski R et al (2018) Candida albicans biofilm-induced vesicles confer drug resistance through matrix biogenesis. PLoS Biol 16:e2006872. https://doi.org/10.1371/journal.pbio.2006872
Zarnowski R et al (2021) Coordination of fungal biofilm development by extracellular vesicle cargo. Nat Commun 12:6235. https://doi.org/10.1038/s41467-021-26525-z
Zarnowski R et al (2022) A common vesicle proteome drives fungal biofilm development. Proc Natl Acad Sci USA 119:e2211424119. https://doi.org/10.1073/pnas.2211424119
Zhao M et al (2021) Turbinmicin inhibits Candida biofilm growth by disrupting fungal vesicle-mediated trafficking. J Clin Invest 131:e145123. https://doi.org/10.1172/jci145123

Chapter 10
Candida Biofilms and Immunity

Jeniel E. Nett and Ashley M. Holt

Abstract *Candida* spp. exhibit strong capacities for biofilm formation on medical devices and other surfaces. Biofilm formation complicates medical management as the adherent communities are difficult to eradicate with antifungal management alone. The presence of these cohesive fungal conglomerates alters host recognition and responses, when compared to interactions with free-floating planktonic *Candida*. In this review, we highlight the clinical importance of *Candida* biofilms, host responses to biofilms in vivo, and leukocyte-biofilm interactions. We describe the mechanisms underpinning these interactions, including how biofilm extracellular vesicles, extracellular matrix, secreted products, and dispersion influence immunity.

The Role of *Candida* Biofilms During Infection

Candida spp. frequently cause invasive disease in hospitalized patients, ranking first among causative pathogens of bloodstream infections in a United States point prevalence study (Magill et al. 2014). Care of hospitalized patients often involves the placement of medical devices, including vascular and urinary catheters, which can serve as substrates for *Candida* biofilm proliferation (Magill et al. 2014; Nobile and Johnson 2015; Douglas 2003). Central venous catheters are implicated in just over 80% of bloodstream infections caused by *Candida* spp., and the association appears more frequent for the recently emergent pathogen *Candida auris* (Andes et al. 2012; Alvarez-Moreno et al. 2023). *Candida* biofilms adherent to the artificial surfaces of

J. E. Nett (✉)
Department of Medicine, University of Wisconsin, Madison, WI, USA

Department of Medical Microbiology & Immunology, University of Wisconsin, Madison, WI, USA
e-mail: jenett@medicine.wisc.edu

A. M. Holt
Department of Medicine, University of Wisconsin, Madison, WI, USA

G. Ramage, D. Andes (eds.), *Fungal Biofilms*, Springer Series on Biofilms 15,
https://doi.org/10.1007/978-3-031-93315-8_10

medical devices are particularly resilient in the face of antifungals and host immune responses, rendering them nearly impossible to clear without physical removal (Douglas 2003; Donlan 2001; Chandra et al. 2001a). While catheter removal is recommended and often required to clear candidemia, mortality rates, even in the setting of catheter removal, remain high (Andes et al. 2012; Ala-Houhala and Anttila 2020; Pappas et al. 2016). These findings suggest that biofilm formation perpetuates invasive candidiasis and that additional anti-biofilm strategies could be helpful for these patients.

While *Candida* biofilms are commonly associated with central venous catheter infections, *Candida* spp. similarly infect other devices, including urinary catheters, joint prostheses, dialysis catheters, prosthetic heart valves, cardiac devices, and central nervous system shunts (Kojic and Darouiche 2004). The rates of infection vary by device, with *Candida* spp. estimated to account for approximately 1% to over 60% of infections (Kojic and Darouiche 2004). Like central venous catheter-associated *Candida* bloodstream infection, device removal is recommended and often required to clear these infections (Pappas et al. 2016). Unfortunately, the extraction of many of these devices requires operative intervention which can have accompanying risks that are not trivial.

While much of the *Candida* biofilm literature has focused on biofilms adherent to artificial devices and abiotic surfaces, the importance of oral mucosal biofilms has been increasingly recognized (Douglas 2002; Ganguly and Mitchell 2011; Dongari-Bagtzoglou et al. 2009; Dongari-Bagtzoglou 2008). The clinical relevance of mucosal biofilm formation is less clear for vaginal candidiasis. While murine models of vaginitis reveal biofilm structures, examination of biopsies from patients with vaginal candidiasis show invasive lesions without clear mucosal-associated biofilm (Swidsinski et al. 2019; Harriott et al. 2010). Similarly, ex vivo skin models show biofilm for *C. auris* on the skin surface (Eix et al. 2022). However, little is known about the structures of *C. auris* cells during colonization on the skin of patients.

In most clinical studies, *Candida albicans* remains the most frequently identified species (Pfaller et al. 2019). However, data from the SENTRY antifungal surveillance programs show a pattern of increasing isolation of *Candida glabrata*, *Candida parapsilosis*, and more recently, C. auris (Pfaller et al. 2019; Castanheira et al. 2024). In addition, Latin America and India have seen a climb in the identification of C. tropicalis (Pfaller et al. 2019; Castanheira et al. 2024; Mathur et al. 2024). The prevalence of *C. auris* varies greatly by geography and institution with some studies reporting *C. auris* as the most identified *Candida* species from patients with candidemia, accounting for over 40% of infections (Prayag et al. 2022). Mirroring this pattern, earlier in vitro and in vivo biofilm work frequently utilized *C. albicans* as a model organism (Ramage et al. 2001; Hawser and Douglas 1994; Chandra et al. 2001b; Uppuluri et al. 2010; Andes et al. 2004). However, it has become increasingly clear that other *Candida* spp., including *C. glabrata, C. tropicalis, C. parapsilosis,* and *C. auris,* also form biofilms in these models, consistent with clinical observation of catheter and other device-associated infections for these species

(Hawser and Douglas 1994; Bizerra et al. 2008; Lewis et al. 2002; Sherry et al. 2017; Kuhn et al. 2002; Jain et al. 2007).

Host Responses to *Candida* Biofilms in Vivo

Many animal model systems have been developed to mimic the variety of *Candida* biofilm infections that occur in patients, and these have been integral for the analysis of host interactions (Kojic and Darouiche 2004; Andes et al. 2004; Lazzell et al. 2009; Schinabeck et al. 2004; Ricicova et al. 2010; Kucharikova et al. 2015a; Nett et al. 2014; Johnson et al. 2012; Nett et al. 2010; Wang and Fries 2011; Capote-Bonato et al. 2018). Similar to descriptions of in vitro biofilms, *Candida* spp. grow adherent to these medical devices and form dense biofilms comprised of combinations of yeast, pseudohyphae, and hyphae (Andes et al. 2004; Kucharikova et al. 2015a, b; Nett et al. 2007; Tseng et al. 2024). The observed morphotypes are species- and strain-specific but can also be influenced by the environmental niche (Nett et al. 2007; Tseng et al. 2024). For example, administration of total parenteral nutrition appears to impair the degree of filamentation observed for *C. albicans* (Tseng et al. 2024). Biofilms formed in the setting of nutrient-rich intravenous solution are mainly composed of yeast and pseudohyphal forms. The model systems routinely show dense biofilms on the surface on in vivo vascular catheters, urinary catheters, and dentures infected with *Candida* spp. (Wang and Fries 2011; Kucharikova et al. 2015b; Nett et al. 2015). The in vivo biofilms also entangle host proteins and blood cells, including red blood cells, white blood cells, and platelets (Tseng et al. 2024; Nett et al. 2015). In one model system, over 95% of extracellular matrix proteins were estimated to be of host origin (Nett et al. 2015). Processes regulating the incorporation of host components into *Candida* biofilms remain largely unexplored. It is theorized that the host proteins and other components deposit in biofilm matrix through a combination of non-specific incorporation and regulated processes. For example, the finding that the proteins of in vitro *C. albicans* biofilms largely reflect the *C. albicans* secretome supports non-specific integration of proteins (Thomas et al. 2006). In contrast, some *C. albicans* biofilms can release a toxin (candidalysin) that can trigger neutrophils to form neutrophil extracellular traps (NETs) which serve as a mechanism for the deposition for cell-specific proteins, such as histones from neutrophils (Tseng et al. 2024). This mechanism could presumably lead to the deposition of other NET-associated molecules, including DNA and granular components (Brinkmann et al. 2004).

Similar to animal models that mimic vascular catheter infections, rodent models can be utilized to examine *Candida* biofilms on urinary catheters (Nett et al. 2014; Wang and Fries 2011; Capote-Bonato et al. 2018). These models show *Candida* biofilms adherent to catheters in the urinary environment, as well as involvement of the bladder with *Candida* propagating on the epithelial surface, mirroring the acute cystitis observed for patients with catheter-associated urinary tract infections (Nett et al. 2014; Capote-Bonato et al. 2018). These model systems show that neutrophils recruit to urinary *Candida* biofilms formed by both *C. albicans* and *C. tropicalis* (Capote-Bonato et al. 2018; Nett et al. 2015). However, the inflammatory responses do not appear to be able to eliminate these catheter-associated urinary infections. Likewise, when catheter segments are subcutaneously implanted, pro-inflammatory cells recruit to the biofilm infection site and lack the capacity to eradicate the biofilm (Ricicova et al. 2010; Kucharikova et al. 2015a; Nieminen et al. 2014).

A variety of mouse and rat models have been employed to examine biofilms on oral mucosal surfaces and/or study *Candida* propagating on the surface of oral devices, such as dentures (Johnson et al. 2012; Nett et al. 2010; Lee et al. 2011; Yano et al. 2019; Tobouti et al. 2016; Sultan et al. 2019; Dwivedi et al. 2011). Similar to the catheter models of biofilm infection, *Candida* spp. robustly adhere to denture devices forming dense biofilms, often resulting in associated mucosal inflammation, mirroring the disease of denture stomatitis frequently observed in denture-wearing patients (Johnson et al. 2012; Nett et al. 2010; Tobouti et al. 2016; Sultan et al. 2019; Chen et al. 2011). Analysis of the oral biofilm composition shows that the fungal biofilm cells become encased in fungal-derived extracellular matrix as well as salivary proteins and inflammatory cells (Nett et al. 2015). Like biofilms formed on abiotic surfaces, oral biofilm structures exhibit resistance to killing by innate immune cells (Dwivedi et al. 2011).

Neutrophil Responses to *Candida* Biofilms

Neutrophils recruit rapidly to many invasive fungi, including *Candida*, and are necessary for control of invasive disease (Fidel Jr. 2002). Patients lacking functional numbers of neutrophils are prone to invasive candidiasis (Erwig and Gow 2016; Edwards Jr. et al. 1978). While neutrophils exhibit antifungal activity against *Candida* when growing in logarithmic phase or planktonic culture, biofilms are substantially less susceptible to these antifungal effects (Katragkou et al. 2010, 2011a, b; Johnson et al. 2016; Xie et al. 2012; Kernien et al. 2017) (Fig. 10.1). For example, for *C. albicans*, human peripheral blood neutrophils are approximately five-fold less effective at killing biofilm cells when compared to planktonic cells in ex vivo studies (Katragkou et al. 2010, 2011b; Johnson et al. 2016; Xie et al. 2012). Similarly, biofilms formed by *C. glabrata* and *C. parapsilosis* withstand neutrophil attack (Katragkou et al. 2011a; Johnson et al. 2017). Poor activity against *Candida* biofilms appears to persist even in the setting of stimulating the cells with cytokines such as interferon-γ and granulocyte colony-stimulating factor (Katragkou et al. 2011b).

Neutrophils counter pathogens with a variety of effector functions, including phagocytosis, production of reactive oxygen species (ROS), and the formation of neutrophil extracellular traps (NETs), extracellular lattice-like structures containing DNA, histones, and antimicrobial proteins (Brinkmann et al. 2004). For *Candida* spp., neutrophil responses appear to vary based on fungal morphology. For example, when *C. albicans* grows in yeast-locked form, human neutrophils respond with

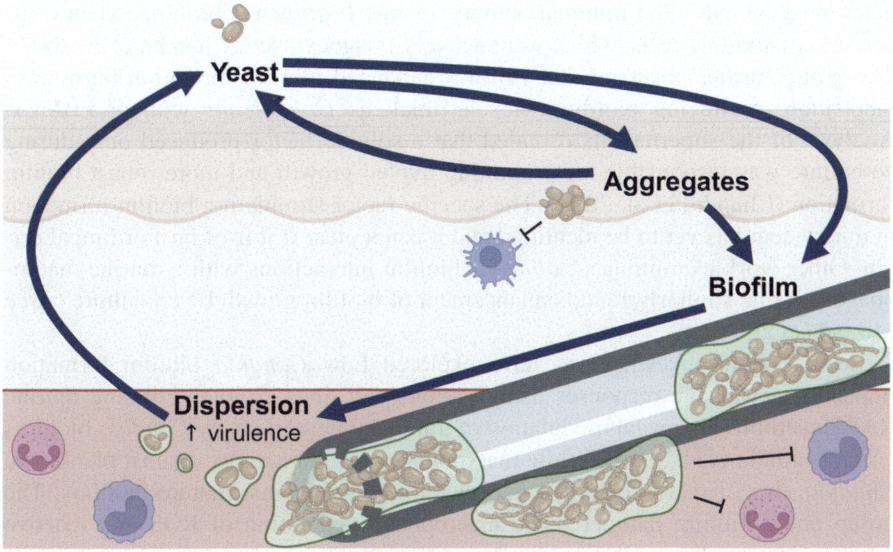

Fig. 10.1 *Candida* biofilm life cycle and host interactions. Biofilms formed by *Candida* spp. alter host cell recognition when compared to free-floating planktonic *Candida*. Biofilm formation can impair phagocyte responses and their capacity for fungal killing. Dispersed biofilm cells retain virulence traits. Aggregative *C. auris* strains display some similarities to biofilm cells. Figure made with Biorender

phagocytosis (Branzk et al. 2014). This contrasts with *C. albicans* hyphae, which are often too large to undergo complete phagocytosis. Upon encountering hyphae and other large organisms, neutrophils generate ROS and form NETs, which provide antifungal activity without the need for phagocytosis (Branzk et al. 2014; Urban et al. 2006). However, biofilms fail to trigger ROS production or NET formation (Brinkmann et al. 2004; Johnson et al. 2016; Xie et al. 2012; Kernien et al. 2017). This inhibition persists even in the setting of a potent inducer of ROS and NETs (phorbol ester) (Johnson et al. 2016). The finding that neutrophils artificially stimulated to produce ROS and NETs ex vivo exhibit anti-biofilm activity suggests that these neutrophil responses may be effective against biofilm if they were not otherwise inhibited (Johnson et al. 2016).

Monocyte and Macrophage Interactions with *Candida* Biofilms

Similar to observations for neutrophil-biofilm interactions, biofilm formation also impacts how monocytes and macrophages recognize and respond to *Candida* (Katragkou et al. 2010; Simitsopoulou et al. 2018; Chandra et al. 2007; Arce Miranda et al. 2019; Alonso et al. 2017) (Fig. 10.1). Chandra et al. found that human peripheral blood mononuclear cells (PBMCs, enriched for monocytes and macrophages) exhibited minimal activity against *C. albicans* biofilms when compared to planktonic cells, which were actively phagocytosed (Chandra et al. 2007). The group further observed that PBMCs enhanced biofilm formation during co-incubation, producing biofilm twice as thick as *C. albicans* without PBMCs. Analysis of the supernatants revealed that a soluble factor produced only during co-culture was responsible for promoting hyphal growth and more robust biofilm formation (Chandra et al. 2007). The specific factor stimulating biofilm formation in this system has yet to be identified and it is not clear if it is of host or fungal origin. Other work examining *C. albicans* biofilm interactions with a murine macrophage cell line similarly found enhancement of biofilm growth by co-culture (Arce Miranda et al. 2019).

Several other investigations have explored how *Candida* biofilm formation impacts macrophage responses using various cell lines. Studies utilizing murine RAW macrophages reveal modulation of immune interactions by *Candida* biofilms (Arce Miranda et al. 2019). While macrophages exhibit activity against planktonic *Candida*, they display negligible antifungal activity against mature biofilms. The group noted biofilm hampered the macrophage generation of ROS and reactive nitrogen intermediates (RNIs), similar to the inhibition observed for neutrophils encountering mature biofilms (Johnson et al. 2016; Xie et al. 2012; Arce Miranda et al. 2019). Studies using macrophage-like cells derived from a human monocytic cell line (THP-1) found different cytokine profiles following co-culture with *C. albicans* biofilms when compared to planktonic cells, with lower induction of TNF-α after biofilm exposure *(*Katragkou et al. 2010*)*. This may be relevant in the clinical setting as recombinant TNF-α can reduce the burden of C. albicans biofilms (FaC et al. 2017). Consistent with the studies using RAW macrophages, macrophages derived from the THP-1 cell line also exhibited less antifungal activity against *C. albicans* biofilms when compared to planktonic cells *(*Katragkou et al. 2010). Macrophages responses to *Candida* biofilms appear to vary by species, as biofilms formed *C. albicans* induce a heightened pro-inflammatory response by THP-1 cells when compared to co-incubation with *C. parapsilosis* biofilms (Simitsopoulou et al. 2018).

In other work using a different murine macrophage line (J774.1), *C. albicans* biofilms were found to impair macrophage migration, with macrophages exhibiting a two-fold lower mean track velocity in response to encounter with biofilms when compared to planktonic cells (Alonso et al. 2017). This impairment of macrophage migration appears to be linked to the physical structure of the biofilm and not necessarily dependent on the cell wall or extracellular matrix composition, as biofilms formed by a mutant with genetic disruption of the mannan cell wall and matrix production exhibited a similar phenotype. It is postulated that the network of hyphae present in *C. albicans* biofilms may prevent fungal capture and lead to a frustrated phagocytosis phenotype (Alonso et al. 2017).

Biofilm Vesicles and Other Secreted Factors

During both biofilm and planktonic growth, *Candida* spp. release extracellular vesicles (EVs), small spherical structures of bioactive molecules encased in a lipid bilayer (Honorato et al. 2022; Zarnowski et al. 2022). EVs can modulate diverse functions, including fungal morphology, fungal resistance, and host interactions (Honorato et al. 2022; Zarnowski et al. 2022; Brown Harding et al. 2024). As the size and cargo of EVs differ between the biofilm and planktonic states, EVs produced during these growth phases can diverge in function and activity (Zarnowski et al. 2018). As may be expected, vesicle cargo also varies among biofilms produced by different *Candida* spp. (Kulig et al. 2022). Despite differences, some biofilm-derived EVs exhibit similar impacts on immunity. For example, EVs collected from *C. tropicalis*, *C. parapsilosis*, and *C. glabrata* biofilms were shown to trigger macrophages (differentiated from THP-1 cells) to boost pro-inflammatory cytokines (TNF-α and IL-8) and blunt production of the anti-inflammatory IL-10 (Kulig et al. 2022).

Recent studies utilizing a murine model shed light on EV interactions with host immunity (Brown Harding et al. 2024). EVs produced by *C. albicans* biofilms were found to modulate macrophage responses, inducing IFN-β production, with a more robust response observed for biofilm EVs over planktonic EVs. The biofilms EVs were shown to be internalized by macrophages, leading to release of EV cargo, including fungal DNA, which in turn activated the cytosolic sensor cGAS (cyclic GMP-AMP synthase) and the downstream STING pathway components to produce a type I interferon response. *Sting−/−* mice with disruption of this pathway exhibited higher fungal burdens in a systemic candidiasis model. The work shows how biofilm EVs can rewire macrophage responses, even prior to engagement and/or phagocytosis with the fungi, altering the global immune response.

C. albicans also secretes a cytotoxic peptide, candidalysin, that can damage host cell membranes and drive a variety of inflammatory responses (Moyes et al. 2016; Kasper et al. 2018; Richardson et al. 2018). This toxin was recently shown to modulate neutrophil interactions during biofilm formation (Tseng et al. 2024). Upon examination of biofilms on catheters from patients with catheter-associated candidiasis, Tseng et al. observed human neutrophil markers consistent with the formation of NETs. However, the presence of these varied by *Candida* isolate. On further examination of the strains, many formed biofilms with minimal hyphae and larger amounts of yeast and pseudohyphae, a phenotype different from the common laboratory strain SC5314 which forms biofilms almost entirely of hyphae. Analysis of filamentation-associated genes showed decreased transcript abundance for *ECE1*, the gene encoding candidalysin, across all strains, while other filamentation-associated gene abundances were not altered. This suggests that *C. albicans* can downregulate *ECE1* to diminish NET formation, promoting fungal persistence on catheters in the bloodstream. This phenotype was shown to be particularly relevant in the setting of total parenteral nutrition administration.

Candida spp. additionally secrete low molecular weight molecules called autoinducers that provide signals for population density in a process termed quorum sensing. The *C. albicans*-secreted quorum-sensing molecule farnesol appears to activate innate immune cells, including neutrophils, but dampen cellular adaptive immunity (Leonhardt et al. 2015; Zawrotniak et al. 2019). The impact of these findings likely varies across the life span of *Candida* biofilms, in line with altering levels of quorum sensing molecules produced during different phases of biofilm formation (Leonhardt et al. 2015; Zawrotniak et al. 2019).

Biofilm Matrix and Dispersion

Recent studies show that several *Candida* biofilm structural properties may influence interactions with neutrophils and other immune cells (Johnson et al. 2016; Xie et al. 2012; Chandra et al. 2007; Alonso et al. 2017) (Fig. 10.2). In contrast to free-floating *Candida* grown in planktonic conditions, *Candida* cells within a biofilm produce EVs that deliver cargo for assembly of a protective extracellular matrix (Zarnowski et al. 2018). This polymeric material promotes cellular interactions and protects from external insults (Chandra et al. 2001a; Uppuluri et al. 2010; Wall et al. 2019). The biofilm extracellular matrix can mask the cell wall structures typically displayed to the host, which can greatly influence how host cells engage, interact, and respond (Johnson et al. 2016; Hawser et al. 1998; Chaffin et al. 1998). This is

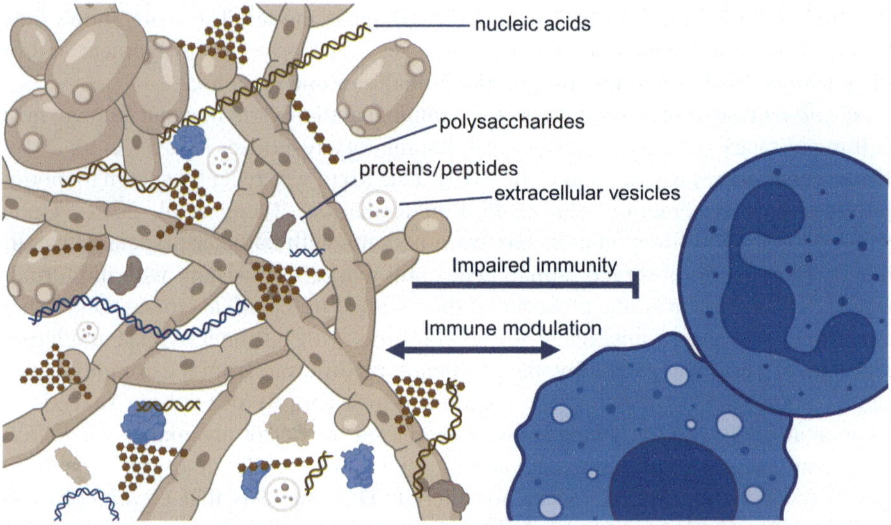

Fig. 10.2 *Candida* biofilm matrix modulates host immunity. The extracellular matrix of *Candida* biofilms is delivered by EVs and is comprised of unique polysaccharides, nucleic acids, proteins, and peptides. The matrix can mask fungal cell wall structures, impair immune responses, and modulate host interactions. Figure made with Biorender

particularly relevant given the variation in polysaccharides and proteins between the biofilm cell wall and biofilm extracellular matrix (Thomas et al. 2006; Taff et al. 2012; Silva et al. 2009; Baillie and Douglas 2000). For example, the extracellular matrix of *C. albicans* contains an abundant unique high molecular weight α-1,2-branched α-1,6 mannan which interacts with other polysaccharides to form a mannan-glucan complex not present in the cell wall (Chaffin et al. 1998; Zarnowski et al. 2014; Mitchell et al. 2015). Genetic disruption of the α-mannan production diminishes mannan-glucan complex assembly in the matrix, altering host recognition (Johnson et al. 2016). While wildtype *C. albicans* biofilms typically blunt neutrophil responses, biofilms with disruption of this complex activate neutrophils and trigger NET formation. As the formation of a mannan-glucan complex is a hallmark of many *Candida* biofilms, including *C. tropicalis*, *C. parapsilosis*, *C. glabrata*, and *C. auris*, biofilms formed by these species may similarly modulate host responses through this mechanism (Dominguez et al. 2018, 2019).

Many other individual biofilm components have the capacity to modulate immune responses to *Candida* biofilms. For example, DNA and RNA isolated from the extracellular matrix of *C. albicans* can induce neutrophils to form NETs ex vivo following the endocytosis of nucleic acid by neutrophils and TLR8 activation (Smolarz et al. 2021). The isolated nucleic acid can also induce human neutrophil chemotaxis ex vivo. The relevance of the isolated free nucleic acid to intact biofilms is somewhat uncertain, but biofilms release nucleic acid as cargo in EVs and presumably from intact biofilms following physical disruption or dispersion (Brown Harding et al. 2024). Furthermore, the activity of the *C. albicans* extracellular nucleic acid can bind the host cationic peptide LL-37, creating a complex that attenuates NET formation (Juszczak et al. 2024). These findings suggest that the nucleic acid of *Candida* biofilms may heighten or dampen neutrophil responses, depending on the context. *C. albicans* cell wall proteins can also influence host interactions. For instance, the GPI-anchored cell wall protein Hyr1 has been shown to be protective against fungal damage caused by the HL60 neutrophil-like cell line (Dwivedi et al. 2011).

It has become increasingly recognized that cells release from mature biofilms in a regulated process of biofilm dispersion (Uppuluri et al. 2010). In the setting of a device-associated infection, biofilm dispersion allows *Candida* to spread throughout the bloodstream and cause disseminated disease at distant sites. Signals for *C. albicans* dispersion include growth in a rich carbon source or at a low pH, and it is likely other factors are involved (Uppuluri et al. 2010; Sellam et al. 2009). Pioneering studies utilizing *C. albicans* biofilms found that cells released from biofilms most closely resemble yeast, often with an elongated form (Uppuluri et al. 2010). However, despite a morphology similar to planktonic *C. albicans*, the dispersed cells exhibit distinct phenotypes relevant to clinical infection. For example, cells dispersed from biofilms are highly adhesive and form more robust biofilms, when compared to planktonic cells (Uppuluri et al. 2010). Furthermore, they are more toxic to endothelial cells and result in increased mortality in a murine model of invasive candidiasis (Uppuluri et al. 2010). These findings show how cells exiting the biofilm morphology continue to harbor virulence traits that can be advantageous for promoting infection at a distant site, helping to explain why patients without *Candida* biofilm removal have poor outcomes (Pappas et al. 2016). Newly dispersed biofilm cells appear to be transcriptionally programmed for increased pathogenicity, particularly in the setting of bloodstream infections involving vascular catheter infections (Uppuluri et al. 2010, 2018; Sellam et al. 2009).

C. auris Aggregates

Some *C. auris* strains display a multicellular aggregative or clumping phenotype that is distinct from other *Candida* spp. and nonaggregating *C. auris* strains (Borman et al. 2016). Many of these aggregative strains also form more robust biofilms in vitro (Bing et al. 2023; Brown et al. 2020). Studies show that aggregative phenotypes can emerge during the course of infection, resulting in clones with competitive advantages (Bing et al. 2024). For example, *C. auris* aggregative strains resist phagocytosis by a murine macrophage cell line (RAW264.7), but once disrupted by sonication, individual yeast cells could be phagocytosed (Bing et al. 2024). In addition, aggregation promotes resistance to host antimicrobial peptides, including LL-37 and pituitary adenylate cyclase-activating peptide (PACAP) (Bing et al. 2024). In other work using wounding models of skin epithelial cells, an aggregative *C. auris* displayed higher epithelial cytotoxicity (Brown et al. 2020). Together, studies suggest that aggregative phenotypes may emerge during infection and have the potential to alter virulence. *C. auris* aggregation appears to serve as a mechanism of immune evasion under certain clinical scenarios (Bing et al. 2024).

Conclusions

Candida biofilm formation commonly occurs during clinical infection, particularly in the setting of catheters and other medical devices. While many of the original investigations have utilized *C. albicans* as a model organism, the vast majority of *Candida* spp. have been shown to proliferate as biofilms adherent to medical devices or epithelial surfaces. Compared to single-cell planktonic organisms, the formation of biofilm conglomerates influences both host recognition and the resilience of *Candida* to host defenses. Across spp., *Candida* biofilms display resistance to phagocytosis by both neutrophils and macrophages. Even upon recognition, biofilm formation alters cellular responses and cytokine production, impacting downstream immunity. Biofilm EVs deposit cargo that accumulates to form an extracellular matrix, which decorates fungal cell walls, obscuring cell wall motifs from immune cells. The biofilm extracellular matrix also provides protection from host cells and antimicrobial peptides. Even following release from biofilm, dispersed cells are programmed for enhancement of virulence traits, likely contributing to the pathogenesis observed for disseminated candidiasis. As discoveries of fungal-immune interactions emerge, it will remain important to analyze how these findings interface with biofilm formation, as the host clearly responds differently to biofilm and planktonic *Candida* spp.

Conflict of Interest The authors declare that the research was conducted in the absence of any commercial or financial relationships that could be construed as a potential conflict of interest.

Funding JEN is supported by the National Institutes of Health (R**01** AI**145939** and R**21** AAI**8678**). AMH is supported by the National Institutes of Health (T**32**GM**140935**).

References

Ala-Houhala M, Anttila VJ (2020) Persistent versus non-persistent candidaemia in adult patients in 2007–2016: a retrospective cohort study. Mycoses 63:617

Alonso MF, NaR G, Erwig LP et al (2017) Macrophage migration is impaired within *Candida albicans* biofilms. J Fungi (Basel) 3:31

Alvarez-Moreno CA, Morales-Lopez S, Rodriguez GJ et al (2023) The mortality attributable to candidemia in *C. auris* is higher than that in other *Candida* species: myth or reality? J fungi (Basel) 9:430

Andes D, Nett J, Oschel P et al (2004) Development and characterization of an in vivo central venous catheter *Candida albicans* biofilm model. Infect Immun 72:6023–6031

Andes DR, Safdar N, Baddley JW et al (2012) Impact of treatment strategy on outcomes in patients with candidemia and other forms of invasive candidiasis: a patient-level quantitative review of randomized trials. Clin Infect Dis 54:1110–1122

Arce Miranda JE, Baronetti JL, Sotomayor CE et al (2019) Oxidative and nitrosative stress responses during macrophage-*Candida albicans* biofilm interaction. Med Mycol 57:101–113

Baillie GS, Douglas LJ (2000) Matrix polymers of *Candida biofilms* and their possible role in biofilm resistance to antifungal agents. J Antimicrob Chemother 46:397–403

Bing J, Guan Z, Zheng T et al (2023) Clinical isolates of *Candida auris* with enhanced adherence and biofilm formation due to genomic amplification of *ALS4*. PLoS Pathog 19:e1011239

Bing J, Guan Z, Zheng T et al (2024) Rapid evolution of an adaptive multicellular morphology of *Candida auris* during systemic infection. Nat Commun 15:2381

Bizerra FC, Nakamura CV, De Poersch C et al (2008) Characteristics of biofilm formation by *Candida tropicalis* and antifungal resistance. FEMS Yeast Res 8:442–450

Borman AM, Szekely A, Johnson EM (2016) Comparative pathogenicity of United Kingdom isolates of the emerging pathogen *Candida auris* and other key pathogenic *Candida* species. mSphere 1

Branzk N, Lubojemska A, Hardison SE et al (2014) Neutrophils sense microbe size and selectively release neutrophil extracellular traps in response to large pathogens. Nat Immunol 15:1017–1025

Brinkmann V, Reichard U, Goosmann C et al (2004) Neutrophil extracellular traps kill bacteria. Science 303:1532–1535

Brown Harding H, Kwaku GN, Reardon CM et al (2024) *Candida albicans* extracellular vesicles trigger type I IFN signalling via cGAS and STING. Nat Microbiol 9:95–107

Brown JL, Delaney C, Short B et al (2020) *Candida auris* phenotypic heterogeneity determines pathogenicity *in vitro*. mSphere 5:e00371-20

Capote-Bonato F, Bonato DV, Ayer IM et al (2018) Murine model for the evaluation of candiduria caused by *Candida tropicalis* from biofilm. Microb Pathog 117:170–174

Castanheira M, Deshpande LM, Rhomberg PR et al (2024) Recent increase in *Candida auris* frequency in the SENTRY surveillance program: antifungal activity and genotypic characterization. Antimicrob Agents Chemother 68:e0057024

Chaffin WL, Lopez-Ribot JL, Casanova M et al (1998) Cell wall and secreted proteins of *Candida albicans*: identification, function, and expression. Microbiol Mol Biol Rev 62:130–180

Chandra J, Kuhn DM, Mukherjee PK et al (2001a) Biofilm formation by the fungal pathogen *Candida albicans*: development, architecture, and drug resistance. J Bacteriol 183:5385–5394

Chandra J, Mukherjee PK, Leidich SD et al (2001b) Antifungal resistance of candidal biofilms formed on denture acrylic in vitro. J Dent Res 80:903–908

Chandra J, Mccormick TS, Imamura Y et al (2007) Interaction of *Candida albicans* with adherent human peripheral blood mononuclear cells increases *C. albicans* biofilm formation and results in differential expression of pro- and anti-inflammatory cytokines. Infect Immun 75:2612–2620

Chen YL, Brand A, Morrison EL et al (2011) Calcineurin controls drug tolerance, hyphal growth, and virulence in *Candida dubliniensis*. Eukaryot Cell 10:803–819

Dominguez E, Zarnowski R, Sanchez H et al (2018) Conservation and divergence in the *Candida* species biofilm matrix mannan-glucan complex structure, function, and genetic control. mBio 9:e00451-18

Dominguez EG, Zarnowski R, Choy HL et al (2019) Conserved role for biofilm matrix polysaccharides in *Candida auris* drug resistance. mSphere 4:e00680-18

Dongari-Bagtzoglou A (2008) Mucosal biofilms: challenges and future directions. Expert Rev Anti-Infect Ther 6:141–144

Dongari-Bagtzoglou A, Kashleva H, Dwivedi P et al (2009) Characterization of mucosal *Candida albicans* biofilms. PLoS One 4:e7967

Donlan RM (2001) Biofilm formation: a clinically relevant microbiological process. Clin Infect Dis 33:1387–1392

Douglas LJ (2002) Medical importance of biofilms in *Candida infections*. Rev Iberoam Micol 19:139–143

Douglas LJ (2003) *Candida* biofilms and their role in infection. Trends Microbiol 11:30–36

Dwivedi P, Thompson A, Xie Z et al (2011) Role of Bcr1-activated genes Hwp1 and Hyr1 in *Candida albicans* oral mucosal biofilms and neutrophil evasion. PLoS One 6:e16218

Edwards JE Jr, Lehrer RI, Stiehm ER et al (1978) Severe candidal infections: clinical perspective, immune defense mechanisms, and current concepts of therapy. Ann Intern Med 89:91–106

Eix EF, Johnson CJ, Wartman KM et al (2022) Ex vivo human and porcine skin effectively model *Candida auris* colonization, differentiating robust and poor fungal colonizers. J Infect Dis 225:1791–1795

Erwig LP, Gow NA (2016) Interactions of fungal pathogens with phagocytes. Nat Rev 14:163–176

FaC R, Alves A, Rocha MFG et al (2017) Tumor necrosis factor prevents *Candida albicans* biofilm formation. Sci Rep 7:1206

Fidel PL Jr (2002) Immunity to *Candida*. Oral Dis 8(Suppl 2):69–75

Ganguly S, Mitchell AP (2011) Mucosal biofilms of *Candida albicans*. Curr Opin Microbiol 14:380–385

Harriott MM, Lilly EA, Rodriguez TE et al (2010) *Candida albicans* forms biofilms on the vaginal mucosa. Microbiology 156:3635–3644

Hawser SP, Douglas LJ (1994) Biofilm formation by *Candida* species on the surface of catheter materials in vitro. Infect Immun 62:915–921

Hawser SP, Baillie GS, Douglas LJ (1998) Production of extracellular matrix by *Candida albicans* biofilms. J Med Microbiol 47:253–256

Honorato L, De Araujo JFD, Ellis CC et al (2022) Extracellular vesicles regulate biofilm formation and yeast-to-hypha differentiation in Candida albicans. MBio 13:e0030122

Jain N, Kohli R, Cook E et al (2007) Biofilm formation by and antifungal susceptibility of *Candida* isolates from urine. Appl Environ Microbiol 73:1697–1703

Johnson CC, Yu A, Lee H et al (2012) Development of a contemporary animal model of *Candida albicans*-associated denture stomatitis using a novel intraoral denture system. Infect Immun 80:1736–1743

Johnson CJ, Cabezas-Olcoz J, Kernien JF et al (2016) The extracellular matrix of *Candida albicans* biofilms impairs formation of neutrophil extracellular traps. PLoS Pathog 12:e1005884

Johnson CJ, Kernien JF, Hoyer AR et al (2017) Mechanisms involved in the triggering of neutrophil extracellular traps (NETs) by *Candida glabrata* during planktonic and biofilm growth. Sci Rep 7:13065

Juszczak M, Zawrotniak M, Rapala-Kozik M (2024) Complexation of fungal extracellular nucleic acids by host LL-37 peptide shapes neutrophil response to *Candida albicans* biofilm. Front Immunol 15:1295168

Kasper L, Konig A, Koenig PA et al (2018) The fungal peptide toxin Candidalysin activates the NLRP3 inflammasome and causes cytolysis in mononuclear phagocytes. Nat Commun 9:4260

Katragkou A, Kruhlak MJ, Simitsopoulou M et al (2010) Interactions between human phagocytes and *Candida albicans* biofilms alone and in combination with antifungal agents. J Infect Dis 201:1941–1949

Katragkou A, Chatzimoschou A, Simitsopoulou M et al (2011a) Additive antifungal activity of anidulafungin and human neutrophils against *Candida parapsilosis* biofilms. J Antimicrob Chemother 66:588–591

Katragkou A, Simitsopoulou M, Chatzimoschou A et al (2011b) Effects of interferon-gamma and granulocyte colony-stimulating factor on antifungal activity of human polymorphonuclear neutrophils against *Candida albicans* grown as biofilms or planktonic cells. Cytokine 55:330–334

Kernien JF, Johnson CJ, Nett JE (2017) Conserved inhibition of neutrophil extracellular trap release by clinical *Candida albicans* biofilms. J Fungi 3:49

Kojic EM, Darouiche RO (2004) *Candida* infections of medical devices. Clin Microbiol Rev 17:255–267

Kucharikova S, Neirinck B, Sharma N et al (2015a) In vivo *Candida glabrata* biofilm development on foreign bodies in a rat subcutaneous model. J Antimicrob Chemother 70:846–856

Kucharikova S, Vande Velde G, Himmelreich U et al (2015b) Candida albicans biofilm development on medically-relevant foreign bodies in a mouse subcutaneous model followed by bioluminescence imaging. J Vis Exp 95:52239

Kuhn DM, Chandra J, Mukherjee PK et al (2002) Comparison of biofilms formed by *Candida albicans* and *Candida parapsilosis* on bioprosthetic surfaces. Infect Immun 70:878–888

Kulig K, Karnas E, Woznicka O et al (2022) Insight into the properties and immunoregulatory effect of extracellular vesicles produced by *Candida glabrata*, *Candida parapsilosis*, and *Candida tropicalis* biofilms. Front Cell Infect Microbiol 12:879237

Lazzell AL, Chaturvedi AK, Pierce CG et al (2009) Treatment and prevention of *Candida albicans* biofilms with caspofungin in a novel central venous catheter murine model of candidiasis. J Antimicrob Chemother 64:567–570

Lee H, Yu A, Johnson CC et al (2011) Fabrication of a multi-applicable removable intraoral denture system for rodent research. J Oral Rehabil 38:686–690

Leonhardt I, Spielberg S, Weber M et al (2015) The fungal quorum-sensing molecule farnesol activates innate immune cells but suppresses cellular adaptive immunity. MBio 6:e00143

Lewis RE, Kontoyiannis DP, Darouiche RO et al (2002) Antifungal activity of amphotericin B, fluconazole, and voriconazole in an in vitro model of *Candida* catheter-related bloodstream infection. Antimicrob Agents Chemother 46:3499–3505

Magill SS, Edwards JR, Bamberg W et al (2014) Multistate point-prevalence survey of health care-associated infections. N Engl J Med 370:1198–1208

Mathur P, Srivastav S, Thakur AK et al (2024) Candidaemia and central line-associated Candidaemia in a network of Indian ICUs: impact of COVID-19 pandemic. Mycoses 67:e13790

Mitchell KF, Zarnowski R, Sanchez H et al (2015) Community participation in biofilm matrix assembly and function. Proc Natl Acad Sci USA 112:4092–4097

Moyes DL, Wilson D, Richardson JP et al (2016) Candidalysin is a fungal peptide toxin critical for mucosal infection. Nature 532:64–68

Nett J, Lincoln L, Marchillo K et al (2007) Beta −1,3 glucan as a test for central venous catheter biofilm infection. J Infect Dis 195:1705–1712

Nett JE, Marchillo K, Spiegel CA et al (2010) Development and validation of an in vivo *Candida albicans* biofilm denture model. Infect Immun 78:3650–3659

Nett JE, Brooks EG, Cabezas-Olcoz J et al (2014) Rat indwelling urinary catheter model of *Candida albicans* biofilm infection. Infect Immun 82:4931–4940

Nett JE, Zarnowski R, Cabezas-Olcoz J et al (2015) Host contributions to construction of three device-associated *Candida albicans* biofilms. Infect Immun 83:4630–4638

Nieminen MT, Hernandez M, Novak-Frazer L et al (2014) DL-2-hydroxyisocaproic acid attenuates inflammatory responses in a murine *Candida albicans* biofilm model. Clin Vaccine Immunol 21:1240–1245

Nobile CJ, Johnson AD (2015) *Candida albicans* biofilms and human disease. Ann Rev Microbiol 69:71–92

Pappas PG, Kauffman CA, Andes DR et al (2016) Clinical practice guideline for the Management of Candidiasis: 2016 update by the Infectious Diseases Society of America. Clin Infect Dis 62:e1–e50

Pfaller MA, Diekema DJ, Turnidge JD et al (2019) Twenty years of the SENTRY antifungal surveillance program: results for *Candida* species from 1997–2016. Open Forum Infect Dis 6:S79–S94

Prayag PS, Patwardhan S, Panchakshari S et al (2022) The dominance of *Candida auris:* a single-center experience of 79 episodes of Candidemia from Western India. Indian J Crit Care Med 26:560–563

Ramage G, Vandewalle K, Wickes BL et al (2001) Characteristics of biofilm formation by *Candida albicans*. Rev Iberoam Micol 18:163–170

Richardson JP, Willems HME, Moyes DL et al (2018) Candidalysin drives epithelial signaling, neutrophil recruitment, and immunopathology at the vaginal mucosa. Infect Immun 86:e00645-17

Ricicova M, Kucharikova S, Tournu H et al (2010) *Candida albicans* biofilm formation in a new in vivo rat model. Microbiology 156:909–919

Schinabeck MK, Long LA, Hossain MA et al (2004) Rabbit model of *Candida albicans* biofilm infection: liposomal amphotericin B antifungal lock therapy. Antimicrob Agents Chemother 48:1727–1732

Sellam A, Al-Niemi T, Mcinnerney K et al (2009) A *Candida albicans* early stage biofilm detachment event in rich medium. BMC Microbiol 9:25

Sherry L, Ramage G, Kean R et al (2017) Biofilm-forming capability of highly virulent, multidrug-resistant *Candida auris*. Emerg Infect Dis 23:328–331

Silva S, Henriques M, Martins A et al (2009) Biofilms of non-*Candida albicans Candida* species: quantification, structure and matrix composition. Med Mycol 47:681–689

Simitsopoulou M, Chlichlia K, Kyrpitzi D et al (2018) Pharmacodynamic and immunomodulatory effects of micafungin on host responses against biofilms of *Candida parapsilosis* in comparison to those of *Candida albicans*. Antimicrob Agents Chemother 62:e00478-18

Smolarz M, Zawrotniak M, Satala D et al (2021) Extracellular nucleic acids present in the Candida albicans biofilm trigger the release of neutrophil extracellular traps. Front Cell Infect Microbiol 11:681030

Sultan AS, Rizk AM, Vila T et al (2019) Digital design of a universal rat intraoral device for therapeutic evaluation of a topical formulation against *Candida*-associated denture stomatitis. Infect Immun 87:e00617-19

Swidsinski A, Guschin A, Tang Q et al (2019) Vulvovaginal candidiasis: histologic lesions are primarily polymicrobial and invasive and do not contain biofilms. Am J Obstet Gynecol 220:91 e91–91 e98

Taff HT, Nett JE, Zarnowski R et al (2012) A *Candida* biofilm-induced pathway for matrix glucan delivery: implications for drug resistance. PLoS Pathog 8:e1002848

Thomas DP, Bachmann SP, Lopez-Ribot JL (2006) Proteomics for the analysis of the *Candida albicans* biofilm lifestyle. Proteomics 6:5795–5804

Tobouti PL, Casaroto AR, De Almeida RS et al (2016) Expression of secreted aspartyl proteinases in an experimental model of *Candida albicans*-associated denture stomatitis. J Prosthodont 25:127–134

Tseng KY, Huang YT, Huang YT et al (2024) Regulation of candidalysin underlies *Candida albicans* persistence in intravascular catheters by modulating NETosis. PLoS Pathog 20:e1012319

Uppuluri P, Chaturvedi AK, Srinivasan A et al (2010) Dispersion as an important step in the *Candida albicans* biofilm developmental cycle. PLoS Pathog 6:e1000828

Uppuluri P, Acosta Zaldivar M, Anderson MZ et al (2018) *Candida albicans* dispersed cells are developmentally distinct from biofilm and planktonic cells. mBio 9:e01338-18

Urban CF, Reichard U, Brinkmann V et al (2006) Neutrophil extracellular traps capture and kill *Candida albicans* yeast and hyphal forms. Cell Microbiol 8:668–676

Wall G, Montelongo-Jauregui D, Vidal Bonifacio B et al (2019) *Candida albicans* biofilm growth and dispersal: contributions to pathogenesis. Curr Opin Microbiol 52:1–6

Wang X, Fries BC (2011) A murine model for catheter-associated candiduria. J Med Microbiol 60:1523–1529

Xie Z, Thompson A, Sobue T et al (2012) *Candida albicans* biofilms do not trigger reactive oxygen species and evade neutrophil killing. J Infect Dis 206:1936–1945

Yano J, Yu A, Fidel PL Jr et al (2019) *Candida glabrata* has no enhancing role in the pathogenesis of *Candida*-associated denture stomatitis in a rat model. mSphere 4:e00191-19

Zarnowski R, Westler WM, Lacmbouh GA et al (2014) Novel entries in a fungal biofilm matrix encyclopedia. MBio 5:e01333–e01314

Zarnowski R, Sanchez H, Covelli AS et al (2018) *Candida albicans* biofilm-induced vesicles confer drug resistance through matrix biogenesis. PLoS Biol 16:e2006872

Zarnowski R, Sanchez H, Jaromin A et al (2022) A common vesicle proteome drives fungal biofilm development. Proc Natl Acad Sci USA 119:e2211424119

Zawrotniak M, Wojtalik K, Rapala-Kozik M (2019) Farnesol, a quorum-sensing molecule of Candida Albicans triggers the release of neutrophil extracellular traps. Cells 8:1611

Chapter 11
Novel Therapeutic Approaches to Control Fungal Biofilms

Luigi Pedrini Guisso, Natália Pereira Ribeiro, Wilmer Ramírez Carmona, Ana Carolyna Becher Roseno, Juliano Pelim Pessan, and Douglas Roberto Monteiro

Abstract Fungal biofilms pose a significant threat to healthcare services globally due to their distinct pathogenic features, as numerous fungal species lead to local and/or systemic diseases, primarily in immunocompromised individuals. Some of these fungal infections vary from illnesses compromising skin and nails to candidiasis (oral or vaginal) and pulmonary infections. Tolerance to major antifungal drugs by some fungal biofilms is reported worldwide, without discrimination between low- and high-income countries, which complicates the management of biofilm-related diseases. This scenario has stimulated the study and development of novel therapeutic approaches to control fungal biofilms, which include new antifungals, drug combinations, photodynamic therapy, immunotherapy with antibodies and vaccines, antimicrobial peptides, phage therapies, nanotherapies, and the recent use of natural compounds. This chapter discusses these therapies in detail and presents updated data on their efficacy, safety, mechanisms of action, and possible clinical applications.

Introduction

Numerous fungal species are responsible for producing dangerous invasive infections, primarily in immunocompromised individuals, and those that form fungal biofilms pose a significant threat to healthcare services globally due to their distinct pathogenic features. For instance, *Aspergillus fumigatus* is responsible for invasive pulmonary disease in patients with cystic fibrosis, with a fatal outcome of 30%

L. P. Guisso · N. P. Ribeiro · W. R. Carmona · A. C. B. Roseno · J. P. Pessan
São Paulo State University (UNESP), School of Dentistry, Araçatuba,
Department of Pediatric Dentistry and Public Health, Araçatuba, São Paulo, Brazil

D. R. Monteiro (✉)
São Paulo State University (UNESP), School of Dentistry, Araçatuba,
Department of Diagnosis and Surgery, Araçatuba, São Paulo, Brazil
e-mail: douglas.monteiro@unesp.br

G. Ramage, D. Andes (eds.), *Fungal Biofilms*, Springer Series on Biofilms 15,
https://doi.org/10.1007/978-3-031-93315-8_11

(Maertens et al. 2021). Its biofilm is characterized by multinucleated, branched, and spatially interconnected hyphae and vertical growth (Liu et al. 2022). On the other hand, invasive fungal infections linked to *Candida* spp. and the use of medical devices, such as intravascular catheters, have a mortality rate of 35–60% for neonates, 10–35% for children, 20–67% for adults, and 54–66% for patients with septic shock (Wijaya et al. 2023). The pathogens *Candida albicans* and *Candida parapsilosis* are responsible for the formation of most biofilms associated with the use of these devices. *Candida* spp. biofilms differ in structure and extracellular matrix composition and are closely related to antifungal resistance (Wijaya et al. 2023).

Resistance to major antifungal drugs by some fungal species is reported worldwide, without discrimination between low- and high-income countries. A meta-analysis of over 260 clinical isolates of *Candida auris* from 16 countries showed 44%, 15%, and nearly 13% resistance to fluconazole, amphotericin B (AmB), and voriconazole, respectively (Osei Sekyere 2018). Furthermore, specific biofilm characteristics of *C. auris* appear to be a key factor in antifungal resistance (Černáková et al. 2021). Mutations in transcription factors Mrr1 and Tac1 also cause resistance to fluconazole. Azole resistance is also primarily associated with the efflux pump Cdr1. In addition, *erg11* gene mutations are related to *C. auris* resistance to fluconazole and increased expression of efflux pump-related genes (Černáková et al. 2021). In turn, mutations in the *fks1* gene in *Candida* spp., *Cryptococcus*, and *Aspergillus* are responsible for resistance to echinocandins (Lee et al. 2023). Similarly, mutations in ergosterol synthesis genes cause sterol accumulation and are responsible for much of the resistance to polyenes. Resistance to AmB, for example, depends on Hsp90 in the presence of stress (Lee et al. 2023).

In this challenging scenario, there are many proposals for the management of invasive fungal diseases to overcome resistance to current therapies. Antifungal resistance is mainly mediated by the unique pathogenic characteristics of fungal biofilms. New antifungals, drug combinations, photodynamic therapy, immunotherapy with antibodies and vaccines, antimicrobial peptides, phage therapies, nanotherapies, and the recent use of natural compounds, have been proposed as alternative therapies to combat fungal biofilms in severe fungal diseases. In light of the above context, this chapter discusses these therapies in detail and presents updated data on their efficacy, safety, mechanisms of action, and possible clinical applications.

New Antifungals

Opelconazole (PC-945)

Aspergillus infections, especially those involving *A. fumigatus*, are mediated by the formation of filamentous mycelia, currently classified as a type of biofilm (Harding et al. 2009; Mowat et al. 2009). A common approach to treating *A. fumigatus* infections involves the use of azoles, such as itraconazole, posaconazole, and voriconazole orally (Thompson 3rd and Lewis 2nd 2010). However, their systemic

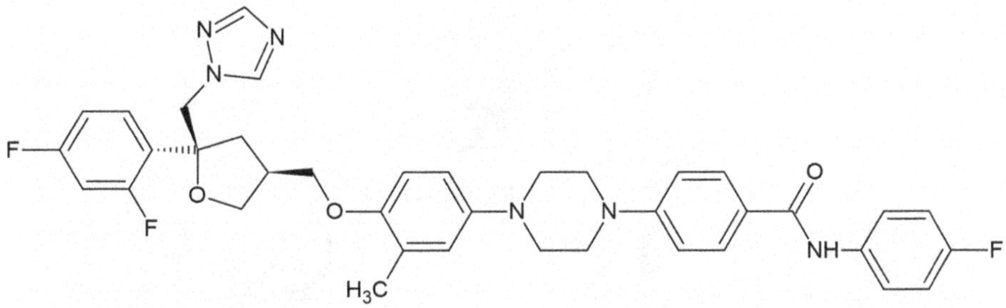

Fig. 11.1 Molecular structure of Opelconazole (4-[4-(4-{[(3R,5R)-5-(2,4-Difluorophenyl)-5-(1H-1,2,4-triazol-1-ylmethyl)oxolan-3-yl]methoxy}-3-methylphenyl)piperazin-1-yl]-N-(4 fluorophenyl)benzamide) (PC945)

pharmacological side effects on tissues, such as temporary changes in vision and notorious abnormalities in liver function (Xiong et al. 2015; Tan et al. 2006), as well as difficulties in oral exposure and delivery (Rodvold et al. 2011), make their use challenging. New drugs derived from azoles, such as Opelconazole, were developed as alternatives.

Opelconazole (PC-945, Pulmocide Ltd.; Fig. 11.1) is a triazole compound of the azole class, specifically used against *Aspergillus* lung infections. The triazole group in its molecule inhibits the enzyme CYP51A1, which is responsible for the biosynthesis of the fungal membrane, converting lanosterol into ergosterol (a component of this membrane), thus causing impairment of fungal growth (Fig. 11.2) (Hoenigl et al. 2021).

Unlike other azoles, which are related to undesirable systemic effects, Opelconazole can be administered via inhalation by a nebulizer. This delivers the drug directly to the lungs, thus markedly reducing the risk of systemic side effects and promoting a greater concentration of the drug at the injection site. Given that its particles are deposited directly on the surface of the biofilm, Opelconazole inhibits biofilm growth and isolates the infectious agent topically, with maximum local therapeutic activity and minimum possible systemic exposure (Kimura et al. 2017; Murray et al. 2020).

Under in vitro conditions involving a human alveolar bilayer model, Opelconazole acts synergistically with systemic drugs against *A. fumigatus* infection (Colley et al. 2019). Also, Opelconazole leads to greater inhibitions of fungal burden and inflammation related to *A. fumigatus* compared to posaconazole and voriconazole in rats (Kimura et al. 2017). Concerning survival of mice infected with *A. fumigatus*, rates of up to 50% were observed in those treated with PC-945 versus 44% for the treatment with posaconazole (Colley et al. 2017).

In addition to *A. fumigatus*, this antifungal drug also presents in vitro antifungal activity against cultures of other species of *Aspergillus*, such as *Aspergillus carbonarius* and *Aspergillus flavus*, but azole-resistant *A. fumigatus* still poses a challenge, showing resistance against the new drug. Different genera of *Candida* spp., such as *C. albicans*, *Candida glabrata*, *Candida krusei*, *Cryptococcus neoformans*, and *Cryptococcus gatti*, also presented good in vitro activity (Hoenigl et al. 2021).

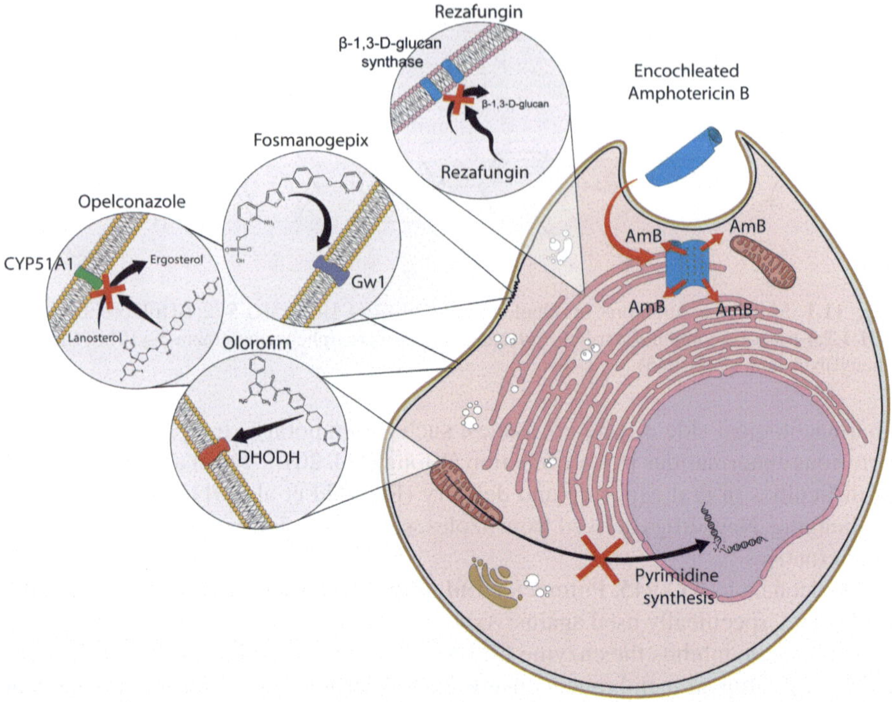

Fig. 11.2 Mechanisms of new drugs in fungal cell membranes. (Based on Hoenigl et al. 2021)

Despite most studies focusing on fungi in their planktonic state, conditions such as invasive and pulmonary aspergillosis are examples of infections involving the formation of biofilms on the cells of lung epithelial tissues (Seidler et al. 2008), so it is reasonable to assume that new antifungals such as Opelconazole also act on microbial consortia in vivo. However, the factors that directly influence the structure and resistance of these biofilms in new treatments are not explicitly mentioned or even studied.

Olorofim

With the increasing use of azole-based fungicides in agriculture, emerging strains resistant to triazoles, such as *A. fumigatus* and other classes of compounds have been described as threats to human health worldwide (Chowdhary et al. 2013), in addition to acute toxicity and drug interactions of some treatments (Denning and Hope 2010). This context has stimulated the development of compounds of different chemical classes and functions, such as Olorofim (F901318, F2G, Inc.; Fig. 11.3). It is a drug from the orthomid class, being an experimental antifungal still in the clinical phase developed by F2G Inc.

Fig. 11.3 Molecular structure of Olorofim (F901318)

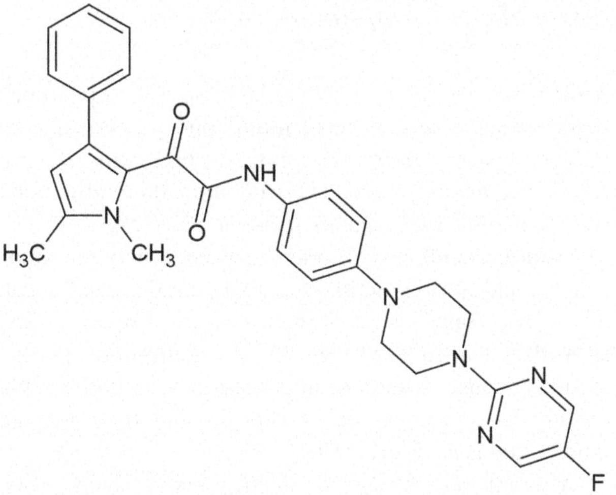

This antifungal works by inhibiting dihydroorotate dehydrogenase (DHODH), the enzyme responsible for the synthesis of pyrimidine in fungi, essential in the synthesis of deoxyribonucleic acids but without cross-reactivity with human DHODH (Negri et al. 2018) (Fig. 11.2). Olorofim affects the biosynthesis of pyrimidine, which is essential for the formation of chitin and β-1,3-glucan, components of the fungal cell wall, despite the lack of evidence of a decrease in β-1,3-glucan in the mycelium. On the other hand, there was an increase in chitin, which can affect electronegativity and hydrophobicity, thus decreasing fungal surface adhesion. This can affect the colonization and formation of biofilms of *A. fumigatus* when treated with Olorofim, thus inhibiting its growth (du Pré et al. 2020; Kirchhoff et al. 2022).

Olorofim still has a short spectrum of reported activity (Hoenigl et al. 2021), showing little or no activity against the order *Mucorales* (Georgacopoulos et al. 2021) and against *Exophiala dermatitidis* (Kirchhoff et al. 2020), but it has interesting effects against specific infections (aspergillosis) and, as it is a new class of drugs, it may present good alternatives against resistant strains in the future.

Under in vitro conditions, Olorofim has strong effects against dysmorphic and filamentous fungi, including more than 100 strains of *Aspergillus* spp. (Oliver et al. 2016), such as *Aspergillus nidulans*, *Aspergillus tubingensis*, *Aspergillus lentulus*, and *Aspergillus calidoustus* (Oliver et al. 2016; Buil et al. 2017). In azole-resistant strains of *A. fumigatus*, however, this new antifungal has similar effects against planktonic forms and at the initial phases of biofilm development compared with AmB and voriconazole, with lower effect on mature biofilms than AmB.

Besides *Aspergillus* spp., Olorofim has also been shown to be effective against *Lomentospora prolificans*, which is a rare, emerging pathogen responsible for several different infections. This fungus has a high biofilm-forming capacity in lung epithelial cells and catheters, being mostly fatal due to its high resistance to common antifungals (Konsoula et al. 2022). Olorofim has a similar antifungal effect to voriconazole and superior to AmB and micafungin in combating *L. prolificans*.

Encochleated Amphotericin B

AmB is one of the most prescribed and potent polyene antifungals, having a broad spectrum against a range of molds and yeasts, such as *Blastomyces dermatitidis*, *C. neoformans*, *Sporothrix* spp., *A. fumigatus*, *C. albicans* (Akinosoglou et al. 2024), and even protozoa (*Leishmania*). Its treatment has been established and prescribed against these agents for over 50 years.

In turn, cochleates are nanoparticle structures composed of negatively charged foliated and spiral lipid bilayers that contain metallic cations, such as calcium (Ca^{+2}) inside. Its properties allow the transport of hydrophilic or hydrophobic drugs, such as AmB, encapsulating them in the calcium phospholipid structure, providing protection against oxidation, degradation, and undesirable environmental impacts to the environment, without compromising their availability (Shende et al. 2019; Aigner and Lass-Flörl 2020).

After encochleated AmB (CAmB) is ingested orally, it is absorbed in the intestine and discharged into the bloodstream, where it is easily ingested by macrophages, opening its spiral structures by the calcium gradient in relation to the cytoplasm and releasing the carried compound (Aigner and Lass-Flörl 2020; Verekar et al. 2024). The mechanism of action of CAmB is essentially the same as AmB, i.e., the drug binds to ergosterol in the fungal cell membrane, opening pores through which cytoplasmic ions are dispersed and lost (Baran et al. 2009) (Fig. 11.2).

The main advantage of encochleation is the increase in the concentrated availability of the drug, while maintaining low toxicity, as it is not toxic to macrophages (Sesana et al. 2011) and is not hemolytic, even at high concentrations and when administered orally, in comparison to AmB deoxycholate (Zarif et al. 2000; Zarif 2005).

CAmB has proven activity against *A. fumigatus* and *Candida* spp., with reduced mortality in aspergillosis, presenting 70% survival in infected rats and a reduction in colonies in the lungs, liver, and kidneys (Delmas et al. 2002). At low doses, CAmB was shown to be more effective against *C. albicans* in rats than Fungizone and AmBisome, providing a 100% survival rate compared to 60% for Fungizone (Zarif 2005). Drug tolerance, especially in the kidneys of HIV-positive patients who survived cryptococcal meningitis, has also been reported (Skipper et al. 2020). All these conditions may involve biofilm formation despite most of the studies focused on planktonic cells.

Rezafungin

Rezafungin is a second-generation echinocandin drug and, like other compounds of this class, its activity covers fungi such as *Pneumocystis* spp., *Aspergillus* spp., and *Candida* spp., including *C. auris* (Tóth et al. 2019), with little or no activity against *C. neoformans*, *Fusarium* spp., and *Zygomycetes* (Denning 2003; Carvalhaes et al. 2022).

Echinocandins are composed of large lipopeptide molecules, including caspofungin and anidulafungin, in addition to rezafungin. The latter has optimized pharmacokinetic and dynamic properties in relation to those of the first generation, having a longer half-life, antifungal activity, and penetration into tissues, with reduced hepatoxicity (Garcia-Effron 2020). This class of compounds inhibits the enzyme β-1,3-d-glucan synthase, a catalyst for the polymer β-D-1,3-glucan, an essential component of the fungal cell wall, representing at least 50% of the dry weight of the microorganism (Fig. 11.2). Glucan inhibition generates morphological changes that cause osmotic instability and cell death (Garcia-Effron 2020).

The US Food and Drug Administration (FDA), in 2023, approved rezafungin in cases of candidemia (an infection of the *Candida* genus in the bloodstream) and candidiasis. A recent study demonstrated 73% eradication for rezafungin versus 65% for caspofungin in cases of candidemia and invasive candidiasis within 5 days (Thompson 3rd et al. 2024). Other studies have demonstrated efficacy in preventing Pneumocystis pneumonia and its reactivation in rats (Ong et al. 2020) and against *A. fumigatus* in vitro biofilm, including those resistant to azole drugs. It also presents activity against biofilm isolates of *A. lentulus*, *A. calidoustus*, *Aspergillus thermomutatus*, and *Aspergillus udzagawae*, with a similar effect compared with micafungin and other drugs in the echinocandin class (Wiederhold et al. 2018). In addition to its efficiency, it has low hepatotoxicity even in patients with liver failure, providing a safe profile for its use (Flanagan et al. 2024), but despite presenting a broad spectrum of activity, as an echinocandin, rezafungin still presents resistance against the *Cryptococcus* genus (Pfaller et al. 2017; Hoenigl et al. 2021).

Fosmanogepix

Fosmanogepix (APX001) is a prodrug of Manogepix (MGX), which is a medication that is initially less active and, after metabolized, is converted into its active pharmacological form, MGX (Fig. 11.4). It has a spectrum of activity against fungi capable of forming biofilm, such as *Coccidioides*, *Aspergillus*, *Fusarium*, *Cryptococcus*, *Scedosporium*, and *Candida* (Viriyakosol et al. 2019; Zhao et al. 2019; Alkhazraji et al. 2020; Hoenigl et al. 2021; Almajid et al. 2024), and can be administered either orally or intravenously.

The mechanism of MGX consists of the inhibition of the Gw1 enzyme in the cell membrane and wall, which transports and anchors mannoproteins, the latter being the sources of biomass for the fungus (Orlean and Menon 2007; McLellan et al. 2012) (Fig. 11.2). Inhibition of Gw1 in *Saccharomyces cerevisiae* and *C. albicans* was shown to affect their growth and prevent protein and biofilm maturation (Watanabe et al. 2012; Shaw and Ibrahim 2020), demonstrating the importance of mannoproteins in cellular integrity. Likewise, during inhibition, high stress in the endoplasmic reticulum, reduction of biofilm, and reduction of hyphae were also observed.

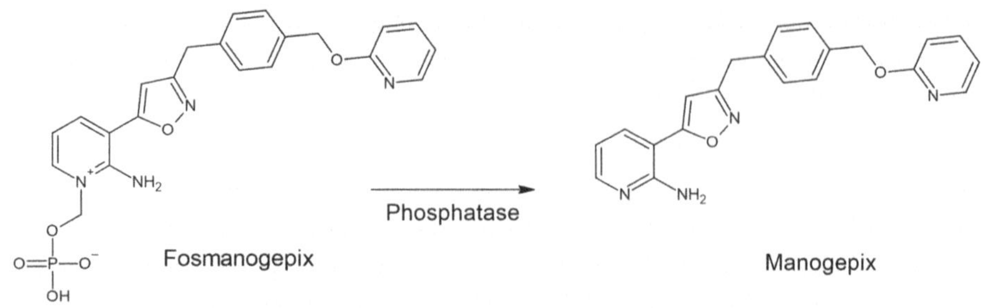

Fig. 11.4 Conversion of Fosmanogepix to Manogepix after systemic phosphatases. (Based on Shaw and Ibrahim 2020)

Phase I clinical trials demonstrate good tolerability and no significant side effects in healthy patients, with transient headaches being the most common effects (Hodges et al. 2017). In phase II, the drug was shown to be well tolerated by patients with invasive candidiasis (Pappas et al. 2020) and candidemia, removing *Candida* from blood cultures in an average of 2.4 days, with an 80% success rate (Kullberg et al. 2021).

For *C. auris*, MGX has greater efficacy against invasive candidiasis in rats infected with fluconazole-resistant *C. auris* compared with caspofungin and fluconazole (Wiederhold et al. 2019), proving to be an effective treatment for this severe condition in the future. Another growing problem is related to echinocandins, azoles, and AmB resistance, especially in *Candida* spp. and *Aspergillus* spp. biofilms. In this context, fosmanogepix can be an effective alternative, as it has demonstrated similar levels of activity regardless of the organism's resistance to these compounds (Arendrup et al. 2018; Jørgensen et al. 2020; Shaw and Ibrahim 2020).

Like MGX, Fosmanogepix, despite presenting good results against several species of *Candida*, had no effect against *C. krusei*, already known for its resistance against this class of drugs (Miyazaki et al. 2011; Hoenigl et al. 2021; Pfaller et al. 2024).

Drug Combinations

Until recently, only four classes of systemic medications were prescribed against fungal diseases related to biofilm, namely azoles, polyenes, echinocandins, and flucytosine, each with their limitations and known adverse effects. Simultaneously, the use of pesticides with these classes of compounds and their inappropriate use in medicine generates resistant microorganisms that have been increasingly linked to fatal infections, representing a growing health risk, especially for immunosuppressed patients (Chowdhary et al. 2013; Zhu et al. 2023).

The development of novel drugs and the combination of their activities, which has already been applied against viruses and bacteria, are among the strategies to overcome this emerging problem. Combinatorial treatment regimens may even involve non-antifungal medications that directly improve the activity of antifungals or alter susceptibility by circumventing the resistance mechanism (Spitzer et al. 2017).

Combinatorial approaches can be applied from mycotic infections on the skin that can eventually manifest complications, with or without the formation of biofilms, affecting the CNS and lungs, such as Phaeohyphomycoses and Mucormycosis, to infections located specifically in these regions, such as Cryptococcosis and Aspergillosis, as it will be further described. The relatively high variety of species that cause these diseases presents a challenge in conventional treatments as they present different resistances to them, therefore, combinatorial procedures and their mechanisms of action must be studied, as they can solve present limitations.

Several advantages were observed in combining drugs. Due to the large reduction in the pathogen population, possible resistance development can be prevented, as the fungus would need to resist simultaneously two or more compounds (Spitzer et al. 2017). Other strategies include the synergistic action of two or more drugs, as in the case of co-administration of caspofungin (an echinocandin) with voriconazole (an azole) (Perea et al. 2002; Wood et al. 2014), in addition to the phenomenon of collateral sensitivity, which occurs when resistance to one compound confers greater sensitivity to the other (Imamovic and Sommer 2013). All the above emphasize the potential role of combinatorial strategies for biofilm-related infections with their broad spectrums of activity against the wide diversity of bacteria and fungi regulated by *quorum sensing*.

Mechanisms

In general, there are three main classifications of mechanisms regarding drug combinations. First, there is a mechanism for increasing the intracellular bioavailability of the drugs through the alteration of the cell membrane by one medication, while the other drug acts on intracellular components more efficiently. The combination of AmB with flucytosine (5-FC) is the most common example of this type of mechanism, in which AmB disrupts the membrane by binding with ergosterol, enhancing the transport of 5-FC into the cell and leading to greater concentration and availability of this antifungal, in addition to AmB's own effect (Zhu et al. 2023).

Another known mechanism is the vertical or parallel inhibition of pathways, which blocks the intended biosynthesis pathway in two different ways. The union of terbinafine with azoles, for example, generates a double inhibition of the ergosterol synthesis in the cell wall, further increasing fungicidal activity. In turn, the action of caspofungin is improved by the synergistic action of nikkomycin Z against *A. fumigatus* by its inhibition of the synthesis of cell wall components (Verwer et al. 2012; Zhu et al. 2023).

Finally, there is the mechanism for suppressing cellular stress responses. Cellular stress is a microorganism reaction induced by the antifungal medication for its own protection, which can even cause the expression of resistance genes so that agents that inhibit the response to this stress can prevent the fungus from reacting to other drugs when co-administered (Zhu et al. 2023). Examples are inhibitors of calcium-regulated phosphatases (calcineurin), which work synergistically with azoles and echinocandins, reducing gene expression associated with their resistance and cell membrane stress, acting mainly against *C. albicans*, *C. neoformans*, and *A. fumigatus* (Uppuluri et al. 2008; Chen et al. 2011; Zhu et al. 2023). Other types of drugs reported are inhibitors of the heat shock 90 (Hsp90) protein, which is responsible for stabilizing proteins and regulating cellular stress, generating tolerance and resistance to antifungals in *Candida* and *Aspergillus* (Cowen et al. 2009). Another example is CMLD013075, which is capable of inhibiting the growth of *C. albicans* in a comparable way to fluconazole and, when combined with it, significantly increases its fungicidal capacity (Whitesell et al. 2019; Zhu et al. 2023).

The strategies above enable the development of therapies with potentially lower doses (and consequently lower toxicity) and greater efficacy, despite adverse effects, interactions between drugs and biofilm penetration must still be monitored (Zimmermann et al. 2007; Spitzer et al. 2017). Table 11.1 summarizes the main drug combinations and their antifungal results compared to monotreatments.

Phaeohyphomycoses

Phaeohyphomycoses are infections caused by dematiaceous fungi, which contain melanin in their cell wall that provide color at the site of colonization, and whose symptoms range from subcutaneous lesions to pneumonia and brain abscesses (Li and de Hoog 2009). Infections of the brain parenchyma caused by dematiaceous are called cerebral phaeohyphomycosis (CPHM), which equally affect healthy and immunosuppressed patients (Chakrabarti et al. 2016; Samaddar et al. 2024). Most drug monotherapies in severe disseminated infections, like in CPHM, fail due to biofilm formation, drug resistance, and cancer progression. In these cases, drug exposure during several months or combination therapy is often needed (Chowdhary et al. 2014).

Most fungal infections in the central nervous system (CNS), such as CPHM, were shown to achieve remission by the combination of AmB and 5-FC compared to monotherapy (Li and de Hoog 2009). It is noteworthy that 5-FC usually shows good biofilm penetration due to low molecular weight, water solubility, and fewer reactions with proteins, being preferable in combination therapies in severe fungal infections of the eye, urinal tract, and CNS (Sigera and Denning 2023). In studies with isolated fungi responsible for CPHM, this combination showed well in vitro synergistic response against cultures of *Rhinocladiella mackenziei*, *Cladophialophora bantiana*, *Exophiala dermatitidis*, and *Fonsecaea monophora* (Coelho et al. 2018; Sigera and Denning 2023).

E. dermatitidis is a black yeast dermatophyte known to form biofilm and linked to phaeohyphomycoses and other skin and lung infections. Like *Candida* spp.,

Table 11.1 Summary of combinations, types of infections, and their results compared to monotreatments

Infection	Drug combination	Results compared to monotreatments	References
Phaeohyphomycoses	5-FC + AmB	Good synergistic response in vitro	Sigera and Denning (2023)
	Everolimus (EVL) + AmB	Showed greater efficacy compared to further combinations	Jia et al. (2023)
Cryptococcal infections	5-FC + AmB	"Gold standard" to treat cryptococcal infections; greater and faster resolutions	Maziarz and Perfect (2016)
	Sertraline + fluconazole	Showed synergy; reduced fungal burden and biomass of the biofilm	Zhai et al. (2012)
	Sertraline + fluconazole + AmB	Faster infection resolution	Rhein et al. (2016)
Mucormycosis	Isavuconazonium sulfate (IS) + LAmB	Increase in the survival time and reduced the fungal load in rats	Gebremariam et al. (2021)
	Deferasirox + LAmB	Higher survival rates in disseminated mucormycosis in rats	Ibrahim et al. (2007
	Deferasirox + LAmB	Higher 90-day mortality rate in 20 patients; low efficacy	Spellberg et al. (2012)
	Calcineurin + LAmB	Prevent the development of more virulent mucormycosis under in vitro and in vivo	Lewis et al. (2013)
Aspergillosis	Fosmanogepix + LAmB	Higher survival rates	Gebremariam et al. (2022)
	Ibrexafungerp + Isavuconazole	Significant reduction in residual fungal load in the lungs	Petraitis et al. (2020)
	Voriconazole + Anidulafungin	Lower mortality rate in humans	Marr et al. (2015)

virulence factors are related to the ability to switch morphologically from yeast to hyphae state, being linked to biofilm formation. Based on the characteristics of this biofilm, as well as its susceptibility to micafungin and voriconazole during the adhesion when premature, preventive surface treatment may be regarded as the best approach (Kirchhoff et al. 2017).

Another compound that deserves attention is everolimus (EVL), which is able to regulate several processes, such as proliferation, development, aggregation, transcription, homeostasis, and autophagy, among others, normally used as an antitumor. For *E. dermatitidis* cultures, a strong synergistic effect of EVL with itraconazole and EVL with posaconazole was observed, highlighting the effect of EVL with AmB, which showed greater efficacy compared to monotreatment and the further combinations (Jia et al. 2023). The mechanism of EVL possibly involves the suppression of genes that confer drug resistance (Jia et al. 2023), making EVL a promising therapeutic alternative for fungal *E. dermatitidis* infections although EVL-biofilm interactions are still unknown.

Cryptococcal Infections

In the same perspective, cryptococcal meningitis is always fatal if left untreated, with around 30% of AIDS patients dying each year as a result of this disease, especially in sub-Saharan Africa (Park et al. 2009). As *C. neoformans* crosses the blood-brain barrier, it disperses and colonizes optimal areas where biofilm formation is possible, such as those with surface support, ideal temperature, neutral pH, oxygen-rich blood, and ideal hydrodynamics (Martinez and Casadevall 2007). After maturing, the biofilms, also called cryptococcomas, can progress with many resistance mechanisms, such as antioxidant melanin synthesis and morphological yeast cell changes, which is directly linked to drug resistance and phagocytic cells protection, aggravating the CNS infection (Martinez and Casadevall 2007; Aslanyan et al. 2017). In this context, classical and alternative novel treatments for cryptococcal meningitis have emerged.

Similar to phaeohyphomycosis, cryptococcal meningitis is also commonly treated with AmB and 5-FC combined, considered the "gold standard" by the Infectious Diseases Society of America (IDSA) (Maziarz and Perfect 2016). Despite concerns about the adverse nephrotoxic effects of polyenes, the low dose and short period of administration of AmB in cryptococcal meningitis has been linked to low toxicity (Spitzer et al. 2017). In immunosuppressed HIV patients in Africa, treatment within 1 week with AmB combined with 5-FC was shown to lead to the lowest mortality rates (24.2%) and was associated with greater and faster resolutions in the infection (Molloy et al. 2018). In the same study, it was also suggested that in the absence of AmB, fluconazole can be used as an effective and sustainable alternative to be combined with 5-FC (Molloy et al. 2018). It is noteworthy that 5-FC has fungicidal and antibiofilm activity even at low concentrations by its inherent properties, as previously mentioned (O'Connor et al. 2013).

Interestingly, one of the most prescribed antidepressants, sertraline, was shown to display antifungal activity against *Cryptococcus* despite the mechanism involved is still not completely explained (Zhai et al. 2012). Administered alone, sertraline initially reduces fungal burden in murine models with systemic cryptococcosis but also works synergically with fluconazole, which further reduces the fungal burden in the brain, kidneys, and spleen, having a marked effect on the reduction of fungal biofilm biomass (Zhai et al. 2012). In *Candida* strains, however, sertraline and its combination with fluconazole was shown to be less effective and even have an antagonistic effect (Zhai et al. 2012). In a study with HIV-positive patients with cryptococcal meningitis, administration of triple-combined AmB, fluconazole, and sertraline resulted in faster cerebrospinal fluid clearance and decreased inflammatory syndromes, with no relapse for 12 weeks (Rhein et al. 2016). However, more recent findings from the same research group in a phase III study noted that the mortality rate was not reduced with this same triple combination, linking its inefficiency to multiple factors and warning that HIV-positive patients with cryptococcal meningitis should not be treated with sertraline, alone or combined, and should ideally be treated with the "gold standard" therapy (i.e., AmB and 5-FC) (Rhein et al. 2019).

Aspergillosis

Invasive aspergillosis is an aggressive infection located mainly in the lungs, which spreads to other organs in people with weakened immune systems and is related to a high mortality rate in this population. Although different from the bacterial biofilm, the *Aspergillus* mycelia can conceptually be considered a biofilm, as it is a dense, complex, multicellular, and three-dimensional structure, surrounded by an extracellular matrix adhered to surfaces and directly related to recalcitrance and drug resistance (Morelli et al. 2021). *Aspergillus* conidia can germinate after entering the respiratory tract, subsequently maturating into biofilms with mycelium that adhere to bronchial epithelial cells and generate toxins that damage the epithelial cell surface, causing and aggravating invasive aspergillosis. The *Aspergillus* genus is responsible not only for invasive aspergillosis but also for several other conditions, such as aspergilloma, hypersensitivity pneumonitis, chronic necrotizing pneumonia, and allergic bronchopulmonary aspergillosis (Seidler et al. 2008).

Most of the studies involving invasive aspergillosis and *A. fumigatus* biofilm were carried out in vitro or in animal models, which suggests higher doses or combinatorial therapy for better penetration (Seidler et al. 2008). In immunosuppressed rats with fungal infections, pulmonary aspergillosis, invasive trichinellosis, and invasive fusariosis, the prodrug Fosmanogepix was tested either alone or combined with liposomal AmB (LAmB), which were shown to provide higher survival rates compared with the drugs administered alone (Gebremariam et al. 2022).

Ibrexafungerp is another new antifungal, known for its action against *Candida* and *Aspergillus* species. Rabbits infected with pulmonary aspergillosis and treated with Ibrexafungerp and Isavuconazole were shown to have a significant reduction in residual fungal load, lung weight, and pulmonary infarction scores, in addition to higher survival rates compared to the Isavuconazole group, due to the synergistic effect between the two drugs (Petraitis et al. 2020). In humans, patients with suspected or confirmed invasive aspergillosis and treated with the combined therapy of voriconazole and anidulafungin had a lower mortality rate (19.3%) compared to voriconazole monotherapy (27.5%), demonstrating the effectiveness of combined therapy compared to monotherapy (Marr et al. 2015).

Mucormycosis

Mucorales biofilms are known for having robust biomass, with filamentous and highly adherent structures (Singh et al. 2011). Although *Mucorales* are usually linked to vascular and cardiac infections, biofilms of these fungi were also found in catheters, implants, and artificial heart valves and are related to osteomyelitis and paranasal fungal balls (Singh et al. 2011). Mucormycosis is transmitted by spores of the order *Mucorales*, causing serious infections in immunosuppressed patients that can affect the respiratory system, skin, nervous system, and other organs, often fatal in severe disseminated cases.

Despite azoles having almost no effect against mucormycosis pathogens, isavuconazole is one of the few that does (Borman et al. 2021). Isavuconazonium sulfate (IS) is a prodrug of isavuconazole that acts against mucormycosis. In neutropenic rats infected with *Rhizopus delemar* or *Mucor circinelloides*, combined therapy of IS and LAmB was more effective than IS monotherapy, with an increase in the general survival time of the animals and reducing the fungal load in their organs (Gebremariam et al. 2021).

Another strategy involves the administration of Deferasirox, an FDA-approved iron chelator that can prevent fungi such as *Trichoderma* or *Mucorales* genera from taking up iron, thus inhibiting their development (Spellberg et al. 2012). In mice with disseminated mucormycosis, deferasirox was shown to be as effective as LAmB, and their combined therapies demonstrated synergism, resulting in higher survival rates (Ibrahim et al. 2007). In a clinical study, however, 20 patients with proven or probable mucormycosis received treatment with LAmB and deferasirox or LAmB and placebo, with a higher 90-day mortality rate in those who received the deferasirox-LAmB treatment, therefore being inconsistent with the evidence from mice. The mechanisms and reasons behind such discrepancy are still unclear. Despite the absence of toxicity, deferasirox cannot be recommended in a combinatorial treatment due to excess mortality, emphasizing the need for larger, less limited, phase III clinical studies to elucidate the true efficacy of deferasirox against mucormycosis (Spellberg et al. 2012).

As previously mentioned, calcineurin is a specific phosphatase protein essential in fungal virulence and growth, so the use of calcineurin inhibitors in monotherapy or combined therapy was shown to prevent the development of mucormycosis under in vitro and in vivo conditions (Thakur and Revankar 2011; Lewis et al. 2013). Such inhibition forces the *Mucor* genus to produce smaller spores and prevents the more virulent dysmorphic transition, characteristics related to the fungal cell wall. As this is the site where other drugs also act, such as LAmB, the coadministration of calcineurin can be an adjuvant therapy.

Despite the onset of new medications, the most common combination strategies still involve the four already known classes of antifungals, through the combination of two different classes. The combination of three classes, on the other hand, requires a very large population for a reliable comparison, so the evidence is still scarce and obtained from studies with a low number of patients (Meena et al. 2023).

In mucormycosis, echinocandins are not as effective as the other classes because they bind to β-(1,3)-d-glucan synthase, while in *Mucorales*, there is a predominance of β-(1,6)-d-glucan synthase in their cell walls (Mroczyńska and Brillowska-Dąbrowska 2020; Meena et al. 2023). Mixed results have been reported regarding combined therapy involving echinocandins. For instance, a study with 106 patients with hematological malignancies and mucormycosis demonstrated no benefits and differences in mortality in the combined treatment of echinocandins, posaconazole, and LAmB in relation to monotreatments (43% vs 41%) (Kyvernitakis et al. 2016). Similar results were reported for the triple combination.

Photodynamic Therapy

Skin fungal infections are common conditions in humans, present mainly in areas rich in keratin, such as skin, hair, and nails, and which may affect the physical and mental health of patients. With the emergence of fungi biofilm resistance to conventional drugs, such as oral and intravenous azoles, standard treatments may become less effective, especially in deep and severe infections, in addition to undesirable side effects on the kidneys and liver and drug interference. Therefore, it becomes necessary to resort to new strategies and alternative techniques or even in combination with drugs.

Photodynamic therapy (PDT) is a consolidated technique, discovered in 1900, after the elimination of microorganisms exposed to photosensitizing dyes (PDs) and sunlight. Since then, PDT has been explored in cancer treatments, destroying tumor-associated tissues and vasculature, as well as in bacterial, fungal, and viral infections (Hamblin 2016; Dąbrowski 2017; Correia et al. 2021). Table 11.2 shows the possible PDTs available for managing fungal infections of biofilm aetiology.

Table 11.2 Summary of treatment outcome for each infection

Infection	Results of PDT	References
Onychomycosis	Biofilm reduction; antibiofilm action; increased susceptibility to conventional antifungals	Chen et al. (2019); Fernandes et al. (2023)
Tinea pedis	High recurrence rates; low efficacy	Sotiriou et al. (2009)
Tinea cruris	Low efficacy alone; better efficacy when combined with azoles	Ji et al. (2024)
Tinea capitis	Better efficacy compared to oral treatment	Chen et al. (2024)
Chromoblastomycosis	Biofilm inhibition; drug synergism with azoles; improvements in lesions	Hu et al. (2019); Lan et al. (2021)
Sporotrichosis	Remission of infection; good response combined with oral treatment	Legabão et al. (2022)
Phaeohyphomycosis	Biofilm reduction; synergism with oral treatment	Gao et al. (2016); Gil-Pallares et al. (2024)
Pityriasis versicolor	Lack of studies; remission of infection with no recurrences	Alberdi and Gómez (2020)

PDs are chemical compounds with the ability to absorb light energy and transfer it to another nearby molecule through photochemical reactions (Plaetzer et al. 2009), which can be administered either intravenously or topically. While PDT may face obstacles in deeper infections and tumors due to the lack of available oxygen and low light penetration into tissue with PDs, in cutaneous and subcutaneous infections, such as mycoses, it can have considerable effects (Dąbrowski 2017).

The effects of PDT can also be further enhanced by combining it with known antifungals. Due to the nature of the treatment, as it requires oxygen and radiation, most of the studies in the literature involving this technique were carried out on cutaneous and subcutaneous diseases that favor its application, among them, Chromoblastomycosis, Sporotrichosis, Phaeohyphomycosis, and Pityriasis can be mentioned.

A wide variety of light sources with different powers can be applied in PDT, such as halogen lamps, light-emitting diodes (LED), and incandescent lamps, although lasers have become more widely used due to their portability, relatively low cost, and ease of handling (Hu et al. 2023). Low potency and localized concentrations of radiation are ideal for reducing possible injuries to adjacent tissues in the patient.

Fractional CO_2 laser treatment can be administered before PDT to eliminate damaged tissue (whether in nail or skin), allowing greater penetration and optimization of the photodynamic treatment (de Oliveira et al. 2015), or after PDT to stimulate collagen formation, which eliminates or reduces scars left by mycoses and other infections.

Several advantages can be mentioned in relation to conventional treatments, including the low incidence of side effects and toxicity, the possibility of multiple applications in the same location, with little treatment time, the ability of the technique to be non-invasive, and to have satisfactory aesthetic results to the patient and the relatively low cost (Correia et al. 2021).

Adverse effects of PDT typically include discomfort, local pain, erythema, and edema, particularly in patients who have received multiple treatments in larger areas and deeper infections (Ibbotson et al. 2019). Nonetheless, there is no evidence of cumulative toxicity or photocarcinogenic risk (Ibbotson et al. 2019), and most effects cease around 2 weeks (Wollina et al. 2019). Also, the need for oxygen present in the photochemical reaction may be challenging under certain circumstances, such as in deeper infections. Other disadvantages and limitations include the need for adjuvant systemic drugs in severe recalcitrant cases or when faced with resistant fungal biofilm, and issues related to patient photosensitivity after treatment and the compatibility of PDT for microorganisms despite their wide operating range (Correia et al. 2021).

Mechanisms of Action

The mechanisms by which PDT acts involve the absorption of light energy (photons) at wavelengths compatible with the PD absorption bands, exciting the dye to an unstable short-lived singlet state (fluorescence). This can transition to a long-lived triplet state (phosphorescence), allowing its reaction with O_2 available in the

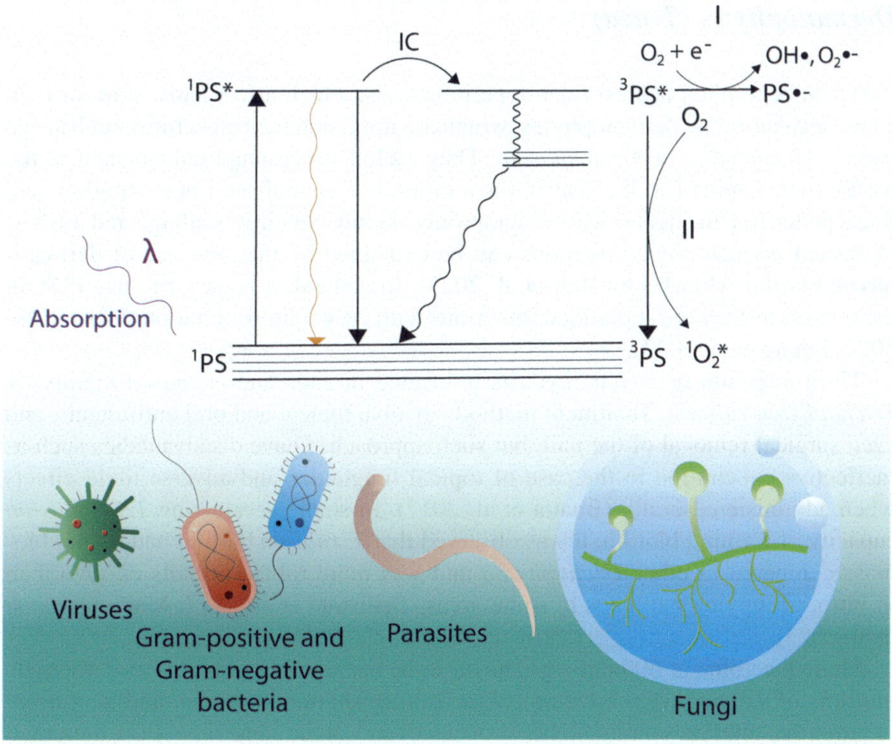

Fig. 11.5 Simplified Jablonski diagram representing the formation of highly reactive oxygen species upon excitation of photosensitizing dye (PD). PD is excited from its ground energy state (^{1}PS) to the excited singlet state (^{1}PS*) and performs the intersystem crossing (IC) passing to the triplet state (^{3}PS*), which, in turn, can transfer electrons to O_2, forming free radicals (I) or providing energy to form O_2 in an excited singlet state (II). (Based on Hamblin 2016)

environment in two possible ways (types 1 or 2), generating chemical species that are harmful to microorganisms (Broekgaarden et al. 2015; Hamblin 2016).

In type 1, an electron (e^-) is transferred in the production of hydroxyl free radical ($HO^{\bullet}$), superoxide ions ($O_2^{\bullet-}$), and hydrogen peroxide (H_2O_2), while in type 2, energy is transported to the O_2, producing O_2 in an excited singlet state ($^1O_2$). Both types form highly reactive oxygen species (ROS) and damage cellular structures and biomolecules in the process of oxidative stress, which can occur simultaneously (Broekgaarden et al. 2015) (Fig. 11.5).

PDs have effects against viruses, bacteria, fungi, and parasites, damaging their DNA or cell membrane (Hamblin and Hasan 2004) despite some PDs being more efficient on certain microorganisms than on others. Specifically in fungi, ROS acts by altering the cell membrane and wall, reducing biofilm formation and interfering with enzymatic processes. Fungal cell walls are composed of thick layers of beta-glucan and chitin, forming impermeable barriers to PDs, similar to the physical barrier to PDT of biofilms. Therefore, to optimize interactions with ROS inside the cells and biofilms, and to overcome these barriers, the dyes can be combined with the nanopeptide polymyxin or EDTA (Kharkwal et al. 2011).

Dermatophytes (Tinea)

PDT can be applied against tinea-type infections, which affect nails, skin, or hair. These are caused by dermatophytes, which are fungi that feed on keratin, such as the genus *Trichophyton* or *Microsporum*. They include tinea unguium (located in the nails), tinea capitis (on the scalp), tinea cruris (on genitals and inner thighs), and tinea pedis (on the feet), whose symptoms include itching, scaling, and rashes. Persistent dermatophyte infections can be explained by the presence of dermatophytic biofilms (Markantonatou et al. 2023). In general, it is easier to use PDT in these cases as they are superficial infections with easy light penetration (Wu and Hu 2022; Leung et al. 2023).

Tinea unguium or onychomycosis is located in nails and is caused mainly by *Trichophyton rubrum*. Treatment methods involve topical and oral antifungals, and even surgical removal of the nail, but such approaches have disadvantages such as ineffective penetration in the case of topical fungicides and adverse toxic effects when administered orally (Bhatta et al. 2017). In studies evaluating *Trichophyton* capacity of forming biofilm, it was observed that *T. rubrum* biofilm had higher biomass compared to other *Trichophyton* and was more robust in nails ex vivo than in vitro, occupying nail gaps in some areas (Brilhante et al. 2017; Markantonatou et al. 2023).

Methylene blue at 600 nm was shown to be capable of reducing dermatophytic biofilms of *Trichophyton mentagrophytes*, *Microsporum gypseum*, and *T. rubrum* in vitro (the latter presenting higher resistance) and, at the same time, PDT increased susceptibility to conventional antifungals (Chen et al. 2019). Other PDs, like P123-encapsulated hypericin (P-Hyp123), also demonstrated antifungal and antibiofilm against *T. rubrum* in more recent studies, in which P-Hyp123 ruptured and withered hyphae and caused biomass loss of mature biofilms (Fernandes et al. 2023).

In clinical trials, 67% of patients received negative culture and microscopy assay results for *T. rubrum* after treatment (Gupta et al. 2017). Overall, the fight against *T. rubrum* by PDT leads to satisfactory results, in addition to presenting a plethora of possible therapeutic PDs, such as *Citrus aurantifolia* essential oil, which is also highly effective (Fekrazad et al. 2015).

Although PDT is efficient against *T. rubrum* and its biofilms in onychomycosis, the same was not observed in tinea pedis. Popularly known as "athlete's foot," tinea pedis is an infection also caused by the dermatophytes *T. rubrum* and *Trichophyton interdigitale*, usually treated orally or topically with azoles, terbinafine, and benzylamine (Leung et al. 2023). Due to its long treatment time, PDT was tried in order to reduce this condition but with variable success and high recurrence rates (Calzavara-Pinton et al. 2004; Sotiriou et al. 2009). The low efficacy for the treatment of tinea pedis may be related to the protocol with aminolevulinic acid (ALA-PDT), which was not aggressive enough against *Trichophyton*. Also, environmental interferences such as temperature, humidity, and pH can hinder the cellular absorption of ALA-PDT, in addition to the irregular shape of the feet being a challenge during light absorption (Calzavara-Pinton et al. 2004; Sotiriou et al. 2009).

For tinea cruris, low rates of cure (around 40%) were reported after 8 weeks of follow-up with the ALA-PDT protocol, with temperature and humidity also being considered as conditions that affect the cellular adsorption capacity of ALA-PDT (Sotiriou et al. 2009). More recently, however, PDT was shown to be an effective antifungal adjuvant combined with azoles, successfully improving tinea by using ALA-PDT (Ji et al. 2024). Due to the conflicting results around PDT efficacy, its role in mycoses (Qiao et al. 2010) and in biofilm control remains unclear.

Tinea capitis, in turn, causes inflammation of the scalp (i.e., kerion). Most cases are mainly caused by *Trichophyton violaceum*, *Microsporum canis*, and *T. mentagrophytes* and, like other tinea, show little response to oral treatment, mainly due to drug-resistant strains (Chen et al. 2024). Clinical cases of six patients with kerion demonstrated resolution of symptoms within 6 weeks to 3 months without recurrence of the disease with treatment performed by PDs.

Chromoblastomycosis

Chromoblastomycosis is a subcutaneous fungal infection caused, among several fungi, by *F. monophora*, *Fonsecaea pedrosoi*, and *Cladophialophora carrionii*, with a low probability of cure, difficult-to-treat lesions, and a high probability of relapses (Hu et al. 2019; Sousa et al. 2022). Like other fungi, its conidial cells have the ability to adhere to substrates, differentiate into hyphae, and form complex biofilms with robust biomass composed of extracellular polymers (Sousa et al. 2022). Mycotic wounds also favor the formation of polymicrobial biofilms that interfere with treatment, making PDT a viable alternative with its broad spectrum of action against various microorganisms.

The effect of PDT against *F. pedrosoi*, *F. monophora*, and *C. carrionii* biofilms in vitro is related to ALA-PDT damaging and altering the fungi cell wall, which inhibits biofilm formation and/or prevents its maturation (Lyon et al. 2013; Hu et al. 2019). In the same way, ALA-PDT also inactivates *F. monophora* conidias reproductive spores through damage caused by highly reactive ROS, presenting a drug synergism with itraconazole in vitro against the same fungi (Yi et al. 2017; Hu et al. 2019).

In complex cases, with resistant strains and involving biofilms, the same treatment with ALA-PDT can be combined with oral drugs, such as itraconazole, terbinafine, and voriconazole (Hu et al. 2019; Sousa et al. 2022). Clinical cases with the previously mentioned treatments showed negative mycological tests after 6 months, but the majority of patients only experienced improvements in the lesions (Hu et al. 2019). As most patients are diagnosed with moderate or severe cases, the combination of two methods is more appropriate, generating better results (Lan et al. 2021).

Sporotrichosis

Another profound type of mycosis that can be treated by PDT is sporotrichosis, which is an infection caused by the *Sporothrix schenckii* complex, a fungus whose most common vector in humans is currently domestic cats (Legabão et al. 2022). *S. schenckii* and *Sporothrix brasiliensis* have the capacity to form biofilm in cat claws, from where it is transmitted to humans (dos Santos et al. 2022). The infection affects the skin, subcutaneous tissue, and lymphatic vessels and is commonly treated with itraconazole, amphotericin B, or terbinafine despite the toxicity and adverse effects of such medications.

As in other microorganisms, *Sporothrix* biofilm formation is linked to virulence and is a key factor in the pathogenesis and recalcitrant mycosis, being formed both in nature and on host surfaces, in a similar way to *Candida*, *Aspergillus*, and *Cryptococcus* species (Sánchez-Herrera et al. 2021). It has also been suggested that resistance to azoles, AmB, and flucytocin in clinical cases is, in part, due to its structure (Brilhante et al. 2017; Sánchez-Herrera et al. 2021). Despite this, no study has assessed the effects of PDT directly on *Sporothrix* biofilms in vitro, nor the factors that could influence the treatment, PDT has shown successful results in clinical cases where biofilm is present and may be a preferable alternative (Orofino-Costa et al. 2017; Legabão et al. 2022).

For instance, in PDT, treatment with 1% methylene blue combined with itraconazole and terbinafine leads to a good clinical response, without significant adverse effects and complete remission of the infection (Gilaberte et al. 2014). A similar treatment carried out on felines led to similar results in 2 weeks of treatment (Cabral et al. 2022; Legabão et al. 2022).

Phaeohyphomycosis

Phaeohyphomycosis is an infection characterized by the formation of melanized (black) hyphae of the *Alternaria* and *Exophiala* genera in the skin that mainly affects immunosuppressed patients, such as those undergoing organ transplants. These black fungi, such as *E. dermatitidis*, are capable of forming biofilm in the host and are often linked to implantation diseases and infections such as phaeohyphomycosis and chromoblastomycosis (Seneviratne et al. 2015; Sav et al. 2016). Its standard treatment includes strengthening the immune system and prescribing azoles (Pastor and Guarro 2008).

Regarding the use of PDT, methylene blue irradiated by LED leads to biologically relevant antimicrobial activity against isolates of *Exophiala* spp., while in *Fusarium* spp., it presented a disinfectant effect in colony-forming unit (CFU) analysis, being an efficient alternative method against the growth of biofilms (Gao et al. 2016). Treatments with PDT in immunosuppressed patients included also PD 5-ALA and its pro-drug, Methylaminolevulinate (MAL), in combination with voriconazole and itraconazole, resolving symptoms from 1.5 to 6 months without relapses (Gil-Pallares et al. 2024). The synergistic effect between oral antifungals and PDT leads to a reduction in treatment time (Gil-Pallares et al. 2024).

Pityriasis Versicolor

Pityriasis versicolor is a fungal infection characterized by the growth of yeast that causes discolored patches on the skin. Its treatment is systemic, especially in severe recalcitrant cases. Virulence in *Malassezia* spp., especially *Malassezia furfur*, is linked to enzymatic activities and biofilm formation, which may explain skin inflammation and invasive infections of not only pityriasis versicolor but also *Malassezia folliculitis* and seborrheic dermatitis, among other symptoms (Saunte et al. 2020; Chebil et al. 2022).

Under in vitro conditions, PDT effectively inactivates the cells of *M. furfur* colonies in a dose-dependent manner with light and PD concentration, by damaging cell wall and increasing intracellular ROS levels (Cui et al. 2021). Clinically, treatment of female patients over 6 sessions with 2% methylene blue and LED lamp spaced every 2 weeks was shown to promote total remission of the disease in 4 weeks, with no recurrence for 6 months (Alberdi and Gómez 2020). In this sense, despite the lack of studies and protocols involving PDT and pityriasis versicolor in the literature, the results above point out the potential application of this treatment.

Immunotherapy and Vaccines

Immunotherapy may be defined as a biological therapy that uses the natural human immune system to fight cancer or fungal infections as well as to reduce the side effects of some therapies, improving tolerance and efficacy. In addition, this therapy may be applied as a preventive or treatment strategy. Biological treatments with monoclonal antibodies and vaccines have been used to treat or control fungal infections (Feng et al. 2024).

The structure, function, and responses of antibodies and B cells generated by *C. albicans* antigens such as secreted aspartyl proteinase (Sap2), heat shock protein (Hsp90), hyphal-regulated protein 1 (Hyr1), hyphal wall protein 1 (Hwp1), and enolase (Eno1), as well as by the Als3 and Mdh1 genes were recently described (Shukla et al. 2021; Feng et al. 2024). The findings demonstrate that a variety of mechanisms, including opsonization, complement fixation, neutralization, biofilm inhibition, and direct candidacidal activity, are used by antibodies produced in response to the vaccine candidates to prevent systemic candidiasis (Shukla et al. 2021).

It has also been proposed that blocking the expression of complement receptor-related protein 3 (surface protein, CR3-RP) through anti-CR3-RP polyclonal antibody inhibits *C. auris* biofilm development by 36–73% (for established biofilms by

28–46%), which is comparable to commonly used medications such as fluconazole, amphotericin B, or caspofungin (Bandara and Samaranayake 2022). More recently, monoclonal antibodies against the hypha-regulated protein of *C. albicans* (Cal-Hyr1p) were generated, which show the ability to recognize clinical isolates of *C. auris*. The anti-Hyr1p antibody inhibits biofilm formation and promotes phagocytosis of *C. auris* by macrophages. Furthermore, anti-Hyr1p protects 55% of mice against lethal systemic *C. auris* infection with a significantly lower fungal burden (Singh et al. 2023).

In turn, the NDV-3A vaccine (based on the N-terminus of the Als3 protein) is capable of generating anti-Als3p antibodies that recognize *C. auris* in vitro, by blocking biofilm formation and increasing macrophage-mediated phagocytosis (Singh et al. 2019). Anti-Als3p antibodies and CD4+ helper T cells, which stimulate tissue macrophages, are responsible for this action. NDV-3A also improves the efficacy of micafungin against candidemia caused by *C. auris* (Singh et al. 2019). Likewise, active immunization of mice with NDV-3A induces high levels of anti-rAls3p-N antibodies, which inhibit the ability of *C. albicans* to adhere and form biofilms (Alqarihi et al. 2019). NDV-3A also hinders the spread of fungal infection to the kidneys and prevents colonization of central venous catheters in mice (Alqarihi et al. 2019).

Additionally, natural compounds have been tested as strategies to control fungal infections with interesting results. The natural compound cinnamon oil has been investigated as an immunomodulator between *C. albicans* biofilms and phagocytes. Cinnamon oil effectively inhibits *C. albicans* in the planktonic state of forming biofilms. Furthermore, the sessile minimum inhibitory concentration of cinnamon is sufficient to downregulate hyphal regulatory genes compared to untreated biofilms (Tartor et al. 2022).

Antimicrobial Peptides

Peptides are molecules formed by the union of more than two amino acids by peptide bonds to form proteins. Peptides with antifungal action can be classified as peptides that cross the membrane (pore formation or action on β-glucan synthesis) and peptides that do not cross the membrane (action on the cell membrane) (Vanzolini et al. 2022). Some mechanisms of action include inhibition of protein synthesis, apoptosis, permeability and inhibition of cell wall synthesis, and enzymatic activity (Vanzolini et al. 2022). Several peptides have been tested as antimicrobial agents, and some with antifungal effects are discussed below.

PepBiotics exhibit excellent antifungal activity against *C. auris* by inhibiting metabolic activity at concentrations below 1 μM. Likewise, lipopeptides produced by *Bacillus subtilis*, such as AF4, exhibit activity against *Candida* spp. For instance, AF4 at a concentration of 3.48 μg/mL affects the hemolytic activity of *C. auris* (Bandara and Samaranayake 2022). Shellfish-derived peptides such as Cm-p5 are capable of eliminating several *Candida* species as well as *C. neoformans* (Bandara and Samaranayake 2022). A double-layer hydrogel functionalized with Cm-p5 is capable of killing *C. auris* at concentrations below 11 μg/mL. Interestingly, a peptide derived from the venom of *Crotalus durissus terrificus*, known as crotamine, inhibits 50% of *C. auris* isolates resistant to AmB and fluconazole (Bandara and Samaranayake 2022).

The CRAMP peptide also inhibits the formation of fungal biofilms, and its fragment AS10 may hinder the growth of *C. albicans* biofilms (Di Somma et al. 2020). Similarly, *C. albicans* biofilms are eliminated by the ZmD32 peptide (Oshiro et al. 2019). In turn, the mammalian peptides α-defensin-3, β-defensin-1, β-defensin-3, and PG-1 present antifungal activity against *C. neoformans* biofilms, with PG-1 reducing biofilm survival by 50% (Oshiro et al. 2019). Psd1, a plant-derived defensin, was also tested on *Candida*. This peptide destroys the polysaccharide matrix of *C. albicans* biofilms and increases cell death, probably acting on the glycosphingolipid GlcCer (Gonçalves et al. 2017). Likewise, microorganism-derived peptides such as Tn-AFP1, AMP-17, and Lycosin-1 successfully inhibit *Candida tropicalis*, *C. neoformans*, and *C. albicans* biofilms, respectively (Ul Haq et al. 2024).

Phages and Mycovirus

Bacteriophages are viruses that infect bacteria and the most numerous microorganisms known. Phages can integrate their genetic material into the host genome which, after replication, may confer immunity against infections by the same type of phage. Some phages are also capable of disseminating bacterial antibiotic resistance genes and may be allies due to their biofilm-reducing capacity (Elois et al. 2023). Some viruses infect fungi and are known as mycoviruses. Mycoviruses generally have RNA genomes and are obligate intracellular microorganisms and can be used to control fungal infections (van de Sande and Vonk 2019; Battersby et al. 2024; Urayama et al. 2024). Several types of mycoviruses (ambiguivirus 1, mycotimovirus type 2, narnavirus 1, mitovirus, flexivirus 1, fenuivirus 1, mononegavirus 1 and 2, mycophyovirus, and partitivirus 1) have been identified by RNA sequencing in *Fusarium* spp., a fungus that causes damage to wheat crops (Buivydaitė et al. 2024).

When mice are inoculated with *A. fumigatus* infected with a chrysovirus and with virus-free *A. fumigatus*, lower mortality and pulmonary fungal burden are observed for mice inoculated with the virus-infected fungus (Battersby et al. 2024). Furthermore, the virus-infected *A. fumigatus* strain shows lower tolerance to hypoxic stress, nitric oxide stress, and reduced hydrophobicity. Several *A. fumigatus* genes that are normally upregulated under hypoxic conditions, such as those encoding proteins involved in NAD+ regeneration, are suppressed in the virus-infected strain (Battersby et al. 2024). Furthermore, the fungi upregulate nitric oxide detoxification genes for protection (Battersby et al. 2024). Similarly, a strain of polymycovirus reduces mortality during *A. fumigatus* infection, considering that the infected strain reduces mycelial mass and conidia formation, in addition to delaying germination (Battersby et al. 2024). RNA sequencing of the infected strain revealed that gliotoxin synthesis genes are downregulated, while fumagillin synthesis genes are upregulated. Moreover, decreased adherence and increased sensitivity to phagocytosis are observed (Battersby et al. 2024). One explanation for this is that macrophages produce ROS that induce oxidative stress as an immune response mechanism (Battersby et al. 2024).

The bacterial phage Motto from *Pseudomonas aeruginosa* has been tested on mono- and dual-species biofilms of *P. aeruginosa* and *C. albicans*. The combination of this phage with ciprofloxacin had a significant reducing effect on a single biofilm of *P. aeruginosa*. On the other hand, when Motto was used in combination with fluconazole, the combination therapy was effective only against 6-h-old biofilms, with no noticeable effects for 24-h-old biofilms (Manohar et al. 2024).

In the last decade, *P. aeruginosa* phages Pf4 and Pf1 have also been evaluated on *C. albicans* biofilm formation and preformed *C. albicans* biofilms. Both biofilms are inhibited by phages, in addition to planktonic fungal growth. Interestingly, researchers were able to reverse the biofilm inhibition by adding iron, suggesting that the phages suppress iron that would otherwise be used by *C. albicans* (Nazik et al. 2017). Additionally, phage aggregation in the extracellular matrix of the biofilm has been demonstrated (Nazik et al. 2017). In turn, phage Pf4 inhibits the formation and growth of *A. fumigatus* biofilms and can also inhibit the metabolic activity of biofilms of this fungus, with dose-dependent results. Pf4 binds to iron, an essential element for *A. fumigatus*. In contrast, Pf1 is not able to inhibit *Aspergillus* biofilms (Penner et al. 2016).

Metallic Nanoparticles

Particles with dimensions ranging from 1 to 100 nm are classified as nanoparticles. They have distinct physical, chemical, and biological properties due to their small size and large surface area in relation to volume (Benoit et al. 2019; Najahi-Missaoui et al. 2020; Hou et al. 2021). There are numerous advantages of nanotherapy over conventional treatments in combating fungal biofilms, including improved antifungal activity, greater ease of penetration into the most basal layers of the biofilm, greater stability and bioavailability, and action in the targeted and controlled delivery of drugs, contributing to increased release time and reduced frequency of drug administration. The strategy of combining nanoparticles and antifungal drugs can therefore improve drug penetration into the biofilm and increase its antimicrobial activity (Baptista et al. 2018; Horvat et al. 2020; Najahi-Missaoui et al. 2020; Vera-González and Shukla 2020; Hou et al. 2021; Monteiro and Pessan 2022).

On the other hand, there are some disadvantages and challenges related to nanoparticle application. The first is the risk of toxicity to the host's cells, which ends up restricting the dosage and frequency of use of the nanoparticles. Also, some synthesis processes may require more complex techniques, increasing the cost of their production. Finally, there are regulatory barriers related to their safety and efficacy that make the supervision of nanotherapies rigorous, delaying their use in

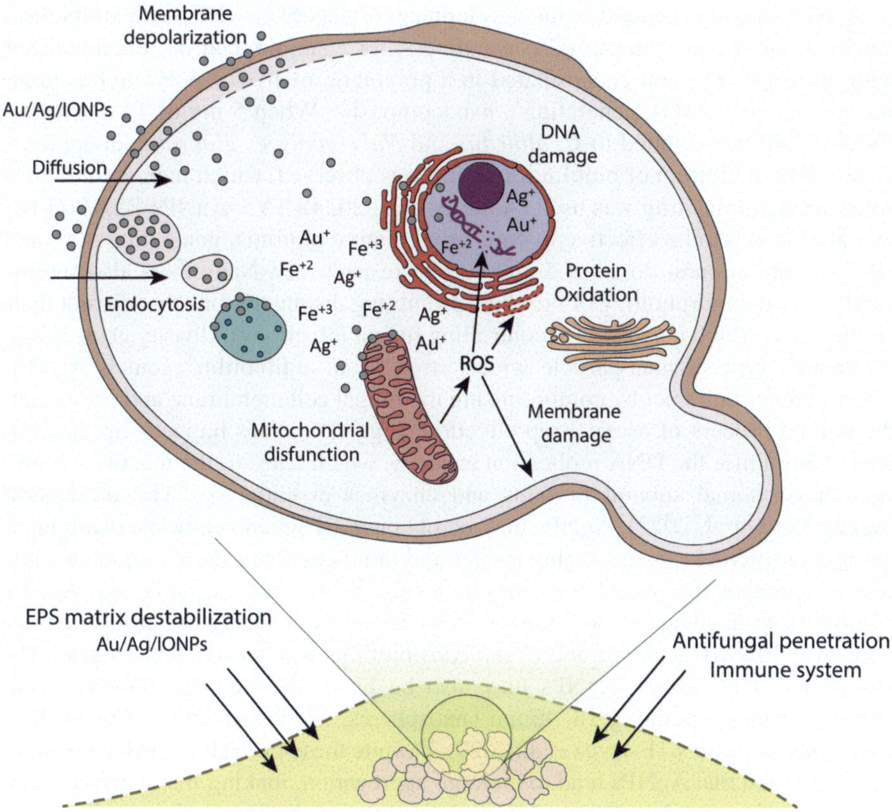

Fig. 11.6 Main targets of antifungal and antibiofilm actions of metallic nanoparticles

clinical applications (Baptista et al. 2018; Horvat et al. 2020; Najahi-Missaoui et al. 2020; Vera-González and Shukla 2020; Hou et al. 2021; Monteiro and Pessan 2022).

In general, the mechanisms of antimicrobial/antibiofilm action of metallic nanoparticles are varied and include binding to the extracellular matrix, causing its destabilization and favoring both the penetration of antifungal agents and the exposure of microbial cells to the immune system (Fig. 11.6). They also generate ROS that leads to oxidative stress in fungal cells, causing damage to cell membranes, proteins, and nucleic acids (Chen and Feng 2022).

Among the main metal nanoparticles applied against fungal biofilms, those made of gold (AuNPs), silver (AgNPs), and iron oxide (IONPs) stand out. AuNPs stabilized with cetyltrimethylammonium bromide (CTAB-AuNPs) or conjugated with cysteine (cys-AuNPs) have been associated with AmB and tested on *C. tropicalis* biofilms. After exposure to CTAB-AuNPs, cys-AuNPs, and AmB, yeasts showed average sizes of 5.00, 5.4, and 3.8 μm, respectively. Furthermore, the combination of CTAB-AuNPs with AmB resulted in a 97.3% reduction in biofilm, with a fractional inhibitory concentration index (FICI) of 0.05, indicating synergism. In turn, the combination of cys-AuNPs with AmB promoted a 91.7% reduction in biofilm, with an additive effect (FICI = 0.75) (Dasilva et al. 2024).

AuNPs may also be used in the development of nanofibers (NF) with antibiofilm potential. Gold and chitosan (CS) nanofibers, for example, can be functionalized with tyrosol (TYS) and encapsulated in a polymeric matrix of polyvinyl alcohol/acacia gum (PVA/AG), generating a nanocomposite. When 5 mg of TYS-AuNPs-PVA/AG-NF was applied to *C. albicans* and *Nakeseomyces glabrata*, an approximately 50% inhibition of biofilm formation was observed, which increased to 80% when 10 mg of the drug was used (Alhariry et al. 2024). TYS-AuNPs-PVA/AG-NF was also shown to be effective in combating mature biofilms, generating 40% and 60–70% inhibition at doses of 5 and 10 mg, respectively. Nanofibers also significantly alter the morphology of biofilms, promoting disintegrations that disrupt their architecture, which is crucial for controlling fungal infections (Alhariry et al. 2024).

Another type of nanoparticle widely used as an antibiofilm agent is AgNPs. These nanoparticles act by compromising the fungal cell membrane and preventing the normal process of asexual reproduction (Fig. 11.6). This happens because Ag ions compromise the DNA replication capacity, which leads to the inactive expression of ribosomal subunit proteins and enzymes essential for ATP production (Sakthi Devi et al. 2022). AgNPs may be obtained by green synthesis, using plant extracts or microorganisms, being a safer and more economical technique, in addition to reducing the use of harmful chemicals. In this context, spherical AgNPs (approximate diameter of 66.7 nm) biosynthesized by *A. fumigatus* lead to reductions over 70% for *A. flavus* and *C. albicans* biofilms when tested at 10 mg/mL (El Dougdoug et al. 2024). AgNPs may also be biosynthesized by *Phanerochaete chrysosporium*, generating 18–46 nm nanospheres, which reduce *C. albicans* biofilm biomass by 80% (Estevez et al. 2020). Despite these favorable results, it should be highlighted that AgNPs tend to undergo aggregation, making their surface functionalization essential to prevent impairment of the antibiofilm action.

Recently, a probiotic-based silver secnidazole nanohybrid emulsion (Ag-Secn-NHE) was investigated as an alternative treatment for the control of vaginal pathogens, with promising results. It was found that the most effective microorganism in the production of probiotics is *S. cerevisiae*, which showed a positive effect when combined with Ag-Secn-NHE. Ag-Secn-NHE generated inhibition halos against *C. albicans*, *C. tropicalis*, *C. glabrata*, *Candida lipolytica*, *C. auris*, *C. parapsilosis*, *Trichosporon inkin*, *C. neoformans*, and *Cypripedium candidum* ranging from 12 to 24 mm, respectively. However, after combination with *S. cerevisiae* probiotic, the halos increased in the range of 17–35 mm. Furthermore, Ag-Secn-NHE associated with the probiotic led to 75% inhibition of *C. auris* biofilm formation and 60% inhibition of *C. neoformans* and *C. albicans* biofilms, compared to the probiotic-free nanocomposite (Mosallam et al. 2024).

In turn, IONPs have been used in the medical field mainly as drug carriers, enhancing the effects of the carried compounds (Arias et al. 2018). These nanoparticles have magnetic properties and may be coated with natural polymers, such as CS. This conjugation favors the stability of the nanoparticles and enhances their antimicrobial effects (Arias et al. 2018). IONPs may penetrate fungal cells by simple diffusion or endocytosis or cause a depolarization of the plasma membrane, affecting the integrity of the yeast. Once these particles reach the intracellular compartment, they may release iron ions, leading to mitochondrial dysfunction (with ROS production) and protein oxidation, in addition to causing DNA damage (Arias et al. 2018) (Fig. 11.6).

Miconazole (MCZ), fluconazole (FLZ), chlorhexidine (CHX), and cetylpyridinium chloride (CPC) have been carried by CS-coated IONPs, and the generated nanocarriers have been tested against different models of fungal biofilms, aiming at the control of oral candidiasis. The administration of IONPs-CS as carriers of MCZ at 78 µg/mL against mono- and dual-species biofilms of *C. albicans* and *C. glabrata* showed great promise. This nanocarrier was able to overcome the antibiofilm effects of free MCZ in reducing metabolic activity and the number of CFUs (Arias et al. 2020b). In turn, when tested on inter-kingdom biofilm models of dental caries, denture, and gingivitis, IONPs-CS-MCZ is capable of promoting reductions in total biomass, metabolic activity, and numbers of CFUs and colony-forming equivalents, in addition to affecting the structure of polymicrobial biofilms and the interactions between fungi and bacteria in these communities (Arias et al. 2020a). On the other hand, IONPs-CS carrying 1250 µg/mL FLZ presents a similar effect to free FLZ on single biofilms of *C. albicans* and *C. glabrata*, and both compounds promote significant reductions in biomass, metabolic activity, and CFU load (de Lima et al. 2020). Exposure of *Candida* biofilms to the IONPs-CS-FLZ system also generates significant increases in the protein, nucleic acid, and carbohydrate loads of the extracellular matrix, which is related to the antimicrobial effect on fungal cells (de Lima et al. 2020). Scanning electron microscopy analysis associated with energy dispersive spectroscopy confirms the ability of the IONPs-CS-FLZ nanocarrier to penetrate the most basal layers of *Candida* biofilms (de Lima et al. 2020).

CS-coated IONPs carrying MCZ or FLZ also show interesting results when applied to mixed biofilms of three *Candida* species (*C. albicans*, *C. glabrata*, and *C. tropicalis*). For this biofilm model, nanocarriers significantly reduce some biofilm parameters (biomass, metabolism, and number of cultivable cells) and may present the same antibiofilm effect as free drugs, however, at a concentration twice lower (Caldeirão et al. 2021a). By increasing the complexity of the biofilm model, the effects of the nanocarriers of MCZ or FLZ are maintained. In this context, the IONPs-CS-MCZ and IONPs-CS-FLZ carrier systems exhibit similar or superior reducing effects to free antifungals on polymicrobial biofilms that simulate an oral candidiasis microcosm (Caldeirão et al. 2021b).

In addition to the aforementioned antifungals, traditional antimicrobials used in dentistry such as CHX and CPC can also be carried by CS-coated IONPs. In this sense, the IONPs-CS-CHX nanocarrier with dimensions less than 40 nm and carrying two different concentrations of CHX (39 or 78 µg/mL) were shown to have significant reducing effects on mono- and dual-species biofilms of *C. albicans* and *Streptococcus mutans* (Vieira et al. 2019). These effects are equivalent to or even surpass those found for CHX in free form at 78 µg/mL (Vieira et al. 2019). For *C. glabrata* mono-species biofilms, the same trend has been reported, considering that the nanocarrier containing CHX at 156 µg/mL is more effective in reducing biofilm biomass and metabolism than free CHX and presents the same reducing effect on the number of CFUs as the antimicrobial administered alone (Araujo et al.

2020b). Furthermore, at low concentrations (≤3.9 μg/mL), the IONPs-CS-CHX nanocarrier is less cytotoxic to murine fibroblasts than free CHX (Araujo et al. 2020b). On the other hand, when the IONPs-CS-CPC nanocarrier is applied to single and mixed biofilms of *C. albicans* and *C. glabrata*, its reducing effect on the number of CFUs significantly surpasses that promoted by CPC in free form (Araujo et al. 2020a).

IONPs-CS may also be used to simultaneously carry two antimicrobial agents. More recently, a dual nanocarrier of CHX and FLZ showed very promising results on polymicrobial biofilms simulating an oral candidiasis microcosm. This dual nanocarrier was more effective in reducing loads of total microorganisms, *S. mutans*, *Lactobacillus* spp., *C. albicans*, and *C. glabrata* than the combination of CHX with FLZ without nanocarrier, in addition to promoting a significant decrease in lactic acid production (Araujo et al. 2023).

Natural Compounds

Phytotherapeutics are medicinal substances extracted from plants, including extracts and essential oils, used for their antimicrobial, anti-inflammatory, and antioxidant properties. These compounds are being explored as effective alternatives to traditional antimicrobial therapies, especially for treating infections caused by drug-resistant pathogens such as *Candida* species. The effectiveness of phytotherapeutics is based on their multiple mechanisms of action and low toxic potential for host cells, in addition to the ability to combine these compounds with other technologies (Sardi et al. 2013; Newman 2017).

The mechanism of action of phytotherapeutics is multifactorial and is related to the action of terpenes, terpenoids, flavonoids, and phenolic acids. Mechanisms include disruption of the biofilm extracellular matrix, alteration of cell permeability, inhibition of fungal growth, disruption of cell wall or nucleic acid synthesis, inhibition of ergosterol synthesis, induction of apoptosis through generation of ROS, disruption of mitochondrial function, and increased antiadhesive properties (Nazzaro et al. 2017; Aboody and Mickymaray 2020; Mishra et al. 2020).

The main advantages related to the use of phytotherapeutics/natural compounds in the control of fungal biofilms include (i) potential to overcome drug resistance, considering their multiple mechanisms of action; (ii) synergistic effect in combination with other antifungal drugs, which may increase their efficiency; (iii) lower toxicity compared to usual antifungals; and (iv) greater accessibility to the population since natural compounds tend to have a low cost. On the other hand, the lack of clinical research and scientific evidence that proves the efficacy and safety of this therapy, in addition to possible allergic reactions and interactions with conventional medications are highlighted as the main disadvantages and limitations of this type of therapy (Sardi et al. 2013).

Several natural compounds can act in the treatment of fungal biofilms, and the most promising are essential oils, polyphenols, and other plant extracts, which are discussed below.

Essential Oils

Essential oils are volatile and highly concentrated compounds with therapeutic properties that include antimicrobial, anti-inflammatory, antioxidant, and antifungal actions. They are applied in the fields of cosmetics, medicine, and aromatherapy. The main categories of essential oils are terpenes and terpenoids. Terpenes are simple molecules of a class of natural hydrocarbons derived from isoprene, composed of carbon and hydrogen atoms. Terpenoids, also derived from isoprenes, undergo oxidation, which results in the addition of functional groups such as hydroxyls, carbonyls, or oxygen (Nazzaro et al. 2017).

Cinnamomum verum essential oil, for instance, has been associated with sertraline in combating *Candida* species in the planktonic or sessile states. For planktonic cells, the best results were obtained with the combination of the two compounds, generating a synergistic effect due to the reductions (88–97%) in the minimum inhibitory concentration (MIC) values of each component compared to the MIC of the isolated substances. For the sessile state (biofilms), the same trend of synergistic effect was observed, considering that the combination of compounds promoted inhibitions in metabolic activity ranging from 60% to 86% (Barbarossa et al. 2024). Although these results are favorable in the control of *Candida* biofilms, further investigations are necessary to clarify the exact mechanism of antibiofilm action and evaluate its efficacy in the clinical setting.

The use of *Melaleuca alternifolia* in the fight against fungi has also been explored. This is a plant native to Australia, also called tea tree, whose extraction is done through steam distillation of its leaves, producing tea tree oil (TTO). Its composition is a combination of several components, with terpinen-4-ol being the main component. When these compounds are evaluated on planktonic *C. albicans*, MIC values of 9.0 and 4.5 mg/mL are observed, respectively, for TTO and terpinen-4-ol. Regarding *Candida* biofilm, 8.86 mg/mL terpinen-4-ol has a significantly greater reducing effect on cell viability (Francisconi et al. 2020). These findings reinforce that TTO and terpinen-4-ol could be applied in both the prevention and control of *Candida* infections.

Polyphenols

Polyphenols are also natural compounds derived from plants, with antimicrobial and antioxidant properties, and potential against fungal biofilms. Curcumin is a polyphenol from the flavonoid class, with antioxidant, anti-inflammatory, antimutagenic, and anticancer effects. When curcumin is tested on *C. albicans*, MIC values of 100 and ≥200 μg/mL are found, respectively, for planktonic and sessile cells. For biofilms, curcumin at 50 and 100 μg/mL has moderate impacts on overall metabolism. However, when applied at 200 μg/mL, a significant reduction in biofilm formation is observed, reaching an inhibition of over 90% compared to the control group (Sun et al. 2015a, b; Raorane et al. 2019).

Another compound belonging to the class of polyphenols with antifungal action is called magnolol, which is present in the bark of *Magnolia officinalis* (Behbehani et al. 2017). The antifungal effect of magnolol was evaluated on 64 *Candida* spp. strains using MIC assays and growth curves. The MIC values for magnolol ranged from 16 to 64 μg/mL, while those for AmB ranged from 0.06 to 1 μg/mL. In the growth curve assays, a remarkable sensitivity to magnolol was observed compared to the control, with reductions of 96.04% for magnolol and 95.95% for AmB. Regarding the biofilm assays, the MTT reduction results showed that magnolol at MIC, MIC/2, and MIC/4 promoted biofilm reductions of 69.5%, 46.7%, and 35.6%, respectively (Behbehani et al. 2017).

Phenolic compounds have also been tested on biofilms of *Pichia anomala*, *S. cerevisiae*, *Schizosaccharomyces pombe*, and *Debaryomyces hansenii*. As for biofilm formation, vanillin shows high activity, inhibiting 77% of biofilm formation in *P. anomala* and 82% in *S. pombe*. For *S. cerevisiae*, the most effective compound is 4-hydroxybenzaldehyde, with 64% inhibition. In turn, biofilms of *D. hansenii*, *P. anomala*, and *S. pombe* are more sensitive to phenolic compounds, while *S. cerevisiae* is more resistant. Cinnamic acid has the greatest anti-adhesion activity against the yeasts mentioned above, with inhibitions ranging from 50 to 91% (Kimani et al. 2021). Therefore, phenolic compounds such as cinnamic acid and vanillin exhibit considerable antibiofilm effects against important food spoilage yeasts (Kimani et al. 2021).

Other Plant Extracts

Plant extracts are important natural resources with the potential to combat pathogenic microorganisms and biofilms. For instance, the antimicrobial and antibiofilm effects of medicinal plants from Turkey have been evaluated against *C. albicans*. While the methanolic extract of *Allium paniculatum* flower has a MIC of 9.75 μg/mL, extracts of *Primula vulgaris* root and leaves of *Cistus laurifolius*, *Cistus salviifolius*, and *Achillea nobilis* show MICs ranging from 156 to 312 μg/mL. Specifically for biofilms, *A. paniculatum* extract significantly reduces the *C. albicans* load in mature biofilms at concentrations equal to or greater than 32 × MIC (Oyardı et al. 2024), being a promising product for the treatment of candidiasis.

Chelerythrine is an alkaloid extracted from *Bocconia cordata*, known for its anticancer, anti-inflammatory, and antimicrobial properties, and which has shown strong antimicrobial activity against *C. albicans* and *Staphylococcus aureus* in single or mixed cultures. Chelerythrine is capable of efficiently inhibiting biofilm formation in both species, with minimum biofilm inhibitory concentration values of 2 μg/mL for single cultures and 3 μg/mL for mixed cultures. Furthermore, when mixed biofilms are exposed to chelerythrine at MIC/4, the numbers of cultivable cells of *S. aureus* and *C. albicans* are substantially reduced (Qian et al. 2020).

Conclusion

Several studies have demonstrated a clear advantage of the alternative approaches listed in this chapter compared with conventional therapies for some biofilm-associated fungal infections. Common limitations, however, include frequent incompatibility between in vitro and in vivo results, either due to pharmacokinetic factors or multiple factors related to the host. The growing resistance to classic fungicides and studies including a low number of strains can also restrict the effectiveness of the treatment and limit the external validity of the studies. Furthermore, translational research from in vitro or animal studies to in vivo clinical conditions is still lacking for some of the therapies covered here. Despite these drawbacks, the development of new therapies associated with artificial intelligence and the use of bioinformatics may open new avenues in the fight against fungal biofilms.

References

Aboody MSA, Mickymaray S (2020) Anti-fungal efficacy and mechanisms of flavonoids. Antibiotics (Basel) 9(2):45

Aigner M, Lass-Flörl C (2020) Encochleated amphotericin B: is the oral availability of amphotericin B finally reached? J Fungi (Basel) 6(2):66

Akinosoglou K, Rigopoulos EA, Papageorgiou D, Schinas G, Polyzou E, Dimopoulou E et al (2024) Amphotericin B in the era of new antifungals: where will it stand? J Fungi (Basel) 10(4):278

Alberdi E, Gómez C (2020) Successful treatment of Pityriasis Versicolor by photodynamic therapy mediated by methylene blue. Photodermatol Photoimmunol Photomed 36(4):308–312

Alhariry J, Kumar A, Yadav TC, Yadav E, Prasad R, Poluri KM et al (2024) Tyrosol-gold nanoparticle functionalized acacia gum-PVA nanofibers for mitigation of *Candida* biofilm. Microb Pathog 193:106763

Alkhazraji S, Gebremariam T, Alqarihi A, Gu Y, Mamouei Z, Singh S et al (2020) Fosmanogepix (APX001) is effective in the treatment of immunocompromised mice infected with invasive pulmonary scedosporiosis or disseminated fusariosis. Antimicrob Agents Chemother 64(3):e01735–e01719

Almajid A, Bazroon A, Al-Awami HM, Albarbari H, Alqahtani I, Almutairi R et al (2024) Fosmanogepix: the novel anti-fungal agent's comprehensive review of *in vitro*, *in vivo*, and current insights from advancing clinical trials. Cureus 16(4):e59210

Alqarihi A, Singh S, Edwards JE Jr, Ibrahim AS, Uppuluri P (2019) NDV-3A vaccination prevents *C. albicans* colonization of jugular vein catheters in mice. Sci Rep 9(1):6194

Araujo HC, Arias LS, Caldeirão ACM, Assumpção LCF, Morceli MG, de Souza Neto FN et al (2020a) Novel colloidal nanocarrier of cetylpyridinium chloride: antifungal activities on *Candida* species and cytotoxic potential on murine fibroblasts. J Fungi (Basel) 6(4):218

Araujo HC, da Silva ACG, Paião LI, Magario MKW, Frasnelli SCT, Oliveira SHP et al (2020b) Antimicrobial, antibiofilm and cytotoxic effects of a colloidal nanocarrier composed by chitosan-coated iron oxide nanoparticles loaded with chlorhexidine. J Dent 101:103453

Araujo HC, Pessan JP, Caldeirão ACM, Sampaio C, Oliveira MJS, Sales DH et al (2023) Dual nanocarrier of chlorhexidine and fluconazole: physicochemical characterization and effects on microcosm biofilms and oral keratinocytes. J Dent 138:104699

Arendrup MC, Chowdhary A, Astvad KMT, Jørgensen KM (2018) APX001A in vitro activity against contemporary blood isolates and *Candida auris* determined by the EUCAST reference method. Antimicrob Agents Chemother 62(10):e01225–e01218

Arias LS, Brown JL, Butcher MC, Delaney C, Monteiro DR, Ramage G (2020a) A nanocarrier system that potentiates the effect of miconazole within different interkingdom biofilms. J Oral Microbiol 12(1):1771071

Arias LS, Pessan JP, de Souza Neto FN, Lima BHR, de Camargo ER, Ramage G et al (2020b) Novel nanocarrier of miconazole based on chitosan-coated iron oxide nanoparticles as a nanotherapy to fight Candida biofilms. Colloids Surf B Biointerfaces 192:111080

Arias LS, Pessan JP, Vieira APM, Lima TMT, Delbem ACB, Monteiro DR (2018) Iron oxide nanoparticles for biomedical applications: a perspective on synthesis, drugs, antimicrobial activity, and toxicity. Antibiotics (Basel) 7(2):46

Aslanyan L, Sanchez DA, Valdebenito S, Eugenin EA, Ramos RL, Martinez LR (2017) The crucial role of biofilms in cryptococcus neoformans survival within macrophages and colonization of the central nervous system. J Fungi (Basel) 3(1):10

Bandara N, Samaranayake L (2022) Emerging and future strategies in the management of recalcitrant Candida auris. Med Mycol 60(4):myac008

Baptista PV, McCusker MP, Carvalho A, Ferreira DA, Mohan NM, Martins M et al (2018) Nanostrategies to fight multidrug resistant bacteria – "A Battle of the Titans". Front Microbiol 9:1441

Baran M, Borowski E, Mazerski J (2009) Molecular modeling of amphotericin B-ergosterol primary complex in water II. Biophys Chem 141(2–3):162–168

Barbarossa A, Rosato A, Carrieri A, Fumarola L, Tardugno R, Corbo F et al (2024) Exploring the antibiofilm effect of sertraline in synergy with *Cinnamomum verum* essential oil to counteract *Candida* species. Pharmaceuticals (Basel) 17(9):1109

Battersby JL, Stevens DA, Coutts RHA, Havlíček V, Hsu JL, Sass G et al (2024) The expanding mycovirome of *Aspergilli*. J Fungi (Basel) 10(8):585

Behbehani J, Shreaz S, Irshad M, Karched M (2017) The natural compound magnolol affects growth, biofilm formation, and ultrastructure of oral *Candida* isolates. Microb Pathog 113:209–217

Benoit DSW, Sims KR Jr, Fraser D (2019) Nanoparticles for oral biofilm treatments. ACS Nano 13(5):4869–4875

Bhatta AK, Keyal U, Wang X, Gellén E (2017) A review of the mechanism of action of lasers and photodynamic therapy for onychomycosis. Lasers Med Sci 32(2):469–474

Borman AM, Fraser M, Patterson Z, Palmer MD, Johnson EM (2021) *In vitro* antifungal drug resistance profiles of clinically relevant members of the Mucorales (Mucoromycota) especially with the newer triazoles. J Fungi (Basel) 7(4):271

Brilhante RSN, Correia EEM, Guedes GMM, Pereira VS, Oliveira JS, Bandeira SP et al (2017) Quantitative and structural analyses of the *in vitro* and *ex vivo* biofilm-forming ability of dermatophytes. J Med Microbiol 66(7):1045–1052

Broekgaarden M, Weijer R, van Gulik TM, Hamblin MR, Heger M (2015) Tumor cell survival pathways activated by photodynamic therapy: a molecular basis for pharmacological inhibition strategies. Cancer Metastasis Rev 34(4):643–690

Buil JB, Rijs AJMM, Meis JF, Birch M, Law D, Melchers WJG et al (2017) *In vitro* activity of the novel antifungal compound F901318 against difficult-to-treat *Aspergillus* isolates. J Antimicrob Chemother 72(9):2548–2552

Buivydaitė Ž, Winding A, Jørgensen LN, Zervas A, Sapkota R (2024) New insights into RNA mycoviruses of fungal pathogens causing *Fusarium* head blight. Virus Res 349:199462

Cabral FV, Sellera FP, Ribeiro MS (2022) Feline sporotrichosis successfully treated with methylene blue-mediated antimicrobial photodynamic therapy and low doses of itraconazole. Photodiagn Photodyn Ther 40:103154

Caldeirão ACM, Araujo HC, Arias LS, Ramírez Carmona W, Miranda GP, Oliveira SHP et al (2021a) Nanocarriers of Miconazole or Fluconazole: effects on three-species *Candida* biofilms and cytotoxic effects *in vitro*. J Fungi (Basel) 7(7):500

Caldeirão ACM, Araujo HC, Tomasella CM, Sampaio C, dos Santos Oliveira MJ, Ramage G et al (2021b) Effects of antifungal carriers based on chitosan-coated iron oxide nanoparticles on microcosm biofilms. Antibiotics (Basel) 10(5):588

Calzavara-Pinton PG, Venturini M, Capezzera R, Sala R, Zane C (2004) Photodynamic therapy of interdigital mycoses of the feet with topical application of 5-aminolevulinic acid. Photodermatol Photoimmunol Photomed 20(3):144–147

Carvalhaes CG, Rhomberg P, Rhomberg P, Strand G, Klauer AL, Castanheira M (2022) 1731. Rezafungin activity against *Candida spp.* and *Aspergillus spp.* isolates causing invasive infections worldwide in 2021. Open Forum Infect Dis 9(Suppl 2):ofac492.1361

Černáková L, Roudbary M, Brás S, Tafaj S, Rodrigues CF (2021) *Candida auris*: a quick review on identification, current treatments, and challenges. Int J Mol Sci 22(9):4470

Chakrabarti A, Kaur H, Rudramurthy SM, Appannanavar SB, Patel A, Mukherjee KK et al (2016) Brain abscess due to *Cladophialophora bantiana*: a review of 124 cases. Med Mycol 54(2):111–119

Chebil W, Rhimi W, Haouas N, Romano V, Belgacem S, Belhadj Ali H et al (2022) Virulence factors of *Malassezia* strains isolated from pityriasis versicolor patients and healthy individuals. Med Mycol 60(8):myac060

Chen B, Sun Y, Zhang J, Chen R, Zhong X, Wu X et al (2019) *In vitro* evaluation of photodynamic effects against biofilms of dermatophytes involved in onychomycosis. Front Microbiol 10:1228

Chen X, Liang Y, Guo Z, Huang R, Zheng W, Wu W et al (2024) Photodynamic therapy combined with antifungal drugs against kerion: a report of six cases and literature review. Photodiagn Photodyn Ther 46:104039

Chen Y, Feng X (2022) Gold nanoparticles for skin drug delivery. Int J Pharm 625:122122

Chen YL, Brand A, Morrison EL, Silao FG, Bigol UG, Malbas FF Jr et al (2011) Calcineurin controls drug tolerance, hyphal growth, and virulence in *Candida dubliniensis*. Eukaryot Cell 10(6):803–819

Chowdhary A, Kathuria S, Xu J, Meis JF (2013) Emergence of azole-resistant aspergillus fumigatus strains due to agricultural azole use creates an increasing threat to human health. PLoS Pathog 9(10):e1003633

Chowdhary A, Perfect J, de Hoog GS (2014) Black molds and melanized yeasts pathogenic to humans. Cold Spring Harb Perspect Med 5(8):a019570

Coelho RA, Brito-Santos F, Figueiredo-Carvalho MHG, Silva JVDS, Gutierrez-Galhardo MC, do Valle ACF et al (2018) Molecular identification and antifungal susceptibility profiles of clinical strains of *Fonsecaea spp.* isolated from patients with chromoblastomycosis in Rio de Janeiro, Brazil. PLoS Negl Trop Dis 12(7):e0006675

Colley T, Alanio A, Kelly SL, Sehra G, Kizawa Y, Warrilow AGS et al (2017) *In vitro* and *in vivo* antifungal profile of a novel and long-acting inhaled azole, PC945, on *Aspergillus fumigatus* infection. Antimicrob Agents Chemother 61(5):e02280–e02216

Colley T, Sehra G, Daly L, Kimura G, Nakaoki T, Nishimoto Y et al (2019) Antifungal synergy of a topical triazole, PC945, with a systemic triazole against respiratory *Aspergillus fumigatus* infection. Sci Rep 9(1):9482

Correia JH, Rodrigues JA, Pimenta S, Dong T, Yang Z (2021) Photodynamic therapy review: principles, photosensitizers, applications, and future directions. Pharmaceutics 13(9):1332

Cowen LE, Singh SD, Köhler JR, Collins C, Zaas AK, Schell WA et al (2009) Harnessing Hsp90 function as a powerful, broadly effective therapeutic strategy for fungal infectious disease. Proc Natl Acad Sci USA 106(8):2818–2823

Cui Z, Zhang M, Geng S, Niu X, Wang X, Zhu Y et al (2021) Antifungal effect of antimicrobial photodynamic therapy mediated by haematoporphyrin monomethyl ether and aloe emodin on *Malassezia furfur*. Front Microbiol 12:749106

Dąbrowski JM (2017) Chapter nine – Reactive oxygen species in photodynamic therapy: mechanisms of their generation and potentiation. In: van Eldik R, Hubbard CD (eds) Advances in inorganic chemistry, vol 70. Academic Press, pp 343–394

Dasilva MA, Andrada KFC, Torales MM, Hughes IM, Pez P, García-Martínez JC et al (2024) Synergistic activity of gold nanoparticles with amphotericin B on persister cells of *Candida tropicalis* biofilms. J Nanobiotechnol 22(1):254

de Lima TM, Arias LS, Afanaci LF, Ferraresse RF, de S Neto FN, de Lima BH et al (2020) Assembly and antifungal effect of a new fluconazole-carrier nanosystem. Future Microbiol 15:273–285

de Oliveira GB, Antonio JR, Antonio CR, Tomé FA (2015) The association of fractional CO_2 laser 10.600nm and photodynamic therapy in the treatment of onychomycosis. An Bras Dermatol 90(4):468–471

Delmas G, Park S, Chen ZW, Tan F, Kashiwazaki R, Zarif L et al (2002) Efficacy of orally delivered cochleates containing amphotericin B in a murine model of aspergillosis. Antimicrob Agents Chemother 46(8):2704–2707

Denning DW, Hope WW (2010) Therapy for fungal diseases: opportunities and priorities. Trends Microbiol 18(5):195–204

Denning DW (2003) Echinocandin antifungal drugs. Lancet 362(9390):1142–1151

Di Somma A, Moretta A, Canè C, Cirillo A, Duilio A (2020) Antimicrobial and antibiofilm peptides. Biomolecules 10(4):652

dos Santos GMP, Borba-Santos LP, Vila T, Ferreira Gremião ID, Pereira SA, De Souza W et al (2022) *Sporothrix* spp. biofilms impact in the zoonotic transmission route: feline claws associated biofilms, itraconazole tolerance, and potential repurposing for miltefosine. Pathogens 11(2):206

du Pré S, Birch M, Law D, Beckmann N, Sibley GEM, Bromley MJ et al (2020) The dynamic influence of Olorofim (F901318) on the cell morphology and organization of living cells of *Aspergillus fumigatus*. J Fungi (Basel) 6(2):47

El Dougdoug NK, Attia MS, Malash MN, Abdel-Maksoud MA, Malik A, Kiani BH et al (2024) *Aspergillus fumigatus*-induced biogenic silver nanoparticles' efficacy as antimicrobial and antibiofilm agents with potential anticancer activity: an *in vitro* investigation. Microb Pathog:106950

Elois MA, Silva RD, Pilati GVT, Rodríguez-Lázaro D, Fongaro G (2023) Bacteriophages as biotechnological tools. Viruses 15(2):349

Estevez MB, Raffaelli S, Mitchell SG, Faccio R, Alborés S (2020) Biofilm eradication using biogenic silver nanoparticles. Molecules 25(9):2023

Fekrazad R, Poorsattar Bejeh Mir A, Ghasemi Barghi V, Shams-Ghahfarokhi M (2015) Eradication of *C. albicans* and *T. rubrum* with photoactivated indocyanine green, *Citrus aurantifolia* essential oil and fluconazole. Photodiagn Photodyn Ther 12(2):289–297

Feng Z, Lu H, Jiang Y (2024) Promising immunotherapeutic targets for treating candidiasis. Front Cell Infect Microbiol 14:1339501

Fernandes JA, Conrado PCV, Perina BS, de Oliveira ACV, Arita GS, Capoci IRG et al (2023) Photodynamic inactivation by hypericin-P123 on azole-resistant isolates of the *Trichophyton rubrum* complex as planktonic cells and biofilm. Photodiagn Photodyn Ther 44:103875

Flanagan S, Ong V, Marbury T, Jandourek A, Gandhi RG, Sandison T (2024) Phase I study of the pharmacokinetics and safety of rezafungin in subjects with moderate/severe hepatic impairment and matched control subjects. Pharmacotherapy 44(6):435–443

Francisconi RS, Huacho PMM, Tonon CC, Bordini EAF, Correia MF, de Cássia Orlandi Sardi J et al (2020) Antibiofilm efficacy of tea tree oil and of its main component terpinen-4-ol against *Candida albicans*. Braz Oral Res 34:e050

Gao L, Jiang S, Sun Y, Deng M, Wu Q, Li M et al (2016) Evaluation of the effects of photodynamic therapy alone and combined with standard antifungal therapy on planktonic cells and biofilms of *Fusarium* spp. and *Exophiala* spp. Front Microbiol 7:617

Garcia-Effron G (2020) Rezafungin-mechanisms of action, susceptibility and resistance: similarities and differences with the other echinocandins. J Fungi (Basel) 6(4):262

Gebremariam T, Gu Y, Alkhazraji S, Youssef E, Shaw KJ, Ibrahim AS (2022) The combination treatment of fosmanogepix and liposomal amphotericin B is superior to monotherapy in treating experimental invasive mold infections. Antimicrob Agents Chemother 66(7):e0038022

Gebremariam T, Gu Y, Singh S, Kitt TM, Ibrahim AS (2021) Combination treatment of liposomal amphotericin B and isavuconazole is synergistic in treating experimental mucormycosis. J Antimicrob Chemother 76(10):2636–2639

Georgacopoulos O, Nunnally NS, Ransom EM, Law D, Birch M, Lockhart SR, Berkow EL (2021) In vitro activity of novel antifungal olorofim against filamentous fungi and comparison to eight other antifungal agents. J Fungi (Basel) 7(5):378

Gilaberte Y, Aspiroz C, Alejandre MC, Andres-Ciriano E, Fortuño B, Charlez L et al (2014) Cutaneous sporotrichosis treated with photodynamic therapy: an *in vitro* and *in vivo* study. Photomed Laser Surg 32(1):54–57

Gil-Pallares P, Gracia-Cazaña T, Álvarez-Salafranca M, Gorgojo MA, García-García M, Beltrán-Rosel A et al (2024) Cutaneous alternariosis in immunosuppressed patients treated with photodynamic therapy and oral antifungals, a synergistic strategy. Pharmaceuticals (Basel) 17(2):245

Gonçalves S, Silva PM, Felício MR, de Medeiros LN, Kurtenbach E, Santos NC (2017) Psd1 effects on *Candida albicans* planktonic cells and biofilms. Front Cell Infect Microbiol 7:249

Gupta AK, Versteeg SG, Shear NH (2017) Onychomycosis in the 21st century: an update on diagnosis, epidemiology, and treatment. J Cutan Med Surg 21(6):525–539

Hamblin MR, Hasan T (2004) Photodynamic therapy: a new antimicrobial approach to infectious disease? Photochem Photobiol Sci 3(5):436–450

Hamblin MR (2016) Antimicrobial photodynamic inactivation: a bright new technique to kill resistant microbes. Curr Opin Microbiol 33:67–73

Harding MW, Marques LL, Howard RJ, Olson ME (2009) Can filamentous fungi form biofilms? Trends Microbiol 17(11):475–480

Hodges MR, Ople E, Shaw KJ, Mansbach R, Van Marle SJ, Van Hoogdalem E et al (2017) First-in-human study to assess safety, tolerability and pharmacokinetics of APX001 administered by intravenous infusion to healthy subjects. OFID 4:S526

Hoenigl M, Sprute R, Egger M, Arastehfar A, Cornely OA, Krause R et al (2021) The antifungal pipeline: Fosmanogepix, Ibrexafungerp, Olorofim, Opelconazole, and Rezafungin. Drugs 81(15):1703–1729

Horvat S, Yu Y, Böjte S, Teßmer I, Lowman DW, Ma Z et al (2020) Engineering nanogels for drug delivery to pathogenic fungi *Aspergillus fumigatus* by tuning polymer amphiphilicity. Biomacromolecules 21(8):3112–3121

Hou Y, Yang M, Li J, Bi X, Li G, Xu J et al (2021) The enhancing antifungal effect of AD1 aptamer-functionalized amphotericin B-loaded PLGA-PEG nanoparticles with a low-frequency and low-intensity ultrasound exposure on *C. albicans* biofilm through targeted effect. NanoImpact 21:100275

Hu Q, Li T, Yang J, Peng Y, Liu Q, Liu N (2023) Efficacy of photodynamic therapy in the treatment of oral candidiasis: a systematic review and meta-analysis. BMC Oral Health 23(1):802

Hu Y, Qi X, Sun H, Lu Y, Hu Y, Chen X et al (2019) Photodynamic therapy combined with antifungal drugs against chromoblastomycosis and the effect of ALA-PDT on *Fonsecaea in vitro*. PLoS Negl Trop Dis 13(10):e0007849

Ibbotson SH, Wong TH, Morton CA, Collier NJ, Haylett A, McKenna KE et al (2019) Adverse effects of topical photodynamic therapy: a consensus review and approach to management. Br J Dermatol 180(4):715–729

Ibrahim AS, Gebermariam T, Fu Y, Lin L, Husseiny MI, French SW et al (2007) The iron chelator deferasirox protects mice from mucormycosis through iron starvation. J Clin Invest 117(9):2649–2657

Imamovic L, Sommer MO (2013) Use of collateral sensitivity networks to design drug cycling protocols that avoid resistance development. Sci Transl Med 5(204):204ra132

Ji J, Liang CT, Zhong JJ, Kong X, Xu HX, Xu CC et al (2024) 5-aminolevulinic acid-based photodynamic therapy in combination with antifungal agents for adult kerion and facial ulcer caused by *Trichophyton rubrum*. Photodiagn Photodyn Ther 45:103954

Jia G, Hu J, Tan L, Li L, Gao L, Sun Y (2023) *In vitro* and *in vivo* evaluation of synergistic effects of Everolimus in combination with antifungal agents on *Exophiala dermatitidis*. Microbiol Spectr 11(3):e0530222

Jørgensen KM, Astvad KMT, Arendrup MC (2020) *In vitro* activity of Manogepix (APX001A) and comparators against contemporary molds: MEC comparison and preliminary experience with colorimetric MIC determination. Antimicrob Agents Chemother 64(8):e00730–e00720

Kharkwal GB, Sharma SK, Huang YY, Dai T, Hamblin MR (2011) Photodynamic therapy for infections: clinical applications. Lasers Surg Med 43(7):755–567

Kimani BG, Kerekes EB, Szebenyi C, Krisch J, Vágvölgyi C, Papp T et al (2021) *In vitro* activity of selected phenolic compounds against planktonic and biofilm cells of food-contaminating yeasts. Food Secur 10(7):1652

Kimura G, Nakaoki T, Colley T, Rapeport G, Strong P, Ito K et al (2017) In vivo biomarker analysis of the effects of intranasally dosed PC945, a novel antifungal triazole, on *Aspergillus fumigatus* infection in immunocompromised mice. Antimicrob Agents Chemother 61(9):e00124–e00117

Kirchhoff L, Dittmer S, Buer J, Rath PM, Steinmann J (2020) *In vitro* activity of olorofim (F901318) against fungi of the genus, *Scedosporium* and *Rasamsonia* as well as against *Lomentospora prolificans*, *Exophiala dermatitidis* and azole-resistant *Aspergillus fumigatus*. Int J Antimicrob Agents 56(3):106105

Kirchhoff L, Dittmer S, Furnica DT, Buer J, Steinmann E, Rath PM et al (2022) Inhibition of azole-resistant *Aspergillus fumigatus* biofilm at various formation stages by antifungal drugs, including olorofim. J Antimicrob Chemother 77(6):1645–1654

Kirchhoff L, Olsowski M, Zilmans K, Dittmer S, Haase G, Sedlacek L et al (2017) Biofilm formation of the black yeast-like fungus *Exophiala dermatitidis* and its susceptibility to antiinfective agents. Sci Rep 7:42886

Konsoula A, Tsioutis C, Markaki I, Papadakis M, Agouridis AP, Spernovasilis N (2022) *Lomentospora prolificans*: an emerging opportunistic fungal pathogen. Microorganisms 10(7):1317

Kullberg BB, Pappas L, Vazquez PG, Goosen JA, Arauz J, Paruk AB et al (2021) Clinical efficacy and safety of the novel antifungal fosmanogepix in patients with candidaemia and/or invasive candidiasis caused by Candida auris: results from a phase II proof of concept trial. 31st ECCMID, Vienna

Kyvernitakis A, Torres HA, Jiang Y, Chamilos G, Lewis RE, Kontoyiannis DP (2016) Initial use of combination treatment does not impact survival of 106 patients with haematologic malignancies and mucormycosis: a propensity score analysis. Clin Microbiol Infect 22(9):811.e1–811.e8

Lan Y, Lu S, Zhang J (2021) Retinoid combined with photodynamic therapy against hyperkeratotic chromoblastomycosis: a case report and literature review. Mycoses 64(1):18–23

Lee Y, Robbins N, Cowen LE (2023) Molecular mechanisms governing antifungal drug resistance. NPJ Antimicrob Resist 1(1):5

Legabão BC, Fernandes JA, de Oliveira Barbosa GF, Bonfim-Mendonça PS, Svidzinski TIE (2022) The zoonosis sporotrichosis can be successfully treated by photodynamic therapy: a scoping review. Acta Trop 228:106341

Leung AK, Barankin B, Lam JM, Leong KF, Hon KL (2023) Tinea pedis: an updated review. Drugs Context 12:2023-5-1

Lewis RE, Ben-Ami R, Best L, Albert N, Walsh TJ, Kontoyiannis DP (2013) Tacrolimus enhances the potency of posaconazole against *Rhizopus oryzae in vitro* and in an experimental model of mucormycosis. J Infect Dis 207(5):834–841

Li DM, de Hoog GS (2009) Cerebral phaeohyphomycosis–a cure at what lengths? Lancet Infect Dis 9(6):376–383

Liu S, Le Mauff F, Sheppard DC, Zhang S (2022) Filamentous fungal biofilms: conserved and unique aspects of extracellular matrix composition, mechanisms of drug resistance and regulatory networks in *Aspergillus fumigatus*. NPJ Biofilms Microbiomes 8(1):83

Lyon JP, Moreira LM, de Carvalho VS, dos Santos FV, de Lima CJ, de Resende MA (2013) *In vitro* photodynamic therapy against *Foncecaea pedrosoi* and *Cladophialophora carrionii*. Mycoses 56(2):157–161

Maertens JA, Rahav G, Lee DG, Ponce-de-León A, Ramírez Sánchez IC, Klimko N et al (2021) Posaconazole versus voriconazole for primary treatment of invasive aspergillosis: a phase 3, randomised, controlled, non-inferiority trial. Lancet 397(10273):499–509

Manohar P, Loh B, Nachimuthu R, Leptihn S (2024) Phage-antibiotic combinations to control *Pseudomonas aeruginosa-Candida* two-species biofilms. Sci Rep 14(1):9354

Markantonatou AM, Samaras K, Vyzantiadis TA (2023) Dermatophytic biofilms: characteristics, significance and treatment approaches. J Fungi (Basel) 9(2):228

Marr KA, Schlamm HT, Herbrecht R, Rottinghaus ST, Bow EJ, Cornely OA et al (2015) Combination antifungal therapy for invasive aspergillosis: a randomized trial. Ann Intern Med 162(2):81–89

Martinez LR, Casadevall A (2007) *Cryptococcus neoformans* biofilm formation depends on surface support and carbon source and reduces fungal cell susceptibility to heat, cold, and UV light. Appl Environ Microbiol 73(14):4592–4601

Maziarz EK, Perfect JR (2016) Cryptococcosis. Infect Dis Clin N Am 30(1):179–206

McLellan CA, Whitesell L, King OD, Lancaster AK, Mazitschek R, Lindquist S (2012) Inhibiting GPI anchor biosynthesis in fungi stresses the endoplasmic reticulum and enhances immunogenicity. ACS Chem Biol 7(9):1520–1528

Meena DS, Kumar D, Bohra GK (2023) Combination therapy in Mucormycosis: current evidence from the world literature, a mini review. J Mycol Med 33(1):101332

Mishra R, Panda AK, De Mandal S, Shakeel M, Bisht SS, Khan J (2020) Natural anti-biofilm agents: strategies to control biofilm-forming pathogens. Front Microbiol 11:566325

Miyazaki M, Horii T, Hata K, Watanabe NA, Nakamoto K, Tanaka K, Shirotori S, Murai N, Inoue S, Matsukura M, Abe S, Yoshimatsu K, Asada M (2011) In vitro activity of E1210, a novel antifungal, against clinically important yeasts and molds. Antimicrob Agents Chemother 55(10):4652–4658

Molloy SF, Kanyama C, Heyderman RS, Loyse A, Kouanfack C, Chanda D et al (2018) Antifungal combinations for treatment of cryptococcal meningitis in Africa. N Engl J Med 378(11):1004–1017

Monteiro DR, Pessan JP (2022) Nanocarriers of antifungal drugs: possible strategies to overcome antimicrobial resistance. Future Microbiol 17:1199–1202

Morelli KA, Kerkaert JD, Cramer RA (2021) *Aspergillus fumigatus* biofilms: toward understanding how growth as a multicellular network increases antifungal resistance and disease progression. PLoS Pathog 17(8):e1009794

Mosallam FM, Helmy EA, El-Bastawisy HS, El-Batal AI (2024) Silver secnidazole nano-hybrid emulsion-based probiotics as a novel antifungal formula against multidrug-resistant vaginal pathogens. Biotechnol Appl Biochem 72:295. https://doi.org/10.1002/bab.2663. Epub ahead of print

Mowat E, Williams C, Jones B, McChlery S, Ramage G (2009) The characteristics of *Aspergillus fumigatus* mycetoma development: is this a biofilm? Med Mycol 47(Suppl 1):S120–S126

Mroczyńska M, Brillowska-Dąbrowska A (2020) Review on current status of echinocandins use. Antibiotics (Basel) 9(5):227

Murray A, Cass L, Ito K, Pagani N, Armstrong-James D, Dalal P et al (2020) PC945, a novel inhaled antifungal agent, for the treatment of respiratory fungal infections. J Fungi (Basel) 6(4):373

Najahi-Missaoui W, Arnold RD, Cummings BS (2020) Safe nanoparticles: are we there yet? Int J Mol Sci 22(1):385

Nazik H, Joubert LM, Secor PR, Sweere JM, Bollyky PL, Sass G et al (2017) *Pseudomonas* phage inhibition of *Candida albicans*. Microbiology (Reading) 163(11):1568–1577

Nazzaro F, Fratianni F, Coppola R, Feo VD (2017) Essential oils and antifungal activity. Pharmaceuticals (Basel) 10(4):86

Negri CE, Johnson A, McEntee L, Box H, Whalley S, Schwartz JA et al (2018) Pharmacodynamics of the novel antifungal agent F901318 for acute sinopulmonary aspergillosis caused by *Aspergillus flavus*. J Infect Dis 217(7):1118–1127

Newman DJ (2017) The influence of brazilian biodiversity on searching for human use pharmaceuticals. J Braz Chem Soc 28(3):402–414

O'Connor L, Livermore J, Sharp AD, Goodwin J, Gregson L, Howard SJ et al (2013) Pharmacodynamics of liposomal amphotericin B and flucytosine for cryptococcal meningoencephalitis: safe and effective regimens for immunocompromised patients. J Infect Dis 208(2):351–361

Oliver JD, Sibley GEM, Beckmann N, Dobb KS, Slater MJ, McEntee L et al (2016) F901318 represents a novel class of antifungal drug that inhibits dihydroorotate dehydrogenase. Proc Natl Acad Sci USA 113(45):12809–12814

Ong V, Ashbaugh A, Cushion MT (2020) Rezafungin prevention of pneumocystis pneumonia and pneumocystis reactivation using different doses and durations of prophylaxis in a mouse model. Biol Blood Marrow Transplant 26(3):328–329

Orlean P, Menon AK (2007) Thematic review series: lipid posttranslational modifications. GPI anchoring of protein in yeast and mammalian cells, or: how we learned to stop worrying and love glycophospholipids. J Lipid Res 48(5):993–1011

Orofino-Costa R, Macedo PM, Rodrigues AM, Bernardes-Engemann AR (2017) Sporotrichosis: an update on epidemiology, etiopathogenesis, laboratory and clinical therapeutics. An Bras Dermatol 92(5):606–620

Osei Sekyere J (2018) *Candida auris*: a systematic review and meta-analysis of current updates on an emerging multidrug-resistant pathogen. Microbiology 7(4):e00578

Oshiro KGN, Rodrigues G, Monges BED, Cardoso MH, Franco OL (2019) Bioactive peptides against fungal biofilms. Front Microbiol 10:2169

Oyardı Ö, Hacıoğlu M, Özdemir E, Erbay MŞ, Kültür Ş, Bozkurt GÇ (2024) Screening of antimicrobial, antibiofilm, and cytotoxic activities of some medicinal plants from Balıkesir Province, Türkiye: potential effects of *Allium paniculatum* flower. Turk J Pharm Sci 21(3):252–258

Pappas PG, Kullberg BJ, Vazquez JA, Oren I, Rahav G, Aoun M et al (2020) Clinical safety and efficacy of novel antifungal, Fosmanogepix, in the treatment of candidemia: results from a phase 2 proof of concept trial. IDWeek 2020. In: Programs and Abstracts of the 55th Annual Infectious Diseases Society of America (IDSA) Meeting

Park BJ, Wannemuehler KA, Marston BJ, Govender N, Pappas PG, Chiller TM (2009) Estimation of the current global burden of cryptococcal meningitis among persons living with HIV/AIDS. AIDS 23(4):525–530

Pastor FJ, Guarro J (2008) *Alternaria infections*: laboratory diagnosis and relevant clinical features. Clin Microbiol Infect 14(8):734–746

Penner JC, Ferreira JAG, Secor PR, Sweere JM, Birukova MK, Joubert LM et al (2016) Pf4 bacteriophage produced by *Pseudomonas aeruginosa* inhibits *Aspergillus fumigatus* metabolism via iron sequestration. Microbiology (Reading) 162(9):1583–1594

Perea S, Gonzalez G, Fothergill AW, Kirkpatrick WR, Rinaldi MG, Patterson TF (2002) *In vitro* interaction of caspofungin acetate with voriconazole against clinical isolates of *Aspergillus* spp. Antimicrob Agents Chemother 46(9):3039–3041

Petraitis V, Petraitiene R, Katragkou A, Maung BBW, Naing E, Kavaliauskas P et al (2020) Combination therapy with Ibrexafungerp (formerly SCY-078), a first-in-class triterpenoid inhibitor of (1→3)-β-d-glucan synthesis, and Isavuconazole for treatment of experimental invasive pulmonary aspergillosis. Antimicrob Agents Chemother 64(6):e02429–e02419

Pfaller M, Huband M, Bien PA, Carvalhaes CG, Klauer A, Castanheira M (2024) *In vitro* activity of manogepix and comparators against infrequently encountered yeast and mold isolates from the SENTRY Surveillance Program (2017-2022). Antimicrob Agents Chemother 68(2):e0113223

Pfaller MA, Messer SA, Rhomberg PR, Castanheira M (2017 Sep) CD101, a long-acting echinocandin, and comparator antifungal agents tested against a global collection of invasive fungal isolates in the SENTRY 2015 Antifungal Surveillance Program. Int J Antimicrob Agents 50(3):352–358

Plaetzer K, Krammer B, Berlanda J, Berr F, Kiesslich T (2009) Photophysics and photochemistry of photodynamic therapy: fundamental aspects. Lasers Med Sci 24(2):259–268

Qian W, Zhang J, Wang W, Liu M, Fu Y, Li X et al (2020) Efficacy of chelerythrine against mono- and dual-species biofilms of *Candida albicans* and *Staphylococcus aureus* and its properties of inducing hypha-to-yeast transition of *C. albicans*. J Fungi (Basel) 6(2):45

Qiao J, Li R, Ding Y, Fang H (2010) Photodynamic therapy in the treatment of superficial mycoses: an evidence-based evaluation. Mycopathologia 170(5):339–343

Raorane CJ, Lee J-H, Kim Y-G, Rajasekharan SK, García-Contreras R, Lee J (2019) Antibiofilm and antivirulence efficacies of flavonoids and curcumin against *Acinetobacter baumannii*. Front Microbiol 10:990

Rhein J, Huppler Hullsiek K, Tugume L, Nuwagira E, Mpoza E, Evans EE et al (2019) Adjunctive sertraline for HIV-associated cryptococcal meningitis: a randomised, placebo-controlled, double-blind phase 3 trial. Lancet Infect Dis 19(8):843–851

Rhein J, Morawski BM, Hullsiek KH, Nabeta HW, Kiggundu R, Tugume L et al (2016) Efficacy of adjunctive sertraline for the treatment of HIV-associated cryptococcal meningitis: an open-label dose-ranging study. Lancet Infect Dis 16(7):809–818

Rodvold KA, Yoo L, George JM (2011) Penetration of anti-infective agents into pulmonary epithelial lining fluid: focus on antifungal, antitubercular and miscellaneous anti-infective agents. Clin Pharmacokinet 50(11):689–704

Sakthi Devi R, Girigoswami A, Siddharth M, Girigoswami K (2022) Applications of gold and silver nanoparticles in theranostics. Appl Biochem Biotechnol 194(9):4187–4219

Samaddar A, Mendonsa JM, Sudhaharan S, Nagarathna S, Kindo AJ, Shetty A et al (2024) *In vitro* evaluation of antifungal combinations against neurotropic dematiaceous fungi associated with primary cerebral phaeohyphomycosis. Microbiol Spectr 12(8):e0078124

Sánchez-Herrera R, Flores-Villavicencio LL, Pichardo-Molina JL, Castruita-Domínguez JP, Aparicio-Fernández X, Sabanero López M et al (2021) Analysis of biofilm formation by *Sporothrix schenckii*. Med Mycol 59(1):31–40

Sardi JCO, Scorzoni L, Bernardi T, Fusco-Almeida AM, Mendes Giannini MJS (2013) *Candida* species: current epidemiology, pathogenicity, biofilm formation, natural antifungal products and new therapeutic options. J Med Microbiol 62(Pt 1):10–24

Saunte DML, Gaitanis G, Hay RJ (2020) *Malassezia*-associated skin diseases, the use of diagnostics and treatment. Front Cell Infect Microbiol 10:112

Sav H, Ozakkas F, Altınbas R, Kiraz N, Tümgör A, Gümral R et al (2016) Virulence markers of opportunistic black yeast in *Exophiala*. Mycoses 59(6):343–350

Seidler MJ, Salvenmoser S, Müller FM (2008) *Aspergillus fumigatus* forms biofilms with reduced antifungal drug susceptibility on bronchial epithelial cells. Antimicrob Agents Chemother 52(11):4130–4136

Seneviratne CJ, Fong PH, Wong SS, Lee VH (2015) Antifungal susceptibility and phenotypic characterization of oral isolates of a black fungus from a nasopharyngeal carcinoma patient under radiotherapy. BMC Oral Health 15:39

Sesana AM, Monti-Rocha R, Vinhas SA, Morais CG, Dietze R, Lemos EM (2011) *In vitro* activity of amphotericin B cochleates against *Leishmania chagasi*. Mem Inst Oswaldo Cruz 106(2):251–253

Shaw KJ, Ibrahim AS (2020) Fosmanogepix: a review of the first-in-class broad spectrum agent for the treatment of invasive fungal infections. J Fungi (Basel) 6(4):239

Shende P, Khair R, Gaud RS (2019) Nanostructured cochleates: a multi-layered platform for cellular transportation of therapeutics. Drug Dev Ind Pharm 45(6):869–881

Shukla M, Chandley P, Rohatgi S (2021) The role of B-cells and antibodies against *Candida* vaccine antigens in invasive candidiasis. Vaccines (Basel) 9(10):1159

Sigera LSM, Denning DW (2023) Flucytosine and its clinical usage. Ther Adv Infect Dis 10:20499361231161387

Singh R, Shivaprakash MR, Chakrabarti A (2011) Biofilm formation by zygomycetes: quantification, structure and matrix composition. Microbiology (Reading) 157(Pt 9):2611–2618

Singh S, Barbarino A, Youssef EG, Coleman D, Gebremariam T, Ibrahim AS (2023) Protective efficacy of anti-hyr1p monoclonal antibody against systemic candidiasis due to multi-drug-resistant *Candida auris*. J Fungi (Basel) 9(1):103. https://doi.org/10.3390/jof9010103

Singh S, Uppuluri P, Mamouei Z, Alqarihi A, Elhassan H, French S et al (2019) The NDV-3A vaccine protects mice from multidrug resistant *Candida auris* infection. PLoS Pathog 15(8):e1007460

Skipper CP, Atukunda M, Stadelman A, Engen NW, Bangdiwala AS, Hullsiek KH et al (2020) Phase I EnACT trial of the safety and tolerability of a novel oral formulation of Amphotericin B. Antimicrob Agents Chemother 64(10):e00838–e00820

Sotiriou E, Koussidou T, Patsatsi A, Apalla Z, Ioannides D (2009) 5-Aminolevulinic acid-photodynamic treatment for dermatophytic tinea pedis of interdigital type: a small clinical study. J Eur Acad Dermatol Venereol 23(2):203–204

Sousa IS, Mello TP, Pereira EP, Granato MQ, Alviano CS, Santos ALS et al (2022) Biofilm Formation by Chromoblastomycosis fungi *Fonsecaea pedrosoi* and *Phialophora verrucosa*: involvement with antifungal resistance. J Fungi (Basel) 8(9):963

Spellberg B, Ibrahim AS, Chin-Hong PV, Kontoyiannis DP, Morris MI, Perfect JR et al (2012) The deferasirox-AmBisome therapy for mucormycosis (DEFEAT Mucor) study: a randomized, double-blinded, placebo-controlled trial. J Antimicrob Chemother 67(3):715–722

Spitzer M, Robbins N, Wright GD (2017) Combinatorial strategies for combating invasive fungal infections. Virulence 8(2):169–185

Sun L, Liao K, Wang D (2015a) Effects of magnolol and honokiol on adhesion, yeast-hyphal transition, and formation of biofilm by *Candida albicans*. PLoS One 10(2):e0117695

Sun LM, Liao K, Liang S, Yu PH, Wang DY (2015b) Synergistic activity of magnolol with azoles and its possible antifungal mechanism against *Candida albicans*. J Appl Microbiol 118(4):826–838

Tan K, Brayshaw N, Tomaszewski K, Troke P, Wood N (2006) Investigation of the potential relationships between plasma voriconazole concentrations and visual adverse events or liver function test abnormalities. J Clin Pharmacol 46(2):235–243

Tartor YH, Elmowalid GA, Hassan MN, Shaker A, Ashour DF, Saber T (2022) Promising anti-biofilm agents and phagocytes enhancers for the treatment of *Candida albicans* biofilm-associated infections. Front Cell Infect Microbiol 12:807218

Thakur M, Revankar SG (2011) *In vitro* interaction of caspofungin and immunosuppressives against agents of mucormycosis. J Antimicrob Chemother 66(10):2312–2314

Thompson GR 3rd, Lewis JS 2nd (2010) Pharmacology and clinical use of voriconazole. Expert Opin Drug Metab Toxicol 6(1):83–94

Thompson GR 3rd, Soriano A, Honore PM, Bassetti M, Cornely OA, Kollef M et al (2024) Efficacy and safety of rezafungin and caspofungin in candidaemia and invasive candidiasis: pooled data from two prospective randomised controlled trials. Lancet Infect Dis 24(3):319–328

Tóth Z, Forgács L, Locke JB, Kardos G, Nagy F, Kovács R et al (2019) *In vitro* activity of rezafungin against common and rare Candida species and *Saccharomyces cerevisiae*. J Antimicrob Chemother 74(12):3505–3510

Ul Haq I, Maryam S, Shyntum DY, Khan TA, Li F (2024) Exploring the frontiers of therapeutic breadth of antifungal peptides: a new avenue in antifungal drugs. J Ind Microbiol Biotechnol 51:kuae018

Uppuluri P, Nett J, Heitman J, Andes D (2008) Synergistic effect of calcineurin inhibitors and fluconazole against *Candida albicans* biofilms. Antimicrob Agents Chemother 52(3):1127–1132

Urayama SI, Zhao YJ, Kuroki M, Chiba Y, Ninomiya A, Hagiwara D (2024) Greetings from virologists to mycologists: a review outlining viruses that live in fungi. Mycoscience 65(1):1–11

van de Sande WWJ, Vonk AG (2019) Mycovirus therapy for invasive pulmonary aspergillosis? Med Mycol 57(Suppl_2):S179–S188

Vanzolini T, Bruschi M, Rinaldi AC, Magnani M, Fraternale A (2022) Multitalented synthetic antimicrobial peptides and their antibacterial, antifungal and antiviral mechanisms. Int J Mol Sci 23(1):545

Vera-González N, Shukla A (2020) Advances in biomaterials for the prevention and disruption of *Candida* biofilms. Front Microbiol 11:538602

Verekar R, Desai S, Ayyanar M, Nadaf S, Gurav S (2024) Nanocochleates: revolutionizing lipid-based drug delivery with enhanced bioavailability, a review. Hybrid Adv 6:100215

Verwer PE, van Duijn ML, Tavakol M, Bakker-Woudenberg IA, van de Sande WW (2012) Reshuffling of Aspergillus fumigatus cell wall components chitin and β-glucan under the influence of caspofungin or nikkomycin Z alone or in combination. Antimicrob Agents Chemother 56(3):1595–1598

Vieira APM, Arias LS, de Souza Neto FN, Kubo AM, Lima BHR, de Camargo ER et al (2019) Antibiofilm effect of chlorhexidine-carrier nanosystem based on iron oxide magnetic nanoparticles and chitosan. Colloids Surf B Biointerfaces 174:224–231

Viriyakosol S, Kapoor M, Okamoto S, Covel J, Soltow QA, Trzoss M et al (2019) APX001 and other Gwt1 inhibitor prodrugs are effective in experimental *Coccidioides immitis* pneumonia. Antimicrob Agents Chemother 63(2):e01715–e01718

Watanabe NA, Miyazaki M, Horii T, Sagane K, Tsukahara K, Hata K (2012) E1210, a new broad-spectrum antifungal, suppresses *Candida albicans* hyphal growth through inhibition of glycosylphosphatidylinositol biosynthesis. Antimicrob Agents Chemother 56(2):960–971

Whitesell L, Robbins N, Huang DS, McLellan CA, Shekhar-Guturja T, LeBlanc EV et al (2019) Structural basis for species-selective targeting of Hsp90 in a pathogenic fungus. Nat Commun 10(1):402

Wiederhold NP, Locke JB, Daruwala P, Bartizal K (2018) Rezafungin (CD101) demonstrates potent *in vitro* activity against *Aspergillus*, including azole-resistant *Aspergillus fumigatus* isolates and cryptic species. J Antimicrob Chemother 73(11):3063–3067

Wiederhold NP, Najvar LK, Shaw KJ, Jaramillo R, Patterson H, Olivo M et al (2019) Efficacy of delayed therapy with fosmanogepix (APX001) in a murine model of *Candida auris* invasive candidiasis. Antimicrob Agents Chemother 63(11):e01120–e01119

Wijaya M, Halleyantoro R, Kalumpiu JF (2023) Biofilm: the invisible culprit in catheter-induced candidemia. AIMS Microbiol 9(3):467–485

Wollina U, Bitel A, Vojvodic A, Lotti T (2019) Rosacea flare - up after photodynamic therapy (PDT) for field cancerization and a review on adverse events with PDT in general. Open Access Maced J Med Sci 7(18):2998–3001

Wood KB, Wood KC, Nishida S, Cluzel P (2014) Uncovering scaling laws to infer multidrug response of resistant microbes and cancer cells. Cell Rep 6(6):1073–1084

Wu X, Hu Y (2022) Photodynamic therapy for the treatment of fungal infections. Infect Drug Resist 15:3251–3266

Xiong WH, Brown RL, Reed B, Burke NS, Duvoisin RM, Morgans CW (2015) Voriconazole, an antifungal triazol that causes visual side effects, is an inhibitor of TRPM1 and TRPM3 channels. Invest Ophthalmol Vis Sci 56(2):1367–1373

Yi X, Fransisca C, He Y, Liu Y, Lu S, He L et al (2017) Photodynamic effects on *Fonsecaea monophora* conidia and RAW264.7 *in vitro*. J Photochem Photobiol B 176:112–117

Zarif L, Graybill JR, Perlin D, Mannino RJ (2000) Cochleates: new lipid-based drug delivery system. J Liposome Res 10(4):523–538

Zarif L (2005) Drug delivery by lipid cochleates. Methods Enzymol 391:314–329

Zhai B, Wu C, Wang L, Sachs MS, Lin X (2012) The antidepressant sertraline provides a promising therapeutic option for neurotropic cryptococcal infections. Antimicrob Agents Chemother 56(7):3758–3766

Zhao M, Lepak AJ, Marchillo K, Vanhecker J, Sanchez H, Ambrose PG et al (2019) APX001 Pharmacokinetic/pharmacodynamic target determination against *Aspergillus fumigatus* in an *in vivo* model of invasive pulmonary aspergillosis. Antimicrob Agents Chemother 63(4):e02372–e02318

Zhu P, Li Y, Guo T, Liu S, Tancer RJ, Hu C et al (2023) New antifungal strategies: drug combination and co-delivery. Adv Drug Deliv Rev 198:114874

Zimmermann GR, Lehár J, Keith CT (2007) Multi-target therapeutics: when the whole is greater than the sum of the parts. Drug Discov Today 12(1–2):34–42

Chapter 12
Large-Scale Phenotypic Screenings of Repurposing Libraries to Identify Drugs with Novel Antifungal Activity Against *Candida* Biofilms

Augusto Vazquez-Rodriguez, Celia F. Rodrigues, Gordon Ramage, and Jose L. Lopez-Ribot

Abstract Biofilm formation is a defining feature of many different clinical manifestations of candidiasis, as a majority of *Candida* species are fully capable of forming adherent cell populations. The high levels of mortality detected with *Candida* infections associated with a biofilm etiology make it imperative to develop new therapeutic options. Repurposing, finding new uses for old drugs, can significantly reduce the effort, time and money required in drug development, and these repositioning approaches are especially appealing in disease areas with high unmet needs and a scarcity of novel drug candidates, which is the case in the antifungal space. The availability of repurposing libraries has greatly facilitated repurposing studies by allowing the large-scale screening of hundreds to thousands of drugs in an efficient manner. Here, we present a comprehensive review of these efforts in the search for drugs with novel anti-biofilm activity against *Candida* species.

A. Vazquez-Rodriguez · J. L. Lopez-Ribot (✉)
Department of Molecular Microbiology & Immunology and South Texas Center for Emerging Infectious Diseases, The University of Texas at San Antonio, San Antonio, TX, USA
e-mail: jesus.vazquezrodriguez@utsa.edu; jose.lopezribot@utsa.edu

C. F. Rodrigues
Associate Laboratory i4HB-Institute for Health and Bioeconomy, University Institute of Health Sciences-CESPU, Gandra, Portugal

UCIBIO-Applied Molecular Biosciences Unit, Translational Toxicology Research Laboratory, University Institute of Health Sciences (1H-TOXRUN, IUCS-CESPU), Porto, Portugal

ALiCE-LEPABE-Laboratory for Process Engineering, Environment, Biotechnology and Energy, University of Porto, Porto, Portugal

G. Ramage
Safeguarding Health through Infection Prevention (SHIP) Research Group, Research Centre for Health, School of Health and Life Sciences, Glasgow Caledonian University, Glasgow, UK
e-mail: gordon.ramage@gcu.ac.uk

G. Ramage, D. Andes (eds.), *Fungal Biofilms*, Springer Series on Biofilms 15,
https://doi.org/10.1007/978-3-031-93315-8_12

Introduction: *Candida* Species as Etiological Agents of Candidiasis, and Candidiasis as a Biofilm-Associated Infection

Candida species constitute the most frequent etiological agents of opportunistic fungal invasive infections, and candidiasis now represents the third-to-fourth most frequent nosocomial infection worldwide (Brown et al. 2012; Kohler et al. 2017; Quindos et al. 2018; Tsay et al. 2020), carrying unacceptably high mortality rates of 30–60%, which have alarmingly remained mostly unchanged during the last few decades (Gudlaugsson et al. 2003; Koehler et al. 2019). As normal commensals of humans, *Candida* species are typically considered opportunistic pathogenic yeasts but can take advantage of medically and immune-compromised patients (Thomas-Ruddel et al. 2022). Although *Candida albicans* remains the most common causative agent of candidiasis, in recent years, there has been an alarming trend of an increasing number of infections caused by non-*albicans Candida* spp. such as *C. glabrata* (now *Nakaseomyces glabratus*), *C. parapsilosis*, *C. tropicalis*, *C. krusei* and most recently, the emerging pathogen *C. auris*, which together now account for approximately half the cases of candidiasis, also depending on the geographical area, and many times are associated with increased levels of resistance (Quindos et al. 2018; Kean et al. 2020; Pfaller and Diekema 2007).

Understanding the risk factors for candidiasis is essential for identifying individuals who may be at an increased risk of developing these infections, and efforts to prevent and treat these infections should be tailored to these high-risk populations, including the appropriate use of preventive and infection control measures. One of the most significant risk factors for candidiasis is immunosuppression. Individuals with weakened immune systems, such as those with HIV/AIDS, cancer, diabetes, corticosteroid use or undergoing organ transplantation, are more susceptible to these opportunistic infections. The breakdown of anatomical barriers such as the skin, a number of surgical procedures, and staying in the intensive care unit (ICU) also increase susceptibility to invasive candidiasis. Furthermore, the use of broad-spectrum antibiotics can lead to the overgrowth of opportunistic *Candida* species (Thomas-Ruddel et al. 2022).

Besides the expanding number of immune- and medically compromised patients, major factors contributing to the high morbidity and mortality rates associated with candidiasis are problems with its (quick) diagnosis, the limited arsenal of antifungal drugs (see below), as well as the increasing levels of antifungal drug resistance (Wall and Lopez-Ribot 2020a). In addition, most infections attributed to *Candida* species involve biofilms, which further contribute to the pathogenesis of candidiasis and greatly complicates treatment of these patients (Ramage et al. 2006, 2023). Over the past few decades, research has consistently shown that species in this genus can adhere to both biological and non-biological surfaces, leading to robust biofilm formation. Among pathogenic fungi, *Candida* species are most often associated with biofilm development, with *C. albicans* displaying the absolute highest biofilm-forming ability (Ramage et al. 2023; Cuellar-Cruz et al. 2012). *Candida* biofilms consist of dense cell clusters anchored to a surface and surrounded by a matrix of self-secreted exo-polymeric substances, creating an ideal microenvironment for nutrient uptake and waste removal (Ramage et al. 2023). Moreover, cells dispersed from biofilms are responsible for candidemia and dissemination to deep organs leading to the establishment of invasive disease (Uppuluri et al. 2010; Uppuluri and Lopez-Ribot 2016). Overall, *Candida* biofilms significantly contribute to the persistence and severity of candidiasis, making treatment much more challenging (Ramage et al. 2006; Rajendran et al. 2016).

Current Antifungal Drugs for the Treatment of Candidiasis and Their Activity Against *Candida* Biofilms

Fungi are eukaryotic organisms, and as such, there is a paucity of selective targets for antifungal drug development (Wall and Lopez-Ribot 2020a). Therefore, the current antifungal arsenal is exceedingly limited, and for the clinical treatment of candidiasis (and other invasive fungal infections) effectively restricted to three classes of antifungals: polyenes, azoles and echinocandins (Wall and Lopez-Ribot 2020a).

Polyenes are the oldest antifungal class, with their origins tracing back to the 1950s, when amphotericin B became the first FDA-approved drug to treat invasive fungal infections (Odds et al. 2003). Polyenes act as "sponges," removing ergosterol from the fungal cell membrane, leading to leakage of cellular contents and ultimately cell death (Anderson et al. 2014). These drugs are fungicidal and exhibit broad-spectrum antifungal activity. However, polyenes are known for their toxicity, particularly nephrotoxicity, severely limiting their therapeutic use. The azole class emerged as the next major development in antifungals over the next few decades (Odds et al. 2003). Azoles, which block ergosterol synthesis by inhibiting the cytochrome P-450 14-alpha lanosterol demethylase, are primarily fungistatic. Introduced in the 1980s, fluconazole quickly became a preferred treatment for *Candida* infections, favored for its safety profile over amphotericin B (Wall and Lopez-Ribot 2020a; Odds et al. 2003). Yet, resistance issues soon arose, particularly with the emergence of non-*albicans Candida* species, complicating fluconazole's use in the treatment of candidiasis. Finally, the echinocandins entered clinical use in the early 2000s, with caspofungin being the first FDA-approved drug in this class (Odds et al. 2003). These semisynthetic lipopeptide antibiotics work by inhibiting 1,3-β-D-glucan synthase, an enzyme crucial for the biosynthesis glucan, a key structural component of the *Candida* cell wall and absent in mammalian cells. As such, echinocandins display strong fungicidal activity against most *Candida* species and an excellent safety profile, making them first-line treatment for candidemia (Wall and Lopez-Ribot 2020a).

From the early days of biofilm research, it became evident that *Candida* biofilms exhibit increased recalcitrance to most antifungal drugs (Ajetunmobi et al. 2023a). Studies revealed the intrinsic resistance to azoles, particularly fluconazole, with minimal inhibitory concentrations up to 100–1000 times higher for biofilm than for planktonic cells (Ramage et al. 2001). Amphotericin B also showed around a ten-fold reduction in effectiveness against biofilms compared to planktonic cells, with the relatively high doses required to target biofilms being considered toxic and unsafe (Ramage et al. 2001). Later, research found that liposomal formulations of amphotericin B improved its biofilm efficacy, likely due to enhanced penetration through the biofilm matrix (Kuhn et al. 2002; Rodrigues and Henriques 2017). The development of echinocandins coincided with early biofilm research, where laboratory studies demonstrated that therapeutic levels of echinocandins could effectively combat *C. albicans* biofilms, findings that were then extended to additional *Candida* species (Kuhn et al. 2002; Bachmann et al. 2002). Overall, the decreased susceptibility of *Candida* biofilms to a majority of currently available antifungal agents makes it imperative to search for and develop alternative therapeutic strategies to combat the threat of biofilm resistance (Ajetunmobi et al. 2023a).

Drug Repurposing Compared to Traditional Drug Discovery and Development

The traditional route of discovering and developing new drugs by searching for entirely new chemical matter—referred to as de novo drug discovery—remains a dominant but highly resource-intensive and time-consuming pathway, often with overall low success due to high attrition rates (Hughes et al. 2011). Typically, advancing a drug from discovery to full market approval can span up to two decades and incur estimated costs approaching $2 billion (Sertkaya et al. 2024) (Fig. 12.1). Furthermore, the probability of achieving success through clinical trials remains quite low, with only about 5% of drug candidates successfully completing clinical trials. Largely due to these high costs and risks, most recently the pace of antifungal research and development (R&D) has slowed significantly within large pharmaceutical corporations (Wall and Lopez-Ribot 2020a). Instead, the responsibility of innovating new antifungal treatments has largely shifted to smaller biotech companies, which have more restricted budgets and only a limited selection of agents at various stages of the research and development pipeline (Wall and Lopez-Ribot 2020a). Instead, and in stark contrast, drug repurposing (or repositioning) has emerged as a cost-effective, faster approach to uncover new uses for established or previously shelved compounds, including through completely new indications or through novel commercial prospects for already commercialized drugs (Farha and Brown 2019). Repurposing shortens development timelines by leveraging existing safety, pharmacodynamics and pharmacokinetic data, especially for already approved drugs as well as for clinically evaluated compounds (those having completed multiple clinical trial stages) (Pushpakom et al. 2019). Therefore, with known safety profiles and established pharmacodynamics, repurposed drugs can often bypass much of the discovery, development and preclinical stages and promote a

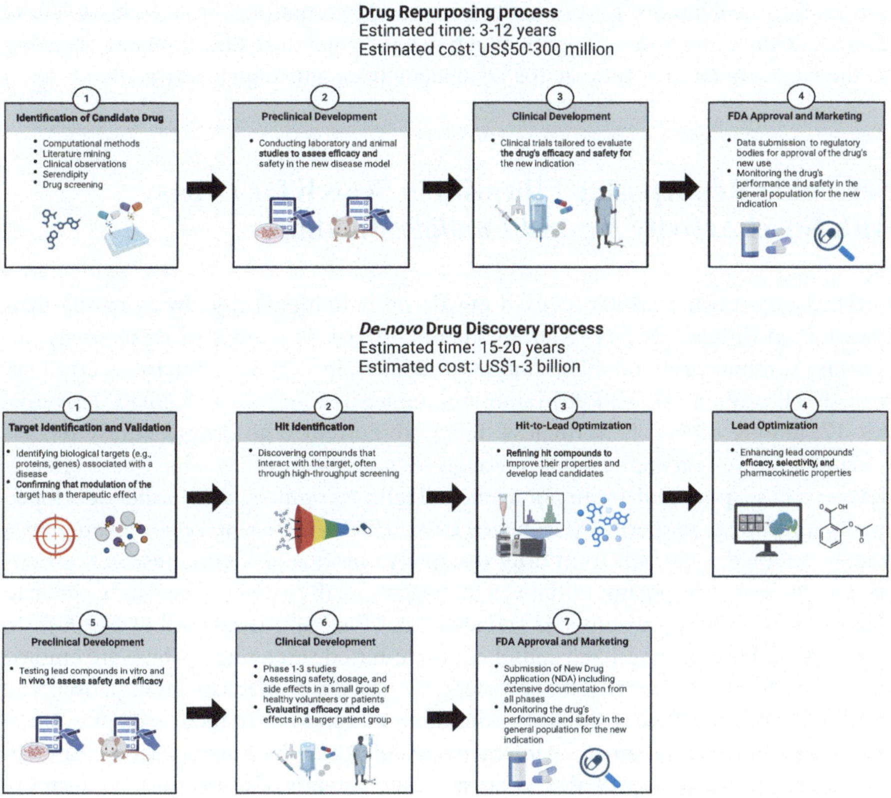

Fig. 12.1 Comparison between de novo drug discovery/development and drug repurposing

quick transition to clinical trial phases. Moreover, a leading repositionable candidate can often avoid the need for extensive Phase I safety studies and enter clinical trials already at Phase II. If the subsequent Phase III is successful, it should allow for expedited review and, if approved, rapid integration into healthcare, thereby creating a potentially life-saving impact for patients (Pushpakom et al. 2019). This could be specially appealing in case of multi-drug resistant and emerging pathogens such as *C. auris (*Kean et al. 2020*;* Sherry et al. 2017). Although repurposing approaches have been common practice and are by no means new, more recently repurposing efforts, strategies and successes in different therapeutic fields have been increasing in numbers, which has brought drug repurposing to the forefront as a promising strategy. Thus, overall, drug repurposing should represent a tempting and practical route for antifungal drug development, where the need for new therapeutic options is undeniable. Notably, repositioning is gaining traction as a strategic pathway in the antifungal arena with the potential to reduce costs while at the same time expedite development and approval timelines (Wall and Lopez-Ribot 2020b). Beyond the scientific benefits, drug repurposing offers an excellent opportunity for fostering collaboration and establishing new partnerships among academia, non-profit organizations, government entities and international organizations. These alliances can help address the critical gap in antifungal drug development, focusing on the urgent need to accelerate the advancement of antifungal treatments.

Screening Repurposing Libraries in Search for Drugs with Novel Activity Against *Candida* Biofilms

Original repurposing efforts focused mostly on individual drugs by assessing their potential antifungal activity (Katragkou et al. 2016). Examples of these are cyclosporine A, rapamycin and other "target of rapamycin" (TOR) inhibitors, sertraline, heat shock protein 90 (HSP90) inhibitors, tamoxifen and the celecoxib derivative AR-12, among others. Nevertheless, there are currently no drugs entirely repositioned for antifungal applications although we note that 5-flucytosine was originally developed as a potential antitumor agent and later recognized and further developed for its antifungal properties (Grunberg et al. 1963). In recent years, leveraging a widely adopted approach from drug discovery, various academic research groups have launched repurposing initiatives to systematically screen libraries in order to uncover new antifungal activities (reviewed in (Wall and Lopez-Ribot 2020b)). In most cases, these have utilized cell/phenotypic-based screenings, which are empirical by nature and do not require knowledge of targets or mechanism of action. The availability of repurposing libraries, consisting of hundreds to thousands of existing drugs and often accompanied by interactive databases, which are accessible through commercial vendors or provided as open-source compound collections by a variety of organizations (i.e. institutional facilities, governmental labs, non-for-profit organizations and even biotechnology and pharmaceutical companies), has greatly supported and facilitated these repurposing efforts (Wall and Lopez-Ribot 2020b). Among these, some have screened specifically for compounds with activity against *Candida* biofilms. Although early studies mostly screened for *C. albicans* biofilm inhibitors (Siles et al. 2013), recent screening efforts have also been directed to the emergent *C. auris*, due to the urgency in identifying new compounds effective against this multi-drug resistant species (Kean et al. 2020; Sherry et al. 2017).

Screening Techniques

Besides the availability of repurposing libraries, another major factor that has made it possible to implement these large-screening techniques has been the availability of relatively simple and highly reproducible methodologies for the formation of *Candida* biofilms. Although other techniques exist, typically these assays are based on the use of multi-well plates, where multiple, essentially equivalent, biofilms can be formed with relative ease. These are mostly adapted from the original description

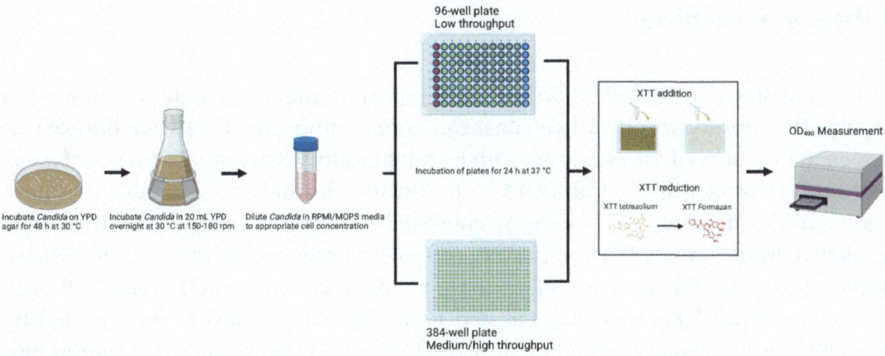

Fig. 12.2 Schematic diagram of multi-well-based models for Candida biofilm formation and their use in phenotype-based screenings

of a 96-well microtiter plate-based platform, which constituted a significant advance toward standardizing *Candida* biofilm formation and susceptibility testing (Fig. 12.2) (Ramage et al. 2001; Pierce et al. 2008). Briefly, in this method, biofilms are formed on the bottom of the wells of 96-well microtiter plates, which are then combined with a colorimetric readout to estimate the extent of biofilm formation. This technique is highly reproducible, fast, simple and cost-effective and can be easily adapted for its use in large-scale screenings. More recently, and mostly due to the inherent low throughput associated with the 96-well microtiter plate model, a similar methodology has been developed using 384-well microtiter plates which allows for true high throughput screening (Ajetunmobi et al. 2023b).

For the actual screening, individual library compounds are deposited within single wells of the microtiter plates, with final concentrations typically in the 10–20 micromolar range. The wells are seeded with a suspension of *Candida* yeast cells and incubated under suitable conditions to favor biofilm formation. Appropriate positive and negative controls are also included in each screening plate. At the end of the incubation period, the extent of biofilm formation is estimated using a colorimetric assay based on the metabolic activity of sessile cells within the biofilms, with results typically expressed as percent biofilm inhibition relative to the average of the control (uninhibited) wells within the same plate (Ajetunmobi et al. 2023b). Hit compounds are identified as those with inhibitory activity above a certain threshold, which can be set arbitrarily (i.e. >70% inhibition), or by using more advanced data analysis and statistical methods (i.e. Z- and B-scores). Validation of initial hits from the primary screening is typically performed by concentration-response assays using the same multi-well plate methodology, in order to confirm their anti-biofilm activity and at the same time establish their potency by calculating their IC_{50} value (defined as the concentration of the compound leading to 50% inhibition of biofilm formation) (Ajetunmobi et al. 2023b).

Library Screenings

In the first study in the scientific literature reporting the large-scale screening of a repurposing library specifically against *Candida* biofilms, the Prestwick library was screened in search of known drugs with novel inhibitory activity against *C. albicans* biofilm formation (Siles et al. 2013). The Prestwick library is available commercially, and at the time of this study, consisted of approximately 1200 diversified marketed drugs, mostly FDA-approved, off-patent drugs, covering a comprehensive range of targets, mechanisms of action and therapeutic/research areas, and with well-characterized pharmacological and toxicological properties. As reported by the authors, the primary screen identified a total of 38 compounds that inhibit biofilm formation by >50% in duplicate screenings. These initial hits were classified into three different classes including known antifungals, antibacterials/antiseptics and multifunctional drugs, which they described as drugs that have been used for other applications but with no previously known antifungal activity, and which represented the most interesting hits from a repurposing point of view. Among these, Auranofin, a gold thiol compound used to treat rheumatoid arthritis, constituted the most attractive repositionable drug. We note that the same Prestwick library was subsequently screened in search of inhibitors of *C. auris (*Wall et al. 2018*)*. Although the initial screen was performed under planktonic conditions, follow-up experiments focused on the anti-biofilm activity of some of the most promising confirmed hits. Among these, it was demonstrated that ebselen, a synthetic organoselenium compound, was able to inhibit biofilm formation in different clinical isolates of *C. auris* at concentrations that were comparable to those that inhibit planktonic cells (Wall et al. 2018).

Vila et al. screened The Pathogen Box for small molecule inhibitors of *C. albicans* biofilm formation (Vila and Lopez-Ribot 2017). The Pathogen Box is a diverse collection of approximately 400 drug-like compounds assembled by Medicines for Malaria Venture (MMV) with the objective to speed up drug discovery against neglected diseases. The initial screen identified seven hits that inhibit biofilm >50% formation, of which six were confirmed during subsequent dose-response confirmatory assays. Follow-up experiments indicated that three compounds, MMV688768, MMV687273 and MMV687807, also displayed activity against preformed biofilms. Of particular interest was the fact that compound MMV688768 exhibited increased activity against mature biofilms as compared to planktonic cultures, suggesting that this drug-like compound may affect processes with a preponderant function during the biofilm life-style. The Pathogen Box was subsequently also screened in search for drugs with activity against *C. auris*, under both biofilm and planktonic growth (Wall et al. 2019). These studies led to the identification of iodoquinol and miltefosine as the most promising compounds inhibiting growth of this emerging pathogen although only miltefosine was active against *C. auris* biofilms.

More recently, another collection from MMV, the "Pandemic Response Box" containing 400 diverse drug-like molecules active against bacteria, viruses or fungi, was screened in search for inhibitors of both *C. albicans* and *C. auris* biofilm formation (Ajetunmobi et al. 2023c). Initial hits were identified as those inhibiting biofilm growth by 70% or more in at least one of two duplicate assays, leading to the identification of 14 and 12 bioactive compounds against *C. albicans* and *C. auris*, respectively, several of which were active against both species. However, not surprisingly, a majority of hits were identified among the known antifungals present in this collection. Among the hit compounds with an original classification as antibacterials or antivirals (but not antifungals), MMV1633966 and MMV1593537 as well as everolimus were selected as the main repositionable leading compounds for further characterization of their antifungal activities.

In another study, the Repurposing, Focused Rescue, and Accelerated Medchem (ReFRAME) library from Calibr at Scripps Research was screened in search for existing compounds with novel antifungal activity and efficacy to prevent biofilm formation by *C. auris (*Wall et al. 2020*)*. Compounds within this library have been assembled by the combination of three databases (GVK Excelra GoStar, Clarivate Integrity and Citeline Pharmaprojects) for fast-track drug discovery, and the library contains approximately 13,000 high-value molecules, representing the largest collection of its kind. Included in this library are investigational new drugs at several stages of clinical development (~58%), US Food and Drug Administration (FDA)-approved/registered drugs (~35%) and approximately 3% of preclinical compounds with available safety data (Janes et al. 2018). As such, most of the compounds in this collection are non-approved therapeutics which offer excellent opportunities for repurposing. The initial screen was performed at a relatively low concentration of 5 μM, and after normalization of the data, the initial threshold for hit identification was established at 40%. A total of 100 compounds satisfied this criterion and were initially identified as hits, which resulted in an initial hit rate of 0.83%. Only 26 out of 100 compounds were confirmed in subsequent concentration-response validation experiments. As expected, a majority of confirmed hits were drugs with already known activity as antifungals or antiseptics. Only five drugs, Miltefosine, AM-24 (2,4,6-triiodophenol), Provecta (rose bengal disodium), Tazomeline and Lonafarnib, were identified as potentially repositionable compounds. More recently, in studies that remain unpublished at the time of this writing, the same ReFRAME library was screened by our group in search of drugs with novel anti-biofilm activity specifically against *C. albicans*. The primary screening was conducted in true High Throughput Screening (HTS) fashion utilizing the newly developed 384-well microtiter plate model of *Candida* biofilm formation. Plates for the initial screening contained individual library compounds each pre-spotted in single wells of the 384 plates. The smaller volume required as compared to the regular 96-well plates, allowed as to screen at a higher initial concentration of 10 micromolar. These "assay-ready" plates were then processed in the laboratory by adding the suspension of yeast cells, incubating under biofilm-forming conditions and developing with the XTT-colorimetric assay. Confirmation of the inhibitory activity of initial hits was performed using the same 384-well model but using an eight-point, 1:3 dilution dose-response format. Additional information on the screening conditions, initial and confirmed hits from this particular screening can be found by searching the associated ReFRAME Database, an open and extendable drug repurposing data, accessible through https://reframedb.org/assays/A00354.

The Ramage group screened the commercially available Tocriscreen 2.0 Micro library, consisting of a total of 1280 biologically active molecules from the Tocris catalog, in search of inhibitors of both *C. albicans* and *C. auris* biofilm formation (Abduljalil et al. 2022). The primary screen identified 90 initial hits against *C. albicans* and 87 against *C. auris*, with 64 overlapping hits against both species. A total of five leading compounds, Polygodial, Darapladib, KHS101, M62812 and Toyocamycin, were selected for further characterization of their biofilm inhibitory efficacy, which revealed their activity against biofilms formed by other *Candida* species, including *C. glabrata* and *C. krusei*. Toyocomycin was identified as the most potent overall inhibitor of *Candida* biofilms. Follow-up in vitro studies also demonstrated the potential of some of the leading hits to synergize with currently available antifungals, with a notable anti-biofilm synergy detected for the combination of darapladib with caspofungin.

Most recently, a high throughput screening of the Repurposing Hub Library using the 384-well microtiter plate platform was conducted to identify drugs with novel inhibitory activity against both *C. albicans* and *C. auris* biofilms (Ajetunmobi et al. 2023d). The Repurposing Hub library, from the Broad Institute, is a comprehensive repurposing library and accompanying interactive and curated database, consisting of over 6000 different compounds with 663 different therapeutic indications and over 2000 different targets (Corsello et al. 2017). Most drugs in this library are approved by the FDA, with others at different stages of clinical development (i.e. have undergone at least one phase of clinical trials). Because of this, pharmacodynamics, pharmacokinetics, safety and toxicity for humans have been characterized for most of these compounds (Corsello et al. 2017). Using an arbitrary threshold of 70% inhibition or higher, the primary screen identified 57 and 33 initial hits against *C. albicans* and *C. auris*, respectively. Subsequent concentration-dependent assays were used to validate the anti-biofilm activity of the initial hits and to establish their potency via IC_{50} determinations. The accompanying Repurposing Hub database was used to uncover the identity of the confirmed hits. There were 19 confirmed compounds which were common to both *C. albicans* and *C. auris*. Of these, five were classified as antifungals, eight were antiseptics, and six had different primary therapeutic indications, among which ebselen, temsirolimus and compound BAY 11–7082 emerged as the most promising compounds from a repositioning point of view.

A complementary approach has been the screening of libraries in search of compounds that can potentiate the activity of current antifungal agents against *Candida* biofilms. An example is the screen performed by the Thevissen group in order to identify compounds that potentiate the anti-biofilm activity of miconazole (De Cremer et al. 2015). This screen used the Pharmakon 1600 repositioning library and resulted in the identification of artesunate, from the artemisinin family of antimalarials, as the most effective compound synergizing with miconazole against highly resistant *C. albicans* biofilms.

Leading Repositionable Drugs

In this section, we would like to present a summary and discussion on what we believe to be some of the most promising compounds, potentially repositionable as antifungals, emanating from these screenings. We note that they are all still at the early stages of drug development, with most studies to date having examined their antifungal activity in vitro; although in some instances, there have been a limited number of mostly preliminary preclinical studies evaluating their in vivo antifungal activity. None of them have advanced to clinical trials as of the time of this writing. As such, additional studies will be needed for their repositioning as antifungal agents for the treatment of biofilm-associated *Candida* infections.

Auranofin, consisting of a gold(I) center coordinated to a thiosugar and triethylphosphine, has been FDA-approved for clinical use since 1985 as a disease-modifying antirheumatic drug due to its anti-inflammatory properties, whose main targets in mammalian cells are thioredoxin reductases (Madeira et al. 2012). After the initial identification of its antifungal activity against *C. albicans* biofilms (Siles et al. 2013), a subsequent study demonstrated that Auranofin displayed a relatively broad spectrum of antifungal activity against a variety of medically important fungi, including against multiple other *Candida* species and molds (Wiederhold et al. 2017). Since then, many different studies have investigated and further characterized the antifungal activity of auranofin against a variety of pathogenic fungi, with emphasis on those resistant to current conventional antifungals (i.e. azole-resistant *Aspergillus*, *Scedosporium* and *Lomentospora* species, *Histoplasma capsulatum* and agents of chromoblastomycosis) (Chen et al. 2023; Coelho et al. 2020; de Abreu Almeida et al. 2024; Rollin-Pinheiro et al. 2021; Yaakoub et al. 2021). One study indicated that the mitochondrial protein Mia40 represents the potential target of Auranofin (Thangamani et al. 2017). The same study also demonstrated the in vivo efficacy of Auranofin treatment against *C. neoformans* in a *Caenorhabditis elegans* model (Thangamani et al. 2017).

The synthetic organoselenium compound Ebselen was identified as a hit, particularly against *C. auris*, during the screenings of multiple repurposing libraries (Wall et al. 2018; Ajetunmobi et al. 2023d). Ebselen is recognized for its anti-inflammatory and antioxidant properties, acting as a mimic of glutathione peroxidase resulting in the prevention of damage from reactive oxygen species. Although not currently FDA-approved, Ebselen is part of the NIH Clinical Collection, considered clinically safe and currently undergoing clinical trials for multiple indications (Azad and Tomar 2014; Kil et al. 2017). Follow-up studies on its antifungal activity indicated that Ebselen almost completely abolishes the growth of *C. auris* under planktonic growth conditions at concentrations as low as 2.5 μM although higher concentrations are required for biofilm inhibition (Wall et al. 2018). Furthermore, follow-up studies indicated that, besides *C. auris*, Ebselen displays broad antifungal spectrum of action based on its activity against a variety of medically important fungi, including yeasts and molds, with the notable exception of mucorales (Wall et al. 2018). Of note, Ebselen was also effective in vivo in a clinically relevant murine model of hematogenously disseminated *C. albicans* infection (Ajetunmobi et al. 2023d), and, although a trend toward improved survival against *C. auris* was detected, the protective effects against infection with this species did not reach statistical significance (Ajetunmobi et al. 2023d).

Interestingly, both Auranofin and Ebselen also display potent antibacterial activity, in particular against *Staphylococcus aureus*. Thus, it is possible that they could also be repurposed for the treatment of mixed fungal/bacterial infections for which novel therapeutic strategies are urgently needed as they are particularly difficult to treat and often lead to unacceptably high mortality rates (Rodrigues et al. 2019; She et al. 2020).

During the screening of most repurposing libraries, Alexidine dihydrochloride, an antiseptic, antibacterial and antiplaque agent with good safety characteristics, was consistently identified among the top hits displaying some of the highest levels of anti-biofilm inhibitory activity. Confirming these results, Alexidine was also identified in a different screen by the Ibrahim and Uppuluri labs (Mamouei et al. 2018). Although their initial screen was performed under planktonic conditions, in subsequent studies, they were able to demonstrate the killing activity of Alexidine against preformed *Candida* biofilms. Moreover, treatment with Alexidine was able to eradicate *Candida* biofilms in a mouse model of central venous catheter candidiasis, further highlighting its potential to be repositioned as an antifungal for the treatment of biofilm-associated candidiasis (Mamouei et al. 2018). The authors were also able to demonstrate its in vitro activity against a broad spectrum of medically important fungi, an observation that was further corroborated by a different group that demonstrated its activity against highly resistant pathogens (i.e. *Aspergillus*, *Fusarium*, *Scedosporium*, *Rhizopus* and *Lichtheimia* species) (Yousfi et al. 2020).

Rapamycin (sirolimus) and different rapamycin analogs ("rapalogs") were frequent hits virtually in almost every single screening performed with the different repurposing libraries. These are highly specific mTOR inhibitors (for "mammalian target of rapamycin") (Li et al. 2014) and have been mostly studied because of their anticancer properties; although we note that rapamycin was originally discovered because of its antifungal activity but subsequently demonstrated to exert antiproliferative and immunosuppressive effects (Sehgal et al. 1975; Vezina et al. 1975). Also, rapalogs have been known to synergize with antifungals, including against *Candida* biofilms (Uppuluri et al. 2008). Among these rapalogs, Temsirolimus and Everolimus were common hits to both *C. albicans* and *C. auris* and seemed to exhibit the most potent anti-biofilm activity, with IC_{50} values determined from confirmatory dose-response experiments approaching the picomolar range (Ajetunmobi et al. 2023c, d). Follow-up experiments demonstrated their in vitro activity against multiple pathogenic species within the *Candida* genus. Of note, treatment with Temsirolimus and Everolimus increased the survival of mice in the murine model of hematogenously disseminated *C. albicans* infections (Ajetunmobi et al. 2023c, d) although no statistically significant differences between treated and untreated groups of mice were observed when used in a similar model of *C. auris* infection (Ajetunmobi et al. 2023c, d). Temsirolimus is an ester analog of rapamycin with improved solubility and pharmacokinetic properties, which after administration is converted to rapamycin via enzymatic hydrolysis (Gore 2007; Zanardi et al. 2015). Everolimus is a second-generation "rapalog," constituting the 40-O-(2-hydroxyethyl) derivative of sirolimus, which results in improved pharmadynamic and pharmacokinetic characteristics as well as enhanced drug-target interactions (Klawitter et al. 2015). As such, it shows higher potency compared to rapamycin. It is FDA-approved for the treatment of different types of cancer, is also used in solid organ transplantation and also displays antiviral properties. Overall, both Temsirolimus and Everolimus display more potent antifungal properties compared to other rapalogs as well as lower immunosuppressive effects (Bravo-San Pedro and Senovilla 2013), offering reassurances about their potential to be repurposed as antifungals.

The compound BAY 11-7082 was identified during the screens of the Repurposing Hub Library as one of the top hits with inhibitory activity against both *C. albicans* and *C. auris* biofilms, whereas it is closely related analog BAY 11-7085 was a hit in the *C. albicans* screen only. These investigational compounds regulate cytokine activity and inflammation via their inhibition of IκB-α phosphorylation in mammalian cells (Pierce et al. 1997); as a result, they exhibit broad-spectrum anti-inflammatory activity against multiple targets (Lee et al. 2012). However, they are phenyl vinyl sulfone-related compounds and these compounds are also irreversible inhibitors of cysteine proteases. Their biological activities range from anticancer to anti-inflammatory and neuroprotective effects, but they have been primarily used as investigational bioactive small molecules for research in the gene regulation field (Irrera et al. 2017; Kim et al. 2010; White and Burchill 2008). BAY 11-7082 has been evaluated in preclinical studies of inflammation after hematopoietic stem cell transplants in mice, anticancer activity and psoriasis (Irrera et al. 2017; Kim et al. 2010; White and Burchill 2008). Interestingly, a screen of the LOPAC library for antifungal activity against *C. albicans* also identified compounds BAY 11-7082 and BAY 11-7085 among the main hits (Watamoto et al. 2015); although the initial screen was performed under planktonic conditions, follow-up studies demonstrated their more limited activity against both adhered cells and pre-formed biofilms. Another study reported on the activity of BAY 11–7085 against *C. albicans* mono-species biofilms as well as *C. albicans/Staphylococcus aureus* dual-species biofilms (Escobar et al. 2021). Most recently, the O'Meara group used a high throughput imaging screen to probe an FDA drug repurposing library of 2017 compounds, assembled in-house by the University of Michigan Center for Drug Repurposing and largely sourced from the SelleckChem bioactive compound library, in search for inhibitors of *C. albicans* filamentation (Metzner et al. 2023). Both BAY 11-7085 and BAY 11-7082 were identified as hits during this screen, and the authors hypothesized that the effect on biofilm formation may be in turn due to the inhibition of filamentation. Moreover, their studies indicate that this family of compounds prevent filamentation by modifying translation through the eIF3 complex. As one of the main repositionable leads from the Repurposing Hub Library screening, the in vivo activity of compound BAY 11-7082 against *C. albicans* was evaluated in the murine model of hematogenously disseminated candidiasis; unfortunately, mice treated with compound BAY 11-7082 did not exhibit any significant differences in survival as compared to the control group, at least under the conditions used in this study (Ajetunmobi et al. 2023d).

Niclosamide, which is FDA-approved and has a long history of safety and efficacy as an anthelmintic, was a hit in an initial screen for inhibitors of *C. albicans* filamentation, with follow-up experiments demonstrating its inhibitory activity against both *C. albicans* and *C. auris* biofilm formation (Garcia et al. 2018). It was subsequently identified as the top repositionable compound in a target-based screen for inhibitors of *C. albicans* Ndu1p, a mitochondrial protein that controls biofilm growth and virulence in this opportunistic pathogenic fungus (Sutar et al. 2022). The authors demonstrated that Niclosamide targets mitochondrial respiration and induces biofilm detachment. Because this drug suffers from poor bioavailability after oral administration, in very clever studies, the authors developed a nanoparticle formulation for the encapsulation of Niclosamide and demonstrated its in vitro efficacy of the resulting nanoformulation against *C. albicans* and *C. auris*, both under planktonic- and biofilm-growing conditions (Sutar et al. 2022). Moreover, in vivo studies demonstrated the efficacy of this nanoformulation of Niclosamide (incorporated in a gel) in the established mouse model of oropharyngeal candidiasis caused by *C. albicans*, which is associated with biofilm formation on the oral mucosa (Sutar et al. 2022).

Finally, Miltefosine represented one of the top hits after the screening of the ReFRAME library and the MMV's Pathogen Box against *C. auris*, with follow-up studies demonstrating its activity against multiple *C. auris* strains and other *Candida* spp. (Wall et al. 2019, 2020). This was not entirely surprising as the antifungal activity of miltefosine has been described previously (Vila et al. 2015). Miltefosine is an alkylphosphocholine drug originally developed as an anticancer drug. It was shown to possess activity against various parasites, and it has been approved as the first oral therapy for visceral and cutaneous leishmaniasis (Dorlo et al. 2012). Miltefosine has dose-limiting gastrointestinal toxicity, but side effects are normally moderate in severity with the caveat that at very high doses it may display hepatotoxicity and nephrotoxicity (Dorlo et al. 2012).

Conclusions and Future Perspectives

Antifungal therapeutic options are scarce, and current clinically used antifungals display limited efficacy against biofilms, contributing to high morbidity and mortality rates. As such, the need for new and effective antifungals with activity against highly recalcitrant *Candida* biofilms is undeniable. There has been an increasing interest in repurposing as a potential expedited path to drug development, with the potential to reduce costs and accelerate approval timelines in the antifungal field in general and in particular in the search for novel drugs with activity against *Candida* biofilms. The work summarized here on the screening of repurposing libraries, and the resulting identification of potentially repositionable compounds represents the initial stage of efforts in repurposing programs and will necessitate much additional work to confirm the promise of these approaches. We also expect an increase in the number of repurposing screens to be conducted in the near future.

Compliance and Ethical Standards.

The authors confirm that they have no conflicts and have complied with all ethical standards.

References

Abduljalil H et al (2022) Screening the Tocriscreen bioactive compound library in search for inhibitors of Candida biofilm formation. APMIS 130(9):568–577

Ajetunmobi OH et al (2023a) Antifungal therapy of Candida biofilms: past, present and future. Biofilms 5:100126

Ajetunmobi OH et al (2023b) A 384-well microtiter plate model for Candida biofilm formation and its application to high-throughput screening. Methods Mol Biol 2658:53–64

Ajetunmobi OH et al (2023c) Screening the medicine for malaria venture's pandemic response box to identify novel inhibitors of Candida albicans and Candida auris biofilm formation. APMIS 131:613

Ajetunmobi OH et al (2023d) High-throughput screening of the repurposing hub library to identify drugs with novel inhibitory activity against Candida albicans and Candida auris biofilms. J Fungi (Basel) 9(9):879

Anderson TM et al (2014) Amphotericin forms an extramembranous and fungicidal sterol sponge. Nat Chem Biol 10(5):400–406

Azad GK, Tomar RS (2014) Ebselen, a promising antioxidant drug: mechanisms of action and targets of biological pathways. Mol Biol Rep 41(8):4865–4879

Bachmann SP et al (2002) In vitro activity of caspofungin against Candida albicans biofilms. Antimicrob Agents Chemother 46(11):3591–3596

Bravo-San Pedro JM, Senovilla L (2013) Immunostimulatory activity of lifespan-extending agents. Aging (Albany NY) 5(11):793–801

Brown GD et al (2012) Hidden killers: human fungal infections. Sci Transl Med 4(165):165rv13

Chen P et al (2023) In vitro antifungal and antibiofilm activities of auranofin against itraconazole-resistant aspergillus fumigatus. J Mycol Med 33(2):101381

Coelho RA et al (2020) A screening of the MMV pathogen box(R) reveals new potential antifungal drugs against the etiologic agents of chromoblastomycosis. PLoS One 15(5):e0229630

Corsello SM et al (2017) The drug repurposing hub: a next-generation drug library and information resource. Nat Med 23(4):405–408

Cuellar-Cruz M et al (2012) Candida species: new insights into biofilm formation. Future Microbiol 7(6):755–771

de Abreu Almeida M et al (2024) Auranofin is active against Histoplasma capsulatum and reduces the expression of virulence-related genes. PLoS Negl Trop Dis 18(10):e0012586

De Cremer K et al (2015) Artemisinins, new miconazole potentiators resulting in increased activity against Candida albicans biofilms. Antimicrob Agents Chemother 59(1):421–426

Dorlo TP et al (2012) Miltefosine: a review of its pharmacology and therapeutic efficacy in the treatment of leishmaniasis. J Antimicrob Chemother 67(11):2576–2597

Escobar IE et al (2021) Repurposing kinase Inhibitor Bay 11-7085 to combat Staphylococcus aureus and Candida albicans biofilms. Front Pharmacol 12:675300

Farha MA, Brown ED (2019) Drug repurposing for antimicrobial discovery. Nat Microbiol 4(4):565–577

Garcia C et al (2018) A phenotypic small-molecule screen identifies halogenated salicylanilides as inhibitors of fungal morphogenesis, biofilm formation and host cell invasion. Sci Rep 8(1):11559

Gore ME (2007) Temsirolimus in the treatment of advanced renal cell carcinoma. Ann Oncol 18(Suppl 9):ix87-8

Grunberg E, Titsworth E, Bennett M (1963) Chemotherapeutic activity of 5-Fluorocytosine. Antimicrob Agents Chemother (Bethesda) 161:566–568

Gudlaugsson O et al (2003) Attributable mortality of nosocomial candidemia, revisited. Clin Infect Dis 37(9):1172–1177

Hughes JP et al (2011) Principles of early drug discovery. Br J Pharmacol 162(6):1239–1249

Irrera N et al (2017) BAY 11-7082 inhibits the NF-kappaB and NLRP3 inflammasome pathways and protects against IMQ-induced psoriasis. Clin Sci (Lond) 131(6):487–498

Janes J et al (2018) The ReFRAME library as a comprehensive drug repurposing library and its application to the treatment of cryptosporidiosis. Proc Natl Acad Sci USA 115(42):10750–10755

Katragkou A, Roilides E, Walsh TJ (2016) Can repurposing of existing drugs provide more effective therapies for invasive fungal infections? Expert Opin Pharmacother 17(9):1179–1182

Kean R et al (2020) Candida auris: a decade of understanding of an enigmatic pathogenic yeast. J Fungi (Basel) 6(1):30

Kil J et al (2017) Safety and efficacy of ebselen for the prevention of noise-induced hearing loss: a randomised, double-blind, placebo-controlled, phase 2 trial. Lancet 390(10098):969–979

Kim YS et al (2010) BAY 11-7082, a nuclear factor-kappaB inhibitor, reduces inflammation and apoptosis in a rat cardiac ischemia-reperfusion injury model. Int Heart J 51(5):348–353

Klawitter J, Nashan B, Christians U (2015) Everolimus and sirolimus in transplantation-related but different. Expert Opin Drug Saf 14(7):1055–1070

Koehler P et al (2019) Morbidity and mortality of candidaemia in Europe: an epidemiologic meta-analysis. Clin Microbiol Infect 25(10):1200–1212

Kohler JR et al (2017) Fungi that Infect Humans. Microbiol Spectr 5:3

Kuhn DM et al (2002) Antifungal susceptibility of Candida biofilms: unique efficacy of amphotericin B lipid formulations and echinocandins. Antimicrob Agents Chemother 46(6):1773–1780

Lee J et al (2012) BAY 11-7082 is a broad-spectrum inhibitor with anti-inflammatory activity against multiple targets. Mediat Inflamm 2012:416036

Li J, Kim SG, Blenis J (2014) Rapamycin: one drug, many effects. Cell Metab 19(3):373–379

Madeira JM et al (2012) The biological activity of auranofin: implications for novel treatment of diseases. Inflammopharmacology 20(6):297–306

Mamouei Z et al (2018) Alexidine Dihydrochloride has broad-Spectrum activities against diverse fungal pathogens. mSphere 3(5):e00539-18

Metzner K et al (2023) Imaging-based screening identifies modulators of the eIF3 translation initiation factor complex in Candida albicans. Antimicrob Agents Chemother 67(7):e0050323

Odds FC, Brown AJ, Gow NA (2003) Antifungal agents: mechanisms of action. Trends Microbiol 11(6):272–279

Pfaller MA, Diekema DJ (2007) Epidemiology of invasive candidiasis: a persistent public health problem. Clin Microbiol Rev 20(1):133–163

Pierce JW et al (1997) Novel inhibitors of cytokine-induced IkappaBalpha phosphorylation and endothelial cell adhesion molecule expression show anti-inflammatory effects in vivo. J Biol Chem 272(34):21096–21103

Pierce CG et al (2008) A simple and reproducible 96-well plate-based method for the formation of fungal biofilms and its application to antifungal susceptibility testing. Nat Protoc 3(9):1494–1500

Pushpakom S et al (2019) Drug repurposing: progress, challenges and recommendations. Nat Rev Drug Discov 18(1):41–58

Quindos G et al (2018) The continuous changes in the aetiology and epidemiology of invasive candidiasis: from familiar Candida albicans to multiresistant Candida auris. Int Microbiol 21(3):107–119

Rajendran R et al (2016) Biofilm formation is a risk factor for mortality in patients with Candida albicans bloodstream infection-Scotland, 2012–2013. Clin Microbiol Infect 22(1):87–93

Ramage G et al (2001) Standardized method for in vitro antifungal susceptibility testing of Candida albicans biofilms. Antimicrob Agents Chemother 45(9):2475–2479

Ramage G, Martinez JP, Lopez-Ribot JL (2006) Candida biofilms on implanted biomaterials: a clinically significant problem. FEMS Yeast Res 6(7):979–986

Ramage G et al (2023) Our current clinical understanding of Candida biofilms: where are we two decades on? APMIS 131:636

Rodrigues CF, Henriques M (2017) Liposomal and Deoxycholate amphotericin B formulations: effectiveness against biofilm infections of Candida spp. Pathogens 6(4)

Rodrigues ME, Gomes F, Rodrigues CF (2019) Candida spp./bacteria mixed biofilms. J Fungi (Basel) 6(1):5

Rollin-Pinheiro R et al (2021) Identification of promising antifungal drugs against Scedosporium and Lomentospora species after screening of pathogen box library. J Fungi (Basel) 7(10):803

Sehgal SN, Baker H, Vezina C (1975) Rapamycin (AY-22,989), a new antifungal antibiotic. II. Fermentation, isolation and characterization. J Antibiot (Tokyo) 28(10):727–732

Sertkaya A et al (2024) Costs of drug development and Research and Development intensity in the US, 2000–2018. JAMA Netw Open 7(6):e2415445

She P et al (2020) Antibiofilm efficacy of the gold compound auranofin on dual species biofilms of Staphylococcus aureus and Candida sp. J Appl Microbiol 128(1):88–101

Sherry L et al (2017) Biofilm-forming capability of highly virulent, multidrug-resistant Candida auris. Emerg Infect Dis 23(2):328–331

Siles SA et al (2013) High-throughput screening of a collection of known pharmacologically active small compounds for identification of Candida albicans biofilm inhibitors. Antimicrob Agents Chemother 57(8):3681–3687

Sutar Y et al (2022) Niclosamide-loaded nanoparticles disrupt Candida biofilms and protect mice from mucosal candidiasis. PLoS Biol 20(8):e3001762

Thangamani S et al (2017) Repurposing approach identifies Auranofin with broad Spectrum antifungal activity that targets Mia40-Erv1 pathway. Front Cell Infect Microbiol 7:4

Thomas-Ruddel DO et al (2022) Risk factors for invasive Candida infection in critically ill patients: a systematic review and meta-analysis. Chest 161(2):345–355

Tsay SV et al (2020) Burden of Candidemia in the United States, 2017. Clin Infect Dis 71(9):e449–e453

Uppuluri P, Lopez-Ribot JL (2016) Go forth and colonize: dispersal from clinically important microbial biofilms. PLoS Pathog 12(2):e1005397

Uppuluri P et al (2008) Synergistic effect of calcineurin inhibitors and fluconazole against Candida albicans biofilms. Antimicrob Agents Chemother 52(3):1127–1132

Uppuluri P et al (2010) Dispersion as an important step in the Candida albicans biofilm developmental cycle. PLoS Pathog 6(3):e1000828

Vezina C, Kudelski A, Sehgal SN (1975) Rapamycin (AY-22,989), a new antifungal antibiotic. I. Taxonomy of the producing streptomycete and isolation of the active principle. J Antibiot (Tokyo) 28(10):721–726

Vila T, Lopez-Ribot JL (2017) Screening the pathogen box for identification of Candida albicans biofilm inhibitors. Antimicrob Agents Chemother 61(1)

Vila TV et al (2015) In vitro activity of Miltefosine against Candida albicans under planktonic and biofilm growth conditions and in vivo efficacy in a murine model of oral candidiasis. Antimicrob Agents Chemother 59(12):7611–7620

Wall G, Lopez-Ribot JL (2020a) Current Antimycotics, new prospects, and future approaches to antifungal therapy. Antibiotics (Basel) 9(8)

Wall G, Lopez-Ribot JL (2020b) Screening repurposing libraries for identification of drugs with novel antifungal activity. Antimicrob Agents Chemother 64(9)

Wall G et al (2018) Screening a repurposing library for inhibitors of multidrug-resistant Candida auris identifies Ebselen as a repositionable candidate for antifungal drug development. Antimicrob Agents Chemother 62(10)

Wall G, Herrera N, Lopez-Ribot JL (2019) Repositionable compounds with antifungal activity against multidrug resistant Candida auris identified in the Medicines for Malaria Venture's pathogen box. J Fungi (Basel) 5(4):92

Wall G et al (2020) Screening the CALIBR ReFRAME library in search for inhibitors of Candida auris biofilm formation. Front Cell Infect Microbiol 10:597931

Watamoto T et al (2015) Screening of pharmacologically active small molecule compounds identifies antifungal agents against Candida biofilms. Front Microbiol 6:1453

White DE, Burchill SA (2008) BAY 11-7082 induces cell death through NF-kappaB-independent mechanisms in the Ewing's sarcoma family of tumours. Cancer Lett 268(2):212–224

Wiederhold NP et al (2017) Repurposing auranofin as an antifungal: in vitro activity against a variety of medically important fungi. Virulence 8(2):138–142

Yaakoub H et al (2021) Repurposing of auranofin and honokiol as antifungals against Scedosporium species and the related fungus Lomentospora prolificans. Virulence 12(1):1076–1090

Yousfi H et al (2020) Identification of repositionable drugs with novel antimycotic activity by screening the Prestwick chemical library against emerging invasive moulds. J Glob Antimicrob Resist 21:314–317

Zanardi E et al (2015) Clinical experience with temsirolimus in the treatment of advanced renal cell carcinoma. Ther Adv Urol 7(3):152–161

Chapter 13
Aspergillus Biofilms

Jose A. Ferreira and David A. Stevens

Abstract The formation of a biofilm assists microbes in survival in their environments, defends against stresses, and may enable the extension of its range. In this chapter, we review various aspects of the biofilms of Aspergillus, an important ubiquitous global fungus, found in many environments, and capable of causing diseases, decay and/or intoxications. The extracellular matrix is a very important component of these biofilms. Biofilm formation is affected by the environmental situation and affects the biology and physiology of the fungus. Here, we review the course of Aspergillus biofilm formation, its composition, its architecture, the interactions with the hosts (particularly patients), in intermicrobial competition, in nature, and its effects in the industry. Mixed biofilms with other microbes are explored, as is the interaction with antifungal molecules. All the features just enumerated and reviewed have implications for understanding fungal biology and the development of treatment strategies.

Introduction

As with other aerial fungi, Aspergillus species can form a community of attached microbes, and with Aspergilli, these may be encased, in whole or in part, in a non-crystalline extracellular matrix (ECM) (Mitchell et al. 2016). The whole is referred to here as biofilm, with the ECM, as a dominant component. The present review is focused almost exclusively on *Aspergillus fumigatus* (Af), in most areas of the world the dominant species causing disease and also frequently found in the

J. A. Ferreira
Department of Medicine, Infectious Diseases and Public Health Unit,
Unifenas University School of Medicine, Belo Horizonte, Brazil

D. A. Stevens (✉)
Department of Medicine, Division of Infectious Diseases and Geographic Medicine,
Stanford University Medical School, California, USA
e-mail: stevens@stanford.edu

G. Ramage, D. Andes (eds.), *Fungal Biofilms*, Springer Series on Biofilms 15,
https://doi.org/10.1007/978-3-031-93315-8_13

environment (commonly on decaying vegetation). The formation of a biofilm creates a favorable environment for the fungus, where the generation of new cells is enhanced, and the effect of host defenses, environmental threats, and mechanical stresses is reduced (Morelli et al. 2021). The stability of the fungus in environmental degradation, even in environmental remediation, is assisted.

Course of Aspergillus Biofilm

In culture, Af may be grown in a static or shaken liquid medium or on a solid substrate, such as agar. In liquid medium, shaken cultures, and the submerged component in static cultures, there will be a paucity of ECM (Loussert et al. 2010) although this amount can be modulated by the contents of the medium used (e.g., glucose). The formation of a biofilm and the speed of its development in those milieus will depend on the seeding density. The amount of ECM, in any other type of nonliquid culture, will depend on the medium used, its nutrients, temperature, pO_2, and pH (Roudbary et al. 2022). The amount of ECM as well as the ECM composition will vary with the time of incubation. Variations in content may be expected in biofilms obtained from cultures vs. from in vivo settings, as discussed further below. Metabolomic studies reveal distinct profiles in biofilm cells, highlighting metabolic pathways that differ significantly from planktonic cells, underscoring adaptive biofilm physiology (Muszkieta et al. 2013).

If in vitro cultures are started from conidia, at temperatures from 22 to 37 °C, germination occurs (Fig. 13.1). If the cultures were made in conjunction with abiotic surfaces or mammalian cells, attachment to those cell surfaces would have preceded this (Seidler et al. 2008; Singhal et al. 2011), mediated by hydrophobic interactions and surface adhesins (molecules such as RodA hydrophobin). Generally, hyphae will be formed, with or without mammalian cells present, in 8–10 h. Hyphae will intertwine, genes involved in biofilm formation will be differentially regulated, and ECM formation starts in 10–16 h (Mowat et al. 2007; Kaur and Singh 2014; Van Acker and Coenye 2016). Biofilms formed on surfaces will be hydrophobic at the apex. The ECM can be visualized with safranin stain (Seidler et al. 2008). The hyphal mass can be quantitated simply with crystal violet (Mowat et al. 2007) although mass is commonly inferred from oxidative metabolism via the XTT assay (Beauvais and Muller 2009). Colony-forming unit assays are unreliable, in part because of the difficulties in disarticulating cells in hyphae and preserving viability and because of the nature of Af structures, where hyphal cells can communicate with each other through open septal pores (Manavathu et al. 2014). With further maturation, seeding of individual cells to new sites can occur, sometimes prompted by changes in the local environment, restarting the developmental process elsewhere.

Induced mutations in the key developmental genes disrupt the formation of the final biofilm product, as can epigenetic modifications and small RNA post-transcriptional regulators (Lin et al. 2020). Nucleic acid sequencing has enlarged our understanding of the key genetic pathways (Muszkieta et al. 2013). Some

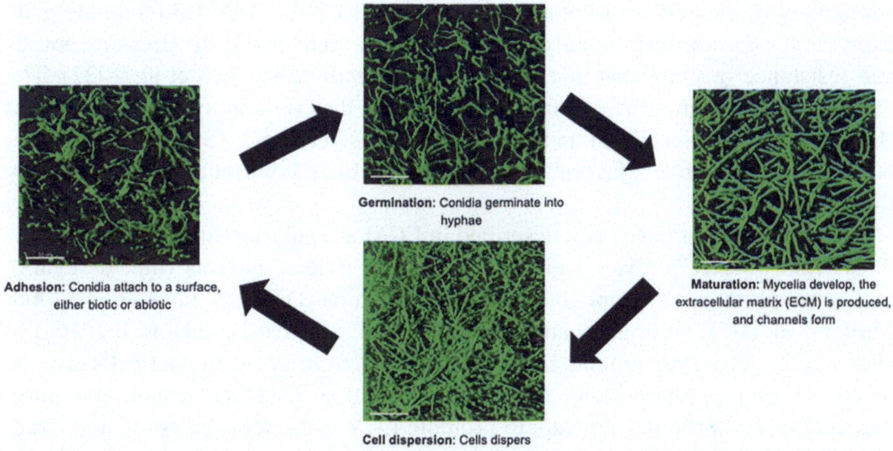

Fig. 13.1 Various stages in the formation of *Aspergillus fumigatus* biofilm. Mature biofilms, embedded within a dense extracellular matrix (ECM), develop <48–72 h, following the adhesion and germination of fungal conidia. Multiple fungal factors, including components of the cell wall and secondary metabolites, are activated and play a role in biofilm formation. The *A. fumigatus* biofilm exhibits a heterogeneous structure in terms of the distribution of fungal cells and ECM. In the first part of the figure, germination has already begun from the seeded conidia (the spherical tiny objects; from these, the slender mycelia emerge). Summary of each stage of biofilm development showing the representative fluorescence confocal microscopy image of *A. fumigatus* biofilms. Scale Bar: 50 μm

mutations in these pathways occur endogenously, under selective pressure, to enable the fungal community to adapt to its environment, such as choosing to assimilate different nutrients, based on scarcity in the environment, or resist stresses. Introducing inhibitors of key developmental pathways can modulate biofilm development and integrity.

Composition of Aspergillus Biofilm

As the biofilm matures, there is increased production of secondary metabolites (including gliotoxin, discussed below) in the hyphal cells (Shopova et al. 2013), upregulation of genes, production of proteins (and their phosphorylation) involved in membrane composition (which may help in resistance to osmotic changes and oxidative pressures) (Mitchell et al. 2016; Beauvais et al. 2014), cellular transport, iron metabolism, sulfur metabolism, nutrient utilization, the oxidative and the hypoxia stress responses and responses to other stresses, and, specifically, production of surface hydrophobins and alcohol dehydrogenases (Lin et al. 2020; Beauvais et al. 2014; Bruns et al. 2010; Zheng et al. 2015; Valsecchi et al. 2017). There may be signaling molecules released to affect the behavior of other cells in the community (Barriuso et al. 2018; Wongsuk et al. 2016). Pathways involved in strengthening the cell wall come into play. Small RNAs (sRNAs) modulate gene expression post-transcriptionally within biofilms, contributing to stress responses and resistance mechanisms that fortify biofilm resilience (Chen et al. 2022). The most proximal (to the surface attachment) cells in the community become dormant (Beauvais and Muller 2009), possibly related to lowered pO_2 (Stewart et al. 2019) but may revive under favorable conditions, which contributes to community longevity.

Components that have been described in ECM are galactomannan, galactosaminogalactan (GAG), alpha-1,3-glucans, monosaccharides, polyols (mixed sugars), melanins, proteins, hydrophobins, and extracellular DNA (Seidler et al. 2008; Muller et al. 2011; Guimarães and Takahashi 2014; Sheppard and Howell 2016; Liu et al. 2022). The extracellular DNA, presumably from lysed hyphal cells, can be detected with propidium iodide (Shopova et al. 2013). This DNA can bind to polysaccharides and proteins, appears to promote ECM adhesion, diffusion, and structural integrity of the community, and contributes to resistance to antifungal drugs; application of DNAase enzyme can disrupt the biofilm integrity (Shopova et al. 2013; Krappman and Ramage 2013; Nijland et al. 2010). When Af biofilms are found in mammalian hosts and likely when grown in vitro with mammalian cells, mammalian extracellular DNA may be found in the ECM too. In culture, and presumably in vivo, serum (apparently serum glycoproteins) promotes more hyphal branching, polysaccharide presence, biofilm strength, and ECM localization at hyphal tips (Manavathu et al. 2014; Shopova et al. 2013; Bugli et al. 2014). The biofilm matrix produced by *Aspergillus* can contain several extracellular enzymes, such as hydrolases and proteases, which facilitate nutrient acquisition and biofilm maintenance. These enzymes can also play a role in microbial interactions by degrading host tissues, competing microbial cells, or environmental substrates. There is a paucity of beta-1,3-glucan and chitin in ECM although these are important components in the Af cell wall (Beauvais and Muller 2009).

Attempts to generate quantitative descriptions of the Af ECM have been somewhat limited by the inability to completely solubilize ECM components, possible disruption of material that can occur during such sample preparation, and possible loss of components in that process, frustrating quantitative analysis. As mentioned above, ECM composition will vary with time and culture milieu. In studies by our group on ECM composition (Reichhardt et al. 2015, 2016, 2019), we used a solid-state nuclear magnetic resonance (NMR) approach, which avoids such problems (Fig. 13.2). With ^{15}N,^{13}C, and ^{31}P spectra, NMR gives an atomic-level chemical accounting of the types of these molecules, uses intact biofilms, and can characterize insoluble materials. This NMR approach used cross-polarization magic-angle spinning and rotational-echo double-resonance. NMR was integrated with complementary biochemical analysis (e.g., protein electrophoresis) and scanning electron microscopy to study the structure. In growth in RPMI, carbonyls accounted for 12% of the total carbon mass and alpha carbons 7%. The ECM consisted of 40% protein, which had previously been underestimated by other methods (and includes catalase B, which protects against reactive oxygen species, and Asp f2, which damages host

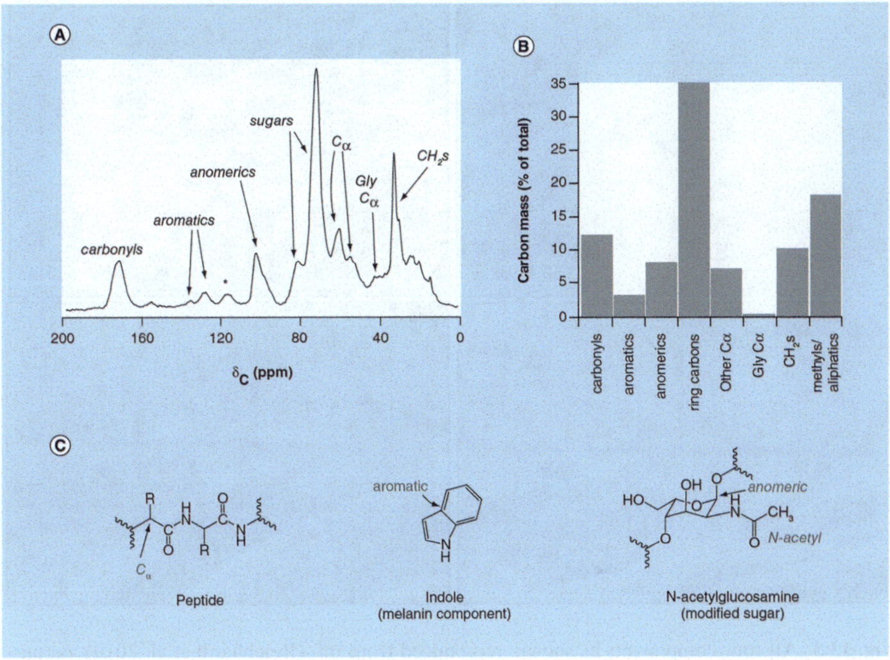

Fig. 13.2 Figure reproduced from ref. (Reichhardt et al. 2016), with permission from Taylor and Francis, publishers. *Aspergillus fumigatus* extracellular matrix composition was determined using solid-state nuclear magnetic resonance. (**a**) The 13C cross-polarization magic-angle spinning spectrum of *A. fumigatus* extracellular matrix showed carbon contributions from a range of biomolecules including proteins, carbohydrates, and lipids. (**b**) The contributions of specific carbon pools to the total carbon mass were annotated using cross-polarization magic angle spinning and spectral editing via 13C{15 N} and 13C{31P}REDOR nuclear magnetic resonance experiments, and these contributions are summarized in the graph. (**c**) Chemical structures of representative biomolecules that were identified in the *A. fumigatus* extracellular matrix

epithelium, as major proteins), 43% polysaccharide, 3% aromatic-containing components (such as melanin), and up to 14% lipid (including some phospholipids).

Architecture of Aspergillus Biofilms

The architecture of biofilms was delineated in some detail when the aforementioned ["Composition of Aspergillus Biofilm" section] specific molecules were described above (Beauvais et al. 2007; Shin et al. 2023; Nett and Andes 2015). Scanning electron microscopy (SEM) analysis of the three-dimensional architecture of biofilms is often affected by standard fixation and drying techniques. Variations in sample preparation protocols, microscope selection, and image acquisition can influence the observed architecture. Some loss of architecture can be associated with

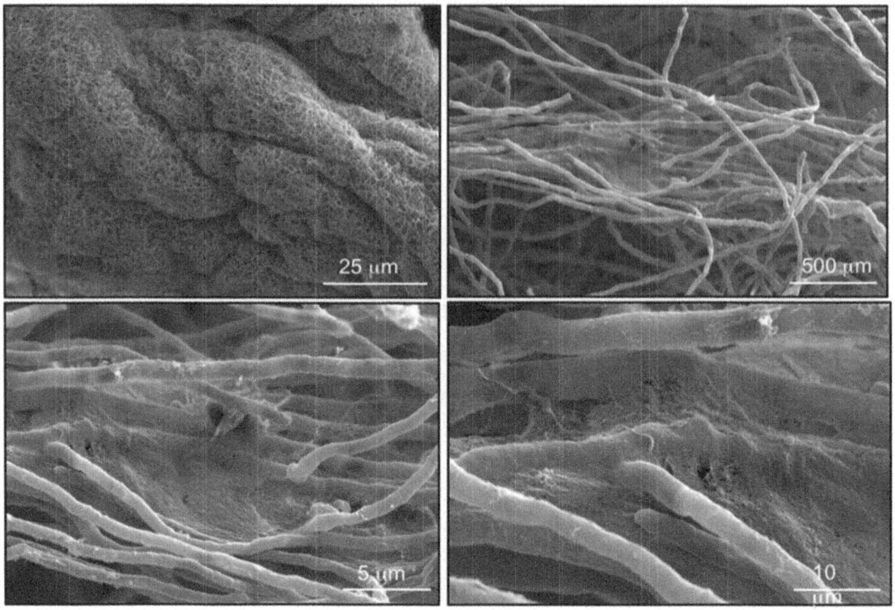

Fig. 13.3 All four components as shown, reproduced from ref. (Reichhardt et al. 2019), permission from Oxford University Press. High-resolution ultrastructural SEM analysis of *A. fumigatus* biofilms grown in RPMI 1640 nutrient broth. Micrographs resulted from optimized sample preparation and processing parameters as well as SEM detection methods to enable high-resolution visualization of the fibrous ultrastructure of ECM on and between hyphae. Samples were harvested from flasks and fixed for limited time periods in 4% paraformaldehyde with 2% glutaraldehyde in 0.1 M sodium cacodylate buffer, rinsed in the same buffer, and post-fixed with 1% OsO_4. Samples were then rinsed in water and gradually dehydrated in increasing concentrations of ethanol (50, 70, 90, and 100%, 5 min each) and critically point dried with liquid CO_2. Dried samples were sputter-coated (50A°) and imaged with a Hitachi S-3400 N SEM operated at 10 kV under a high vacuum, using an Everhart-Thornley (ET) SE detector. Top left micrograph was originally published in ref. (Joubert et al. 2017). Bottom right micrograph was originally published in ref. (Reichhardt et al. 2015)

dehydration required for conventional SEM. We tried to optimize visualizing Af biofilms and develop optimal SEM protocols (Reichhardt et al. 2019; Joubert et al. 2015, 2017), using conventional SEM, field emission SEM (FESEM), and variable pressure (VP)-SEM (Fig. 13.3). Visualization by high-resolution SEM and FESEM techniques using a secondary electron detector proved favorable conditions. VP-SEM enables the observation of biofilms in their natural hydrated state. Important factors are aldehyde fixation of samples, heavy metal contrasting, drying techniques, and ionic liquid treatment. We found achieving consistent high-resolution ultrastructural SEM analysis of cellular features, and ECM of Af biofilms relied on quite short fixation times, up to 1 h is ideal for sample preparation. OsO_4 post-fixation yields improved contrast and visualization of cellular vs. extracellular regions, followed by critical point drying. Other heavy-metal reagents can provide specific contrasting of biofilm, depending on the desire to visualize different components (cells and ECM surface features). Specimen preparation with cryo-fixation and lyophilization provides a useful and rapid alternative to conventional chemical fixation and drying.

Interaction of Aspergillus Biofilm with the Host

Af biofilm interacts with host defenses (Kernien et al. 2018; Mannan et al. 2024; Hernández-Chávez et al. 2017). ECM induces resistance to host cell phagocytosis (Beauvais and Muller 2009). Restricting host cell access to the fungus is a factor. Alpha-1,3-glucans are commonly found at the cell wall surface in ECM (Loussert et al. 2010; Beauvais et al. 2014). These glucans appear important in hyphal agglutination, in resisting host defenses in infection, and in masking fungal immunogens (Kaur and Singh 2014; Beauvais and Latge 2015), and, with GAG, are responsible for adherence to host epithelium (Beauvais and Muller 2009; Gravelat et al. 2013).

GAG is a heteropolysaccharide composed of galactopyranose and partially deacetylated N-acetylgalactosamine, made by polymerization by a glucosyl transferase enzyme. GAG is a promoter of host Th2 responses, decreases leukocyte recruitment, causes leukocyte apoptosis, apparently by activating NK cells, enhances resistance to leukocyte extracellular traps, induces the host receptor IL-1Ra, which blocks IL-1 signaling, inhibiting Th1 and Th17 responses, including production of their cytokines involved in the host response to Aspergillus (Shankar et al. 2024), such as interferon-gamma (a pleiotropic and possibly universal antifungal) (Stevens et al. 2006), IL-17, and IL-22, activates platelets, and suppresses the host inflammatory response by masking beta-glucans, and those glucans would otherwise trigger dectin-1 for the host response (Beauvais et al. 2014; Beauvais and Latge 2015; Gravelat et al. 2013; Lee et al. 2015). The resistance to leukocytes is mediated by resistance to NADPH oxidase-dependent killing, whereas resistance to reactive oxygen species is not increased. This may be explained by NADPH oxidase also coordinating the release of DNA and antimicrobial peptides from neutrophil granules, to form the leukocyte extracellular traps (Lee et al. 2015). The production of GAG is dependent on the gene *uge3*, a UDP-glucose-4-epimerase (Gravelat et al. 2013). GAG-less gene mutants induce more inflammation, but this appears to provide minimal added protection, possibly because fungal adherence is decreased as well (Gravelat et al. 2013). Other biofilm components than GAG may help in resistance to reactive oxygen species. The mycotoxin, gliotoxin (Gomez-Lopez et al. 2022), which has several analogues, has been mentioned as a secondary metabolite, whose production increases as the biofilm matures (Kaur and Singh 2014; Bastos et al. 2022; Bruns et al. 2010; Bugli et al. 2014). Gliotoxin is a virulence factor in mammalian infections and also helps the fungus, in any milieu, by its deleterious effects on microbial competitors (Sugui et al. 2017). Its disulfide bridge is key to these actions (Scharf et al. 2016). It breaks down bronchial epithelial barriers (Illek et al. 2023; Dunne et al. 2023). It inhibits leukocyte influx to infection sites (König et al. 2019) and does this by blocking leukotriene B4 formation, activating thrombocytes (Speth et al. 2013), and impairing the blood-brain barrier (Patel et al. 2018).

Aspergillus Biofilms in Patients

In vivo, aspergillomas are rich in ECM, and the same polysaccharides are present; ECM appears to have a role in the development of fungal balls (Mowat et al. 2009; Williams et al. 2016). Biofilm and ECM likely help perpetuate infection, cause fibrosis, resist host defenses, and help resist therapeutic attempts, in cystic fibrosis airways, chronic pulmonary aspergillosis, and allergic bronchopulmonary aspergillosis (ABPA) (King et al. 2016; Denning et al. 2016; Sehgal et al. 2024; Chatterjee et al. 2024). Chronic pulmonary aspergillosis is a progressive lung disease caused by *Aspergillus* infection, often in individuals with pre-existing lung conditions, such as tuberculosis or chronic obstructive pulmonary disease (Denning et al. 2016). In ABPA, biofilms contribute to the persistence of fungal antigens in the lungs, which triggers recurrent allergic reactions. This can contribute to bronchiectasis and fibrosis. Managing ABPA requires a combination of antifungal and anti-inflammatory treatments although biofilm presence complicates treatment success (Sehgal et al. 2024). In all these entities, the ECM protects the fungal community, including host defenses and drug therapy, and leads to persistence (Roudbary et al. 2022; Chandrasekar and Manavathu 2008; Boisvert et al. 2016; Imbert 2018; Latge 2001; Martinez and Fries 2010; Wu et al. 2017; Paulussen et al. 2016). In some hosts, the persistence can lead to invasive aspergillosis. With combination biofilms, different therapeutic strategies may be needed.

In invasive aspergillosis, there is usually a paucity of ECM, the hyphae are not agglutinated, alpha-1,3-glucans are lacking, and the host contribution to any biofilm is much more prominent (Kaur and Singh 2014; Muller et al. 2011; Binder and Lass-Flörl 2013). In invasive disease, some of the assets to the fungus in the war between host and fungus provided by biofilm are substituted by other fungal assets, such as the ability to invade the host microvasculature and successful competition in acquiring iron (Hsu et al. 2018). In both in vivo settings, aspergilloma and invasive aspergillosis, major secreted antigens, such as Aspf1p, are sparse in the ECM, compared to the in vitro setting (Loussert et al. 2010).

Aspergillus biofilms are notorious for their ability to colonize medical devices, including catheters, prosthetic valves, and pacemakers. Biofilm formation on these surfaces leads to device-related infections that are difficult to eradicate owing to their resistance to both host defenses and antifungal drugs, leading to chronicity. In many cases, device removal is the only effective treatment for biofilm-associated infections. Research into the use of enzymes to break down the ECM and biofilm could be productive. Improved infection control practices in hospitals are needed to prevent *Aspergillus* colonization in the hospital environment (e.g., ceiling tiles and air vents) and on medical devices (Mitchell et al. 2016).

Aspergillus Biofilm in Intermicrobial Competition

Great interest has been in intermicrobial competition with Af (Debourgogne et al. 2023), and most of the data has come from studies of *Pseudomonas aeruginosa* with Af because these are commonly found together in the airways of patients with cystic fibrosis or in immunocompromised patients and because early in vitro studies, and then an extensive literature, documented how Pseudomonas, and specific Pseudomonas molecules, can potently inhibit Af (reviewed in (Sass et al. 2019a)). This inhibition can be studied by adding (the more common avenue of study) Pseudomonas culture filtrates, or specific Pseudomonas molecules, to pure Af planktonic cultures or Af biofilms (Mowat et al. 2010; Manavathu et al. 2015), or in the more complicated situation of mixed bacterial-fungal biofilms (discussed in more detail below) (Ferreira et al. 2015).

Our laboratory has explored the effects of Pseudomonas cultures and specific Pseudomonas molecules, using Af pure biofilm formation, or formed Af biofilm, as the targets (Fig. 13.4). The strength of inhibition by Pseudomonas is dependent on the source (e.g., from CF patients), phenotype, and growth conditions of the bacterium (Ferreira et al. 2015) although Af isolates from CF and non-CF sources were equally susceptible (Nazik et al. 2017) and inhibition is diminished in hypoxic or anaerobic conditions (Anand et al. 2017). Although many *P. aeruginosa* molecules had been shown to inhibit Af (Mangan 1969; Blyth and Forey 1971; Briard et al. 2017), with fungal apoptosis defined as the mechanism (Shirazi et al. 2016) (although the field of fungal cell death is complex and controversial (Galluzzi et al. 2018)), there hadn't been a systematic approach to define the principal causative molecule. Although these molecules had been studied individually, their relative contributions in a mixture of Pseudomonas and Af molecules are unclear, as is the relevance to concentrations actually present in sites in vivo. We, therefore, studied 24 *P. aeruginosa* mutants, with deletions in genes important for virulence, iron acquisition, or quorum sensing, for their effect on Af biofilms in RPMI medium, which has undetectable iron content (Sass et al. 2018). Four *P. aeruginosa* mutants were noninhibitory. The common defect of these mutants was deficient production of the *P. aeruginosa* siderophore pyoverdine. Pure pyoverdine inhibited Af, at clinically relevant concentrations (Martin et al. 2011), even in hypoxic conditions and restored inhibition by the mutants. The key inhibitory mechanism for pyoverdine was iron chelation and denial of iron, a critical nutrient for both fungal and bacterial species to Af (Matthaiou et al. 2018; Bairwa et al. 2017). Similarly, iron chelators can inhibit Af biofilm formation (Nazik et al. 2015). The Pseudomonas quorum-sensing molecule PQS (Pseudomonas quinolone signal; 2-heptyl-3,4-dihydroxyquinoline) has antifungal activity under low iron conditions (Bredenbruch et al. 2006), but in higher amounts of iron, and dependent on Af siderophores, PQS supports fungal metabolism to a greater degree than iron alone (Nazik et al. 2020a). It was curious that phenazine deletion mutants were inhibitory because phenazines had been shown to inhibit Af (Mangan 1969; Kerr et al. 1999). However, when the intermicrobial competition studies were carried out in high iron concentrations,

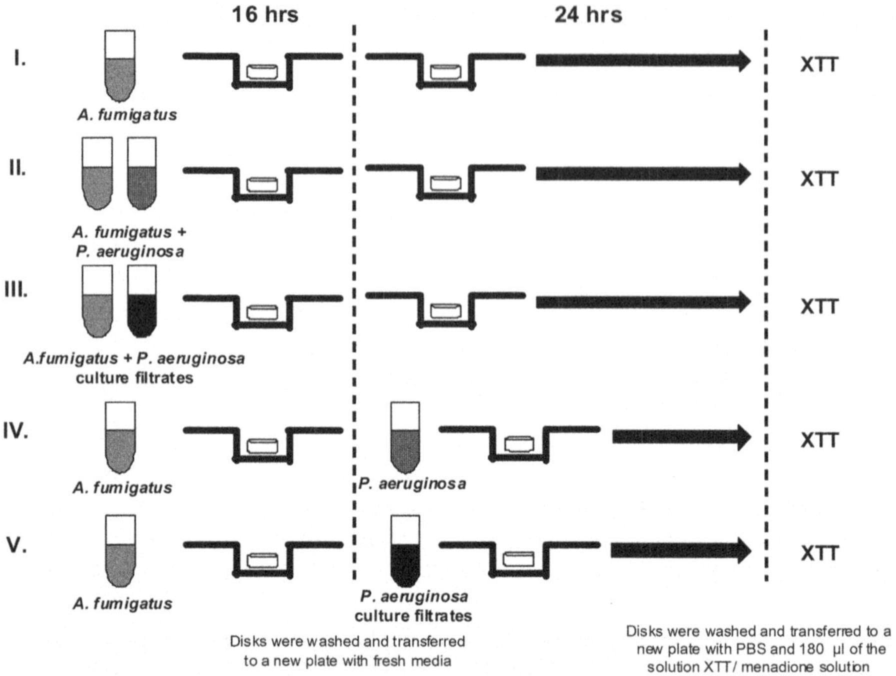

Fig. 13.4 Experimental design. To form an Af biofilm, polystyrene disks were placed in tissue culture plates with conidia and media. Disks were incubated to allow the conidia to attach. Following the attachment phase, disks were transferred to new plates containing and incubated for an additional 24 h (line I). Direct interaction of live *Pseudomonas aeruginosa* (Pa) on Af biofilm formation: suspensions of conidia and bacteria were combined in tissue culture plates for 16 h. Disks were then rinsed gently, transferred to new plates, and incubated for an additional 24 h (line II). Direct interaction of live Pa on preformed Af biofilm: fungal biofilms were formed as described. After 16 h the disks were rinsed, transferred to new plates containing Pa suspension, and incubated for an additional 24 h (lines II and IV). Pa planktonic supernatant assay: to obtain planktonic culture filtrates, Pa suspension was incubated in conical tubes for 24 h. The spent medium was centrifuged to remove suspended cells or debris. The supernatant was filter sterilized and added to wells of a tissue culture plate previously inoculated with Af suspension. Fungal biofilms are attached and formed. Disks containing biofilms were washed, transferred to a new plate, and incubated for an additional 24 h (line III). For the preformed biofilm assay, filtered supernatant was added to wells. Disks containing Af-preformed biofilms were washed, transferred to the plate containing the bacterial filtrates, and incubated for an additional 24 h. Af wells without bacterial supernatant were included as controls (line V). To obtain Pa biofilm filtrates, a suspension of Pa adhered to tissue culture flasks for 2 h (attachment phase). The flasks were rinsed, fresh RPMI-1640 was added to the flask, and adhered cells formed Pa biofilms for 22 h. The spent medium was then removed. Af conidia forming biofilms or preformed Af biofilms were challenged with the Pa biofilm culture filtrate (lines III and V). Figure reproduced from ref. (Ferreira et al. 2015)

where pyoverdine iron denial is neutralized, the phenazine pyocyanin then becomes the major inhibitor of Af biofilms (Sass et al. 2021a). The pseudomonal inhibitory molecules can synergize in inhibition against Af with antifungal drugs *(*Sass et al. 2021b*)*, but some anti-Pseudomonal drugs can enhance Af growth or metabolism

(Sass et al. 2022). All the studies summarized above are relevant to Af biofilms forming in a liquid medium, but it turns out that other Pseudomonas weapons (PQS, HHQ, rhamnolipids, and elastase) predominate for Pseudomonas' offense when Af is growing on solid medium (Sass et al. 2021c).

Further experiments reveal a counteracting, self-protective mechanism by Af against Pseudomonas (Reece et al. 2018; Margalit et al. 2020) although inhibition of the bacterium may not always be the case. The Af defense is largely based on Af hydroxamate siderophore production, dependent on the Af SidA gene (Sass et al. 2019b). Small organic molecules, released from Pseudomonas as volatiles, can inhibit Af as well (Nazik et al. 2020b).

Aspergillus in Nature

Aspergillus biofilms play an important role in environmental ecosystems: decomposition of organic material, such as dead plant matter, contributes to nutrient cycling and soil health. The biofilm mode of growth can enhance the fungal ability to degrade complex carbohydrates, important for the breakdown of lignocellulose in natural and agricultural environments. In soil, essential nutrients such as carbon, nitrogen, and phosphorus are released back into the environment, important for maintaining soil fertility and supporting plant growth. Biofilms also protect *Aspergillus* from environmental stresses such as drought and temperature fluctuations, enhancing its ability to persist. Their metabolic versatility makes them possible agents for bioremediation, the use of organisms to clean up environmental pollutants. *Aspergillus* species have been shown to degrade hydrocarbons, pesticides, and heavy metals in contaminated environments, such as polluted soils and water (Ramage et al. 2009; Richardson and Rautemaa-Richardson 2019).

In agriculture, controlling *Aspergillus* biofilms is important for ecosystem balance and Aspergillus toxin contamination. Strategies include monitoring moisture levels in agricultural fields to control fungal overgrowth and using biocontrol agents (of note, fungal viruses) to inhibit fungi.

Aspergillus in Industry

In industrial settings, *Aspergillus* biofilms can have both beneficial and detrimental impacts (Ramage et al. 2011). On the one hand, biofilms are harnessed for biotechnological applications, such as the production of industrial enzymes and bioactive compounds. On the other hand, *Aspergillus* biofilms can cause biofouling, where they adhere to surfaces in fermentation tanks, water pipelines, and cooling towers, leading to contamination and process inefficiency. *Aspergillus* biofilms can contribute to biocorrosion, where microbial activity accelerates the degradation of metal surfaces. Biofilm formation on metallic surfaces creates localized areas of chemical activity, which can lead to the breakdown of protective coatings and metal integrity. This process has significant implications for industries reliant on metal infrastructure, such as oil and gas, water treatment, and manufacturing. Physical barriers may be needed to prevent biofilm development on surfaces in industrial settings.

Mixed Biofilms

A special case is when biofilms might be mixed structures (Barac et al. 2024; Luo et al. 2022). Biofilms of *Aspergillus* spp. in nature, including on decaying plant matter, in water, or in animal species, may consist of multicellular aggregates, composed of fungal cells but also include bacteria, other fungal species, and sometimes even host cells. The microbial interactions within these biofilms may influence their formation, structure, and resistance to external stressors, including antimicrobial treatments. If biofilms with Aspergillus are created in vitro with a bacterial component, such as *Pseudomonas aeruginosa* or *Stenotrophomonas maltophilia,* both the fungal and bacterial components can be inhibited in growth and metabolism, fungal secondary metabolites decreased, the fungal cell walls increased, immune cell recruitment affected, and susceptibility of the bacterial component to antibacterials altered (Bastos et al. 2022; Melloul et al. 2016; Benoit et al. 2015; Ramírez Granillo et al. 2015). Bacteriophages can play a role in inhibiting the fungus (Penner et al. 2016).

The ratios of the two components in mixed biofilms, and whether hypoxia is present (or its degree) would affect the amount of growth or metabolic inhibition of the components (Manavathu et al. 2014; Bastos et al. 2022). Synergistic interactions also may occur within biofilms. Some bacterial species produce enzymes, such as proteases and lipases, that degrade host tissue, releasing nutrients that Aspergillus can utilize. These bacterial enzymes can indirectly promote fungal virulence and biofilm stability, illustrating a mutualistic relationship. Other microbes within the biofilm can release enzymes or toxins that facilitate biofilm dispersal, thereby enhancing the spread of fungal and bacterial pathogens.

Aspergillus spp. can secrete secondary metabolites, which may interfere with bacterial signaling pathways (Moree et al. 2012). Of particular note, gliotoxin, whose production is already increased as biofilm develops in pure Aspergillus biofilm, can be increased further by a Pseudomonas component (Bastos et al. 2022). This observation is intriguing because it could help explain why mixed infections in vivo, involving Aspergillus, are associated with greater virulence, and/or why Aspergillus colonization in patients can transition to Aspergillus invasion (Bastos et al. 2022; Svahn et al. 2014).

In vivo models could provide further insights into the role of microbial interactions in Aspergillus biofilm-associated infections (Sass and Stevens 2023). Research into affecting the bacterial members of mixed biofilms, such as affecting bacterial signaling molecules and observing the results of such efforts on the viability of the mixed biofilms, could be fruitful.

In nature, environmental factors, such as temperature and humidity, can modulate the interactions within Aspergillus biofilms. Biofilms of *Pseudomonas aeruginosa* can co-exist with *Aspergillus* spp. in both environmental and clinical settings. Aspergillus biofilms have been shown to interact with protozoa and other eukaryotic microorganisms in aquatic environments. These interactions can influence biofilm structure and microbial diversity. Predation by protozoa can selectively remove certain microbial species, altering the community composition within the biofilm (Wuertz et al. 2004).

Effect of Antifungals

The biofilm is an important factor in drug susceptibility (Roudbary et al. 2022; Jabra-Rizk et al. 2004; Perlin et al. 2017; Chandra et al. 2005; Shishodia et al. 2019; Rajendran et al. 2015; Borghi et al. 2016; Desai et al. 2014), enough so as to confound the wisdom of routine susceptibility testing with planktonic organisms (Simitsopoulou et al. 2016). Drug resistance increases with time of the biofilm development and hyphal impaction (Beauvais and Latge 2015; Mowat et al. 2008). Decreased drug diffusion, owing to physical barriers or chemical sequestering or binding of antifungals by the ECM, repulsion of antifungals due to electrostatic or hydrophobicity issues, or decreased pO_2 may be factors (Manavathu et al. 2014; Kowalski et al. 2020). Increased activity of efflux pumps occurs early in the biofilm development process (Kowalski et al. 2020; Rajendran et al. 2011; Holmes et al. 2016). ATP-binding cassette transporters are upregulated and the major facilitator superfamily (Kaur and Singh 2014). All three major antifungal drug classes (polyenes, azoles, and echinocandins) (Denning and Stevens 2023) can be resisted by the fungus (Kowalski et al. 2020). Amphotericin B, a drug where several mechanisms of antifungal action can now be postulated (Stevens and Hope n.d.), seems to be the most active and most rapidly active in this situation (Mowat et al. 2008). However, in mixed biofilms (preceding section of this review), *Pseudomonas aeruginosa* has been shown to enhance fungal resistance to amphotericin B, likely due to the production of bacterial virulence factors that modulate fungal biofilm structure (Manavathu et al. 2015).

The role of ECM in inducing drug resistance can be demonstrated by washing the biofilm, resulting in a decline in resistance (Rajendran et al. 2011). As the biofilm reaches full maturity, drug resistance can decline (Mowat et al. 2008). However, the basal biofilm layers, in particular, can provide a reservoir of cells with lowered metabolism, cells that are drug-resistant, that can reseed the biofilm (Manavathu et al. 2014; Kowalski et al. 2020).

The development of novel antifungal agents that target specific microbial interactions within *Aspergillus* biofilms is an active area of research (Ribeiro et al. 2016). For example, drugs, including host defense peptides (or their mimics), or probiotics, that disrupt microbial communication or biofilm formation are being explored as potential therapeutic agents (Pletzer et al. 2016; Woodburn et al. 2019; Larwood and

Stevens 2024). Bacteriophages have been proposed as potential treatments for biofilm-associated infections. These biological agents can selectively target bacterial co-infectors within the biofilm, potentially weakening the biofilm structure and enhancing the effectiveness of antifungal therapies.

Conclusion

The formation of biofilms by *Aspergillus* species has important implications for human health, industry, and environmental ecosystems. Efforts to inhibit the production of the ECM or inhibit the physical interactions between secreted components have emerged as an attractive anti-virulence and anti-drug-resistance approach. Knowing the ECM components and how they are produced and assembled (Muszkieta et al. 2013; Wongsuk et al. 2016), would be key, for example, in drug design, including in targeting, drug combinations and immunotherapy (Choi et al. 2023; Del Pozo 2021), and in new diagnostics. The effects of Aspergillus viruses on biofilm formation are unknown (Battersby et al. 2024). Transcriptional studies would be useful, such as in deciphering the functional roles of genes involved in biofilm formation, maintenance and resilience, and regulatory elements (Manavathu et al. 2014; Frisvad et al. 2009; Coenye et al. 2014). These efforts could lead to novel therapeutic strategies, including immunomodulation or targeting key genetic pathways with RNA interference (Muszkieta et al. 2013).

Acknowledgements We regret being unable to cite all the relevant articles in the literature.

We thank our Stanford University and California Institute for Medical Research collaborators, responsible for their studies cited, for their efforts, expertise, and friendship, which has taught us so much: Gabriele Sass, Lydia-Marie Joubert, Lynette Cegelski, Courtney Reichert, Karl V. Clemons.

References

Anand R, Clemons KV, Stevens DA (2017) Effect of anaerobiasis or hypoxia on Pseudomonas aeruginosa inhibition of Aspergillus fumigatus biofilm. Arch Microbiol 199:881–890

Bairwa G, Hee Jung W, Kronstad JW (2017) Iron acquisition in fungal pathogens of humans. Metallomics 9(3):215–227

Barac A, Vujovic A, Peric J, Tulic I, Stojanovic M, Stjepanovic M (2024) Rethinking aspergillosis in the era of microbiota and mycobiota. Mycopathologia 189(4):49

Barriuso J, Hogan DA, Keshavarz T, Martinez MJ (2018) Role of quorum sensing and chemical communication in fungal biotechnology and pathogenesis. FEMS Microbiol Rev 42(5):627–638

Bastos RW, Akiyama D, dos Reis TF, Colabardini AC, Luperini RS, de Castro PA, Baldini RL, Fill T, Goldman GH (2022) Secondary metabolites produced during *Aspergillus fumigatus* and *Pseudomonas aeruginosa* biofilm formation. MBio 13:4

Battersby J, Stevens DA, Coutts RHA, Havlíček V, Hsu JL, Sass G, Kotta-Loizou I (2024) The expanding mycovirome of *Aspergillus*. J. Fungi 10:585

Beauvais A, Latge J-P (2015) Aspergillus biofilm in vitro and in vivo. Microbiol Spectr 3:MB-0017

Beauvais A, Muller F-M (2009) Biofilm formation in *Aspergillus fumigatus*. In: Latge J-P, Steinbach WJ (eds) *Aspergillus fumigatus* and *Aspergillosis,* chapter 12. ASM Press, Washington, D.C.

Beauvais A, Schmidt C, Guadagnini S, Roux P, Perret E, Henry C, Paris S, Mallet A, Prévost MC, Latgé JP (2007) An extracellular matrix glues together the aerial-grown hyphae of Aspergillus fumigatus. Cell Microbiol 9:1588–1600

Beauvais A, Fontaine T, Aimanianda V, Latgé JP (2014) Aspergillus cell wall and biofilm. Mycopathologia 178(5–6):371–377

Benoit I, van den Esker MH, Patyshakuliyeva A, Mattern DJ, Blei F, Zhou M, Dijksterhuis J, Brakhage A, Kuipers OP, de Vries RP, Kovacs AT (2015) Bacillus subtilis attachment to *Aspergillus niger* hyphae results in mutually altered metabolism. Environ Microbiol 17(6):2099–2113

Binder U, Lass-Flörl C (2013) New insights into invasive aspergillosis–from the pathogen to the disease. Curr Pharm Des 19(20):3679–3688

Blyth W, Forey A (1971) The influence of respiratory bacteria and their biochemical fractions on Aspergillus fumigatus. Sabouraudia 9:273–282

Boisvert AA, Cheng MP, Sheppard DC, Nguyen D (2016) Microbial biofilms in pulmonary and critical care diseases. Ann Am Thorac Soc 13(9):1615–1623

Borghi E, Borgo F, Morace G (2016) Fungal biofilms: update on resistance. Adv Exp Med Biol 931:37–47

Bredenbruch F, Geffers R, Nimtz M, Buer J, Häussler S (2006) The Pseudomonas aeruginosa quinolone signal (PQS) has an iron-chelating activity. Environ Microbiol 8:1318–1329

Briard B, Rasoldier V, Bomme P, ElAouad N, Guerreiro C, Chassagne P, Muszkieta L, Latgé JP, Mulard L, Beauvais A (2017) Dirhamnolipids secreted from Pseudomonas aeruginosa modify antifungal susceptibility of Aspergillus fumigatus by inhibiting β1,3 glucan synthase activity. ISME J 11:1578–1591

Bruns S, Seidler M, Albrecht D, Salvenmoser S, Remme N, Hertweck C, Brakhage AA, Kniemeyer O, Muller F-MC (2010) Functional genomic profiling of *Aspergillus fumigatus* biofilm reveals enhanced production of the mycotoxin gliotoxin. Proteomics 10:3097–3107

Bugli F, Sterbini FP, Cacaci M, Lancelloti S, Stigliano E, Torelli R, Arena V, Caira M, Posterero P, Sanguinetti M, Posterero B (2014) Increased production of gliotoxin is related to the formation of biofilm by *Aspergillus fumigatus*. Pathog Dis 70:379–389

Chandra J, Zhou G, Ghannoum MA (2005) Fungal biofilms and antimycotics. Curr Drug Targets 6(8):887–894

Chandrasekar PH, Manavathu EK (2008) Do Aspergillus species produce biofilm? Future Microbiol 3(1):19–21

Chatterjee P, Moss CT, Omar S, Dhillon E, Borges CDH, Tang AC, Stevens DA, Hsu J (2024) Allergic bronchopulmonary aspergillosis (ABPA) in the era of cystic fibrosis transmembrane conductance regulator (CFTR) modulators. J. Fungi 10:656

Chen Y, Stewart JIP, Liu S, Sheppard DC, Lu L, Zhang S (2022) Spt20, a structural subunit of the SAGA complex, regulates *Aspergillus fumigatus* biofilm formation, asexual development, and virulence. Appl Environ Microbiol 88(1):e0153521

Choi V, Rohn JL, Stoodley P, Carugo D, Stride E (2023) Drug delivery strategies for antibiofilm therapy. Nat Rev Microbiol 21(9):555–572

Coenye T, Van Dijck P, Bjarnsholt T, Forsberg A (2014) Microbial biofilms–the coming of age of a research field. Pathog Dis 70(3):203–204

Debourgogne A, Monpierre L, Sy KA, Valsecchi I, Decousser JW, Botterel F (2023) Interactions between bacteria and *Aspergillus fumigatus* in airways: from the mycobiome to molecular interactions. J Fungi (Basel). 9(9):900

Del Pozo JL (2021) Novel treatment dynamics for biofilm-related infections. Expert Rev Anti-Infect Ther 19(11):1443–1456

Denning DW, Stevens DA (2023) Systemic antifungal agents. In: Goldman L, Cooney K (eds) Chapter 307, Goldman-cecil textbook of medicine, 27th edn. Elsevier

Denning DW, Cadranel J, Beigelman-Aubry C, Ader F, Chakrabarty A, Blot S, Ullmann AJ, Dimopoulos G, Lange C (2016) European Society for Clinical Microbiology and Infectious Diseases and European Respiratory Society. Chronic pulmonary aspergillosis: rationale and clinical guidelines for diagnosis and management. Eur Respir J 47(1):45–68

Desai JV, Mitchell AP, Andes DR (2014) Fungal biofilms, drug resistance, and recurrent infection. Cold Spring Harb Perspect Med 4(10):a019729

Dunne K, Reese E, McClean S, Doyle S, Rogers TR, Murphy P, Renwick J (2023) *Aspergillus fumigatus* supernatants disrupt bronchial epithelial monolayers: potential role for enhanced invasion in cystic fibrosis. J. Fungi 9:490

Ferreira JAG, Penner J, Moss RB, Haagensen JAJ, Clemons KV, Spormann AM, Nazik H, Cohen K, Banaei N, Carolino E, Stevens DA (2015) Inhibition of Aspergillus fumigatus and its biofilm by Pseudomonas aeruginosa is dependent on the source, phenotype and growth conditions of the bacterium. PLoS One 10:e0134692

Frisvad JC, Rank C, Nielsen KF, Larsen TO (2009) Metabolomics of Aspergillus fumigatus. Med Mycol 47(Suppl 1):S53–S71

Galluzzi L, Vitale I, Aaronson SA, Abrams JM, Adam D, Agostinis P, Alnemri ES, Altucci L, Amelio I, Andrews DW, Annicchiarico-Petruzzelli M, Antonov AV, Arama E, Baehrecke EH, Barlev NA, Bazan NG, Bernassola F, Bertrand MJM, Bianchi K, Blagosklonny MV, Blomgren K, Borner C, Boya P, Brenner C, Campanella M, Candi E, Carmona-Gutierrez D, Cecconi F, Chan FK, Chandel NS, Cheng EH, Chipuk JE, Cidlowski JA, Ciechanover A, Cohen GM, Conrad M, Cubillos-Ruiz JR, Czabotar PE, D'Angiolella V, Dawson TM, Dawson VL, De Laurenzi V, De Maria R, Debatin KM, DeBerardinis RJ, Deshmukh M, Di Daniele N, Di Virgilio F, Dixit VM, Dixon SJ, Duckett CS, Dynlacht BD, El-Deiry WS, Elrod JW, Fimia GM, Fulda S, García-Sáez AJ, Garg AD, Garrido C, Gavathiotis E, Golstein P, Gottlieb E, Green DR, Greene LA, Gronemeyer H, Gross A, Hajnoczky G, Hardwick JM, Harris IS, Hengartner MO, Hetz C, Ichijo H, Jäättelä M, Joseph B, Jost PJ, Juin PP, Kaiser WJ, Karin M, Kaufmann T, Kepp O, Kimchi A, Kitsis RN, Klionsky DJ, Knight RA, Kumar S, Lee SW, Lemasters JJ, Levine B, Linkermann A, Lipton SA, Lockshin RA, López-Otín C, Lowe SW, Luedde T, Lugli E, MacFarlane M, Madeo F, Malewicz M, Malorni W, Manic G, Marine JC, Martin SJ, Martinou JC, Medema JP, Mehlen P, Meier P, Melino S, Miao EA, Molkentin JD, Moll UM, Muñoz-Pinedo C, Nagata S, Nuñez G, Oberst A, Oren M, Overholtzer M, Pagano M, Panaretakis T, Pasparakis M, Penninger JM, Pereira DM, Pervaiz S, Peter ME, Piacentini M, Pinton P, Prehn JHM, Puthalakath H, Rabinovich GA, Rehm M, Rizzuto R, Rodrigues CMP, Rubinsztein DC, Rudel T, Ryan KM, Sayan E, Scorrano L, Shao F, Shi Y, Silke J, Simon HU, Sistigu A, Stockwell BR, Strasser A, Szabadkai G, Tait SWG, Tang D, Tavernarakis N, Thorburn A, Tsujimoto Y, Turk B, Vanden Berghe T, Vandenabeele P, Vander Heiden MG, Villunger A, Virgin HW, Vousden KH, Vucic D, Wagner EF, Walczak H, Wallach D, Wang Y, Wells JA, Wood W, Yuan J, Zakeri Z, Zhivotovsky B, Zitvogel L, Melino G, Kroemer G (2018) Molecular mechanisms of cell death: recommendations of the nomenclature committee on cell death. Cell Death Differ 25(3):486–541

Gomez-Lopez A, Rueda C, Pozo RP, Gonzalez LMS (2022) Dynamics of gliotoxin and bis(methylthio) gliotoxin production during the course of *Aspergillus fumigatus* infection. Med Mycol 60(4):myac025

Gravelat FN, Beauvais A, Liu H, Lee MJ, Snarr BD, Chen D, Xu W, Kravtsov I, Hoareau CMQ, Vanier G, Urb M, Campoli P, Al Abdallah QA, Lehoux M, Chabot JC, Ouimet M-C, Baptista SD, Fritz JH, Nierman WC, Latge JP, Mitchell AP, Filler SG, Fontaine T, Sheppard DC (2013) Aspergillus galactoaminogalactan mediates adherence to host constituents and conceals hyphal beta-glucan from the immune system. PLoS Pathog 9:e1003575

Guimarães LL, Takahashi HK (2014) A snapshot of extracellular DNA influence on Aspergillus biofilm. Front Microbiol 5:260

Hernández-Chávez MJ, Pérez-García LA, Niño-Vega GA, Mora-Montes HM (2017) Fungal strategies to evade the host immune recognition. J Fungi (Basel) 3(4):51

Holmes AR, Cardno TS, Strouse JJ, Ivnitski-Steele I, Keniya MV, Lackovic K, Monk BC, Sklar LA, Cannon RD (2016) Targeting efflux pumps to overcome antifungal drug resistance. Future Med Chem 8(12):1485–1501

Hsu JL, Manouvakhova OV, Clemons KV, Inayathullah M, Tu AB, Sobel RA, Tian W, Nazik H, Pothineni VR, Pasupneti S, Jiang X, Dhillion GS, Bedi H, Rajadas J, Haas H, Aurelian L, Stevens DA, Nicolls MR (2018) Microhemorrhage-associated tissue iron enhances the risk for Aspergillus fumigatus invasion in murine airway transplantation. Sci Trans Med 10:eaag2616

Illek B, Fischer H, Machen TE, Hari G, Clemons KV, Sass G, Ferreira JA, Stevens DA (2023) Protective role of CFTR during fungal infection of cystic fibrosis bronchial epithelial cells with *Aspergillus fumigatus*. Front Cell Infect Microbiol 13:1196581

Imbert C (ed) (2018) Fungal biofilms and related infections. Advances in microbiology, infectious diseases and public health volume 3. Springer, Cham

Jabra-Rizk MA, Falkle WA, Meiller TF (2004) Fungal biofilms and drug resistance. Emerg Infect Dis 10(1):14–19

Joubert L-M, Ferreira JAG, Stevens DA, Cegelski L (2015) Aspergillus fumigatus biofilms: a comparison of processing techniques for scanning electron microscopy visualization of fungal mycelium and extracellular matrix. Microsc Microanal 21(Suppl. S3):935–936

Joubert L-M, Ferreira JAG, Stevens DA, Nazik H, McDonald KL, Cegelski L (2017) Visualization of Aspergillus fumigatus biofilms with SEM (scanning electron microscopy) and VP-SEM (variable pressure-SEM): a comparison of processing techniques. J Microbiol Methods 132:46–55

Kaur S, Singh S (2014) Biofilm formation by Aspergillus fumigatus. Med Mycol 52(1):2–9

Kernien JF, Snarr BD, Sheppard DC, Nett JE (2018) The interface between fungal biofilms and innate immunity. Front Immunol 8:1968

Kerr JR, Taylor GW, Rutman A, Høiby N, Cole PJ, Wilson R (1999) *Pseudomonas aeruginosa* pyocyanin and 1-hydroxyphenazine inhibit fungal growth. J Clin Pathol 52:385–387

King J, Brunel SF, Warris A (2016) Aspergillus infections in cystic fibrosis. J Infect 72(Suppl):S50–S55

König S, Pace S, Pein H, Heinekamp T, Kramer J, Romp E, Strassburger M, Troisi F, Proschak A, Dworschak J, Scherlach K, Rossi A, Sautebin L, Haeggström JZ, Hertweck C, Brakhage AA, Gerstmeier J, Proschak E, Werz O (2019) Gliotoxin from *Aspergillus fumigatus* abrogates leukotriene B4 formation through inhibition of leukotriene A4 hydrolase. Cell Chem Biol 26(4):524–534

Kowalski CH, Morelli KA, Schultz D, Nadell CD, Cramer RA (2020) Fungal biofilm architecture produces hypoxic microenvironments that drive antifungal resistance. Proc Nat, Acad Sci 117:22473–22483

Krappman S, Ramage G (2013) A sticky situation: extracellular DNA shapes *Aspergillus fumigatus biofilms*. Front Microbiol. 4:1

Larwood DJ, Stevens DA (2024) Antifungal activity of brilacidin, a nonpeptide host defense molecule. Antibiotics 13:405

Latge JP (2001) The pathobiology of Aspergillus fumigatus. Trends Microbiol 9:382–389

Lee MJ, Liu H, Barker BM, Snarr BD, Gravelat FN, Al Abdallah Q, Gavino C, Baistrocchi SR, Ostapska H, Xiao T, Ralph B, Solis NV, Lehoux M, Baptista SD, Thammahong A, Cerone RP, Kaminsky SG, Guiot MC, Latgé JP, Fontaine T, Vinh DC, Filler SG, Sheppard DC (2015) The fungal exopolysaccharide galactosaminogalactan mediates virulence by enhancing resistance to neutrophil extracellular traps. PLoS Pathog 11:e10065187

Lin CJ, Hou YH, Chen YL (2020) The histone acetyltransferase GcnE regulates conidiation and biofilm formation in Aspergillus fumigatus. Med Mycol 58(2):248–259

Liu S, Le Mauff F, Sheppard DC, Zhang S (2022) Filamentous fungal biofilms: conserved and unique aspects of extracellular matrix composition, mechanisms of drug resistance and regulatory networks in Aspergillus fumigatus. NPJ Biofilms Microbiomes 8(1):83

Loussert C, Schmitt C, Prevost MC, Balloy V, Fadel E, Philippe B, Kauffmann-Lacroix C, Latgé J-P, Beauvais A (2010) In vivo biofilm composition of Aspergillus fumigatus. Cell Microbiol 12(3):405–410

Luo A, Wang F, Sun D, Liu X, Xin B (2022) Formation, development, and cross-species interactions in biofilms. Front Microbiol 12:75732

Manavathu EK, Vager DL, Vazquez JA (2014) Development and antimicrobial susceptibility studies of in vitro monomicrobial and polymicrobial biofilm models, with *Aspergillus fumigatus* and *Pseudomonas aeruginosa*. BMC Microbiol 14:53

Manavathu EK, Vager DL, Vazquez JA (2015) Fungal growth phase dependent fungicidal interaction of *Pseudomonas aeruginosa* and *Aspergillus fumigatus*. J Multidiscip Pathol 2:1–16

Mangan A (1969) Interactions between some aural Aspergillus species and bacteria. J Gen Microbiol 58:261–266

Mannan M, Nabeela S, Mishra R, Uppuluri P (2024) Host immune response against fungal biofilms. Curr Opin Microbiol 81:102520

Margalit A, Carolan JC, Sheehan D, Kavanagh K (2020) The *Aspergillus fumigatus* secretome alters the proteome of *Pseudomonas aeruginosa*. Mol Cell Proteomics 19(8):1346–1359

Martin LW, Reid DW, Sharples KJ, Lamont IL (2011) Pseudomonas siderophores in the sputum of patients with cystic fibrosis. Biometals 24:1059–1067

Martinez LR, Fries BC (2010) Fungal biofilms: relevance in the setting of human disease. Curr Fungal Infect Rep 4(4):266–275

Matthaiou EI, Sass G, Stevens DA, Hsu JL (2018) Iron: an essential nutrient for *Aspergillus fumigatus* and a fulcrum for pathogenesis. Curr Opin Infect Dis 31:506–511

Melloul E, Luigi S, Anais L, Arne P, Costa J-M, Fihman V, Briard B, Dannaoui E, Guillot G, Decousser J-W, Beauvais A, Botterel F (2016) Characteristics of *Aspergillus fumigatus* in association with *Stenotrophomonas maltophilia* in an in vitro model of mixed biofilm. PLoS One 11:e0166325

Mitchell KF, Zarnowski R, Andes DR (2016) The extracellular matrix of fungal biofilms. Adv Exp Med Biol 931:21–35

Moree WJ, Phelan VV, Wu CH, Bandeira N, Cornett DS, Duggan BM, Dorrestein PC (2012) Interkingdom metabolic transformations captured by microbial imaging mass spectrometry. Proc Natl Acad Sci USA 109(34):13811–13816

Morelli KA, Kerkaert JD, Cramer RA (2021) Aspergillus fumigatus biofilms: toward understanding how growth as a multicellular network increases antifungal resistance and disease progression. PLoS Pathog 17(8):e1009794

Mowat E, Butcher J, Lang S, Williams C, Ramage G (2007) Development of a simple model for studying the effects of antifungal agents on multicellular communities of Aspergillus fumigatus. J Med Microbiol 56:1205–1212

Mowat E, Lang S, Williams C, McCulloch E, Jones B, Ramage G (2008) Phase-dependent antifungal activity against *Aspergillus fumigatus* developing multicellular filamentous biofilms. J Antimicrob Chemother 62:1281–1284

Mowat E, Williams C, Jones B, McChlery S, Ramage G (2009) The characteristics of Aspergillus fumigatus mycetoma development: is this a biofilm? Med Mycol 47(Suppl 1):S120–S126

Mowat E, Rajendran R, Williams C, McCulloch E, Jones B, Lang E, Ramage R (2010) *Pseudomonas aeruginosa* and their small diffusible extracellular molecules inhibit *Aspergillus fumigatus* biofilm formation. FEMS Microbiol Lett 313:96–102

Muller F-MC, Seidler M, Beauvais A (2011) Aspergillus fumigatus biofilms in the clinical setting. Med Mycol 49(Suppl. 1):S96–S120

Muszkieta L, Beauvais A, Pähtz V, Gibbons JG, Anton Leberre V, Beau R, Shibuya K, Rokas A, Francois JM, Kniemeyer O, Brakhage AA, Latgé J-P (2013) Investigation of Aspergillus fumigatus biofilm formation by various "omics" approaches. Front Microbiol 4:13

Nazik H, Penner JC, Ferreira JAG, Haagensen JAJ, Cohen K, Spormann AM, Martinez M, Chen V, Hsu JL, Clemons KV, Stevens DA (2015) Effect of iron chelators on the formation and development of Aspergillus fumigatus biofilm. Antimicrob Agents Chemother 59:6514–6520

Nazik H, Moss RB, Karna V, Clemons KV, Banaei N, Cohen K, Choudhary V, Stevens DA (2017) Are cystic fibrosis aspergillus fumigatus isolates different? Intermicrobial interactions with Pseudomonas. Mycopathologia 182:315–318

Nazik H, Sass G, Ansari SR, Ertekin R, Haas H, Deziel E, Stevens DA (2020a) Novel intermicrobial molecular interaction: Pseudomonas aeruginosa Quinolone Signal (PQS) modulates *Aspergillus fumigatus* response to iron. Microbiology 166:44–55

Nazik H, Sass G, Deziel E, Stevens DA (2020b) Aspergillus is inhibited by *Pseudomonas aeruginosa* volatiles. J Fungi 6(3):118. https://doi.org/10.3390/jof6030118

Nett JE, Andes D (2015) Fungal biofilms: in vivo models for discovery of anti-biofilm drugs. Microbiol Spectr 3(3):E30

Nijland R, Hall MJ, Burgess JG (2010) Dispersal of biofilms by secreted, matrix degrading, bacterial DNase. PLoS One 5(12):e15668

Patel R, Hossain MA, German N, Al-Ahmad AJ (2018) Gliotoxin penetrates and impairs the integrity of the human blood-brain barrier in vitro. Mycotoxin Res 34:257–268

Paulussen CE, Hallsworth JE, Sergio Álvarez-Pérez S, Nierman WC, Hamill PG, Blain D, Rediers H, Lievens B (2016) Ecology of aspergillosis: insights into the pathogenic potency of *Aspergillus fumigatus* and some other *Aspergillus* species. Microbial Biotechnol 10(2):296–322

Penner JC, Ferreira JAG, Secor PR, Sweere J, Birukova M, Joubert L-M, Haagensen JAJ, Garcia O, Malkovskiy AV, Kaber G, Nazik H, Manasherob R, Spormann AM, Clemons KV, Stevens DA, Bollyky PL (2016) Pf4 bacteriophage produced by Pseudomonas aeruginosa inhibits *Aspergillus fumigatus* metabolism via iron sequestration. Microbiology 162:1583–1594

Perlin DS, Rautemaa-Richardson R, Alastruey-Izquierdo A (2017) The global problem of antifungal resistance: prevalence, mechanisms, and management. Lancet Infect Dis 17(12):e383–e392

Pletzer D, Coleman SR, Hancock RE (2016) Anti-biofilm peptides as a new weapon in antimicrobial warfare. Curr Opin Microbiol 33:35–40

Rajendran R, Mowat E, McCulloch E, Lappin DF, Jones B, Lang S, Majithiya JB, Warn P, Williams C, Ramage G (2011) Azole resistance of *Aspergillus fumigatus* biofilm is partly associated with efflux pump activity. Antimicrob Agents Chemother 55:2092–2097

Rajendran R, Mowat E, Jones B, Williams C, Ramage G (2015) Prior in vitro exposure to voriconazole confers resistance to amphotericin B in *Aspergillus fumigatus* biofilms. Int J Antimicrob Agents 46(3):342–345

Ramage G, Mowat E, Jones B, Williams C, Lopez-Ribot J (2009) Our current understanding of fungal biofilms. Crit Rev Microbiol 35(4):340–355

Ramage G, Rajendran R, Lambris-Correa M, Jones B, Williams C (2011) Aspergillus biofilms: clinical and industrial significance. FEMS Microbiol Lett 324(2):89–97

Ramírez Granillo A, Canales MG, Espíndola ME, Martínez Rivera MA, de Lucio VM, Tovar AV (2015) Antibiosis interaction of *Staphylococccus aureus* on *Aspergillus fumigatus* assessed in vitro by mixed biofilm formation. BMC Microbiol 15:33

Reece E, Doyle S, Greally P, Renwick J, McClean S (2018) *Aspergillus fumigatus* inhibits *Pseudomonas aeruginosa* in co-culture: implications of a mutually antagonistic relationship on virulence and inflammation in the CF airway. Front Microbiol 9:1205

Reichhardt C, Ferreira JAG, Joubert L, Clemons KV, Stevens DA, Cegelski L (2015) Analysis of the Aspergillus fumigatus biofilm extracellular matrix by solid-state nuclear magnetic resonance. Eukaryot Cell 14:1064–1072

Reichhardt C, Stevens DA, Cegelski L (2016) Fungal biofilm composition and opportunities in drug discovery. Future Med Chem 8:1455–1468

Reichhardt C, Joubert L-M, Clemons KV, Stevens DA, Cegelski L (2019) Integration of electron microscopy and solid-state NMR analysis for new views and compositional parameters of A. fumigatus biofilms. Med Mycol 57:S239–S244

Ribeiro SM, Felício MR, Boas EV, Gonçalves S, Costa FF, Samy RP, Santos NC, Franco OL (2016) New frontiers for anti-biofilm drug development. Pharmacol Ther 160:133–144

Richardson M, Rautemaa-Richardson R (2019) Exposure to *Aspergillus* in home and healthcare facilities' water environments: focus on biofilms. Microorganisms 7(1):7

Roudbary M, Vahedi-Shahandashti R, Santos ALSD, Roudbar Mohammadi S, Aslani P, Lass-Flörl C, Rodrigues CF (2022) Biofilm formation in clinically relevant filamentous fungi: a therapeutic challenge. Crit Rev Microbiol 48(2):197–221

Sass G, Stevens DA (2023) Model of pulmonary co-infection of Aspergillus and Pseudomonas in immunocompetent mice, and review of the literature. Microbiol Res 14:1843–1861

Sass G, Nazik H, Penner J, Shah H, Ansari SR, Clemons KV, Groleau M-C, Dietl A-M, Visca P, Haas H, Déziel E, Stevens DA (2018) Studies of *Pseudomonas aeruginosa* mutants indicate pyoverdine as the central factor in inhibition of *Aspergillus fumigatus* biofilm. J Bacteriol 200(1):e00345–e00317

Sass G, Nazik H, Penner J, Shah H, Ansari SR, Clemons KV, Groleau M-C, Dietl A-M, Visca P, Haas H, Deziel E, Stevens DA (2019a) Aspergillus-Pseudomonas interaction, relevant to competition in airways. Med Mycol 57:S228–S232

Sass G, Ansari SR, Dietl A-M, Deziel E, Haas H, Stevens DA (2019b) Intermicrobial interaction: *Aspergillus fumigatus* siderophores protect against competition by *Pseudomonas aeruginosa*. PLoS One 14(5):e0216085

Sass G, Nazik H, Chatterjee P, Stevens DA (2021a) In high iron environment, pyocyanin is a major anti-fungal molecule; differences in prototype *P. aeruginosa* strains. Med Mycol 59:453–464

Sass G, Shrestha P, Stevens DA (2021b) *Pseudomonas aeruginosa* virulence factors support voriconazole effects on Aspergillus. Pathogens 10:519

Sass G, Nazik H, Chatterjee P, Shrestha P, Groleau M-C, Deziel E, Stevens DA (2021c) Altered Pseudomonas strategies to inhibit surface Aspergillus colonies. Front Cell Infect Microbiol 11:734296

Sass G, Scherpe L, Martinez M, Marsh JJ, Stevens DA (2022) Metrics of anti-fungal effects of ciprofloxacin on Aspergillus fumigatus planktonic growth and biofilm metabolism; effects of iron and siderophores. J. Fungi 8:240

Scharf DH, Brakhage AA, Mukherjee PK (2016) Gliotoxin–bane or boon? Environ Microbiol 18(4):1096–1109

Sehgal IS, Muthu V, Seidel D, Sprute R, Armstrong-James D, Asano K, Chalmers JD, Gangneux JP, Godet C, Salzer HJF, Cornely OA, Agarwal R (2024) EQUAL ABPA score 2024: a tool to measure guideline adherence for managing allergic bronchopulmonary aspergillosis. Mycoses 67(10):e13810

Seidler MJ, Salvenmoser S, Muller F-MC (2008) Aspergillus fumigatus forms biofilms with reduced antifungal drug susceptibility on bronchial epithelial cells. Antimicrob Agents Chemother 52:4130–4136

Shankar J, Thakur R, Clemons KV, Stevens DA (2024) Interplay of cytokines and chemokines in aspergillosis. J. Fungi 10:251

Sheppard DC, Howell PL (2016) Biofilm exopolysaccharides of pathogenic fungi: lessons from bacteria. J Biol Chem 291(24):12529–12537

Shin SH, Ye MK, Lee DW, Geum SY (2023) Immunopathologic role of fungi in chronic rhinosinusitis. Int J Mol Sci 24(3):2366

Shirazi F, Ferreira JAG, Stevens DA, Clemons KV, Kontoyiannis DP (2016) *Pseudomonas aeruginosa* strains isolated from cystic fibrosis patients strongly induce apoptosis and inhibition of preformed *Aspergillus fumigatus* biofilm. PLoS One 11(3):e0150155

Shishodia SK, Tiwari S, Shankar J (2019) Resistance mechanism and proteins in *Aspergillus* species against antifungal agents. Mycology 10(3):151–165

Shopova I, Bruns S, Thywissen A, Kniemeyer O, Brakhage AA, Hillmann F (2013) Extrinsic extracellular DNA leads to biofilm formation and colocalizes with matrix polysaccharides in the human pathogenic fungus *Aspergillus fumigatus*. Front Microbiol. 4:1

Simitsopoulou M, Chatzimoschou A, Roilides E (2016) Biofilms and antifungal susceptibility testing. Methods Mol Biol 1356:183–197

Singhal D, Baker L, Wormald PJ, Tan L (2011) *Aspergillus fumigatus* biofilm on primary human sinonasal epithelial culture. Am J Rhinol Allergy 25(4):219–225

Speth C, Hagleitner M, Ott HW, Wurzner R, Lass-Florl C, Rambach G (2013) *Aspergillus fumigatus* activates thrombocytes by secretion of soluble compounds. J Infect Dis 207:823–833

Stevens DA, Hope W (n.d.) Polyene antifungals. In: Blaser MJ, Cohen JI, Holland SM (eds) Principles and practice of infectious disease, 10th edn. Elsevier, Philadelphia. in press

Stevens DA, Brummer E, Clemons KV (2006) Interferon-gamma as an antifungal. J Infect Dis 194:S33–S37

Stewart PS, White B, Boegli L, Hamerly T, Williamson KS, Franklin MJ, Bothner B, James GA, Fisher S, Vital-Lopez FG, Wallqvist A (2019) Conceptual model of biofilm antibiotic tolerance that integrates phenomena of diffusion, metabolism, gene expression, and physiology. J Bacteriol 201(22):e00307–e00319

Sugui JA, Rose SR, Nardone G, Swamydas M, Lee CR, Kwon-Chung KJ, Lionakis MS (2017) Host immune status-specific production of gliotoxin and bis-methyl-gliotoxin during invasive aspergillosis in mice. Sci Rep 7(1):10977

Svahn KS, Goransson U, Chryssanthou E, Olsen B, Sjolin J, Stromstedt AA (2014) Induction of gliotoxin secretion in *Aspergillus fumigatus* by bacteria-associated molecules. PLoS One 9:e93685

Valsecchi I, Dupres V, Stephen-Victor E, Guijarro JI, Gibbons J, Beau R, Bayry J, Coppee JY, Lafont F, Latgé JP, Beauvais A (2017) Role of hydrophobins in *Aspergillus fumigatus*. J Fungi (Basel) 4(1):2

Van Acker H, Coenye T (2016) The role of efflux and physiological adaptation in biofilm tolerance and resistance. J Biol Chem 291:P12565–P12572

Williams C, Rajendran R, Ramage G (2016) Aspergillus biofilms in human disease. Adv Exp Med Biol 931:1–11

Wongsuk T, Pumeesat P, Luplertlop N (2016) Fungal quorum sensing molecules: role in fungal morphogenesis and pathogenicity. J Basic Microbiol 56(5):440–447

Woodburn KW, Clemens LE, Jaynes J, Joubert L-M, Botha A, Nazik H, Stevens DA (2019) Designed antimicrobial peptides for recurrent vulvovaginal candidiasis treatment. Antimicrob Agents Chemother 63:e02690–e02618

Wu S, Wang Y, Liu N, Dong G, Sheng C (2017) Tackling fungal resistance by biofilm inhibitors. J Med Chem 60(6):2193–2211

Wuertz S, Okabe S, Hausner M (2004) Microbial communities and their interactions in biofilm systems: an overview. Water Sci Technol 49(11–12):327–336

Zheng H, Kim J, Liew M, Yan JK, Herrera O, Bok JW, Kelleher NL, Keller NP, Wang Y (2015) Redox metabolites signal polymicrobial biofilm development via the NapA oxidative stress cascade in Aspergillus. Curr Biol 25(1):29–37

Chapter 14
Assays to Interrogate *Aspergillus fumigatus* Biofilm Fitness

Jane T. Jones, Charles T. S. Puerner, and Robert A. Cramer

Abstract *Aspergillus fumigatus* is a filamentous fungus commonly found in compost piles, but also ubiquitously throughout the environment. Importantly, it is a major causative agent of pulmonary fungal diseases, including invasive aspergillosis (IA), chronic pulmonary aspergillosis (CPA), aspergillomas, and allergic bronchopulmonary aspergillosis (ABPA). Despite overall effectiveness against asexual spores (conidia) in vitro, antifungal treatments often fail to improve overall health outcomes in treated patients. There are many potential reasons behind treatment failures, including a change in the nutritional environment, accessibility of the drug to the fungus, complex microbial communities, and different stages of growth of the fungus in the lung vs. in vitro. In the mammalian lung, *A. fumigatus* grows as a complex network of hyphae that may form biofilms. Features of *A. fumigatus* biofilms include production of an extracellular matrix, adherence to surfaces, and most importantly, reduced susceptibility to antifungal drug treatment. Classical in vitro antifungal drug testing occurs on *A. fumigatus* conidia; however, conidia are not the morphotype of the fungus treated during clinical infection which are highlighted by the formation of biofilms. Therefore, methods to grow and assess damage to *A. fumigatus* biofilms in vitro are critical in determining the effectiveness of both contemporary antifungal drugs and investigational molecules at different stages of drug development. Here, we review novel methods of growing *A. fumigatus* submerged biofilms in vitro and a variety of methods that can be utilized to assess the effect of molecules on the fungal biofilm including metabolic activity, lysis, membrane permeability, biofilm biomass reduction, and biofilm recovery after damage. These assays are expected to better inform potential efficacy of contemporary and novel antifungal treatments.

J. T. Jones · C. T. S. Puerner · R. A. Cramer (✉)
Geisel School of Medicine, Dartmouth, Hanover, NH, USA
e-mail: Robert.a.cramer.jr@dartmouth.edu

G. Ramage, D. Andes (eds.), *Fungal Biofilms*, Springer Series on Biofilms 15,
https://doi.org/10.1007/978-3-031-93315-8_14

Introduction

Aspergillus fumigatus (*A. fumigatus*) is a filamentous fungus that can grow in both diverse environmental niches and clinical settings. As a saprophyte, it utilizes varied sources of organic material found in compost piles, soil, homes, and mammalian hosts (Loussert et al. 2010). In immunocompromised patients, *A. fumigatus* can colonize the mammalian lung and initiate an infection that leads to invasive disease. While there are four classes of antifungals approved for the treatment of fungal infections and three for *A. fumigatus* specifically, antifungal drug efficacy remains sub-optimal (Beauvais and Latgé 2015; Perea and Patterson 2002). In vivo antifungal drug efficacy is determined by complex and as yet fully defined factors. Consequently, it remains unclear why infections caused by strains susceptible to antifungals in vitro may still cause significant morbidity and mortality despite contemporary antifungal therapies. With regard to infections caused by *A. fumigatus* and related molds, one potential mechanism underlying antifungal drug therapy failure is the fungal population-level morphology found at sites of infection in relation to how antifungal efficacy is tested in vitro.

Current antifungal susceptibility testing for molds specifically involves the conidial stage of fungal development (Wayne, PA 2008). Conidia initiate infection but are not the morphological state of the mold found in established infection environments (Loussert et al. 2010). Following germination of an *A. fumigatus* conidia (asexual propagule) or spore (sexual propagule) long, branched, multi-nucleate hyphae form a complex network of cells termed a mycelium. The mycelium adheres to multiple surfaces in part through production of a polysaccharide rich extracellular matrix that can be found in vitro and in vivo (Loussert et al. 2010; Beauvais and Latgé 2015). As these filamentous fungal mycelia are also generally recalcitrant to antifungal treatments in monoculture (Kowalski et al. 2020) and in co-culture with epithelial cells (Seidler et al. 2008), recent literature has described these *A. fumigatus* mycelia as biofilms (Harding et al. 2009; Chandrasekar and Manavathu 2008; Mowat et al. 2009). The initiation development, maintenance, and emergent properties of filamentous fungal biofilms remain to be fully defined.

Currently, there are two main models used to study *A. fumigatus* biofilm development and function in vitro. The first is the use of a solid agar substrate containing a medium of choice. As *A. fumigatus* grows robustly on a diverse array of carbon and nitrogen sources, medium composition used in the literature is highly varied though most utilize non-host relevant concentrations of glucose as the primary carbon source. This method of mold growth can be used to assess susceptibility to a diverse array of chemicals and conditions that induce a stress response including antifungal drugs. For this method, a defined concentration of *A. fumigatus* conidia are inoculated onto the center of the plate and assessed for growth (colony diameter at given time points and/or radial growth rates), conidiation, and colony morphology typically grown at 37 °C 5%CO_2 and ambient oxygen for 24–72 h. These "colony" biofilms have been the traditional approach to study *A. fumigatus* and related molds and has provided a great deal of knowledge and information about filamentous fungal development and stress responses. However, a key feature of this approach is the robust asexual development that colony biofilms undergo in this culture condition. As in vivo biofilms seldom undergo asexual reproduction, the physiology of the fungal cells with this approach is quite distinct from in vivo fungal growth. Moreover, a consideration for interpretation of results is that both conidia, at the initial inoculation, and subsequent hyphae, are exposed to the stress agent under study. Thus, additional experiments and models are ultimately needed to determine the morphology of the fungus impacted by the stress under study. For example, alternative methods are needed to determine if the stress agent under study kills a certain percentage of the inoculation conidia and/or inhibit hyphal growth. Stability of the respective antifungal compounds in the medium is also a consideration in addition to drug diffusion rates in agar. A variation of this approach includes serial conidia dilution assays similar to what is utilized with yeast and bacteria (Wayne, PA 2008).

As it became increasingly clear that hyphae respond differently to important stress-inducing compounds such as antifungal drugs, a second model, the submerged biofilm model, emerged as a method to study *A. fumigatus* biology. Interestingly, this model must utilize a specific range of conidial density to form an adhered biofilm on a polystyrene surface, as conidia concentrations either too low or high fail to form biofilms (Mowat et al. 2007). The mechanisms driving the density-dependent conidia requirement for an in vitro adhered biofilm community to form remain ill-defined. Importantly, submerged biofilms do not conidiate until hyphae reach the air–liquid interface and thus more accurately represent *A. fumigatus* communities observed in infection models and in histopathology samples from patients (Fig. 14.1). A similar alternative to the submerged biofilm model in polystyrene plates is growth in co-culture with mammalian epithelial cells. This model is arguably more clinically relevant; however, the added complexity of an additional

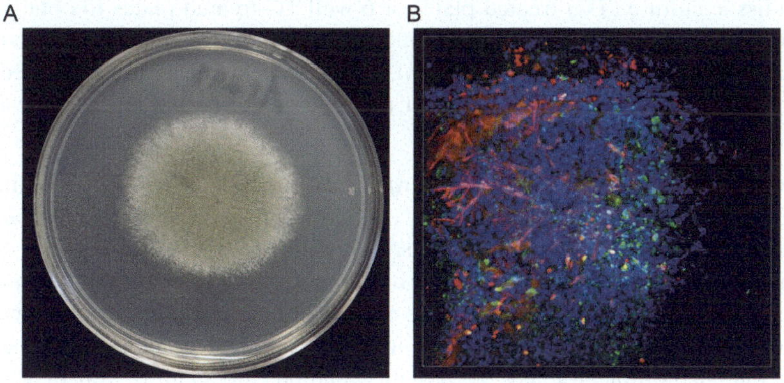

Fig. 14.1 Comparison of agar plate colony biofilms and in vivo biofilms. (**a**) Wild-type strain CEA10 was inoculated onto an agar plate containing GMM (1×10^3 conidia), grown for 72 h and imaged. (**b**) Outbred CD-1 mice were immune suppressed with Kenalog 40 mg/kg for 24 h and subsequently inoculated with laboratory-evolved strain EVOL-20^{tdtomato}. 4 days later, lungs were excised, tissue was cleared by miPACT methods, and stained with DAPI and FITC-SBA. Lungs were imaged using spinning disc confocal microscopy. Red stain indicates biofilm morphology

organism as well as significant damage to host cells induced by the growing fungus can make interpretation of results from these experiments challenging (Seidler et al. 2008; Singhal et al. 2011). Using a variety of models, both in simplicity (in vitro) and complexity (co-culture) is critical to understanding emergent properties of biofilms and how to treat complex *A. fumigatus* infections. This chapter will discuss different methods of characterizing, treating, and troubleshooting submerged *A. fumigatus* in vitro biofilms.

Aspergillus fumigatus Biofilm Growth and Analysis

Previous studies (Mowat et al. 2007; Kowalski et al. 2020; Kerkaert et al. 2022; Jones et al. 2023) have described methods to grow submerged *A. fumigatus* biofilms. *A. fumigatus* conidia are first grown for 3 days as colony biofilms on glucose minimal media, GMM (1% glucose, 6-g/L $NaNO_3$, 0.52-g/L KCl, 0.52-g/L $MgSO_4•7H_2O$, 1.52-g/L KH_2PO_4 monobasic, 2.2-mg/L $ZnSO_4•7H_2O$, 1.1-mg/L H_3BO_3, 0.5-mg/L $MnCl_2•4H_2O$, 0.5-mg/L $FeSO_4•7H_2O$, 0.16-mg/L $CoCl_2•5H_2O$, 0.16-mg/L $CuSO_4•5H_2O$, 0.11-mg/L $(NH_4)_6Mo_7O_{24}•4H_2O$, and 5-mg/L Na_4EDTA; pH 6.5) in 1.5% agar at 37 °C 5% CO_2 and controlled for humidity in incubators. Importantly, there is significant variation in the literature for media compositions used to culture *A. fumigatus* and this has unclear impacts on biofilm formation and experimental results. Conidia are collected in a 0.01% tween-80 solution due to their hydrophobicity. Conidia should be counted in a rigorous manner, most often utilizing a hemacytometer, and seeded into liquid glucose minimal media (L-GMM) at a density of 1×10^5 conidia per mL media. This density is used for all conditions tested in the following sections of this chapter. The volume/size of wells used depends on the assay. For example, when measuring total biomass, inoculate either 10 cm tissue culture (TC)-treated plates or 6 well TC-treated plates to obtain measurable biomass. When assessing treatment of biofilms for metabolic activity or lysis assays (described in later sections), biofilms as small as those formed in a 96-well plate can be used that allows for multiple conditions/strains to be tested at a larger scale.

Submerged A. fumigatus Biofilm Characterization Using the above method to develop *A. fumigatus* biofilms, analysis of the biofilms can subsequently be conducted. A major consideration in contrast to biofilms formed by unicellular organisms is the filamentous nature of mold biofilms. Colony-forming units cannot be utilized with mold biofilms and multiple approaches are thus needed to interrogate the biofilm. A main method to measure mold biofilm development over time is to quantify total biofilm biomass. To use this method, inoculate 6-well plates with 2 ml/well L-GMM + 1 × 10^5/ml conidia. Grow biofilms over a 12–30 h period of time, depending on which developmental stage(s) are of interest. At earlier time points (12–14 h), there is very little biomass and wells may need to be pooled to determine measurable values. At later time points, biofilms will likely reach the air–liquid interface and begin to undergo asexual reproduction that could impact results. For later biofilm time points, additional technical challenges emerge and may best be approached with a microfluidics model. At the time points of interest, remove the medium and scrape biomass into a pre-labeled, pre-weighed, 2 ml screw top tube. Wash 3× with water to remove all salt/sugars from biofilms as to not impact data collection. The significant amount of polysaccharide produced by many mold strains can result in misleading biomass results. After washing, freeze biofilms at −80 °C and subsequently lyophilize to remove water weight, weigh and record as described (Kerkaert et al. 2022; Jones et al. 2023). Quantification of total biomass with and without an antifungal treatment can allow the assessment of growth inhibition in the presence of the drug but cannot distinguish between growth inhibition and cell killing.

Consequently, additional approaches are needed to determine the impact of a given stress on a mold biofilm. Quantitative microscopy is a powerful approach and can be used to interrogate biomass, morphology, and biological processes including cell viability within a biofilm. Fungal reporter strains have been successfully leveraged and can give spatial and temporal insights into fungal cell physiology within the biofilm. For example, to address whether different regions of a biofilm are translationally active, a xylose-inducible GFP reporter strain can be used as a proxy for translation within a biofilm. After growth of a biofilm is established for 18 h, xylose is added to the biofilm for an additional 3 h to test if GFP is detectable by microscopy, indicating protein translation (Kowalski et al. 2020). Reporters can be used for other processes, such as monitoring the hypoxia response by genetically inserting GFP under a promoter that is induced by a given environmental condition such as low oxygen (Kowalski et al. 2020). Intracellular stains can also be used to observe changes in organelles, such as MitoTracker (mitochondria) or Hoechst (nuclei). These can be complemented with fluorescent tagging of organelles using fluorescent protein markers.

Strains either constitutively expressing fluorescent proteins (GFP/mRFP, etc.) or stained with calcofluor white (CFW) 25 μg/mL can be imaged by fluorescence confocal microscopy at different stages of biofilm development and thoroughly characterized. To image biofilms, use an objective with at least 20× magnification and a sufficient working distance to enable 3D image acquisition of the entire biofilm height using a point-scanning or spinning disk confocal microscope. Water immersion objectives with these properties can be used to increase signal capture and resolution. Images can be visually assessed using common image analysis software such as Nikon NIS-Elements Viewer (Kowalski et al. 2019). Images can be quantified for branch length, branch density distribution, and biovolume using the software BiofilmQ as described (Hartmann et al. 2021). All visualization of branches and quantification of biomass can be performed in BiofilmQ in combination with Paraview visualization software, or as shown in Fig. 14.2, a resliced maximum intensity image produced via Fiji image analysis software (Fig. 14.2). Taken together, the use of fungal reporter strains, fluorescent markers, and cell permeable and impermeable dyes can allow quantitative insight into mold biofilm form and function. As mentioned, new technologies such as the use of microfluidics systems are expected to further advance our ability to interrogate the biology of these complex infection-relevant structures formed by molds.

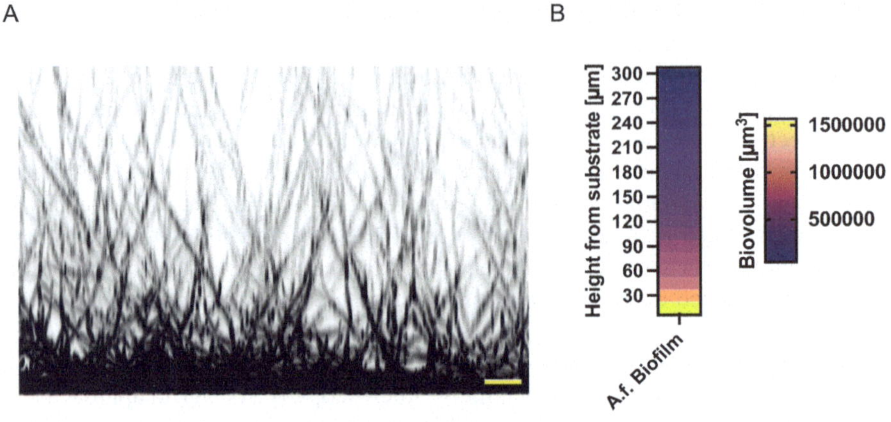

Fig. 14.2 Imaging and quantification of submerged biofilms. (**a**) Reference strain A1160, also known as CEA10, was grown at 1×10^5 conidia/ml of liquid GMM for 24 h. Biofilms were imaged by confocal microscopy and (**b**) quantified using BiofilmQ

Measuring *A. fumigatus* Biofilm Damage

A key goal of investigating mold biofilms is to determine their susceptibility, and the underlying mechanisms, to infection-relevant stress conditions including contemporary antifungal drugs. *A. fumigatus* biofilms are exposed to a variety of stressors in infection environments in the mammalian lung. For example, host immune responses, oxygen and iron limitation, oxidative stress, pH changes, and osmotic/cell wall stress are present at the site of infection. As the submerged biofilm model has the unique advantage to examine any stage of biofilm development, this model can be used to interrogate responses to essentially any host-relevant stress. Moreover, it is relatively straightforward in this model to expose the organism to combinatorial stresses. This contrasts with the earlier described agar colony biofilm, which is generally exposed to a single stressor and requires application of the stress at the point of conidia inoculation. While some studies have utilized hyphal plug assays, which remove a portion of vegetative mycelia from a grown colony biofilm and transfer them onto a plate containing a stressor of interest, these assays prove to be inconsistent.

With unicellular organisms it is relatively straightforward to measure the impact of a stressor/antifungal on an organism by measuring Optical Density (OD) at 600 nm. This approach can be used to measure early growth of *A. fumigatus* from conidia to an immature hyphal stage (0–10 h). However, once a mature biofilm forms and a heterogeneous network of hyphae and extracellular matrix develop, OD measurements are no longer accurate representations of biomass. Their filamentous growth pattern leads to uneven light scattering and inaccurate readings (Miyazawa et al. 2022). Methods measuring only hyphal length in a biofilm are also not an accurate representation of biofilm growth, as they do not take into account density of the biofilm. Here, we discuss arguably more accurate methods to determine damage and loss of viability to an *A. fumigatus* biofilm.

As mentioned, a main advantage of the submerged biofilm model is that the biofilm can be treated at any developmental stage by removing the growth medium and replacing it with fresh medium containing the stressor of interest. Stress agents can also be directly added to the existing biofilm culture. The growth stage at which a biofilm is assayed depends on the question being addressed, and each stage is likely relevant at different points in an active infection. Typically, mature biofilms (24–30 h) are less susceptible to common stressors than developing/immature (12–18 h) biofilms. This is not simply due to there being differences in biomass. Kowalski et al. (2020) demonstrated that a 12 h biofilm was recalcitrant to azole-induced damage when treated in a low oxygen environment, which suggests that other changes that occur in the developing biofilm contribute to drug susceptibility/recalcitrance. General membrane stressors include antifungals amphotericin B (AmB) and voriconazole (0.5–2 μg/ml) and the detergent SDS (0.01–1%). Cell wall stressors that are effective against biofilms are echinocandins (caspofungin, micafungin, and anidulafungin (0.125–2 μg/ml) and CFW (12.5–100 μg/ml) (Kerkaert et al. 2022). Mitochondrial inhibitors/activators include FCCP (1–10 μM), rotenone (10–50 μM), antimycin A (5–20 μg/μl), and KCN (0.5–5mM) and can be used to interrogate general metabolic changes in biofilms. Osmotic stressors include NaCl and sorbitol (0.25–1 M) and ROS stress can be induced by hydrogen peroxide (1–10 mM) and menadione (5–20 μM). The doses selected, time to treat, and endpoints used to test damage all depend on the *A. fumigatus* strains used and maturity of biofilms.

To test synergy between any of these drugs/stressors on a biofilm, a modified checkerboard assay can be used. Biofilms are grown to the desired maturity on a 96 well plate and a dose curve of one stressor (wells 1–11) and the other (A-G) generated with the last well containing no drug. Combine them and add to pre-formed biofilms for an additional 3–6 h. An XTT assay (as described in the next section) can be utilized to assess metabolic activity. Normalize treated samples to untreated control (set to 100%) to determine synergy (<25% metabolic activity), indifference (25–75% metabolic activity), or antagonism (>75% metabolic activity recovered) compared to each individual drug alone (Fig. 14.3) (Jones et al. 2023).

Metabolic Activity A common method to assess the damage caused by antifungal drugs and other stress conditions to mold hyphae utilizes the reduction of various tetrazolium or related dyes to quantitate metabolic activity of the cells before and after treatment. Two main methods to detect changes in metabolic activity are used: the tetrazolium dye XTT (2,3-bis-(2-methoxy-4-nitro-5-sulphenyl)-(2H)-tetrazolium-5-carboxanilide) and resazurin (7-hydroxy-3H-phenoxazin-3-one 10-oxide). In the XTT assay, metabolically active cells reduce yellow salt to an orange formazan dye which is proportional to the number of actively respiring cells. Reduction of tetrazolium dye is dependent on NADPH-dependent oxidoreductase enzymes which are generally localized to the cytosolic compartment, but it remains unclear in filamentous fungi which specific enzymes are responsible for the dye

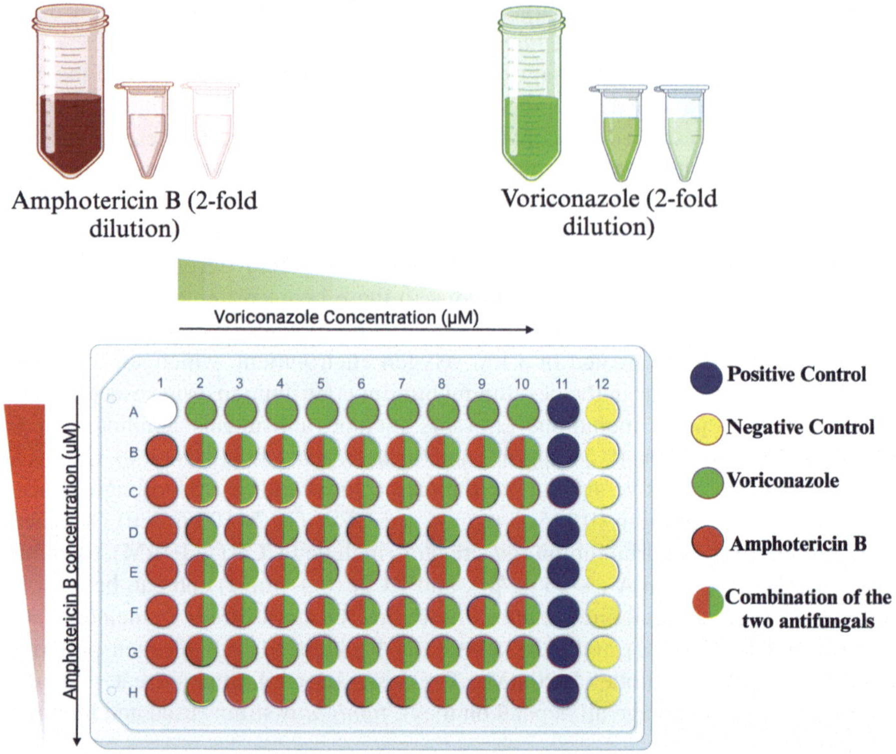

Fig. 14.3 Schematic of checkerboard synergy assay for biofilms (Biorender)

reductions. The uncertainty about the mechanism underlying dye reduction in molds suggests one must be cautious in interpreting the meaning of the assay. Similarly, resazurin (starts as oxidized/blue) reduces to resorufin (pink/red) in metabolically active cells. Other similar compounds are also available for these reduction-based assays. The XTT assay measures metabolic activity of a fungal biofilm at a single time point, whereas the resazurin assay measures changes in metabolic activity in a biofilm over time. Often, these two methods can be used in parallel as complimentary approaches. To perform this assay, after treatment with a stressor of interest, remove medium and add XTT (0.5 mg/ml PBS + 25 μM menadione as terminal electron acceptor) and place plate back in the incubator 37 °C 5% CO_2. Typically, it takes 30–60 min for the reduction to develop; however, it is recommended that once the untreated controls turn a bright orange, to stop the reaction and read the OD at 450 nm. For the resazurin assay, the reagent can be added directly to the media plus stressor (10% solution) and fluorescence (excitation 544, emission 590) measured in a plate-reader every 15 min as desired or until the biofilm reaches air–liquid interface. Data then can be normalized to untreated controls and a percent of control or percent damage can be represented in a graph.

Troubleshooting metabolic activity assays and considerations such as the state of the cell wall and ECM, which can impact uptake of XTT/resazurin reagents, are also important when interpreting results. The XTT/resazurin results are typically variable, as metabolic heterogeneity exists within biofilms and there is well-known lot to lot variation in the tetrazolium salts. Therefore, for each experiment it is recommended that there be at least 6 technical replicates, negative control (media alone: no damage), and a positive control for damage. Typically, polyenes serve as the best positive control in this assay for biofilms, as doses of 0.5–2 ug/ml usually lead to a loss of more than 75% metabolic activity, depending on the maturity of the biofilm (Fig. 14.4a). However, the azole voriconazole does not strongly impact metabolic activity in a more developed biofilm (18–30 h; (Kowalski et al. 2020)). One major consideration for the XTT assay is deciding how long to allow the XTT reagent to develop on a biofilm, which can be subjective, increasing the chance for variable results experiment to experiment. As mentioned, it is important to include a "no damage" control, which should ideally turn a bright orange color. At this point (usually 30–60 min after addition of XTT reagent), the XTT reagent should be removed from the biofilm and OD measurement taken. If the XTT reagent remains on the biofilm too long, this may under-represent the data as the XTT can overdevelop and signal can max out. Fortunately, this is not a concern with the resazurin assay since measurements are taken every 10–15 min throughout the time course. It is recommended that values be used for data analysis at the peak fluorescence of the positive control used or measure the area under the curve for each condition. Consequently, the time it takes an XTT assay to develop depends on many factors, such as the strain of *A. fumigatus*, medium utilized, stress agent concentrations and time of application, and developmental stage of biofilm examined. All these factors must be taken into consideration when designing an experiment. Multiple biological replicates are critical for rigor of conclusions from these assays and additional assays are needed to fully interrogate the impact of a given antifungal agent on a filamentous fungal biofilm.

Membrane Permeability As XTT and resazurin only measure changes in metabolic activity (as measured by reduction of the respective dyes through unclear mechanisms), it is important to consider and explore alternative methods to detect damage to an *A. fumigatus* biofilm. Plasma membrane permeability is another endpoint that can be assessed to determine biofilm damage. Two techniques have been developed to measure membrane damage in a filamentous fungal biofilm specifically: SYTOX staining and adenylate kinase (AK) release into the medium. The first approach utilizes quantitative microscopy with dyes in the SYTOX family that enter cells with compromised membranes but do not enter healthy, live cells (Langsrud and Sundheim 1996). This is based on the premise that biofilms with compromised membranes will uptake SYTOX dyes and then stain nucleic acids, which is observable by microscopy. To use this method, grow strains either constitutively expressing fluorescent proteins (GFP/tomato red, etc.) or stained with CFW 25 μg/mL to a biofilm development stage of choice and treat with the stressor being tested. Optically clear mediums are critical for these experiments and standard glucose

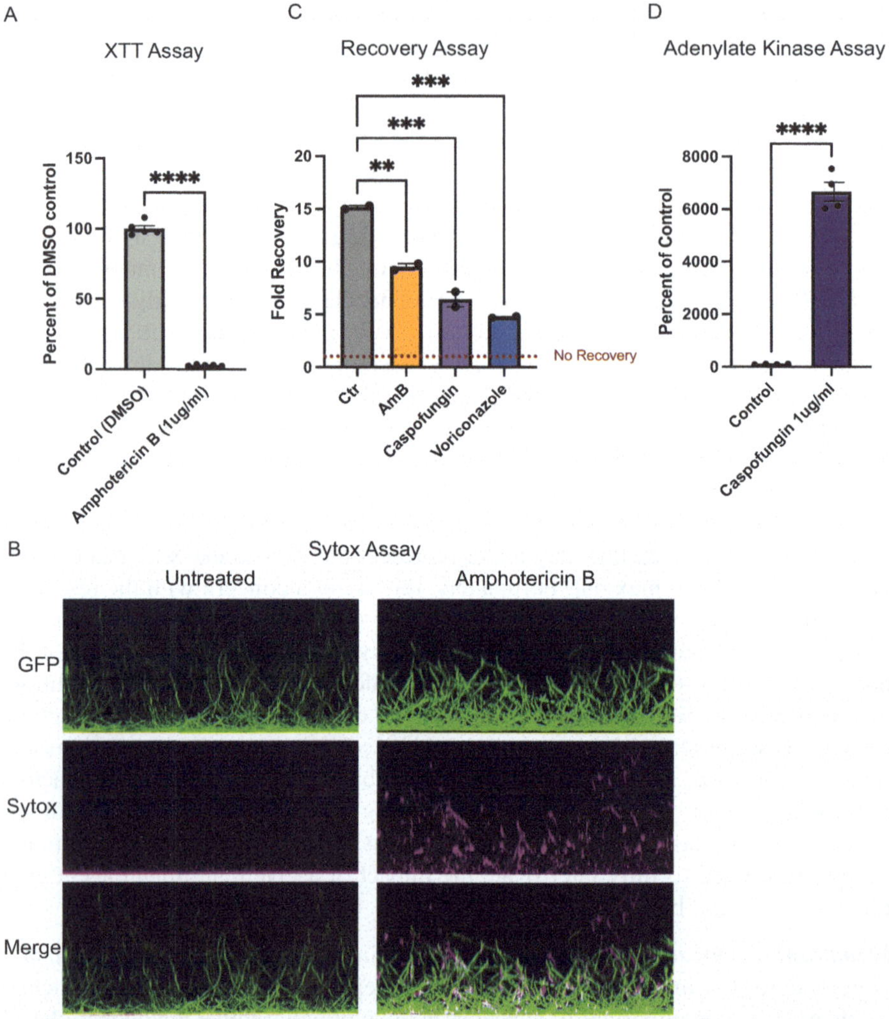

Fig. 14.4 Assessment of biofilm damage. (**a**) Wild-type CEA10 conidia (1×10^5/ml) were grown for 16 h, media removed, and treated with either fresh L-GMM or L-GMM + AmB 1 µg/ml for an additional 3 h. XTT reagent + menadione 25 µM was added to biofilms and when controls turned dark orange, read at OD 450. Data are normalized to untreated control (100%). (**b**) GFP-expressing CEA10 conidia (1×10^5/ml) were grown for 16 h, media removed, and treated with either fresh L-GMM or L-GMM + AmB 1 µg/ml for an additional 3 h. Biofilms were stained with SYTOX Blue and imaged using confocal microscopy. (**c**) WT CEA10 conidia (1×10^5/ml) were grown for 16 h, medium removed, and treated with either fresh L-GMM or L-GMM + Caspofungin 1 µg/ml for an additional 3 h. Media was removed, and total adenylate kinase was measured (data represented as percent of untreated controls). (**d**) Wild-type CEA10 conidia (1×10^5/ml) were grown for 16 h, media removed, and treated with either fresh L-GMM or L-GMM + AmB, Caspofungin, or Voriconazole (all 1 µg/ml) for an additional 3 h. Baseline biomass was collected and in parallel biofilms were given fresh L-GMM and recovered for 24 h. Biomass was collected and quantified. Data are represented as fold change over baseline biomass

minimal medium may be used but will require filtration to remove any insoluble particles that can interfere with microscopy. Total biofilms can be visualized as described above and counterstained with either SYTOX blue (GFP strain) or SYTOX deep red (CFW stained biofilm). Images are acquired as described above using confocal microscopy in two channels. Images are processed by raw image stacks sliced from top down and maximum projections generated (Jones et al. 2023). To analyze the co-localization of cytosolic GFP or CFW with SYTOX signal, the Fiji plugin JaCoP can be used (Rueden et al. 2017; Schindelin et al. 2012; Bolte and Cordelières 2006). Manually set channel thresholds to exclude background signal. Mander's overlap is used to quantify a ratio of SYTOX-stained fungi (damaged) to total fungi (undamaged and damaged) occurs in response to treatment. This will be represented as the fraction of GFP/CFW-positive pixels that overlap with SYTOX-positive pixels as described (Jones et al. 2023). Treatment with antifungal drug AmB (1 ug/ml) (Fig. 14.4b) or hydrogen peroxide (2.5–10 mM) are recommended as positive controls; however, it is important not to treat with too high of a concentration. For this assay to be effective, the biofilm must remain intact for proper staining and imaging. This assay is challenging at later stages of maturity as more dense biofilms are difficult to quantify.

Lysis As an alternative, complementary approach to SYTOX staining, a cell lysis assay can be used. Hyphae presenting with a compromised plasma membrane release AK into the supernatant of the culture and is indicative of lysis (Beattie and Krysan 2021). To quantify AK release, collect supernatants of treated biofilms and keep at 4 °C or on ice. Add the AK detection reagent according to manufacturer's instructions (100-μL Toxilight non-destructive cytotoxicity bioassay, Lonza) to each well of samples (20 μL) and incubate at room temperature for 5 minutes prior to luminescence measurements (1-h, 5-min intervals) as described (Beattie and Krysan 2021). Data are normalized to untreated controls and a percent of control or percent damage can represent the data.

Changes in membrane permeability can often be subtle in response to treatment with stressors and not all stressors cause this type of biofilm damage. Lysis is a very specific manifestation of biofilm damage and not all stressors will lead to this type of response. Therefore, it is important, especially when testing a new stressor, to always include a positive control. Caspofungin and Micafungin are echinocandins that target cell wall synthesis and damage biofilms by causing hyphal tip lysis. Treatment with either of these two antifungal drugs (1–2 ug/ml) serves as a strong, consistent positive control (Fig. 14.4c).

Biofilm Viability Assay A direct way to measure damage to *A. fumigatus* biofilms after antifungal treatment is by measuring its capacity to regrow after treatment (regrowth/recovery assay). It is important to grow biofilms in at least 6-well plates or larger to provide enough measurable biomass. To conduct a regrowth assay, grow biofilms to desired maturity, then treat with stressor of choice for desired time though time courses provide the most rigorous data. At each time point, collect biofilm biomass (baseline sample). In a parallel-treated sample, remove the stressor from the biofilm and replace with fresh medium for an additional 16–24 h. After collecting biomass, washing, lyophilizing, and weighing, the final (recovery) sample can be divided by its baseline sample to determine the fold recovery as described (Jones et al. 2023). Treatment with antifungals AmB, voriconazole, or caspofungin should all have reduced recovery compared to untreated controls (Fig. 14.4d). This is arguably one of the more clinically relevant endpoints because it shows how impactful a treatment is on the viability of the biofilm.

It is important when conducting regrowth experiments that the initial baseline drug-treated measurements are not too different from the control-treated measurements. If there is too much damage in the initial treatment, this could skew the final calculations to make it appear as though the damage was not extensive. Furthermore, as the biofilm is in culture for an extended period, the biofilm could reach the air–liquid interface leading to a potential under-representation of the potential impact of a stressor. Additional medium can be added to biofilms to prevent this from occurring.

Interpreting *A. fumigatus* Biofilm Assay Results

We have discussed methods for assessing biofilm development (biomass quantification, microscopy) and response to stressors metabolic activity (XTT, resazurin), membrane permeability/lysis (SYTOX, AK), and viability (regrowth/recovery assay). However, there are important technical challenges for each assay that must be considered when interpreting the results. It is important that one distinguishes between effects on metabolic activity versus effects on membrane versus overall viability. Changes in permeability can cause reductions in metabolic activity and often a loss of metabolic activity will lead to loss of viability, but these must be directly measured for rigorous conclusions. Alternatively, if no changes in metabolic activity are detected, this does not necessarily mean there is no change in viability.

Summary

In summary, interrogating the structure and function of filamentous fungal biofilms requires multiple approaches. Additional approaches are also needed and should be a focus of future research. For example, a major current limitation in the field is determining the viability of fungal hyphae in vivo at different stages of infection and disease. Overcoming these challenges is needed to help develop the next generation of mold-effective antifungal agents.

References

Beattie SR, Krysan DJ (2021) A unique dual-readout high-throughput screening assay to identify antifungal compounds with Aspergillus fumigatus. mSphere 6:e0053921

Beauvais A, Latgé J-P (2015) Aspergillus biofilm in vitro and in vivo. Microbiol Spectr 3

Bolte S, Cordelières FP (2006) A guided tour into subcellular colocalization analysis in light microscopy. J Microsc 224:213–232

Chandrasekar PH, Manavathu EK (2008) Do Aspergillus species produce biofilm? Future Microbiol 3:19–21

Harding MW, Marques LLR, Howard RJ, Olson ME (2009) Can filamentous fungi form biofilms? Trends Microbiol 17:475–480

Hartmann R, Jeckel H, Jelli E, Singh PK, Vaidya S, Bayer M, Rode DKH, Vidakovic L, Díaz-Pascual F, Fong JCN, Dragoš A, Lamprecht O, Thöming JG, Netter N, Häussler S, Nadell CD, Sourjik V, Kovács ÁT, Yildiz FH, Drescher K (2021) Quantitative image analysis of microbial communities with BiofilmQ. Nat Microbiol 6:151–156

Jones JT, Morelli KA, Vesely EM, Puerner CTS, Pavuluri CK, Ross BS, van Rhijn N, Bromley MJ, Cramer RA (2023) The cystic fibrosis treatment Trikafta affects the growth, viability, and cell wall of Aspergillus fumigatus biofilms. MBio 14

Kerkaert JD, Le Mauff F, Wucher BR, Beattie SR, Vesely EM, Sheppard DC, Nadell CD, Cramer RA (2022) An alanine aminotransferase is required for biofilm-specific resistance of Aspergillus fumigatus to Echinocandin treatment. MBio 13

Kowalski CH, Kerkaert JD, Liu K-W, Bond MC, Hartmann R, Nadell CD, Stajich JE, Cramer RA (2019) Fungal biofilm morphology impacts hypoxia fitness and disease progression. Nat Microbiol 4:2430

Kowalski CH, Morelli KA, Schultz D, Nadell CD, Cramer RA (2020) Fungal biofilm architecture produces hypoxic microenvironments that drive antifungal resistance. Proc Natl Acad Sci USA 117:22473–22483

Langsrud S, Sundheim G (1996) Flow cytometry for rapid assessment of viability after exposure to a quaternary ammonium compound. J Appl Bacteriol 81:41–42

Loussert C, Schmitt C, Prevost MC, Balloy V, Fadel E, Philippe B, Kauffmann-Lacroix C, Latgé JP, Beauvais A (2010) In vivo biofilm composition of Aspergillus fumigatus. Cell Microbiol 12:405–410

Miyazawa K, Umeyama T, Hoshino Y, Abe K, Miyazaki Y (2022) Quantitative monitoring of mycelial growth of Aspergillus fumigatus in liquid culture by optical density. Microbiol Spectr 10:e00063–e00021

Mowat E, Butcher J, Lang S, Williams C, Ramage G (2007) Development of a simple model for studying the effects of antifungal agents on multicellular communities of Aspergillus fumigatus. J Med Microbiol 56:1205–1212

Mowat E, Williams C, Jones B, McChlery S, Ramage G (2009) The characteristics of Aspergillus fumigatus mycetoma development: is this a biofilm? Med Mycol 47(Suppl 1):S120

Perea S, Patterson TF (2002) Antifungal resistance in pathogenic fungi. Clin Infect Dis 35:1073–1080

Rueden CT, Schindelin J, Hiner MC, DeZonia BE, Walter AE, Arena ET, Eliceiri KW (2017) ImageJ2: ImageJ for the next generation of scientific image data. BMC Bioinf 18:1

Schindelin J, Arganda-Carreras I, Frise E, Kaynig V, Longair M, Pietzsch T, Preibisch S, Rueden C, Saalfeld S, Schmid B, Tinevez JY, White DJ, Hartenstein V, Eliceiri K, Tomancak P, Cardona A (2012) Fiji: an open-source platform for biological-image analysis. Nat Methods 9:676–682

Seidler MJ, Salvenmoser S, Müller FMC (2008) Aspergillus fumigatus forms biofilms with reduced antifungal drug susceptibility on bronchial epithelial cells. Antimicrob Agents Chemother 52:4130–4136

Singhal D, Baker L, Wormald PJ, Tan LW (2011) Aspergillus fumigatus biofilm on primary human sinonasal epithelial culture. Am J Rhinol Allergy 25:219–225

Wayne PA (2008) Reference method for broth dilution antifungal susceptibility testing of filamentous fungi. Clinical and Laboratory Standards Institute, Wayne

Chapter 15
Implications of Fungal Biofilms in Microbial Keratitis

Leonie Fingerhut, Kelvin Kah Wai Cheng, Aysha Ali, and Bethany Mills

Abstract Fungal keratitis—an infection of the cornea caused by yeast or filamentous fungi—is a leading cause of visual impairment, blindness and eye-loss worldwide. Infection progression most commonly follows epithelial cell damage, leading to spore adhesion to the ocular surface, germination, hyphal growth and finally biofilm formation within the corneal stroma. Abiotic fungal biofilms may also be introduced to the eye directly via contaminated contact lenses. Interestingly, not all keratitis-causing fungi form biofilm (when grown in vitro), and it is not essential for severe disease manifestation. However, biofilm-forming fungi modulate the immune response, and confer greater resistance to drug and environmental stresses, such as antifungals and UV irradiation. Herein, we report fungal keratitis pathogenesis from a pre-clinical and clinical perspective, with and without contact lens propagation. We further present current and emerging treatment strategies in the context of biofilm eradication, using approaches such as novel drug delivery systems, immunomodulation and light-mediated therapies.

Introduction to Fungal Keratitis

Microbial Keratitis Epidemiology

Microbial keratitis is a painful and sight-threatening infection of the cornea. While the true incidence and burden of disease is not accurately known, conservative estimates suggest that in excess of two million people experience microbial keratitis per year, with the greatest burden found within low- and middle-income countries

Authors Leonie Fingerhut and Kelvin Kah Wai Cheng have equally contributed to this chapter.

L. Fingerhut · K. K. W. Cheng · A. Ali · B. Mills (✉)
Centre for Inflammation Research, Institute for Regeneration and Repair, University of Edinburgh, Edinburgh, UK
e-mail: beth.mills@ed.ac.uk

G. Ramage, D. Andes (eds.), *Fungal Biofilms*, Springer Series on Biofilms 15,
https://doi.org/10.1007/978-3-031-93315-8_15

(LMICs) across the tropics (Hossain 2021; Brown et al. 2021a). After cataract, MK remains the most common cause of unilateral visual impairment, blindness and eye-loss (Ung et al. 2019; Khor et al. 2018, Whitcher et al. 2001). Across Asia, over 60% of microbial keratitis patients are left with moderate (or worse) visual impairment at 6 months, and over 10% of patients require surgical intervention (e.g., therapeutic keratoplasty (TPK)), the success rate of which is only 10–30% (Oliva et al. 2012; Khor et al. 2018).

In the majority of microbial keratitis cases, infection follows an injury to the corneal epithelium, caused either by vegetative matter or through contact lens wear. Other predisposing factors include ocular surface disease, prior corneal surgery and systemic immunosuppression (Ting et al. 2021; Mills et al. 2021). There is a broad spectrum of microorganisms responsible for the aetiology of microbial keratitis; with over 680 species of bacteria, fungi, virus and amoeba implicated (Bartimote et al. 2019). This includes 393 species of fungi across 169 genera; however, 95% of fungal keratitis cases are caused by the filamentous fungi *Aspergillus* spp. and *Fusarium* spp., or the yeast *Candida* spp., with a geographical skew of filamentous fungi towards tropical climates. While yeasts are more prevalent in temperate climates, there has been a marked increase in filamentous fungal keratitis, associated with increased contact lens wear, and potentially due to changing climates (Oliveira Dos Santos et al. 2020; Walther et al. 2017; Ong et al. 2016).

The annual incidence of fungal keratitis is reported as 73 per 100,000 in South Asia and 13.5 per 100,000 across Africa, compared to just 0.02 per 100,000 across Europe (Brown et al. 2021a). Microbial keratitis caused by filamentous fungi typically exhibits worse outcomes compared to corneal infections caused by yeast or bacteria (Mills et al. 2021; Khor et al. 2018; Gopinathan et al. 2009; Prajna et al. 2013b; Shah et al. 2019), with corneal perforation reported in up to 25% of cases, and leading to the removal of an estimated 84,000–167,000 eyes annually (Hoffman et al. 2021; Brown et al. 2021a). Such is the toll of fungal keratitis that the global ophthalmic community are advocating for fungal keratitis to be recognised as a neglected tropical disease across LMICs, where an estimated 600,000 eyes are blinded due to the condition each year (Brown et al. 2022). This leads to significant social and economic difficulties for fungal keratitis patients (Whitcher et al. 2002; Radhakrishnan et al. 2022; Shah et al. 2019), with quality of life negatively impacted beyond the time of active disease, even for patients who recover their vision (Arunga et al. 2019).

There is no doubt that the burden of fungal keratitis resides across LMICs, but fungal keratitis also exhibits significant morbidity and economic impact within high income countries. In the United Kingdom (UK), microbial keratitis caused by any pathogen has an annual incidence of 34–52 cases per 100,000 population (27,000–35,000 people) (Moussa et al. 2021; Ting et al. 2021), and it is the most common non-surgical ophthalmic emergency admission (Ibrahim et al. 2009b). Fungal keratitis accounts for up to 9.3% of the UK microbial keratitis cases, and it is likely increasing (Moledina et al. 2023; Hoffman et al. 2021; Ong et al. 2016). Ting et al reported that hospitalisation was required for 81.2% of UK fungal keratitis patients for an average duration of 18.9 days. Furthermore, over 50% of fungal keratitis patients required additional surgical interventions (Ting et al. 2021), and economic analysis by Moussa et al determined that any hospital stay longer than 4.8 days was financially loss-making for the National Health Service (NHS) (Moussa et al. 2021). In the United States, there are an estimated one million healthcare interactions resultant from microbial keratitis (of which 10–24% are caused by fungi) each year, with an estimated healthcare cost of $175 million USD (Collier et al. 2014; Brown et al. 2021a).

Fungal Pathogenesis in Fungal Keratitis

As per fungal pathogenesis and infection establishment at other sites within the body, development of fungal keratitis progresses through the same key stages, including: (i) immune evasion of conidia/yeast; (ii) adhesion; (iii) invasion; and (iv) toxin production (Mills et al. 2021). The next stage in pathogenesis, biofilm formation, is species and strain dependent, and while it has important implications for treatment susceptibility, is not an essential requirement for manifesting a severe fungal keratitis infection (Fig. 15.1) (Ranjith et al. 2018; Ranjith et al. 2017; Córdova-Alcántara et al. 2019).

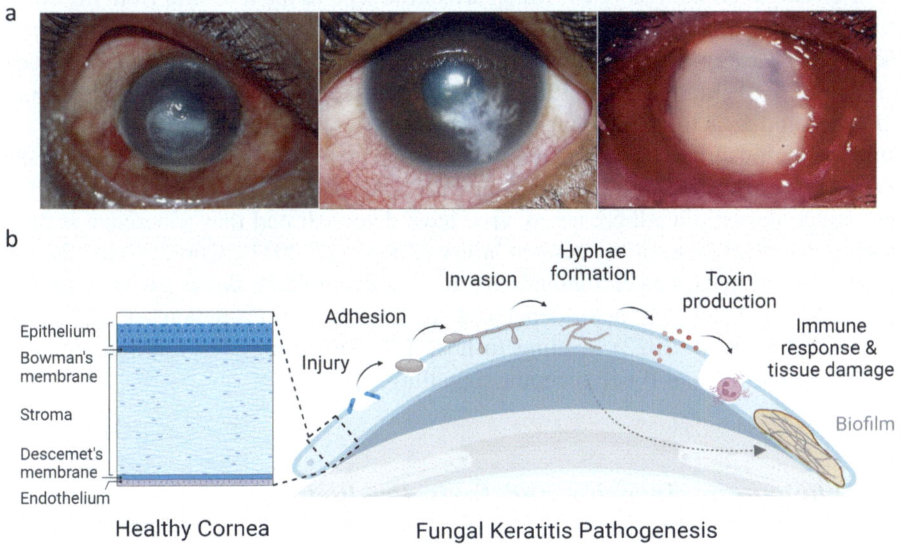

Fig. 15.1 (**a**) Photographs of progressing fungal keratitis. Left: clinically indistinguishable from bacterial keratitis. Centre: feathery margins indicative of growing fungi. Right: advanced fungal keratitis, indistinguishable from bacterial keratitis. Images reproduced from (left) (Brown et al. 2022) (centre, right) (Prajna et al. 2017). (**b**) Schematic of human cornea anatomy and fungal keratitis progression, including hyphae formation, toxin production, immune response and possible biofilm formation

The Cornea

The cornea is a transparent avascular tissue at the front of the eye, consisting of 5 layers, which anterior to posterior, are the corneal epithelium, the Bowmans layer, the corneal stroma, the Descemet's membrane and the corneal endothelium (Fig. 15.1b). The cornea is also protected by a number of physical barriers, such as the eyelid and the tear-film, which acts as a lubricant for the eyelid, provides fluid to wash potential pathogens from the eye via sheer stresses, and enables provision of oxygen, nutrients and soluble defence molecules, such as antimicrobial peptides (AMPs) to the cornea. These can be upregulated in response to infection threats (Mohammed et al. 2021; Mohammed et al. 2017). To-date, over 2500 unique proteins and 100 metabolites have been identified from the tear film of healthy eyes to support and protect the cornea. The cornea is also protected by innate immune cells, such as neutrophils which are resident within closed-eye tears (Mills et al. 2021; Shoham 2015).

Conidia Adhesion

Development of an initial infection requires corneal epithelial damage and the presence of fungal conidia or yeast, which are introduced to the tear-film from the environment or mycobiome (Shivaji et al. 2022; Shivaji et al. 2019). During wound healing, cornea epithelial cells increase expression of cell-surface mannose glycoproteins (Zieske and Gipson 1986), which are recognised by lectin-like proteins (mannan and galactomannan) on the outer fibril layer of fungal conidia of both yeast and filamentous fungi enabling adhesion (Blango et al. 2019), while simultaneously evading immune detection (Aimanianda et al. 2009; Carrion de et al. 2013). Studies investigating conidia adherence ex vivo have demonstrated that adherence is proportional to the extent of the corneal injury (Zhou et al. 2011). Conidia may also be phagocytosed by cornea epithelium directly. For example, *A. flavus* has been shown to be internalised via actin-dependent processes (Arunachalam et al. 2021; Arunachalam et al. 2022), leading to host TNF signalling, Th17 differentiation, NF-kB signalling and B cell receptor signalling.

Morphogenesis, Invasion and Toxin Production

Conidia adhered to the corneal epithelium surface, or internalised within the cell phagosome, undergo morphogenesis to grow hyphal filaments which are able to invade into the corneal stroma (Donovan et al. 2022), causing tissue damage directly, and inciting a pro-inflammatory immunogenic response which leads to further tissue damage (Niu et al. 2020; Yang et al. 2021; Ratitong and Pearlman 2021; Mpakosi and Kaliouli-Antonopoulou 2024). Fungal survival and invasiveness are further propagated by secreted virulence factors which subsequently degrade the corneal stroma, leading to corneal melt and inactivation of host defence factors, such as AMPs (Hohl and Feldmesser 2007; Mellon et al. 2007; Park et al. 2013; Shibuya et al. 2006). In particular, matrix metalloproteases (MMPs), collagenases, serine proteases and cysteine proteases comprise the most abundant secreted proteases (Balakrishnan Sangeetha et al. 2020; Monod et al. 2002; Yike 2011). Together these are responsible for tissue destruction and nutrient acquisition from the microenvironment.

Fungal Keratitis (Biotic) Biofilm Formation

Fungal biofilms are three-dimensional communities of cells. Biofilms of yeasts, like *Candida* spp., are comprised of fungal conidia, hyphae and/or pseudohyphae, whereas biofilms of filamentous fungi, like *Aspergillus* spp., are primarily formed of interconnected branched multinucleate hyphae (Liu et al. 2022). Irrespective of cell morphology composition, all encompass self-produced extracellular matrix (ECM) of polymeric substances (Wang et al. 2024). ECM composition contains macromolecules and small molecules, including polysaccharides, proteins, lipids and extracellular DNA (eDNA) (Zarnowski et al. 2014; Córdova-Alcántara et al. 2019). The proportion of each is dynamic and changeable in relation to the level of biofilm maturity, and the microenvironment. The function of individual macromolecules in biofilm structure and function are not well understood; however, suggested and demonstrated roles are emerging. For example, while polysaccharides are reported to provide the backbone of the biofilm (Pham et al. 2020), the exopolysaccharide galactosaminogalactan (GAG) found in the biofilm matrix of *A. fumigatus* is found both bound to hyphae, and secreted from the cell, with dual roles across adhesion and immune evasion. It is also essential for biofilm formation (Liu et al. 2022). Increased protein content increases the hydrophobicity of the biofilm, and while some reports suggest a stabilising and "gluing" together of the biofilm with hydrophobic ECM, particularly for *Aspergillus*, many others reported conflicting data around the impact that hydrophobicity may have on biofilm formation (Beauvais et al. 2007; Danchik and Casadevall 2021; Wang et al. 2024). eDNA supports the biofilm integrity through cell–cell adhesion and may further contribute to modulation of the immune response (Martins et al. 2010). Cells within biofilms also release extracellular vesicles (EVs), which further provide community coordination, and promote ECM formation and antifungal resistance (Zarnowski et al. 2021; Mannan et al. 2024). Overall, formation of biofilm by fungi renders the pathogen more resistant to environmental stresses, including host immune response, UV irradiation and antifungal drugs, compared to planktonic and non-biofilm growth stages (Fig. 15.2) (Mannan et al. 2024; Morelli et al. 2021).

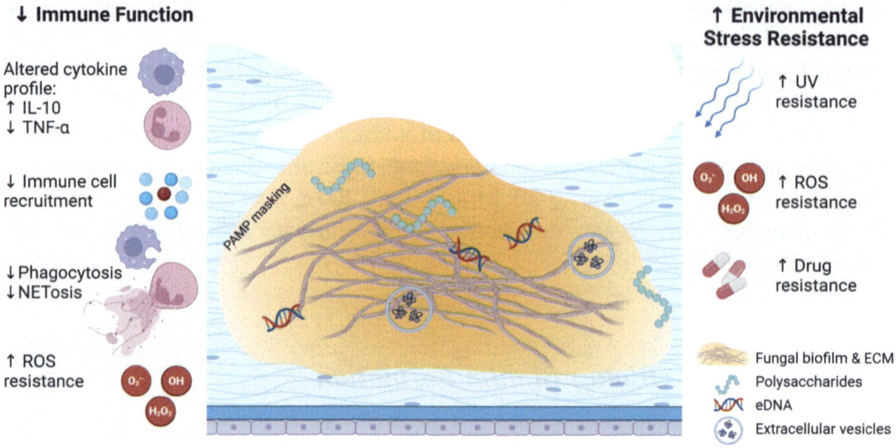

Fig. 15.2 Schematic of fungal biofilm formed of fungal hyphae, with extracellular matrix components (ECM) and indication of enhanced resistance to host and environmental challenges

While mechanisms of fungal biofilm formation and regulation are still not fully understood, particularly for filamentous fungi, even fewer studies have investigated biofilm development in the context of fungal keratitis. The vast majority of conducted studies have relied on assessing the ability of clinical isolates to form biofilm when grown in vitro, or not, and comparing tolerance of biofilms vs planktonic phenotypes against various drugs or other stresses, such as UV irradiation (Córdova-Alcántara et al. 2019; Peiqian et al. 2014). More recent studies have begun to measure gene expression in relation to biofilm formation of these clinical isolates (Ranjith et al. 2018).

Immune Evasion by Fungal Biofilms

Biofilms are increasingly recognised as an important factor in the success of fungal infections, with several reviews covering this in depth (Kernien et al. 2017; Garcia-Perez et al. 2018). Multiple theoretical models have been proposed to explain the mechanism by which fungal biofilms evade the immune system (Garcia-Perez et al. 2018). The first model suggested by Nett et al involves immunological silence by which biofilms shield fungal pathogen-associated molecular patterns (PAMPs) from immune cells (Nett 2016). As the fungi is not recognised, an immune response is not mounted. The second model suggests a deviation in the immune response to fungi. Fungi release factors that shift the immune response to a more tolerant one. For example, *C. albicans* can induce IL-10 expression, which is known to induce a Th2, anti-inflammatory response (Netea et al. 2015). Fungal biofilms are also able to influence the cytokine profile. An example of this is the downregulation of TNF-α which typically stimulates phagocyte activation (Katragkou et al. 2010). Third, the structural or molecular features of the biofilm renders the fungi resistant to immune clearance despite an appropriate immune response (Garcia-Perez et al. 2018). For instance, *Aspergillus* biofilm protects it from neutrophil attack (Lee et al. 2015). Along the same vein, peripheral blood mononuclear cells fail to phagocytose biofilm-associated *C. albicans* (Chandra et al. 2007). Additionally, biofilms have been reported to dampen the generation of reactive oxygen species (ROS) (Xie et al. 2012) and release of neutrophil extracellular traps (NETs) by neutrophils (Johnson et al. 2016).

Evidence of Biofilm Formation by Fungal Keratitis Clinical Isolates

Implications of *Fusarium* spp. Isolates and Biofilm Production

Following a large outbreak (>300 cases) of contact lens-related fungal keratitis caused by *Fusarium* spp. between 2005 and 2007 (Chang et al. 2006) (reported in more detail section "Fungal biofilms linked to contact lenses"), 25 of the clinical isolates and two additional fungal keratitis isolates were investigated for their ability to form biofilm in vitro and in vivo utilising a contact lens model (Mukherjee et al. 2012). Of the isolates, 22 belonged to the *F. solani* species complex, and five belonged to the *F. oxysporum* species complex. All 27 isolates were able to form biofilm, with the *F. solani* strains forming 1.5-fold more biofilm within 48 h, compared to the *F. oxysporum* species complex strains. These data were corroborated by similar findings in a murine model of fungal keratitis, whereby the same strains were inoculated via intrastromal injection, in the absence of a contact lens surface, and "high biofilm former" strains produced thicker and denser biofilms compared to the "low biofilm producer" strains.

The antifungal susceptibility of the strains was next evaluated against natamycin, voriconazole and amphotericin B. Planktonic fungi (conidia) were susceptible to the drug treatments, with minimum inhibitory concentrations (MICs) in the range of 1–8 μg mL^{-1}. However, biofilms of *F. solani* exhibited increased MIC to amphotericin B compared to planktonic, but no increase in MIC for natamycin or voriconazole. Conversely, *F. oxysporum* biofilms demonstrated increased MICs towards voriconazole, but not to the other drugs. This suggests species-dependent differences in biofilm composition and regulation, with depletion of ergosterol in biofilms of *F. solani* and upregulation of efflux pumps in *F. oxysporum* hypothesised to be responsible for increased amphotericin B (Walsh et al. 2003) and voriconazole (Mukherjee et al. 2003) resistance, respectively. Interestingly, the MIC for natamycin was unaffected by biofilm formation, and remained low. These data support the findings of the Mycotic Ulcer Treatment Trial (Prajna et al. 2013a), comparing natamycin and voriconazole for the treatment of fungal keratitis, with natamycin producing a significantly better clinical and microbiological outcome, particularly for patients with *Fusarium* keratitis.

In contrast to the 25 *Fusarium* spp. clinical fungal keratitis isolates evaluated by Mukherjee et al, which all formed biofilm in vitro (Mukherjee et al. 2012), none of the three *Fusarium* keratitis clinical isolates tested by Sav et al were able to form biofilm in vitro (Sav et al. 2018). In this study, the biofilm-forming potential of 38 clinical and three environmental *Fusarium* isolates was investigated. Only seven (18%) were able to form biofilm, and all of the biofilm formers were from two species of the *F. solani* species complex (*F. petroliphilum* and *F. keratoplasticum*); which are species commonly associated with fungal keratitis. However, in this study, none of these biofilm formers were isolated from cornea scrapings. The discrepancy between studies examining *Fusarium* keratitis isolates is somewhat surprising, but could be explained by different growth conditions of the fungi, with biofilms preferentially formed on contact lens materials compared to polystyrene well-plates, or by inconsistent definitions and classifications of the term "biofilm". For example, in the study by Mukherjee (Mukherjee et al. 2012), biofilm was confirmed by imaging the presence of hyphae and polysaccharide extracellular material, whereas the study by Sav (Sav et al. 2018) confirmed biofilm presence through crystal violet assay (which stains cell-surface and ECM polysaccharides) against pre-defined absorbance thresholds.

Similar to the previous study, biofilm formation of the *F. solani* species complex pathogens conferred increased resistance to amphotericin B, but in contrast, the MIC towards voriconazole was also increased compared to planktonic counterparts (Sav et al. 2018). A third study further demonstrated that antifungal resistance increased for *F. solani* biofilm compared to planktonic growth, and while amphotericin B was the most effective antifungal against the planktonic fungi, natamycin was the most effective antifungal against the 48 h biofilm (Zhang et al. 2012). Notably, the MIC of voriconazole remained constant for planktonic and 24 h biofilm (2 mg/L), but was >256 mg/L by 48 h, demonstrating the inherent limitations of directly comparing studies with different methodologies.

In addition to increased antimicrobial resistance, *Fusarium* biofilm grown from fungal keratitis clinical isolates may contribute to increased resistance to other environmental stresses (Córdova-Alcántara et al. 2019; Peiqian et al. 2014), including UV irradiation, with only 35.7% of planktonic fungi surviving compared to 64.8% of fungi within *F. solani* biofilms after 60 min of irradiation in one study (Córdova-Alcántara et al. 2019). Similar protective effects towards UV, heat and cold treatments were found for *F. oxysporum* biofilms (Peiqian et al. 2014).

Implications of *Aspergillus* Isolates and Biofilm Production

Fewer studies have examined the role of *Aspergillus* clinical isolates from fungal keratitis patients in biofilm development, and their potential implications for enhanced stress responses. However, the ability for *A. fumigatus* and *A. flavus* clinical isolates to form biofilm has been demonstrated in vitro and in vivo (Beauvais and Latgé 2015; Kaur and Singh 2014), with biofilm development conferring enhanced antimicrobial resistance compared to planktonic counterparts (Mowat et al. 2008; Fiori et al. 2011). Increased efflux pump activity has been shown to play a significant role in the increased resistance to voriconazole in early biofilm development (Rajendran et al. 2011), with extracellular matrix polysaccharides likely to confer higher resistance in mature biofilms through drug sequestering (Mowat et al. 2008). Progression of a clinical keratitis isolate of *A. fumigatus* to biofilm formation in vitro was captured by scanning electron microscopy (SEM), and confirmed that mature biofilms were formed within 24 h. Interestingly, the matrix deposition was more prevalent when grown at 37 °C, compared to 28 °C (Gonzalez-Ramirez et al. 2016). Micro-hyphae were also observed within the mature biofilm from this fungal keratitis strain (but not a strain isolated from soil, studied under the same conditions), which are associated with high enzymatic activity, maturation and subsequent biofilm dispersal.

Implications of *Candida* Isolates and Biofilm Production

Biofilm formation and implications for increased antimicrobial resistance by *Candida* spp. have been extensively studied in vitro and in vivo (Chandra et al. 2014; Pereira et al. 2021; Malinovska et al. 2023), including with fungal keratitis isolated strains (Ranjith et al. 2017, 2018; Bezerra et al. 2023). A study by Ranjith et al (Ranjith et al. 2017) examined the biofilm-forming capabilities of 43 C*andida* ocular isolates (31 of the isolates were from corneal scrapings; 16 of these were *C. parapsilosis*, 7 were *C. viswanathii,* 5 were *C. albicans* and there was one each of *C. glabarta, C. tropicalis* and *C. orthopsilosis*). 23/43 (53.4%) were able to form biofilm in vitro. Biofilm and non-biofilm producers clustered into two distinct groups based on principal component analysis (PCA) of virulence factor variables, with *C. albicans* remaining more virulent compared to the other strains evaluated. Further studies investigating biofilm formation of *C. albicans* fungal keratitis isolates also demonstrated that 4/7 (57%) were able to form biofilm in vitro, and that antimicrobial resistance was increased 100-200× in biofilm growth stages compared to planktonic yeast (Ranjith et al. 2018). Gene expression analysis confirmed temporal upregulation of the 27 genes evaluated, including those encoding transcription factors, virulence factors, phospholipases, adhesion and the aspartyl proteinase gene family in the biofilm-forming strains, compared to non-biofilm-forming strains. Finally, a recent study comparing 13 fungal keratitis clinical isolates of *Candida* isolated from fungal keratitis patients (11 *C. parapsilosis* and two *C. albicans*) determined that 7 formed very low–low biofilm, and 6 formed medium–high biofilm in vitro (Bezerra et al. 2023). In line with the other reported studies, development of biofilm increased antifungal (amphotericin B, anidulafungin and voriconazole) MIC substantially, compared to the planktonic stage of growth.

Fungal Biofilms Linked to Contact Lenses

Modifications of the Ocular Surface by Contact Lens Wear

As already mentioned, contact lens wear is a major risk factor for developing fungal keratitis. Wearing contact lenses modifies the ocular surface, ultimately influencing the protective mechanisms. For instance, the tear film is affected by decreased tear evaporation and reduction of waste removal facilitated by tear movement across the ocular surface (Liesegang 2002; Bruce and Brennan 1990). In addition, a physical barrier for oxygen delivery from the environment towards the cornea is created, therefore generating a hypoxic and hypercapnic environment (Liesegang 2002; Bruce and Brennan 1990), the extent of which is related to the contact lens type and material. The oxygen permeability of rigid gas-permeable lenses and silicone hydrogel lenses is higher compared to silicone-free hydrogel lenses (Ishihara et al. 2023). Reduced availability of oxygen in the cornea leads to a change in metabolism towards anaerobic glucose metabolism into lactic acid, which then accumulates and drives osmotic stromal swelling (Bruce and Brennan 1990).

Furthermore, other cellular effects of contact lenses, as reviewed elsewhere (Liesegang 2002; Bruce and Brennan 1990; Robertson and Cavanagh 2008), include reduction of epithelial mitosis, cell renewal and metabolism, disruption of epithelial junctions and microtrauma of the corneal epithelium, all of which facilitates adhesion of pathogens. This is important as fungi only adhere to corneal surfaces and consequently induce keratitis if epithelial defects are present (Sun et al. 2010; Rao et al. 1985). In addition, ocular proteins and biomolecules adhering to contact lenses (Luensmann and Jones 2012) provide sources of nutrition for pathogens along with the contact lens substrate itself. In the long term, contact lens wear causes corneal thinning and steepening (Liu and Pflugfelder 2000; Yeniad et al. 2003), as well as reduced sensitivity (Murphy et al. 2001). Moreover, the microbiota of contact lens-wearers is altered to one that more closely resembles skin microbiota (Shin et al. 2016).

The Role of Contact Lens Wear in Microbial Keratitis

These modifications of the corneal surface due to contact lens wear, mutually occurring with the insertion of pathogens into the eye via contaminated contact lens, lead to enhanced numbers of ocular infections. Specifically, contact lens wear is associated with up to 66% of microbial keratitis cases (Dutta et al. 2012), and an eighty-times increase in relative risk (Dart et al. 1991; Jeng et al. 2010), with higher incidences in high-income compared to low-middle income countries, as contact lens are worn more frequently. Contact lens wear was further reported in 37–57% of patients with fungal keratitis in high-income countries (Ong et al. 2016; Keay et al. 2011; Gower et al. 2010), and is the leading cause of fungal keratitis in the United States (Keay et al. 2011).

A transition of abiotic towards biotic biofilm occurs with the transmission of fungal biofilms from the contact lens to the eye (Yi et al. 2023). Abiotic biofilms, using biomaterial rather than tissue as substrate, form on the contact lens. These are then introduced to the ocular surface, where the biofilm protects the fungi from ocular defence mechanisms. Fungi then form biotic biofilms and invade the stroma as described in section "Fungal keratitis (biotic) biofilm formation".

Contact lens-associated fungal keratitis gathered attention in 2005–2006, when worldwide outbreaks occurred in soft contact lens wearers using ReNu with MoistureLoc® contact lens care solution (Khor et al. 2006; Barry et al. 2006). The causative fungi were *F. solani* and *F. oxysporum*. While planktonic fungi were susceptible to the care solution, biofilms on contact lenses were less susceptible (Imamura et al. 2008). A further contributing aspect might be improper use of contact lenses and the care solution, as 81.8% of affected patients in a study from Singapore executed poor lens hygiene (Khor et al. 2006). This presumption is supported by the fact that the concerned cleaning solution proved less effective when applied for shorter times than recommended (Ramani and Chaturvedi 2011).

Biofilm Formation on Contact Lens

Fungi adhere to and can penetrate contact lenses (Simmons et al. 1986; Chen et al. 2000; Ahearn et al. 2007, 2009; Lalitha et al. 2020; Poomany Arul Soundara Rajan et al. 2023), a first step for biofilm formation. Here, they grow in a coiled form within the contact lens matrix, so-called penetration pegs (Simmons et al. 1986; Ahearn et al. 2009). Penetration was seen for a variety of species, including *Acremonium*, *Alternaria*, *Aspergillus*, *Cladosporium*, *Coniochaeta*, *Curvularia*, *Exophiala*, *Fusarium*, *Paecilomyces*, *Penicillium* and *Urocladium* (Simmons et al. 1986; Chen et al. 2000; Ahearn et al. 2007, 2009; Lalitha et al. 2020; Poomany Arul Soundara Rajan et al. 2023).

The biofilms formed on contact lenses have variable morphology depending on the fungal species, showing differences in hyphal morphology, attachment, thickness and metabolic activity (Fig. 15.3). For instance, *Fusarium* biofilms on contact lens are homogeneous and rich in hyphae, while *Candida* biofilms are heterogeneous with a composition of hyphae and conidia (Imamura et al. 2008). Differences were seen even within a fungal genus, as *F. solani* biofilms on contact lens had more robust extracellular matrix than *F. oxysporum* (Retuerto et al. 2012).

Different properties of contact lens material such as surface treatment, ionic state, monomer backbone and water content might influence the formation of fungal

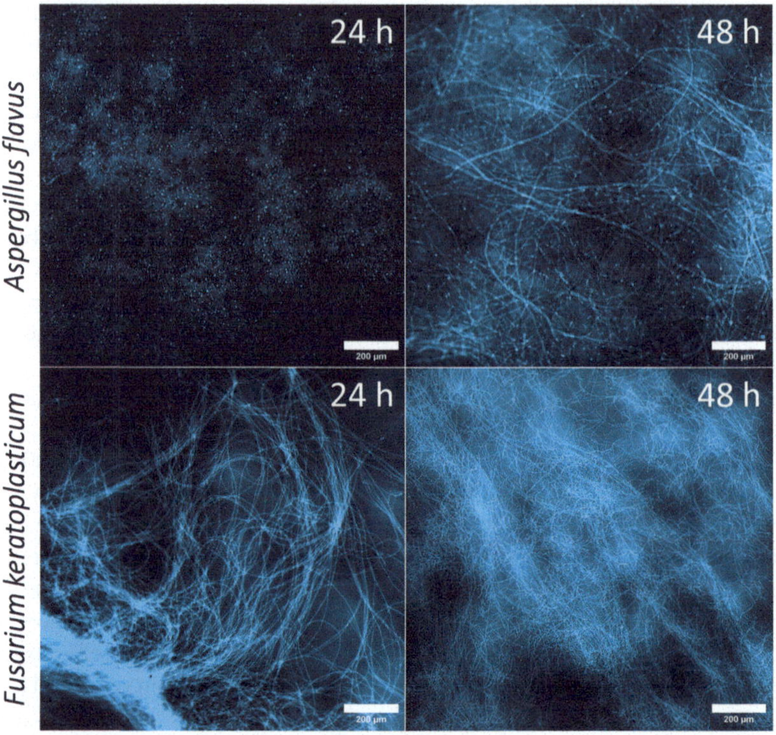

Fig. 15.3 Fungal biofilm progression on contact lenses (CL). *Aspergillus flavus* (ATCC200026) and *Fusarium keratoplasticum* (ATCC36031) spores were added to silicone hydrogel CL (Stenfilcon A) at a concentration of 10^6 mL^{-1} for 90 min at room temperature while rocking at 10 rpm. Non-adhered spores washed away followed by incubation at 28 °C, 5% CO_2 for 24 or 48 h. Biofilms were fixed using formalin, stained with calcofluor white and visualised using an inverted widefield Zeiss Observer microscope, 10× objective. Representative images of *A. flavus* and *F. keratoplasticum* biofilms on CL are shown. Scale bar = 200 µm. Figure is author's own

biofilms. The amount of adherent yeast was different between contact lens types and further elevated in biofilm-positive vs. -negative strains (Onurdag et al. 2011). Some studies report variations in *F. solani/oxysporum* and *C. albicans* biofilm formation depending on the lens type (Imamura et al. 2008) while others found no differences (Retuerto et al. 2012; Kackar et al. 2017). Lens types included rigid gas permeable hard lenses and soft lenses of different materials such as silicone hydrogel or 2-hydroxyethyl methacrylate. There are further contradictory results concerning the influence of water content on fungal growth rate (Simmons et al. 1986; Chen et al. 2000); however, *Fusarium* biofilm was found to be thicker on lenses with higher water content (Imamura et al. 2008).

Another potential influencing factor is the prior usage of the contact lens. *Penicillium* (Chen et al. 2000) and *C. albicans* (Dosler et al. 2020) exhibited enhanced growth or density on used compared to unworn contact lens, while

Fusarium developed penetration pegs earlier on worn lenses (Ahearn et al. 2009). No differences in the ability to form biofilms was found in a study comparing growth of *Fusarium* on worn and unworn contact lens, except for one lens type (etafilcon A) where *F. oxysporum* but not *F. solani* formed more biofilm on worn contact lens (Retuerto et al. 2012).

Further factors enhancing fungal contamination of contact lens are environmental (warm and moist environments), patient-related (e.g. pre-existing corneal disease, broad-spectrum antimicrobial treatment and steroid usage), as well as lens hygiene procedures (e.g. cleaning and extended wear) (Lalitha et al. 2020).

Fungal Contamination of Lens Cases

Fungal contamination and biofilm formation does not only occur on contact lenses, but also in lens cases and cleaning solutions. The feasibility of fungal biofilm formation in lens cases was demonstrated in the laboratory (Kackar et al. 2017). More importantly for contact lens users, contamination rates of contact lens cases with fungi lie between six and 27% in asymptomatic contact lens users (Gray et al. 1995; Willcox et al. 2010; Mela et al. 2015; Wu et al. 2010, 2015b; Kuzman et al. 2014). The rate might be related to compliance with contact lens case handling instructions, as one study found no contaminations in lens cases of fully compliant patients (Kuzman et al. 2014). However, contamination of contact lens cases and contained cleaning solutions with fungi was still as high as 10.2–14% in users complying with manufacturer's instructions for handling, cleaning and replacement in other studies (Willcox et al. 2010; Mela et al. 2015). Some did not further specify fungal specimen (Willcox et al. 2010, Wu et al. 2010, 2015b), while others isolated *F. solani* (Mela et al. 2015; Gray et al. 1995), *Aspergillus* spp. (Mela et al. 2015; Gray et al. 1995), *Candida* spp. (Mela et al. 2015; Gray et al. 1995; Kuzman et al. 2014), *Penicillium* spp. (Kuzman et al. 2014), *Rhodotorula rubra* (Mela et al. 2015), *Cladosporium* spp. (Gray et al. 1995), *Exophiala* spp. (Gray et al. 1995), *Phoma* sp. (Gray et al. 1995) and *Crysosporidium* sp. (Kuzman et al. 2014). One study specified that most fungi were isolated in association with bacteria (Gray et al. 1995). Interestingly, contamination seems to be independent of the contact lens case type when comparing two-well polypropylene cases, silver-impregnated cases and stand-up cases (Wu et al. 2010).

Efficacy of Contact Lens Cleaning Solutions Towards Fungal Biofilms

There are various types of contact lens cleaning solutions, with the main groups of disinfecting systems being hydrogen peroxide, polyhexamethylene biguanide (PHMB) and polyquaternium-1 (Polyquad) without, or in combination with, myristamidopropyl dimethylamine (Aldox). Except for hydrogen peroxide-based systems, these are contained in multipurpose solutions, which serve for cleaning and disinfection in one step and make up the majority of used solutions (Morgan and Efron 2006). Further components include, for instance, biguanide or alexidine as disinfectant/preservative, ethylenediaminetetraacetic acid (EDTA) to reduce protein deposition, hydroxyalkylphosphonate to remove protein deposits, surfactants like poloxamer, poloxamine, polyquaternium 10 and hydroxypropyl methylcellulose and propylene glycol as demulcents. The addition of lactoferricin B was further suggested to enhance the efficacy of lens care solutions against fungal biofilms (Sengupta et al. 2012).

Current regulation for antimicrobial activity testing of multipurpose solutions for contact lens cleaning involves International Standards Organization (ISO) methods ISO 14729 and ISO 18259 (ISO14729:2001 2010; ISO18259:2014 2014). In relation to fungal pathogens, ISO 14729 demands a one $\log_{10}$ reduction at disinfection time as primary criterion, or evidence of disinfection when directions for use are met as secondary criterion. However, this only requires testing of planktonic fungi of a single laboratory strain, thereby omitting variations in efficacy towards multiple clinical and environmental strains (Boost et al. 2010), as well as the known enhanced resistance of fungal hyphae, in particular when present in biofilms. In 2014, ISO 18259 implemented a second step in which contact lens cleaning solutions must be tested in the presence of contact lens and contact lens cases as provided by the manufacturer. This reflects findings of contact lens presence reducing the antimicrobial efficacy of contact lens cleaning solutions against bacteria, *F. solani* and *C. albicans* (Gabriel et al. 2018; McAnally et al. 2021). In addition, contact lens cases were shown to reduce contact lens cleaning solution efficacy due to biocide uptake (Gabriel et al. 2018).

Unfortunately, testing of lens cleaning solutions towards (fungal) biofilms is still not legally required, despite the evidence of reduced efficacy compared to planktonic fungi (Table 15.1). Nevertheless, numerous contact lens cleaning solutions have been tested on fungal biofilms, either for influence on biofilm formation on contact lens in the presence of lens cleaning solution (Dyavaiah et al. 2007; Kackar et al. 2017) or for their influence on viability (Ramani and Chaturvedi 2011; Mendonca et al. 2023), metabolic activity (Retuerto et al. 2012; Sengupta et al. 2012; Imamura et al. 2008) or visual appearance under the microscope (Dyavaiah et al. 2007) of pre-existing biofilms in the presence (Retuerto et al. 2012; Dyavaiah et al. 2007; Imamura et al. 2008) or absence of contact lenses (Ramani and Chaturvedi 2011; Sengupta et al. 2012; Mendonca et al. 2023).

The multitude of variations in testing protocols makes comparisons between studies challenging. Methods vary, as well as tested fungal strains, used contact lenses, and the definition of efficacy (Table 15.1). This might explain why the same contact lens cleaning solution seems to efficiently combat fungal biofilms in one study, but not in another. For instance, ReNu MultiPlus (Bausch & Lomb) was found effective on both *F. solani* and *F. oxysporum* biofilms when tested without contact lenses present, taking viability compared to a control as measure (Ramani and Chaturvedi 2011). Meanwhile, it was reported with reduced efficacy on *F. solani*

Table 15.1 Efficacy of contact lens cleaning solutions against fungal biofilms

Disinfecting system	Contact lens cleaning solution (Manufacturer)	Fungi	Efficacy				Ref.
			Definition	Planktonic	Biofilm formation	Pre-formed biofilm	
Hydrogen peroxide	Clear Care (Alcon)	*F. solani, F. oxysporum*	CFU: 1 $\log_{10}$ reduction for planktonic; ≥50% reduction of metabolic activity for biofilm	**Yes**	/	**Yes**, on CL	Retuerto et al. (2012)
	Oxysept Ultracare Formula (Amo)	*F. solani, F. oxysporum*	CFU: 1 $\log_{10}$ reduction for planktonic; ≥50% reduction of metabolic activity for biofilm	**Yes**	/	**Yes**, on CL	Retuerto et al. (2012)
	Equate (Wal-Mart)	*F. solani, F. oxysporum*	% Viability compared to control	**Yes**	/	**Yes**, without CL	Ramani and Chaturvedi (2011)
Polyhexamethylene biguanide (PHMB)	Complete Moisture Plus (Advanced Medical Optics)	*A. fumigatus, F. solani, C. albicans*	% Metabolic activity for biofilm	/	/	**Reduced**, without CL	Sengupta et al. (2012)
		F. solani, F. oxysporum	CFU: 1 $\log_{10}$ reduction for planktonic; ≥50% reduction of metabolic activity for biofilm	**Yes**	/	**No**, on CL for *F. solani* **Yes**, on CL for *F. oxysporum* on etafilcon A and lotrafilcon A lenses but **not** on balafilcon A lens	Retuerto et al. (2012)
	Aquify (Ciba Vision)	*F. solani, F. oxysporum*	CFU: 1 $\log_{10}$ reduction for planktonic; ≥50% reduction of metabolic activity for biofilm	**Yes**	/	**Yes**, on CL for *F. solani* **Yes**, on CL for *F. oxysporum* on etafilcon A lens but **not** on balafilcon A and lotrafilcon A lenses	Retuerto et al. (2012)
	Bio Soak (Teuto)	*C. albicans*	CFU: >3 $\log_{10}$ reduction (99.9%)	**Yes**	/	**No**, in CL case	Mendonca et al. (2023)
	Ultra Sept (Kley Hertz)	*C. albicans*	CFU: >3 $\log_{10}$ reduction (99.9%)	**Yes**	/	**No**, in CL case	Mendonca et al. (2023)
	Limp Lent (Vita Medic)	*C. albicans*	CFU: >3 $\log_{10}$ reduction (99.9%)	**Yes**	/	**No**, in CL case	Mendonca et al. (2023)

(continued)

Table 15.1 (continued)

Disinfecting system	Contact lens cleaning solution (Manufacturer)	Fungi	Efficacy				Ref.
			Definition	Planktonic	Biofilm formation	Pre-formed biofilm	
	Opto Care (Kley Hertz)	*C. albicans*	CFU: >3 log_{10} reduction (99.9%)	**Yes**	/	**No**, in CL case	Mendonca et al. (2023)
	Unica Sensitive (Avizor)	*C. albicans*	CFU: >3 log_{10} reduction (99.9%)	**Yes**	/	**No**, in CL case	Mendonca et al. (2023)
	Clear Lens Solution Multiuse (Opto Lentes)	*C. albicans*	CFU: >3 log_{10} reduction (99.9%)	**Yes**	/	**No**, in CL case	Mendonca et al. (2023)
	Clear Lens Solution (Opto Lentes)	*C. albicans*	CFU: >3 log_{10} reduction (99.9%)	**Yes**	/	**No**, in CL case	Mendonca et al. (2023)
	Clear Lentes Solution (Opto Lentes)	*C. albicans*	CFU: >3 log_{10} reduction (99.9%)	**Yes**	/	**No**, in CL case	Mendonca et al. (2023)
	All in One Light (Sauflon)	C. albicans	CFU: ≥ 3 log_{10} reduction	/	/	No, on CL after 6 hours	Dosler et al. (2020)
Polyquaternium-1 (Polyquad)	Bio True (Bausch & Lomb)	*C. albicans*	CFU: >3 log_{10} reduction (99.9%)	**Yes**	/	**No**, in CL case	Mendonca et al. (2023)
		C. albicans	CFU: ≥ 3 log_{10} reduction	/	/	**No**, on CL after 6 hours	Dosler et al. (2020)

Polyquaternium-1 (Polyquad) and myristamidopropyl dimethylamine (aldox)	OptiFree Express (Alcon)	*F. solani, F. oxysporum*	% Viability compared to control	**Yes**	/	**Reduced**, without CL, survival up to 6 hours	Ramani and Chaturvedi (2011)
		F. solani, F. oxysporum	CFU: 1 $\log_{10}$ reduction for planktonic; ≥50% reduction of metabolic activity for biofilm	**Yes**	/	**Yes**, on CL	Retuerto et al. (2012)
		C. albicans	CFU: ≥ 3 $\log_{10}$ reduction	/	/	**No**, on CL after 6 hours	Dosler et al. (2020)
	Opti-Free Replenish (Alcon)	*F. solani, F. oxysporum*	CFU: 1 $\log_{10}$ reduction for planktonic; ≥50% reduction of metabolic activity for biofilm	**Yes**	/	**Yes**, on CL for *F. solani* on etafilcon A and lotrafilcon A lenses but **not** on balafilcon A lens No, on CL for *F. oxysporum*	Retuerto et al. (2012)
		C. albicans	CFU: >3 $\log_{10}$ reduction (99.9%)	**Yes**	/	**No**, in CL case	Mendonca et al. (2023)
	Opti-Free PureMoist (Alcon)	*C. albicans*	CFU: >3 $\log_{10}$ reduction (99.9%)	**Yes**	/	**No**, in CL case	Mendonca et al. (2023)
Borax	"Soft lens solution 1"	*C. albicans*	Statistical significance of CFU/OD reduction	/	**Yes**, on CL	/	Kackar et al. (2017)

(continued)

Table 15.1 (continued)

Disinfecting system	Contact lens cleaning solution (Manufacturer)	Fungi	Efficacy				Ref.
			Definition	Planktonic	Biofilm formation	Pre-formed biofilm	
Polyaminopropyl biguanide and boric acid	ReNu MultiPlus (Bausch & Lomb)	*F. solani, F. oxysporum, C. albicans*	CFU: 1 log_{10} reduction for planktonic; % metabolic activity compared to control for biofilm	**Yes**	/	**Reduced** for *Fusarium* on CL; **Ineffective** for *Candida* on CL	Imamura et al. (2008)
		F. solani, F. oxysporum	% Viability compared to control	**Yes**	/	**Yes**, without CL	Ramani and Chaturvedi (2011)
		F. solani, F. oxysporum	CFU: 1 log_{10} reduction for planktonic; ≥50% reduction of metabolic activity for biofilm	**Yes**	/	**Yes**, on CL for *F. solani* on etafilcon A and lotrafilcon A lenses but **not** on balafilcon A lens **Yes**, on CL for *F. oxysporum* on etafilcon A lens but **not** on balafilcon A and lotrafilcon A lenses	Retuerto et al. (2012)
		C. albicans	CFU: ≥ 3 log_{10} reduction	/	/	**No**, on CL after 6 hours	Dosler et al. (2020)
	"Soft lens solution 2"	*C. albicans*	Statistical significance of CFU/OD reduction	/	**Yes**, on CL	/	Kackar et al. (2017)
Polyaminopropyl biguanide	Boston Simplus (Bausch & Lomb)	*C. albicans*	CFU: >3 log_{10} reduction (99.9%)	**Yes**	/	**No**, in CL case	Mendonca et al. (2023)
	Renu Sensitive (Bausch & Lomb)	*C. albicans*	CFU: >3 log_{10} reduction (99.9%)	**Yes**	/	**No**, in CL case	Mendonca et al. (2023)
	Renu Fresh (Bausch & Lomb)	*C. albicans*	CFU: >3 log_{10} reduction (99.9%)	**Yes**	/	**No**, in CL case	Mendonca et al. (2023)
Polymer complex, not further specified	"Rigid gas permeable lens solution"	*C. albicans*	Statistical significance of CFU/OD reduction	/	**No**, on CL	/	Kackar et al. (2017)

Polyquaternium 10 and Poloxamer 407	ReNu with MoistureLoc (Bausch & Lomb)	*F. solani*, *F. oxysporum*	No CFUs recovered for planktonic; microscopic analysis for biofilms	**Yes**; not when applied shorter than indicated	**Yes**, on CL	**Yes**, on CL	Dyavaiah et al. (2007)
		F. solani, F. oxysporum, C. albicans	CFU: 1 $\log_{10}$ reduction for planktonic; % metabolic activity compared to control for biofilm	**Yes**	/	**Reduced** for *Fusarium* on CL; **Ineffective** for *Candida* on CL	Imamura et al. (2008)
		F. solani, F. oxysporum	% Viability compared to control	**Yes**	/	**Reduced**, without CL, survival up to 4 h	Ramani and Chaturvedi (2011)

CFU colony forming unit, *CL* contact lens, *OD* optical density and *F. oxysporum* biofilms on contact lenses (Imamura et al. 2008), and more specifically, effective towards *Fusarium* biofilms on etafilcon A lenses, but not on balafilcon A lenses (Retuerto et al. 2012). Furthermore, it was effective towards *F. solani* but not *F. oxysporum* biofilms on lotrafilcon A lenses (Retuerto et al. 2012). Both studies measured metabolic activity, either compared to untreated controls (Imamura et al. 2008) or with a minimum reduction of 50% required (Retuerto et al. 2012).

While many contact lens solutions appear ineffective towards fungal biofilms, the efficacy of hydrogen peroxide contact lens cleaning systems is contradictory in the literature. All three H_2O_2-containing solutions tested on fungal biofilms successfully eradicated *Fusarium* biofilms (Retuerto et al. 2012; Ramani and Chaturvedi 2011), including from contact lenses (Retuerto et al. 2012). Despite this perceived efficacy towards this more resistant form of fungal growth, solutions with H_2O_2 were less effective in eliminating fungal species from contact lens cases (Mela et al. 2015). In contrast, earlier studies examining contact lens case contamination with fungi had postulated similar (Willcox et al. 2010) or reduced (Wilson et al. 1990) levels of fungal contamination in lens cases of users of a hydrogen peroxide-containing solution compared to other systems. These contradictory results might be influenced by the evolution of the contact lens cleaning systems that they were compared with. Moreover, none of the studies investigating contamination of contact lens cases with fungi specified if biofilms or planktonic forms were present.

Measures to Prevent Microbial Keratitis Caused by Contact Lens Wear

To reduce the number of microbial keratitis cases caused by contaminated contact lenses, various measures have been suggested. One aspect is the improvement of contact lens cleaning, such as scrubbing of contact lens cases to disrupt potential biofilms (Gray et al. 1995), regular replacement of the lens cases (Gray et al. 1995; Devonshire et al. 1993) with cases of the same manufacturer as the disinfection solution (Wu et al. 2015b), and air drying of lens cases between uses (Gray et al. 1995; Wu et al. 2015b). Another key factor to prevent contamination is to wash hands with soap and water when handling cases and lenses (Wu et al. 2015b). Other factors contributing to contact lens-related fungal keratitis are improper cleaning, such as reduced disinfection times or infrequent disinfection, and smoking (Wu et al. 2015a). It is recommended to use two-step systems if H_2O_2 disinfection is preferred, as all pathogens isolated in the context of a study investigating contact lens case contamination had enzyme catalases that break down hydrogen peroxide to oxygen and water, leaving the solution ineffective (Gray et al. 1995). However, the main risk factors for contact lens-related keratitis remains overnight wear (Stapleton et al. 2008).

Even though the risk to develop microbial keratitis is not reduced in patients using daily disposable lenses compared to patients wearing lenses for extended use (Dart et al. 2008), their microbial keratitis severity is lower (Dart et al. 2008; Stapleton et al. 2008), which in turn emphasises the importance of proper contact lens handling and replacement.

Approaches have been made to improve the intrinsic susceptibility of contact lenses and contact lens cases to biofilm formation (Dutta and Willcox 2014). An example targeting fungi is a novel contact lens coating based on immobilised copper ions in metal-phenolic networks, which not only exhibited broad-spectrum antimicrobial activity, but also proved efficient in inhibition of biofilm formation, including by *C. albicans* (Liu et al. 2020). Similarly, a nanocoating of the contact lens surface with zinc oxide nanoparticles, gallic acid and tobramycin resulted in significant antibiofilm performance against not only several multidrug-resistant bacteria, but also *C. albicans*, *A. fumigatus* and *F. solani* (Khan et al. 2021). Biofilm formation on contact lenses by a drug-resistant *C. albicans* strain (CCARM 14007) was inhibited with a peptide series PS1–3 peptide, making it an antifungal candidate for contact lens coating (Lee et al. 2022).

Fungal Biofilm Modelling in Fungal Keratitis Research

In Vitro Models of Fungal Biofilms

On Abiotic Surfaces: Excluding Contact Lenses

As not all fungal isolates are able to form biofilms, a multitude of studies evaluated the ability of ocular isolates to form biofilms in vitro (Ranjith et al. 2018). Other applications for biofilms grown in vitro involve investigation of their appearance and composition (Ranjith et al. 2018; Córdova-Alcántara et al. 2019; Calvillo-Medina et al. 2019), gene or protein expression (Ranjith et al. 2018; Calvillo-Medina et al. 2019), pathogenesis (Bautista-Hernandez et al. 2019; Córdova-Alcántara et al. 2019) or treatment and susceptibility testing (Roberts et al. 2024; Córdova-Alcántara et al. 2019).

Fungal biofilms can be formed in broth or media under static incubation (Ranjith et al. 2018; Roberts et al. 2024; Bautista-Hernandez et al. 2019; Córdova-Alcántara et al. 2019; Calvillo-Medina et al. 2019). For instance, *C. albicans* ocular isolates were induced for biofilm formation during 48 h incubation in Sabouraud dextrose medium. This resulted in four out of seven isolates growing into biofilms. These biofilm-positive isolates had 27 genes associated with biofilm formation and virulence upregulated (Ranjith et al. 2018). Clinical isolates of *A. flavus* and *F. keratoplasticum* were grown in potato dextrose broth for 24 h to enable testing of cold atmospheric plasma treatment (Roberts et al. 2024). Single and mixed biofilms of *F. falciforme* and *S. aureus* ocular isolates formed in DMEM/Ham's F-12 medium over 24 hours (Bautista-Hernandez et al. 2019). Some protocols allow for an initial adherence phase to the well plate, after which non-adherent cells are removed and fresh media is added for the remaining incubation (Córdova-Alcántara et al. 2019; Calvillo-Medina et al. 2019). Examples are biofilms produced by *F. solani* (Córdova-Alcántara et al. 2019) and *F. falciforme* (Calvillo-Medina et al. 2019) in studies where conidia in RPMI (±2% dextrose) were added to well plates for 4 h initially, and then grown for up to 120 h after medium replacement. This method revealed five biofilm formation stages in vitro, including adherence, germination, hyphal development, maturation and cell detachment, and further resulted in biofilms with an extracellular matrix made of carbohydrates, proteins and extracellular DNA (Córdova-Alcántara et al. 2019). Some of the expressed proteins were unique to biofilms and could not be found in planktonic growth stages of the same isolate (Calvillo-Medina et al. 2019). This method further produced biofilms with increased resistance to antifungals and UV light (Córdova-Alcántara et al. 2019).

While static incubation is common for in vitro development of fungal biofilms, they can also be formed under continuous flow. A chamber with flow rates of 180 μL RPMI per minute for 72 h was used to grow *F. oxysporum* and *F. solani* biofilms (Ramani and Chaturvedi 2011). This method required an initial static adhesion phase of 2 h. A homogeneous and compact biofilm with interwoven and mostly viable hyphae was then produced under flow.

On Abiotic Surfaces: Contact Lenses

As already described, an important risk factor for biofilm introduction to the eye are contact lens. Consequently, different models for investigation of fungal biofilms on contact lens or contact lens cases have been developed. Common approaches are either the incubation of contact lens in conidia suspension (Dosler et al. 2020; Simmons et al. 1986; Dyavaiah et al. 2007; Onurdag et al. 2011; Sengupta et al. 2012; Liu et al. 2020) or 48 h broth culture (Kackar et al. 2017), or an initial adherence phase in conidia suspension followed by washes to remove non-adhered spores and subsequent incubation in growth media until biofilms are formed (Ahearn et al. 2007; Imamura et al. 2008; Mukherjee et al. 2012; Retuerto et al. 2012).

Variations are seen regarding incubation times and temperatures, shaking versus static incubation, fungal load, used media and lens types. As an example, *C. albicans* biofilms grew best on used contact lens, in phosphate-buffered saline with 20% bovine serum albumin and 2 g/L lysozyme when incubated at 37 °C without shaking for 48 h (Dosler et al. 2020). Changes in media composition and temperature resulted in less dense biofilms, whereas shaking and incubation time (24 versus 48 h) did not show significant influences (Dosler et al. 2020). *C. albicans* biofilms were also successfully grown on contact lenses using yeast nitrogen base medium (Imamura et al. 2008), Sabouraud liquid medium (Onurdag et al. 2011) or RPMI (Sengupta et al. 2012). Prolonged incubation times were occasionally used with conidia suspensions, e.g., ten conidia of *F. solani* in PBS were cultured with contact lens for 14 days (Ahearn et al. 2009) and *C. albicans* was repeatedly exposed to contact lens over 14 days to test antibiofilm activities of the contact lens (Liu et al.

2020). If an initial adherence phase is included instead of submersion in fungal suspension throughout the growth phase, the timespan for adhesion needs to be adjusted depending on strain and substrate (Ahearn et al. 2007). For *C. albicans*, *F. solani* and *F. oxysporum*, an adherence phase of 90 min was commonly used (Imamura et al. 2008; Mukherjee et al. 2012; Retuerto et al. 2012).

A different method involved contact lens placed on Sabouraud dextrose agar plates concave side up, addition of fungal suspension onto the lens, incision of the lens to flatten it onto the agar and subsequent incubation at room temperature in sealed plastic bags (Chen et al. 2000). It is further possible to first grow biofilms in well plates, to then add contact lens to the pre-formed biofilm for 24 h and thereby achieve infection of the contact lens (Khan et al. 2021).

When drugs or contact lens cleaning solutions are being evaluated, biofilm induction and biofilm reduction need to be differentiated. While the first tests the formation of biofilm on the contact lens in the presence of the treatment to be tested, the latter allows the biofilm to form first before treatment is applied (Lee et al. 2022).

Co-culture with Corneal Cells

Even though modelling of fungal biofilms in vitro on corneal cells promises insights into the occurring host cell–pathogen interactions, most studies in this field use either spores or non-viable hyphae for exposure to a monolayer of corneal cells. This includes co-culture of *A. fumigatus* (Zhao et al. 2012; He et al. 2022; Luan et al. 2022; Qin et al. 2022) and *A. flavus* (Mallela et al. 2021; Ghosh et al. 2023) spores for up to 6 h on an epithelial monolayer, as well as *F. solani* on a fibroblast monolayer culture (Hu et al. 2017). Non-viable hyphal fragments of *A. fumigatus* were used in interaction with epithelial (Guo and Wu 2009; Li et al. 2015; Zhu et al. 2015; Zhao et al. 2016; Yang et al. 2019b; Fan et al. 2020; Tian et al. 2021; Zhang et al. 2021; Chen et al. 2023) and fibroblast monolayers (Li et al. 2013a). While these studies reflect the influence of fungal spores or inactivated hyphae on corneal cells, they fail to investigate fungal occurrence in the complex form of biofilm on the ocular surface.

However, some studies exposed spores to the mammalian cell culture for longer timespans, e.g., *A. fumigatus* for up to 16 h (Zhang et al. 2018; Zhao et al. 2018) and *C. albicans* for up to 48 h (Hua et al. 2017) on a corneal epithelial model. Similarly, *F. solani* spores were incubated in the presence of corneal fibroblasts for 24 h in a monolayer model (Hu et al. 2016), as well as *A. fumigatus* and *F. petroliphilum* for a maximum of 48 h in a three-dimensional model of fibroblasts in extracellular matrix (Brown et al. 2021b). These prolonged timespans allow spores to germinate and grow into hyphae, which might reflect an early step of biofilm formation on the ocular surface. There is one further report of *A. fumigatus* biofilm exposure to limbal fibroblasts (Ramirez-Granillo et al. 2021). Here, single and mixed fungal and bacterial (*S. aureus*) biofilms were grown for 24 h on confluent fibroblasts in RPMI medium with 10% fetal calf serum to induce biofilm formation, replacing medium after an initial adherence phase of 4 h. Interestingly, the resulting biofilm was thicker than biofilms grown on abiotic surfaces.

Ex Vivo and In Vivo Models of Fungal Keratitis Biofilm

Both ex vivo and in vivo models are advantageous over in vitro models as they recapitulate the microenvironment of the tissue of interest, which influences biofilm development (Petrillo et al. 2023). The various characteristics of ex vivo and in vivo models, model design and outputs have been comprehensively reviewed (Cheng et al. 2024; Urwin et al. 2020). While these models have been utilised to investigate novel treatments and to understand host–pathogen interactions, little has been reported regarding their use to study biofilms. Urwin et al reported using an ex vivo porcine infection model to study bacterial biofilm formation and reported bacterial colonisation using scanning election microscopy (Urwin et al. 2020).

Similarly, there are a few reports of biofilm characterisation using in vivo models which have been focused on bacteria, only one report of fungal biofilms was found (Ranjith et al. 2022). Fortunately, a study demonstrated that the standard murine microbial keratitis model (in which murine corneas are infected with bacteria after scratching) developed biofilms from day two onwards. This suggests that all previous models utilising this model have inadvertently been studying microbial keratitis biofilms.

In vivo models have also proven useful in the study of contact- lens-related fungal keratitis.

In Vivo Models for Studying Contact Lens-Related Fungal Keratitis

While feasibility of biofilm formation on contact lenses, as well as testing of novel treatments or preventative methods can be realised in vitro, the host effects are not involved. Therefore, in vivo models of keratitis caused by fungal biofilm-contaminated contact lenses were developed.

Murine models start with biofilm formation on contact lenses in vitro (e.g., 90 min of adhesion of *F. oxysporum* or *F. solani* to a soft contact lens, subsequent washing and incubation in Sabouraud dextrose broth for 48 h on a rocking platform). Most commonly, the corneal epithelium of anesthetised mice is abraded by three parallel scratches, or by abrasion of a 1-mm diameter area of the central corneal epithelium. Then, an infected two-millimetre contact lens punch is placed on the ocular surface for 2 h, before being removed, followed by examination of the mice for 2 days (Sun et al. 2010; Mukherjee et al. 2012; Chandra et al. 2014). This resulted in fungal keratitis induced from the biofilms formed on contact lenses.

Hyphae penetrated through the cornea and into the anterior chamber, and triggered cellular infiltration (Sun et al. 2010). Corneal opacification was independent from biofilm thickness (Mukherjee et al. 2012).

In addition, a rat model was established using immunosuppressed Sprague-Dawley rats for induction of fungal keratitis through infected contact lenses. For this purpose, the rat eyes were prepared by application of antibiotics to prevent bacterial growth. The corneal epithelium was centrally scraped and the stroma incised four times horizontally and vertically. Contact lenses, previously soaked overnight in *F. solani* suspension in sterile saline solution were then fitted onto the rat eyes for 4 h. The rats were then monitored for 7 days, over which they developed clinical signs similar to human patients (Abou Shousha et al. 2013). For this study, hydrogel contact lenses were specifically designed for rat eyes to ensure optimal fitting.

Diagnosis of Fungal Keratitis

Timely diagnosis of fungal keratitis and initiation of appropriate antimicrobials remains elusive in many settings; however, it holds the key to improving patient outcomes. Distinguishing fungi from bacteria early in infection remains the most pressing clinical decision point for initiating treatment, yet this often cannot be determined by clinical presentation alone (Dalmon et al. 2012; Redd et al. 2022b). Often, empirical antibiotics are the first-line treatment prescribed in the absence of a formal diagnosis, further delaying fungal keratitis patients from initiating antifungal therapy, and leading to worsened outcomes (Tuft et al. 2022; Ting et al. 2022). To date, the presence, absence and implication of biofilm during clinical diagnosis is not a specific consideration.

Current diagnosis is invasive and can only be performed within tertiary care. Each step requires skilled personnel to collect the sample (via corneal scraping with a blade or needle), and for sample processing and microbiological evaluation (Ngo et al. 2020). Gold standard culture may take 7–14 days to yield fungal results; however, >50% of cultures remain negative (Tuft et al. 2022; Hicks et al. 2022; Moledina et al. 2023). Smear microscopy of corneal scrapes is able to provide a same-day result; however, suffers from poor sensitivity. Other diagnostic approaches used clinically include in vivo *confocal microscopy* (IVCM) and molecular techniques, such as PCR, although are these are prohibitively costly and require expertise often unavailable in many regions. Current diagnostic procedures for fungal keratitis are reviewed extensively elsewhere (Ting et al. 2022; Mills et al. 2021; Hoffman et al. 2022b; Leck 2009). Simplified and less-invasive sampling techniques, using Corneal Impression Membranes (CIMs) are replacing corneal scraping in some clinical settings, and have been reported to improve culture positivity rate and be better tolerated by patients (Somerville et al. 2021, 2024).

There are a number of emerging approaches under development, aimed with improving diagnostic ease and accuracy of fungal keratitis. These include novel optical imaging methods (Gunasekaran et al. 2020), lateral flow technologies (Gunasekaran et al. 2023) and AI-assisted technologies designed to evaluate corneal ulcer photographs, in order to distinguish bacterial from fungal keratitis (Sarayar et al. 2023; Redd et al. 2022a; Wei et al. 2023). The clinical utility and scale up of these approaches remain to be seen.

Current and Future Treatment Strategies for Fungal Keratitis

Current Treatments

Currently Used Topical Antifungals

Several classes of antifungals are commonly utilised for fungal keratitis, and current evidence for use of these medications is not discriminatory of fungi with or without biofilm formation. These antifungals include azoles, such as voriconazole, and polyene-type antifungals, such as amphotericin B and natamycin. Azoles inhibit ergosterol synthesis in fungal cell membranes. In contrast, polyenes insert into fungal membranes closely associated with ergosterol, resulting in the formation of porin channels that subsequently leads to loss of transmembrane potential and dysregulated cellular function (Kanafani and Perfect 2008). One of the limitations of currently recommended antifungal eye drops is their fungistatic rather than fungicidal nature.

Natamycin is currently the only approved United States Food and Drug Administration (USFDA) drug to treat fungal keratitis. The landmark Mycotic Ulcer Treatment Trial (MUTT) in 2013, one of the main evidences for natamycin use, studied the effectiveness of natamycin versus voriconazole in filamentous keratitis, and demonstrated superiority in patients treated with natamycin. They showed significantly better best spectacle-corrected visual acuity at 3 months, and patients were less likely to have a perforation or require therapeutic keratoplasty (TPK) (Prajna et al. 2013a). In particular, natamycin was reported to be more effective against *Fusarium* keratitis. However, natamycin exhibits a weak-to-moderate action against yeast and is toxic to the corneal epithelium, which may slow healing of the cornea. In addition, suspended particles of natamycin can remain around the wound and eyelid, making assessment of the eye more difficult. On the other hand, amphotericin B is fungicidal against *Candida* but less predictable with moulds (Donovan et al. 2022). Amphotericin B also suffers from the disadvantage of causing punctate epithelial erosions and occasionally discolours the cornea green (Ansari et al. 2013). Both amphotericin B and natamycin have poor corneal penetration in the presence of an intact corneal epithelium (O'Day et al. 1986).

Oral voriconazole has also been investigated as an adjunct treatment as it exhibits excellent ocular penetration and may provide a more stable drug in the deeper stromal layers compared to intermittently applied eye drops (Thiel et al. 2007). However, a trial investigating the utility of oral voriconazole as a supplement to topical antifungals reported that it did not improve patient outcomes. Rather, oral voriconazole was associated with an increase in adverse events such as elevated liver enzymes and visual disturbance (Prajna et al. 2016). Similarly, another randomised controlled trial (RCT) which studied the utility of supplemental oral ketoconazole demonstrated no added benefit (Rajaraman et al. 2015).

Subconjunctival, Intrastromal and Intracameral Antifungal Treatment

In order to overcome the disadvantages of topical antifungals, including surface toxicity and poor ocular drug penetration, other methods of delivering antifungal medications have been investigated, including subconjunctival, intrastromal and intracameral injection. In practice, these techniques have been reserved for severe keratitis, sclerokeratitis or if there are concerns regarding treatment adherence (Said et al. 2011). While subconjunctival injections are easier to perform, they are painful for patients.

The utility of intrastromal injections have been studied in several pre-clinical (Nejabat et al. 2016; Mimouni et al. 2014) and clinical studies. A number of case series reported a high rate of resolution of disease for patients with otherwise refractory fungal keratitis (Sharma et al. 2011; Aydin et al. 2020; Sharma et al. 2019). A randomised controlled trial comparing intrastromal voriconazole, amphotericin B and natamycin as an adjunct to topical natamycin in recalcitrant fungal keratitis showed that although all three agents were equally successful at treating keratitis, intrastromal natamycin showed the fastest healing time, and intrastromal amphotericin B was associated with greater deep-corneal neovascularisation (Saluja et al. 2021). In contrast, in a trial of 40 patients randomised to either topical natamycin +/− adjuvant intrastromal voriconazole, topical treatment alone lead to a significant improvement in best corrected visual acuity at final follow-up, but there was no difference in treatment success or healing rate (Sharma et al. 2013).

Intracameral injections (injections into the anterior chamber) have been explored as a delivery method, but are less commonly reported compared to intrastromal injections. Clinically, they are routinely given to patients undergoing TPK for fungal keratitis after the removal of the diseased cornea to irrigate the anterior chamber, and prevent the spread of fungi into the globe (Said et al. 2011). Several small studies have demonstrated efficacy of intracameral amphotericin B (Dong and Krebs 2022; Yoon et al. 2007; Shao et al. 2010). A case series compared conventional treatment (topical amphotericin B and topical and oral fluconazole) +/− intracameral amphotericin B, and saw improved resolution of hypopyon in the group with intracameral amphotericin B, but no difference in treatment outcomes (Yoon et al. 2007). Similar outcomes were reported in other RCTs (Shao et al. 2010).

Of the three techniques described, intrastromal injections are the most well studied in the context of recalcitrant fungal keratitis, but the evidence of its superiority over topical therapy is lacking. The frequency of injections and agent of choice still requires further validation in large randomised controlled trials of longer duration.

Therapeutic Keratoplasty (TPK)

Despite antifungals, fungal keratitis outcomes are far from ideal, with a sizeable proportion of patients requiring TPK, ranging from 13% in Australia (Thew and Todd 2008), 24.8% in the United Kingdom (Ting et al. 2021) and 38% in Brazil (Ibrahim et al. 2009a). In a smaller proportion of cases (4–7%), the globe itself is removed (evisceration) to limit posterior spread of the fungal infection (Ting et al. 2021; Xie et al. 2007; Kasım and Koçluk 2020). Surgery is more likely in patients with advanced age, greater ulcer size or worse vision at presentation and isolation of fungal pathogen (Kodali et al. 2024).

Multiple studies have demonstrated the effectiveness of TPK (Xie et al. 2007; Sourlis et al. 2022); however, concerns about corneal transplants are manifold. As corneal transplants are considered for refractory keratitis, these eyes are generally severely inflamed and this carries the risk of corneal graft rejection (27–38.5%) (Kasım and Koçluk 2020; Xie et al. 2007). This is particularly important if a larger graft is required. In addition, there are concerns of infection recurrence (15–21%) (Kasım and Koçluk 2020; Xie et al. 2007). As patients who undergo corneal transplants are usually treated with corticosteroids post-surgery to prevent graft rejection, this immunosuppressive state could increase the risk of recurrence (Ansari et al. 2013). Other significant complications include secondary glaucoma (14–43%) and endophthalmitis (4–9%) (Kasım and Koçluk 2020; Xie et al. 2007). It is notable that a significant proportion of patients require multiple operations—in a study by Sourlis et al, 17.5% of patients needed at least one pars plana vitrectomy (PPV) with 15% patients requiring more than three PPVs while 33.3% patients had undergone repeat keratoplasty due to fungal keratitis recurrence or graft failure (Sourlis et al. 2022).

Given the major issue of graft rejection, deep anterior lamellar keratoplasty ((DALK) a partial corneal transplant in which the anterior lamellae of diseased cornea are removed) can be utilised instead, where the infiltrate only involves the anterior lamellae of the cornea. The immune rejection rate and incidence of secondary glaucoma is significantly reduced in DALK compared to TPK (Chen et al. 2021; Uchio et al. 2019).

Emerging Treatment Modalities for Fungal Keratitis

Innovative treatments for fungal keratitis encompass several categories, including: new antifungal agents; immunomodulators; advanced drug delivery methods and light-mediated therapies, such as photodynamic therapy and deep UV.

Novel Antifungals

In terms of novel antifungals, alternatives to natamycin, amphotericin B and voriconazole have been actively sought. The drugs described here have been tested preclinically for fungal keratitis but have yet to progress to clinical stages. Luliconazole, a new imidazole, was tested in an in vivo rabbit model as an alternative to voriconazole, and was demonstrated to be as effective as natamycin 5% for treating a *F. solani* keratitis. Further investigation into the toxicity profile and drug permeation characteristics of the compound would be valuable (Arimoto et al. 2023). Similarly, micofungin, a type of echinocandin, given subconjunctivally was more effective than fluconazole for *C. albicans* keratitis (Hiraoka et al. 2007). However, like natamycin and amphotericin B, drug penetration was negligible with the corneal epithelium present. With denuded epithelium, therapeutic level was achieved when instilled every 0.5 h. Thus, micofungin in this study did not overcome the limitations of current topical antifungal treatments.

Several novel antifungals evaluated against fungal keratitis have demonstrated concurrent immunomodulatory capabilities. For example, pentoxyfyllne (a methylxanthine derivative), which is used clinically for a variety of dermatological conditions (Hassan et al. 2014) and intermittent claudication (McCarty et al. 2016) has anti-inflammatory, anti-fibrosis and antioxidative properties. It also has been shown to disrupt fungal cell wall integrity, and have a role in neutrophil migration inhibition in a murine model of *A. fumigatus* keratitis, while also being less epitheliotoxic than topical natamycin (Chi et al. 2023).

Naturally occurring compounds such as sesquiterpene and phenolic acid have also been investigated. Nerolidol, a naturally occurring sesquiterpene, showed antifungal activity and reduced neutrophil and macrophage infiltration by inhibiting the LOX-1/IL-1β pathway (Yang et al. 2020). Similarly, 4-methoxycinnamic acid, a natural occurring phenolic acid, was effective against *A. fumigatus* and non-toxic to epithelial (HCEC) and macrophage (RAW 264.7) cells in vitro (Wang et al. 2023). In contrast, in vivo, 4-methoxycinnamic acid in combination with natamycin appeared to be more effective than either compound alone.

Novel Immunomodulators

The immune response is widely known to contribute to the final outcome of keratitis. Although infiltration of immune cells is essential in the response to fungi and their elimination, prolonged presence of immune cells leads to release of harmful enzymes, proteases and matrix metalloproteinases which further degrade the tissue. Despite this, there are limited options in terms of immunomodulatory agents. The use of corticosteroids is widely accepted to be detrimental in fungal keratitis given the frequent identification of steroids as a single risk factor for the development of keratitis and multiple studies demonstrating worsening of fungal keratitis with corticosteroids (Knutsson et al. 2021). Nonetheless, a novel immunomodulator that suppress the immune system without compromising the efficacy of antifungal treatments is still being sought.

There is growing evidence of the effectiveness of cyclosporine A (0.1%) in the management of fungal keratitis, due to its growth inhibition of *F. oxysporum*, *F. solani* and *A. fumigatus* compared to vehicle control and methylprednisolone in vitro (Bell et al. 1999). A synergistic effect between fluconazole and cyclosporine in inhibiting *Candida* spp. growth has also been reported (Marchetti et al. 2000). Clinically, cyclosporine A drops have been used in patients who underwent TPK, and it was reported to be effective in reducing postoperative inflammation without increasing risk of recurrence or graft failure (Chatterjee and Agrawal 2022). A randomised clinical trial investigating cyclosporine A as an adjunct to natamycin in treating fungal keratitis is currently ongoing (NCT06658002).

Another experimental treatment approach sought to enhance autophagy (Yang and Klionsky 2010), which is increased in fungal keratitis (Li et al. 2020). Treatment with rapamycin, an autophagy inducer, decreased clinical scores and neutrophil infiltration, while treatment with autophagy inhibitors resulted in increased clinical scores in murine models of fungal keratitis. The rapamycin treatment also decreased fungal CFU counts (Li et al. 2020). Similar observations were made by Xu et al, whereby rapamycin decreased the murine clinical scores, pro-inflammatory cytokine IL-1β and reduced expression of TLR4 and mTOR signalling pathway (Xu et al. 2018). Other in vivo studies investigating enhanced autophagy included the drugs progranulin (PGRN), A20 and flavopiridol (Qi et al. 2024; Han et al. 2023; Gu et al. 2022), with similar reported outcomes.

Others have successfully demonstrated pre-clinical anti-inflammatory activity of drugs and compounds against murine fungal keratitis models, including maresin 1 (Tang et al. 2019) and wedelolactone (Cheng et al. 2019) which were shown to regulate neutrophil recruitment, increasing anti-inflammatory cytokine IL-10 and reducing pro-inflammatory cytokine IL-1, respectively. Lian et al further demonstrated that knockdown of the NLRP3 (NOD-like receptor family pyrin domain containing 3) inflammasome led to less inflammation and lower clinical scores in a murine *C. albicans* fungal keratitis model (Lian et al. 2022). Notably, while these studies demonstrated efficacy of these drugs at modulating inflammation, CFU counts of infected corneas after treatment were not evaluated. Although clinically applicable, a decrease in clinical score does not always correlate with a decreased pathogen load (Meijome et al. 2022). Future studies evaluating novel immunomodulatory agents should not only investigate the in vitro and in vivo safety and efficacy of a drug but also their antifungal ability using CFU counts.

Novel Drug Delivery Modalities

Besides developing new antifungal and anti-inflammatory agents, efforts have also been focused on improving drug delivery and enhancing retention times. This would be beneficial as it could enhance the efficacy of drugs used presently, particularly for biofilm penetration and anti-biofilm activity, while simultaneously decreasing the burden of eye-drop frequency, indirectly improving patient adherence to

Table 15.2 Antifungals and nanocarriers investigated

Antifungal	Nanocarrier	Reference
Amphotericin B	Chitosan-modified lipid carriers	Fu et al. (2017)
Amphotericin B	Lecithin/chitosan nanoparticles	Chhonker et al. (2015)
Voriconazole Miconazole Itraconazole Amphotericin B	Citric acid crosslinked β-cyclodextrin	Soe et al. (2023)
Econazole	Solid lipid nanoparticles	Liang et al. (2023) and Liang et al. (2021)
Clotrimazole	β-Cyclodextrin	Rasool and Salmo (2012)
Fluconazole	Hydroxypropyl-beta-cyclodextrin complex loaded niosomal vesicles and Eudragit nanoparticles	Elmotasem and Awad (2020)
Voriconazole	Self-nanoemulsifying drug delivery system	Rasoanirina et al. (2020)
Voriconazole	Liposomes	de Sá et al. (2015)
Sertaconazole	Nanostructured lipid carriers	Tavakoli et al. (2019)
Sertaconazole	Poly(ethylene glycol)-block-poly(ε-caprolactone) (PEG-b-PCL) polymeric nanomicelles	Soliman et al. (2014)
Natamycin	Poly(ethylene glycol)-block-poly(glycidyl methacrylate) micelles	Guo et al. (2020)
Natamycin	Poly-lactic-co-glycolic acid- poly-ethylene imine-polyethylene glycol triblock polymer	Sha et al. (2022)
Natamycin	Cell penetrating peptides	Rohira et al. (2021)
Natamycin	Nano-cubosomes	Hosny et al. (2021)

treatment. The significant proportion of advances in drug delivery have focused on the use of nanotechnology, listed in Table 15.2.

Hydrogels or emulsions are formulated to enhance the sustained release profile of drugs. Examples include a combination of ketoconazole and ketorolac in mucoadhesive gels made from Carbopol 940, methylcellulose or hyrocypropyl methylcellulose (El-Mofty et al. 2014), and hybrid natural hydrogels composed of decellularised porcine cornea and gelatin integrated with voriconazole-loaded microspheres (Wang et al. 2021). Drug-loaded contact lenses such as a soft contact lenses impregnated with econazole-loaded hydroxypropyl cyclodentrins have also been reported (Wong et al. 2022).

Light-Mediated Treatments

Photodynamic Therapy

Antimicrobial photodynamic therapy (aPDT) is an emerging treatment approach towards the inactivation of pathogens, present within superficial infections; including bacteria, fungi, viruses and protozoa (Wainwright et al. 2017). Initial clinical applications have shown efficacy against infections within the oral cavity (Afrasiabi et al. 2022; Gholami et al. 2022) and of the skin (de Oliveira et al. 2022), causing reductions in pathogen viability and improving time-to-heal of wound sites. Now, several studies have explored the use of aPDT in the treatment of bacterial (de Paiva et al. 2022) and fungal (Hung et al. 2021) keratitis (Altamirano et al. 2020). In relation to aPDT against fungi, studies span from in vitro inactivation of fungal conidia and biofilms, through to exploratory clinical evaluation of riboflavin and rose bengal photosensitisers for the treatment of fungal keratitis.

PDT relies on the activation of a photosensitiser (PS) by a specific wavelength of light (ranging from ultraviolet-A (UVA)—700 nm), the wavelength of light being dictated by the absorbance maxima of the PS (Kwiatkowski et al. 2018). Following activation, the PS enters an excited singlet state, then undergoes intersystem crosslinking into an excited triplet state, where it reacts with oxygen to generate reactive oxygen species (ROS): superoxide anions, hydroxyl radical and hydrogen peroxide via Type I photo processes and/or singlet oxygen via Type II photo processes (Fig. 15.4). The ROS are short lived, but are able to interact with and cause damage to cellular components in close proximity, such as microbial cell envelopes (cell wall and membrane lipids), peptides and nucleic acids, before returning to the ground state (Wainwright et al. 2017).

aPDT confers a number of theoretical advantages over conventional antimicrobials for the treatment of fungal keratitis. These include rapid action, in the order of minutes, and could be efficacious with just a single treatment (Kim et al. 2023), compared to frequent and lengthy treatment regimens for traditional antifungal drugs (Hoffman et al. 2022a). Moreover, a number of secondary, synergistic benefits of aPDT on wound healing (Oyama et al. 2020), protection against corneal collagen degradation (Wollensak et al. 2003), modulation of the immune response (Paziani et al. 2019; Figueiredo-Godoi et al. 2019, Sun et al. 2020) and drug penetration into biofilm/cornea have been reported (Gao et al. 2016), as well as efficacy against multi-drug resistant bacteria and fungi (Lochenie et al. 2024; Benson et al. 2023; Liu et al. 2015). However, in reality, robust aPDT clinical data for treating of fungal keratitis are limited, with studies reported to date using riboflavin or rose bengal as the PS, as described below.

Riboflavin and Microbial Keratitis

Riboflavin (also known as vitamin B2) is a highly efficient PS, activated by UVA (370 nm) irradiation to generate singlet oxygen (Type II photosensitiser) (Crocker et al. 2022). Riboflavin-UVA is currently utilised as a clinical treatment for keratoconus (a type of corneal ectasia whereby the corneal stroma becomes weakened and thinned, leading to corneal bulging) (Meek and Hayes 2013), and is therefore safe to perform on the human cornea. The riboflavin-UVA treatment leads to corneal crosslinking, strengthening stromal fibrils and protection from enzymatic digestion by collagenases (Wollensak et al. 2003). It is hypothesised that riboflavin-UVA aPDT to treat microbial keratitis would benefit from this protective effect against

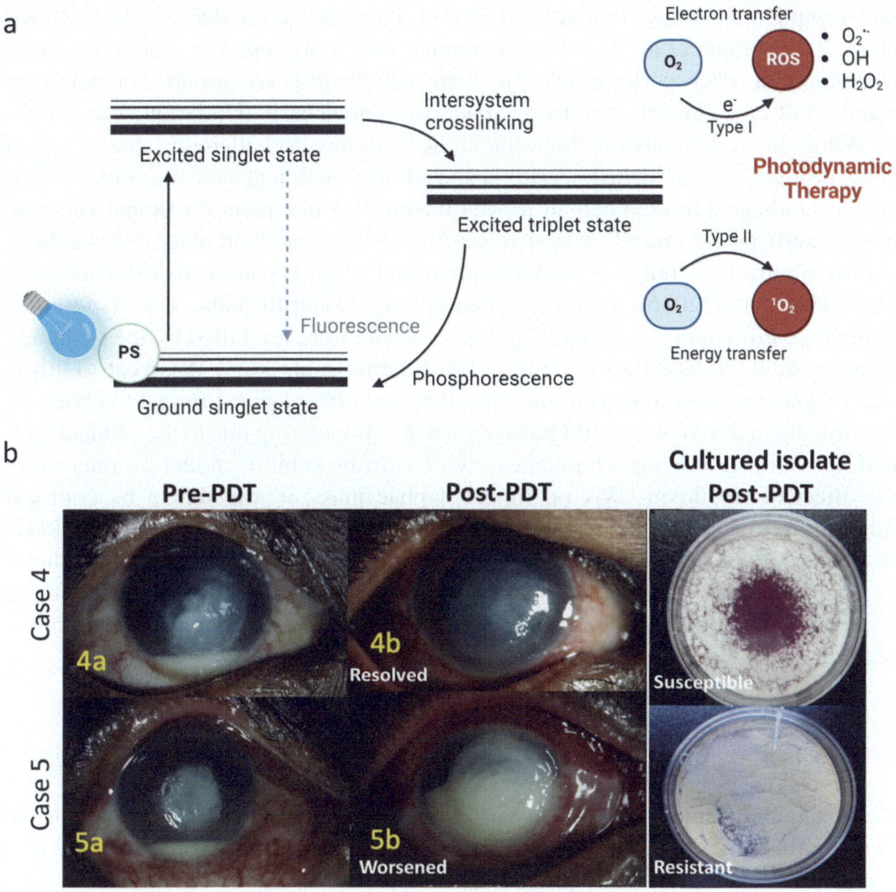

Fig. 15.4 (**a**) Modified Jablonski diagram of photosensitiser (PS) undergoing activation to excited state, and subsequent interaction with molecular oxygen. The PS can undergo electron transfer (Type I photo processes) generating reactive oxygen species (ROS): superoxide, hydroxyl radicals, and hydrogen peroxide, or can undergo energy transfer (Type II reaction) to molecular oxygen producing singlet oxygen. (**b**) Photographs of fungal keratitis patients pre-PDT (photo taken at presenting clinical appointment), and post-PDT treatment with Rose Bengal. Photo taken at time of complete resolution of infection (case 4, Fusarium keratitis. Some cloudiness of cornea still evident), and prior to TPK (case 5, Aspergillus keratitis). Right insert shows in vitro response of the patient clinical isolates to Rose Bengal PDT. (Figure adapted from Bagga et al. 2022)

corneal melt, whilst simultaneously killing the causative pathogen directly (Prajna et al. 2021).

This has been trialled in several clinical studies (Chan et al. 2015). While early studies have suggested a benefit for both bacterial (Makdoumi et al. 2012; Skaat et al. 2014; Price et al. 2012; Bamdad et al. 2015) and fungal (Saglk et al. 2013; Li et al. 2013b) keratitis, including fewer complications (Said et al. 2014), other studies, including randomised control trials have shown limited advantages for healing and complication rates (Prajna et al. 2021), particularly for deeper ulcers (Price et al. 2012; Prajna et al. 2020). In particular, one study reported worse outcomes (increased corneal perforation) for their riboflavin-UVA group compared to standard-of-care controls, and the trial was suspended early (Uddaraju et al. 2015).

While direct comparisons between clinical studies is challenging due to varied study designs, patient inclusion criteria and adjunct treatments, the majority of studies have indicated limited benefit for riboflavin-UVA treatment for fungal keratitis. This is perhaps not entirely surprising; while efficacy of riboflavin-UVA has been shown in vitro for a range of bacterial microbial keratitis pathogens (Martins et al. 2008; Halili et al. 2016), the most common fungal keratitis pathogens, *Aspergillus fumigatus*, *Fusarium solani* and *Candida albicans* were not killed by the treatment (Martins et al. 2008; Arboleda et al. 2014; Sharma et al. 2023). A caveat to this is that fungal spores were used in these studies, and effect against fungal hyphae was not investigated, so care should be taken when extrapolating this to the clinical context. In contrast, utilising a human ex vivo *Fusarium* keratitis model for measuring the effect of riboflavin-UVA on fungal hyphae mass, it was shown by confocal microscopy that there was a significantly lower volume of hyphae in the PDT-treated eyes compared to treatment-free, infected eyes 24 h after treatment (Alshehri et al. 2016). This suggests that the treatment did have a direct impact on the fungi; however, the viability of the remaining hyphae, and subsequent potential re-growth was not measured.

Rose Bengal and Microbial Keratitis

Rose bengal is another dye already used for ophthalmic purposes, such as for the visualisation of corneal lesions, and is activated by blue (455 nm) or green (532 nm) light to undergo type II photo processes. In contrast to riboflavin-UVA, rose bengal PDT does efficiently kill fungal spores and reduce biofilm (as well as bacteria) in vitro (Arboleda et al. 2014; Costa et al. 2012; Amescua et al. 2017; Silva et al. 2016), while also conferring the corneal stiffening and collagenase resistance properties akin to riboflavin (Fadlallah et al. 2016; Cherfan et al. 2013).

Rose bengal aPDT for treatment of microbial keratitis has been investigated clinically (Sepulveda-Beltran et al. 2022). The first case report utilised rose bengal PDT to successfully treat an MDR *Fusarium keratoplasticum* keratitis after two applications (Amescua et al. 2017). A second clinical study, where 5/18 patients had confirmed fungal keratitis (four had *Fusarium*, one had *Curvularia*) unresponsive to standard treatments, demonstrated that three of the fungal keratitis cases had successful treatment outcomes (avoidance of TPK) (Naranjo et al. 2019). A small clinical study by Bagga et al (Bagga et al. 2022) with 7 fungal keratitis patients observed a fungal-species specific difference in patient outcomes (Fig. 15.4b), with all three patients with *Fusarium* infections achieving clinical resolution, whereas the three *Aspergillus* keratitis patients and single *Acremonium* keratitis patient worsened and required therapeutic keratoplasty. Interestingly, the patient outcomes correlated with susceptibility/resistance of the isolated fungi (spores) to rose bengal PDT in vitro. In vitro resistance of six *Aspergillus* spp. keratitis clinical isolates to rose bengal aPDT was also observed by Chou et al, while *Fusarium* spp. isolates were all sensitive to the treatment (Chou et al. 2025). While clinical studies to date have been relatively small, a large, multi-site randomised control trial with 165 participants per arm is underway, aiming to determine if rose bengal PDT is a beneficial adjuvant in the treatment of filamentous fungal or acanthamoeba keratitis (Prajna et al. 2024).

Other Clinically Approved PS and Their Evaluation Against Fungal Biofilm

While both riboflavin and rose bengal have been exploited for aPDT of fungal keratitis clinically, with varying degrees of efficacy, a number of studies have explored the use of alternative, clinically utilised PSs and approaches to treat fungal biofilms, with relevance to fungal keratitis, in vitro (Ziental et al. 2021).

C. albicans biofilms have been shown to have susceptibility to erythrosine, a PS used in dentistry, with biofilm viability reduced by 74% after PDT treatment (Costa et al. 2011). Not only did the clinically approved photosensitiser Photofrin reduce viability, it also reduced the metabolic activity of the *C. albicans* biofilms by ~30%, which was not seen in non-irradiated controls, or biofilms treated with amphotericin B (Chabrier-Roselló et al. 2005). 5-aminolevulinic acid (ALA), a precursor of the photosensitiser protoporphyrin IX (PpIX), was able to be absorbed into *C. albicans* biofilm to reduce viability by >75% with illumination (Shi et al. 2016, Greco et al. 2020). PDT against fungi with photosensitisers such as erythrosine, ALA and methylene blue have also been shown to effect gene expression of *C. albicans*. Sub-lethal treatment of biofilms with these PS was shown to significantly reduce the expression of the genes related to adhesion, hyphae and biofilm formation, thus reducing overall virulence (Freire et al. 2018). Similar effects towards gene expression, including those related to the oxidative stress response have been found by others utilising alternative PS (Jordao et al. 2020; Dias et al. 2023; Shi et al. 2021).

There are few studies investigating the use of clinically approved PS on hyphae or biofilms of *Aspergillus* or *Fusarium* in vitro. ALA activity against mature *Aspergillus* biofilm has been demonstrated (Duan et al. 2023), reducing viability and causing disturbances to structural organisation of hyphae. The PDT activity of methylene blue and toluidine blue O (TBO) against germinated microconidia of three *Fusarium* spp. have been reported (de Menezes et al. 2016). Both PS with illumination resulted in 3–5 $\log_{10}$ reduction in viability, depending on the strain of *Fusarium* used. A similar effect was seen for methylene blue PDT used to treat *Galleria mellonella* larvae infected with *F. keratoplasticum* microconidia, with viability of the microconidia reduced by 62% (Paziani et al. 2019), and prolonged larvae survival. Similar results were also obtained for *C. albicans*-infected *G. mellonella* (Chibebe Junior et al. 2013; Figueiredo-Godoi et al. 2019). Further studies have combined PDT with methylene blue against (antimicrobial resistant) *Fusarium* biofilms with adjunct antimicrobial treatment (Gao et al. 2016). Not only was the PDT highly effective against the biofilm viability, pre-treatment with methylene blue significantly reduced the minimum inhibitory concentration of each of the antifungal drugs, presumably by increasing membrane permeability of the fungal cells. Similar advantages of combining TBO PDT with caspofungin (but not fluconazole) were observed against *C. albicans* biofilm (Huang et al. 2018).

Mixed species biofilms demonstrate higher levels of drug resistance compared to monospecies biofilms. Recently, PDT against a mixed-species biofilm formed of *C. albicans* and *C. tropicalis* with PS Photodithazine (Carmello et al. 2017) and Toluidine Blue O (TBO) (Zhang et al. 2023) have been investigated. Photodithazine was shown to be effective against mono-species biofilms, but not the duo-species biofilms. Conversely, TBO was able to reduce the mixed-species biofilm viability by 99% within 1 min of illumination.

Novel Fluorescent-PS Against Fungal Keratitis In Vitro/In Vivo

An emerging area of interest is the use of imaging-PDT for a theranostic approach, i.e., confirm the infection (e.g., by fluorescent imaging) and then treat with the same compound (e.g., PDT). Three novel photosensitisers and approaches have recently been reported for image-guided PDT.

The first used luminogens with aggregation-induced emission characteristics (AIEgens) to target fungal mitochondria (Zhou et al. 2021). The evaluated compounds were shown to accumulate preferentially within the mitochondria of *C. albicans* over human cornea epithelial cells, allowing for fluorescence imaging followed by robust toxicity against the fungal cells (100% killing of yeast cells within 30 min for compound IQ-TPA), and no impact on the human cell-line. IQ-TPA was also evaluated in an in vivo model of fungal keratitis, infected with *C. albicans* and compared to treatment with rose bengal PDT. Only eyes treated with IQ-TPA and light had resolved infection, limited redness and clear cornea by day 5. Candida were not culturable from the eye. In contrast, the infection was only controlled by rose bengal by day 7, and significant eye-redness and culturable microorganisms remained.

A second approach utilised a novel fluorophore-PS combination conjugated to a pan-microbe binding domain, ubiquicidin$_{29\text{–}41}$ (Lochenie et al. 2024) to enable selective labelling of bacterial and fungal cell envelopes and thus pathogen-specific imaging and killing via PDT. The lead compound, PEP3 led to 99.6% eradication of *C. albicans* biofilms, as well as 100% *F. keratoplasticum* and 96.6% *A. flavus* hyphae. Interestingly, *A. flavus* spores were not labelled by PEP3 (no florescence signature) and were not killed by the PDT action, conversely, *F. keratoplasticum* spores were labelled and completely susceptible to killing via illumination of the PS. PEP3 was also demonstrated to be less cytotoxic against human epithelial cells compared to equimolar rose bengal.

The third approach used AgCu2O-EDTA nanoparticles (AgCuE NPs) developed as part of their study (Ye et al. 2022) to kill fungi (through disruption of the fungal cell wall upon light activation), and treat *C. albicans* in a murine fungal keratitis model. The AgCuE NPs both prevented biofilm formation and destroyed mature biofilms in vitro (82.21% inhibition). When measuring the efficacy of the AgCuE

NPs in vivo, only the NP with illumination group fully resolved by day 15, with animals receiving a clinical score of 0.67. In comparison, amphotericin B treated groups had a final clinical grade of 8, and a significantly greater epithelial defect. Moreover, histology and in vivo laser scanning microscopy determined that no fungal hyphae remained in the AgCuE NP-laser group, while hyphae were still present in the untreated, AgCuE NP-dark and amphotericin B treated groups.

UV Light and Its Mechanism of Action

A general limitation of aPDT for fungal keratitis is the requirement for delivery of the PS, which typically faces the same challenges associated with antimicrobial eye-drops. For example, delivery of riboflavin every 2 min for 30 min prior to illumination (Prajna et al. 2020). Therefore, light-mediated applications, such as ultraviolet (UV) radiation, which do not require the addition of a PS are increasingly attractive.

UV radiation can be categorised into three types based on wavelength: UVA (320–400 nm), UVB (280–320 nm) and UVC (100–280 nm). UVC light is the most effective at killing microorganisms (including bacteria, viruses and fungi (Hockberger 2002)), as the peak germicidal wavelength is in the range of 250–270 nm (Gurzadyan et al. 1995). Absorption of UV photons by DNA or RNA causes the formation of covalent bonds between adjacent pyrimidines, leading to cyclobutene pyrimidine dimers (CPD) and Pyrimidine (6–4) Photoproducts (6-4PPs) (Setlow 1966), and prevention of replication, effectively leading to inactivation or cell death. UVC inactivation is widely used across a range of healthcare and environmental settings (Andersen et al. 2006; Pereira et al. 2023; Bhardwaj et al. 2021; Raeiszadeh and Adeli 2020; Yang et al. 2019a).

UV Light and the Cornea

The cornea is constantly exposed to UV light emitted from the sun, and has evolved to shield and protect the underlying structures against the consequences of UV exposure (Norren and Vos 1974; Sliney 2002). The penetration and associated damage to human cornea differs with the wavelength of UV exposure. For example, while CPD formation was mainly concentrated in the corneal epithelium following UVB and UVC light exposure, UVA induced significant CPD levels throughout the corneal endothelium and into the first layer of the iris, indicating that UVA can reach deeper ocular structures (Mallet and Rochette 2013). Despite this, UVA irradiation is used in several ophthalmic procedures (including corneal cross-linking). These procedures are conducted under strict protocols with a set dosage of UVA irradiation (0.35 mW/cm^2) to corneas with a minimum thickness (>400 μm), thus mitigating the damage to the endothelium or the deeper structures (Spoerl et al. 2011).

More recently, it has been demonstrated that a single 300 s exposure of 265 nm of UVC to primary corneal epithelial cells lead to significant CPD formation (Marasini et al. 2021). However, a 15 s exposure of the same UVC fluence did not result in significant CPD formation, and CPD levels of cells exposed to repeated 15 s of UVC light returned to baseline levels by 72 h after exposure, indicating successful intrinsic DNA damage repair by the cells. Moreover, the UVC inflicted damage was confined to the superficial cells of the cornea in ex vivo porcine models.

Antimicrobial Efficiency of UVC Light

The antimicrobial efficacy of UVC light has successfully been demonstrated in bacterial strains that are implicated in corneal infections by comparing reductions in bacterial counts (CFU) before and after irradiation on agar plates (Dean et al. 2011; Dumpati et al. 2022; Sosnin et al. 2004; Marasini et al. 2022; Song et al. 2022). The conclusion was similar across all the studies; the inhibition zones for *P. aeruginosa* irradiated with UVC were the largest compared to other microorganisms, but all showed significant zones of inhibition. Efficacy of bacterial elimination has also been demonstrated in ex vivo porcine (Marasini et al. 2022) and canine (Turicea et al. 2024) cornea, and in vivo murine microbial keratitis models (Marasini et al. 2022). After UVC (265 nm) treatment of ex vivo porcine microbial keratitis, the bioluminescence of a genetically modified *P. aeruginosa* strain decreased by up to 3.5 $\log_{10}$ CFU mL^{-1} in treated eyes, indicating inhibited metabolism and/or viability, and the overall bacterial burden was higher in the control group than in the UVC treatment group. Treatment with UVC (265 nm) light in the in vivo murine model resulted in the bioluminescence levels of *P. aeruginosa* decreasing below the detection threshold after 8 h, whereas the untreated mice reached a humane endpoint by 24 h.

In comparison to the research on the bactericidal nature of UVC light, there are only a small number of research studies conducted on the efficiency of UVC light on killing fungi. Irradiation of *C. albicans,* dried onto glass slides, exposed to 254 nm for 120 s resulted in a 5-$\log_{10}$ reduction in CFU compared to non-illuminated controls (Guridi et al. 2019). Similar results were obtained for *C. auris* and *A. niger* spores dried onto glass slides, where viability was reduced by a 5-$\log_{10}$ and 4-$\log_{10}$ CFU, respectively (Khan et al. 2022).

However, while these data are promising, the doses (intensity *x* exposure duration) of UVC light for surface sterilisation techniques are much higher (ranging from 830 mJ/cm^2 to 3360 mJ/cm^2) than doses that would be safe to use as treatment options. Doses ranging from 1.5 to 57.9 mJ/cm^2 have been demonstrated as non-lethal to corneal cells but lethal to microorganisms (Sosnin et al. 2004; Dean et al. 2011; Marasini et al. 2021; Song et al. 2022). Dumpati et al evaluated UVC (265 nm) devices against fungal strains that are clinically relevant to keratitis, and demonstrated efficacy against *C. albicans* and *F. solani* spores (Dumpati et al. 2022). The inhibition zones on agar lawns were 5.5 mm for *C. albicans* and 5.0 mm for *F. solani* compared to slightly larger areas of inhibition for the bacterial strains in the study

(7.2 mm for *P. aeruginosa*). Similar studies have shown UVC effectiveness at inhibiting the growth of *C. albicans* and *A. brasiliensis* (Marasini et al. 2022). In these investigations the growth inhibition of fungi as lawns on agar plates were also comparable to the bacteria studied alongside. The doses used in these studies were much lower (1.93 mJ/cm^2) than in doses used for sterilisation of surfaces, therefore likely much safer for the cornea. Irradiation of *C. albicans* at a slightly higher wavelength of 275 nm, but a lower dose of 1.5 mJ/cm^2 also resulted in aseptic zones after 10 s of irradiation (Song et al. 2022).

The studies reported for UVC treatment of fungi are limited in wavelength, experimental design, and strains studied. All of the studies indicated in this review have included antimicrobial efficiency on agar plates, but none have determined whether fungal biofilms on the cornea can be reduced using UVC light. Establishing the depth penetration and absorbance of UVC within the cornea, modelled with fungal infection is an area of study urgently required for assessing the translational potential of UVC as a treatment for fungal keratitis.

Contact Lens Sterilisation Using UVC Light

Another potential application of UVC light is in reducing the bacterial or fungal load on contaminated contact lenses and lens cases (Dumpati et al. 2022). This could align with the treatment of fungal keratitis using UVC light. UVC sterilisation at 265 nm significantly reduced *C. albicans* contamination on contact lens surfaces with up to 2.43-$\log_{10}$ inhibition depending on the type of contact lens material used. A comparable trend was observed against *F. solani*, where UVC showed inhibition of up to 0.90 $\log_{10}$ CFU.

Concluding Remarks

Fungal keratitis is a sight-threatening corneal infection that affects millions of patients per year. The major preventable risk factor is the use of contaminated contact lenses. While this appears easily avoidable by adherence to cleaning regimens, many contact lens cleaning solutions are ineffective against fungal hyphae and testing of such is legally not required. Many ocular isolates from fungal keratitis patients are biofilm-producing fungal strains. This growth form makes treatment more challenging, as biofilms are more resistant to antifungals and present enhanced immune evasion, as well as protection against a number of other environmental- and host-mediated factors.

Prevailing treatment options are limited, and often ineffective towards fungal biofilms for a myriad of reasons, and while invasive methods of drug delivery may be helpful in refractory cases, it does not appear to be superior to topical therapy. Surgical techniques are useful in severe cases, but the post-operative journey is far from easy. Thus, there is a need for novel approaches, with current avenues involving not only novel antifungals, but also immunomodulators, photodynamic therapy and UVC light treatment. Evidently, advanced drug delivery techniques for the management of fungal keratitis are also required to improve dosing of drug compounds, as well as improve the patient experience. Most studies developing and trialling treatments for fungal keratitis did not distinguish between hyphae with or without biofilms, or tested only on fungal spores. These are clearly major limitations and hinder clinical performance. The overarching goal is to specifically target the pathogen, regardless of the presence of extracellular components of biofilms, and without cytotoxic effects towards the corneal cells. Additionally, inhibition of immune responses and the resulting collagenolytic processes which leads to cornea melting would be beneficial. While in vitro biofilm models for efficacy testing of novel drugs are available, the options involving corneal cells or ex vivo corneas are limited. In addition, investigation of the immune responses to fungal keratitis, is to date, only explored using in vivo rodent models of fungal keratitis, and major knowledge gaps persist, both in terms of the immune response directly, and the influence of biofilm in situ in real-time. These impede advancements in the development and evaluation of treatment approaches, which are urgently needed to reduce the burden of visual impairment associated with fungal keratitis globally.

Acknowledgments This work was supported the UK Research and Innovation Future Leaders Fellowship scheme (MR/V026097/1); and Edinburgh Clinical Academic Track—inclusive (ECAT-i)/Wellcome PhD Fellowship for Health Professionals. For the purpose of open access, the author has applied a Creative Commons Attribution (CC BY) licence to any Author Accepted Manuscript version arising from this submission. Figures were prepared with BioRender (BioRender.com/h73m782; BioRender.com/i05v325; BioRender.com/q61h835).

Declaration of Competing Interest The authors jointly do not have any competing interests to declare.

References

Abou Shousha M, Santos AR, Oechsler RA, Iovieno A, Maestre-Mesa J, Ruggeri M, Echegaray JJ, Dubovy SR, Perez VL, Miller D, Alfonso EC, Bajenaru ML (2013) A novel rat contact lens model for Fusarium keratitis. Mol Vis 19:2596–2605

Afrasiabi S, Partoazar A, Chiniforush N, Goudarzi R (2022) The potential application of natural photosensitizers used in antimicrobial photodynamic therapy against oral infections. Pharmaceuticals 15:767. https://doi.org/10.3390/ph15060767

Ahearn DG, Simmons RB, Zhang S, Stulting RD, Crow SA Jr, Schwam BL, Pierce GE (2007) Attachment to and penetration of conventional and silicone hydrogel contact lenses by Fusarium solani and Ulocladium sp. in vitro. Cornea 26:831–839. https://doi.org/10.1097/ICO.0b013e31806c782a

Ahearn DG, Zhang S, Ward MA, Simmons RB, Stulting RD (2009) Hyphal penetration of worn hydrogel contact lenses by Fusarium. Cornea 28:914–917. https://doi.org/10.1097/ICO.0b013e31819c189d

Aimanianda V, Bayry J, Bozza S, Kniemeyer O, Perruccio K, Elluru SR, Clavaud C, Paris S, Brakhage AA, Kaveri SV, Romani L, Latgé J-P (2009) Surface hydrophobin prevents immune recognition of airborne fungal spores. Nature 460:1117–1121. https://doi.org/10.1038/nature08264

Alshehri JM, Caballero-Lima D, Hillarby MC, Shawcross SG, Brahma A, Carley F, Read ND, Radhakrishnan H (2016) Evaluation of corneal cross-linking for treatment of fungal keratitis: using confocal laser scanning microscopy on an ex vivo human corneal model. Investig Ophthalmol Vis Sci 57:6367–6373. https://doi.org/10.1167/iovs.16-20110

Altamirano D, Martinez J, Leviste KD, Parel JM, Amescua G (2020) Photodynamic therapy for infectious Keratitis. Curr Ophthalmol Rep 8:245–251. https://doi.org/10.1007/s40135-020-00252-y

Amescua G, Arboleda A, Nikpoor N, Durkee H, Relhan N, Aguilar MC, Flynn HW, Miller D, Parel J-M (2017) Rose Bengal photodynamic antimicrobial therapy: a novel treatment for resistant Fusarium Keratitis. Cornea 36:1141–1144. https://doi.org/10.1097/ICO.0000000000001265

Andersen BM, Bånrud H, Bøe E, Bjordal O, Drangsholt F (2006) Comparison of UV C light and chemicals for disinfection of surfaces in hospital isolation units. Infect Control Hosp Epidemiol 27:729–734. https://doi.org/10.1086/503643

Ansari Z, Miller D, Galor A (2013) Current thoughts in Fungal Keratitis: diagnosis and treatment. Curr Fungal Infect Rep 7:209–218. https://doi.org/10.1007/s12281-013-0150-110.1007/s12281-013-0150-1

Arboleda A, Miller D, Cabot F, Taneja M, Aguilar MC, Alawa K, Amescua G, Yoo SH, Parel J-M (2014) Assessment of rose bengal versus riboflavin photodynamic therapy for inhibition of fungal keratitis isolates. Am J Ophthalmol 158:64–70.e2. https://doi.org/10.1016/j.ajo.2014.04.007

Arimoto S, Inagaki K, Todokoro D, Suzuki T, Makimura K, Ishino T (2023) Antifungal efficacy of luliconazole in an experimental rabbit model of fungal keratitis caused by Fusarium solani. Mycopathologia 188:775–782. https://doi.org/10.1007/s11046-023-00783-5

Arunachalam D, Namperumalsamy VP, Prajna L, Kuppamuthu D (2021) Human corneal epithelial cells internalize Aspergillus flavus spores by actin-mediated endocytosis. Infect Immun 89. https://doi.org/10.1128/IAI.00794-20

Arunachalam D, Ramanathan SM, Menon A, Madhav L, Ramaswamy G, Namperumalsamy VP, Prajna L, Kuppamuthu D (2022) Expression of immune response genes in human corneal epithelial cells interacting with Aspergillus flavus conidia. BMC Genomics 23:1–21. https://doi.org/10.1186/s12864-021-08218-5

Arunga S, Wiafe G, Habtamu E, Onyango J, Gichuhi S, Leck A, Macleod D, Hu V, Burton M (2019) The impact of microbial keratitis on quality of life in Uganda. BMJ Open Ophthalmol 4:e000351. https://doi.org/10.1136/bmjophth-2019-000351

Aydin B, Cubuk MO, Ucgul A, Ertop M, Ozmen MC, Atalay T, Akata F (2020) Combined Intrastromal Voriconazole and Amphotericin B treatment for persistent fungal keratitis. Eye Contact Lens 46:269–273. https://doi.org/10.1097/ICL.0000000000000723

Bagga B, Sharma S, Ahirwar LK, Sheba E, Vaddavalli PK, Mishra DK (2022) Clinical outcomes of rose bengal mediated photodynamic antimicrobial therapy on fungal keratitis with their microbiological and pathological correlation. Curr Eye Res 47:987–994. https://doi.org/10.1080/02713683.2022.2058019

Balakrishnan Sangeetha A, Abdel-Hadi A, Hassan AS, Shobana CS, Suresh S, Abirami B, Selvam KP, AL-Baradie RS, Banawas S, Alaidarous M, Alshehri B, Dukhyil AAB, Dhanasekaran S, Manikandan P (2020) Evaluation of in vitro activities of extracellular enzymes from Aspergillus species isolated from corneal ulcer/keratitis. Saudi J Biol Sci 27:701–705. https://doi.org/10.1016/j.sjbs.2019.11.023

Bamdad S, Malekhosseini H, Khosravi A (2015) Ultraviolet A/riboflavin collagen cross-linking for treatment of moderate bacterial corneal ulcers. Cornea 34:402–406. https://doi.org/10.1097/ICO.0000000000000375

Barry M, Pendarvis J, Mshar P, Ritger K, Dixon P, Alfonso E, Leguen F, Signs K, Fine A, Nivin B (2006) Update: Fusarium Keratitis–United States, 2005–2006 [Online]. Available: https://www.cdc.gov/mmwr/preview/mmwrhtml/mm5520a5.htm. Accessed June 2024

Bartimote C, Foster J, Watson S (2019) The spectrum of microbial keratitis: an updated review. Open Ophthalmol J 13. https://doi.org/10.2174/1874364101913010100

Bautista-Hernandez LA, Gomez-Olivares JL, Buentello-Volante B, Dominguez-Lopez A, Garfias Y, Acosta-Garcia MC, Calvillo-Medina RP, Bautista-de Lucio VM (2019) Negative interaction of Staphylococcus aureus on Fusarium falciforme growth ocular isolates in an in vitro mixed biofilm. Microb Pathog 135:103644. https://doi.org/10.1016/j.micpath.2019.103644

Beauvais A, Latgé J-P (2015) Aspergillus biofilm in vitro and in vivo. Microb Biofilms 3:149–161. https://doi.org/10.1128/microbiolspec.MB-0017-2015

Beauvais A, Schmidt C, Guadagnini S, Roux P, Perret E, Henry C, Paris S, Mallet A, Prévost MC, Latgé JP (2007) An extracellular matrix glues together the aerial-grown hyphae of Aspergillus fumigatus. Cell Microbiol 9:1588–1600. https://doi.org/10.1111/j.1462-5822.2007.00895.x

Bell NP, Karp CL, Alfonso EC, Schiffman J, Miller D (1999) Effects of methylprednisolone and cyclosporine A on fungal growth in vitro. Cornea 18:306–313. https://doi.org/10.1097/00003226-199905000-00012

Benson S, Kiang A, Lochenie C, Lal N, Mohanan SMP, Williams GO, Dhaliwal K, Mills B, Vendrell M (2023) Environmentally sensitive photosensitizers enable targeted photodynamic ablation of Gram-positive antibiotic resistant bacteria. Theranostics 13:3814. https://doi.org/10.7150/thno.84187

Bezerra FM, Rocchetti TT, Lima SL, Yu MCZ, da Matta DA, Höfling-Lima AL, Melo AS, de Oliveira LA (2023) Candida species causing fungal keratitis: molecular identification, antifungal susceptibility, biofilm formation, and clinical aspects. Braz J Microbiol 54:629–636. https://doi.org/10.1007/s42770-023-00964-w

Bhardwaj SK, Singh H, Deep A, Khatri M, Bhaumik J, Kim K-H, Bhardwaj N (2021) UVC-based photoinactivation as an efficient tool to control the transmission of coronaviruses. Sci Total Environ 792:148548. https://doi.org/10.1016/j.scitotenv.2021.148548

Blango MG, Kniemeyer O, Brakhage AA (2019) Conidial surface proteins at the interface of fungal infections. PLoS Pathogens 15:e1007939. https://doi.org/10.1371/journal.ppat.1007939

Boost M, Lai S, Ma C, Cho P (2010) Do multipurpose contact lens disinfecting solutions work effectively against non-FDA/ISO recommended strains of bacteria and fungi? Ophthalmic Physiol Opt 30:12–19. https://doi.org/10.1111/j.1475-1313.2009.00689.x

Brown L, Leck AK, Gichangi M, Burton MJ, Denning DW (2021a) The global incidence and diagnosis of fungal keratitis. Lancet Infect Dis 21:e49–e57. https://doi.org/10.1016/S1473-3099(20)30448-5

Brown ME, Montgomery ML, Kamath MM, Nicholas S, Liu Y, Karamichos D, Fuller KK (2021b) A novel 3D culture model of fungal keratitis to explore host-pathogen interactions within the stromal environment. Exp Eye Res 207:108581. https://doi.org/10.1016/j.exer.2021.108581

Brown L, Kamwiziku G, Oladele RO, Burton MJ, Prajna NV, Leitman TM, Denning DW (2022) The case for fungal keratitis to be accepted as a neglected tropical disease. J Fungi 8:1047. https://doi.org/10.3390/jof8101047

Bruce AS, Brennan NA (1990) Corneal pathophysiology with contact lens wear. Surv Ophthalmol 35:25–58. https://doi.org/10.1016/0039-6257(90)90046-x

Calvillo-Medina RP, Reyes-Grajeda JP, Barba-Escoto L, Bautista-Hernandez LA, Campos-Guillen J, Jones GH, Bautista-de Lucio VM (2019) Proteome analysis of biofilm produced by a Fusarium falciforme keratitis infectious agent. Microb Pathog 130:232–241. https://doi.org/10.1016/j.micpath.2019.03.001

Carmello JC, Alves F, de Oliveira Mima EG, Jorge JH, Bagnato VS, Pavarina AC (2017) Photoinactivation of single and mixed biofilms of Candida albicans and non-albicans Candida species using Photodithazine®. Photodiagnosis Photodyn Ther 17:194–199. https://doi.org/10.1016/j.pdpdt.2017.08.014

Carrions de J, Leal SM Jr, Ghannoum MA, Aimanianda V, Latgé JP, Pearlman E (2013) The RodA hydrophobin on Aspergillus fumigatus spores masks dectin-1- and dectin-2-dependent responses and enhances fungal survival in vivo. J Immunol 191:2581–2588. https://doi.org/10.4049/jimmunol.1300748

Chabrier-Roselló Y, Foster TH, Pérez-Nazario N, Mitra S, Haidaris CG (2005) Sensitivity of Candida albicans germ tubes and biofilms to photofrin-mediated phototoxicity. Antimicrob Agents Chemother 49:4288–4295. https://doi.org/10.1128/AAC.49.10.4288-4295.2005

Chan TC, Lau TW, Lee JW, Wong IY, Jhanji V, Wong RL (2015) Corneal collagen cross-linking for infectious keratitis: an update of clinical studies. Acta Ophthalmol 93:689–696. https://doi.org/10.1111/aos.12754

Chandra J, Mccormick TS, Imamura Y, Mukherjee PK, Ghannoum MA (2007) Interaction of Candida albicans with adherent human peripheral blood mononuclear cells increases C. albicans biofilm formation and results in differential expression of pro- and anti-inflammatory cytokines. Infect Immun 75:2612–2620. https://doi.org/10.1128/iai.01841-06

Chandra J, Pearlman E, Ghannoum MA (2014) Animal models to investigate fungal biofilm formation. Microbial biofilms: methods and protocols, vol 1147. Springer, New York, pp 141–157. https://doi.org/10.1007/978-1-4939-0467-9_10

Chang DC, Grant GB, O'donnell K, Wannemuehler KA, NOBLE-Wang J, Rao CY, Jacobson LM, Crowell CS, Sneed RS, Lewis FMT, Schaffzin JK, Kainer MA, Genese CA, Alfonso EC, Jones DB, Srinivasan A, Fridkin SK, Park BJ, Fusarium Keratitis Investigation Team, F. T (2006) Multistate outbreak of Fusarium Keratitis associated with use of a contact lens solution. Jama 296:953–963. https://doi.org/10.1001/jama.296.8.953

Chatterjee S, Agrawal D (2022) Use of topical Cyclosporine 0.1% in therapeutic penetrating keratoplasty for Fungal Keratitis. Cornea 41:1116–1121. https://doi.org/10.1097/ico.0000000000002827

Chen J, Fraser T, Fisher D, Khalil Z, Collins M, Hafner L (2000) Characteristics of fungal growth in soft contact lenses. Int Contact Lens Clin 26:84–91. https://doi.org/10.1016/s0892-8967(00)00026-2

Chen X, Li X, Zhang X, Guo X, Qi X, Li S, Shi W, Gao H (2021) Comparison of complications and visual outcomes between big-bubble deep anterior lamellar keratoplasty and penetrating keratoplasty for fungal keratitis. Clin Exp Ophthalmol 49:550–559. https://doi.org/10.1111/ceo.13951

Chen X, Lu D, Liu W, Xie J, Lu Z, Yang H, Yan H, Wang L, Che C (2023) Therapeutic effect of Atractylenolide I on Aspergillus fumigatus keratitis by affecting MyD88/NF-kappaB pathway and IL-1beta, IL-10 expression. Cytokine 162:156112. https://doi.org/10.1016/j.cyto.2022.156112

Cheng M, Lin J, Li C, Zhao W, Yang H, Lv L, Che C (2019) Wedelolactone suppresses IL-1β maturation and neutrophil infiltration in Aspergillus fumigatus keratitis. Int Immunopharmacol 73:17–22. https://doi.org/10.1016/j.intimp.2019.04.050

Cheng KKW, Fingerhut L, Duncan S, Prajna NV, Rossi AG, Mills B (2024) In vitro and ex vivo models of microbial keratitis: present and future. Prog Retin Eye Res 102:101287. https://doi.org/10.1016/j.preteyeres.2024.101287

Cherfan D, Verter EE, Melki S, Gisel TE, Doyle FJ, Scarcelli G, Yun SH, Redmond RW, Kochevar IE (2013) Collagen cross-linking using rose bengal and green light to increase corneal stiffness. Investig Ophthalmol Vis Sci 54:3426–3433. https://doi.org/10.1167/iovs.12-11509

Chhonker YS, Prasad YD, Chandasana H, Vishvkarma A, Mitra K, Shukla PK, Bhatta RS (2015) Amphotericin-B entrapped lecithin/chitosan nanoparticles for prolonged ocular application. Int J Biol Macromol 72:1451–1458. https://doi.org/10.1016/j.ijbiomac.2014.10.014

Chi M, Gu L, Zhang L, Lin J, Xu Q, Jiang N, Wang Y, Qi Y, Diao W, Yi W, Zhao G, Li C (2023) Pentoxifylline treats Aspergillus fumigatus keratitis by reducing fungal burden and suppressing corneal inflammation. Eur J Pharmacol 945:175607. https://doi.org/10.1016/j.ejphar.2023.175607

Chibebe Junior J, Sabino CP, Tan X, Junqueira JC, Wang Y, Fuchs BB, Jorge AO, Tegos GP, Hamblin MR, Mylonakis E (2013) Selective photoinactivation of Candida albicans in the non-vertebrate host infection model Galleria mellonella. BMC Microbiol 13:1–9. https://doi.org/10.1186/1471-2180-13-217

Chou B, Krishna K, Durkee H, Tribin FE, Ahmed A, Lai J, Aguilar MC, Ferreira BC, Leblanc RM, Flynn HW Jr (2025) Photodynamic antimicrobial therapy with Erythrosin B, Eosin Y, and Rose Bengal for the inhibition of fungal keratitis isolates: an in vitro study. J Photochem Photobiol B Biol 263:113090. https://doi.org/10.1016/j.jphotobiol.2024.113090

Collier SA, Gronostaj MP, Macgurn AK, Cope JR, Awsumb KL, Yoder JS, Beach MJ, Centers for Disease, C. & Prevention (2014) Estimated burden of keratitis–United States, 2010 [Online]. Available: https://www.cdc.gov/mmwr/preview/mmwrhtml/mm6345a3.htm. Accessed June 2024

Córdova-Alcántara IM, Venegas-Cortés DL, Martínez-Rivera MÁ, Pérez NO, Rodriguez-Tovar AV (2019) Biofilm characterization of Fusarium solani keratitis isolate: increased resistance to antifungals and UV light. J Microbiol 57:485–497. https://doi.org/10.1007/s12275-019-8637-2

Costa ACBP, de Campos Rasteiro VM, Pereira CA, da Silva Hashimoto ESH, Junior MB, Junqueira JC, Jorge AOC (2011) Susceptibility of Candida albicans and Candida dubliniensis to erythrosine-and LED-mediated photodynamic therapy. Arch Oral Biol 56:1299–1305. https://doi.org/10.1016/j.archoralbio.2011.05.013

Costa ACBP, Rasteiro VMC, Pereira CA, Rossoni RD, Junqueira JC, Jorge AOC (2012) The effects of rose bengal-and erythrosine-mediated photodynamic therapy on Candida albicans. Mycoses 55:56–63. https://doi.org/10.1111/j.1439-0507.2011.02042.x

Crocker LB, Lee JH, Mital S, Mills GC, Schack S, Bistrović-Popov A, Franck CO, Mela I, Kaminski CF, Christie G (2022) Tuning riboflavin derivatives for photodynamic inactivation of pathogens. Sci Rep 12:6580. https://doi.org/10.1038/s41598-022-10394-7

Dalmon C, Porco TC, Lietman TM, Prajna NV, Prajna L, Das MR, Kumar JA, Mascarenhas J, Margolis TP, Whitcher JP (2012) The clinical differentiation of bacterial and fungal keratitis: a photographic survey. Investig Ophthalmol Vis Sci 53:1787–1791. https://doi.org/10.1167/iovs.11-8478

Danchik C, Casadevall A (2021) Role of cell surface hydrophobicity in the pathogenesis of medically-significant fungi. Front Cell Infect Microbiol 10:594973. https://doi.org/10.3389/fcimb.2020.594973

Dart JK, Stapleton F, Minassian D (1991) Contact lenses and other risk factors in microbial keratitis. Lancet 338:650–653. https://doi.org/10.1016/0140-6736(91)91231-i

Dart JK, Radford CF, Minassian D, Verma S, Stapleton F (2008) Risk factors for microbial keratitis with contemporary contact lenses: a case-control study. Ophthalmology 115:1647–1654., 1654 e1-3. https://doi.org/10.1016/j.ophtha.2008.05.003

de Menezes HD, Tonani L, Bachmann L, Wainwright M, Braga GÚL, von Zeska Kress MR (2016) Photodynamic treatment with phenothiazinium photosensitizers kills both ungerminated and germinated microconidia of the pathogenic fungi Fusarium oxysporum, Fusarium moniliforme and Fusarium solani. J Photochem Photobiol B Biol 164:1–12. https://doi.org/10.1016/j.jphotobiol.2016.09.008

de Oliveira AB, Ferrisse TM, Fontana CR, Basso FG, Brighenti FL (2022) Photodynamic therapy for treating infected skin wounds: a systematic review and meta-analysis from randomized clinical trials. Photodiagnosis Photodyn Ther 40:103118. https://doi.org/10.1016/j.pdpdt.2022.103118

de Paiva ACM, Ferreira MDC, da Fonseca AS (2022) Photodynamic therapy for treatment of bacterial keratitis. Photodiagnosis Photodyn Ther 37:102717. https://doi.org/10.1016/j.pdpdt.2022.102717

de Sá FA, Taveira SF, Gelfuso GM, Lima EM, Gratieri T (2015) Liposomal voriconazole (VOR) formulation for improved ocular delivery. Colloids Surf B Biointerfaces 133:331–338. https://doi.org/10.1016/j.colsurfb.2015.06.036

Dean SJ, Petty A, Swift S, Mcghee JJ, Sharma A, Shah S, Craig JP (2011) Efficacy and safety assessment of a novel ultraviolet C device for treating corneal bacterial infections. Clin Exp Ophthalmol 39:156–163. https://doi.org/10.1111/j.1442-9071.2010.02471.x

Devonshire P, Munro FA, Abernethy C, Clark BJ (1993) Microbial contamination of contact lens cases in the west of Scotland. Br J Ophthalmol 77:41–45. https://doi.org/10.1136/bjo.77.1.41

Dias LM, Klein MI, Ferrisse TM, Medeiros KS, Jordao CC, Bellini A, Pavarina AC (2023) The effect of sub-lethal successive applications of photodynamic therapy on Candida albicans biofilm depends on the photosensitizer. J Fungi (Basel) 9:111. https://doi.org/10.3390/jof9010111

Dong LK, Krebs DB (2022) An intracameral approach for recalcitrant fungal keratitis. Am J Ophthalmol Case Rep 25:101369. https://doi.org/10.1016/j.ajoc.2022.101369

Donovan C, Arenas E, Ayyala RS, Margo CE, Espana EM (2022) Fungal keratitis: mechanisms of infection and management strategies. Surv Ophthalmol 67:758–769

Dosler S, Hacioglu M, Yilmaz FN, Oyardi O (2020) Biofilm modelling on the contact lenses and comparison of the in vitro activities of multipurpose lens solutions and antibiotics. PeerJ 8:e9419. https://doi.org/10.7717/peerj.9419

Duan Z, Tong J, Zheng N, Zeng R, Liu Y, Li M (2023) Effect of 5-Aminolevulinic acid photodynamic therapy on Aspergillus fumigatus Biofilms in Vitro. Curr Microbiol 80:334. https://doi.org/10.1007/s00284-023-03351-8

Dumpati S, Naroo SA, Shah S, Dutta D (2022) Antimicrobial efficacy of an ultraviolet-C device against microorganisms related to contact lens adverse events. Antibiotics (Basel) 11. https://doi.org/10.3390/antibiotics11050699

Dutta D, Willcox MD (2014) Antimicrobial contact lenses and lens cases: a review. Eye Contact Lens 40:312–324. https://doi.org/10.1097/ICL.0000000000000056

Dutta D, Cole N, Willcox M (2012) Factors influencing bacterial adhesion to contact lenses. Mol Vis 18:14–21

Dyavaiah M, Ramani R, Chu DS, Ritterband DC, Shah MK, Samsonoff WA, Chaturvedi S, Chaturvedi V (2007) Molecular characterization, biofilm analysis and experimental biofouling study of Fusarium isolates from recent cases of fungal keratitis in New York State. BMC Ophthalmol 7:1. https://doi.org/10.1186/1471-2415-7-1

El-Mofty HM, Abdelhakim MA, EL-Miligi MF, EL-Nabarawi MA, Khalil IA (2014) A new combination formula for treatment of fungal keratitis: an experimental study. J Ophthalmol 2014:173298. https://doi.org/10.1155/2014/173298

Elmotasem H, Awad GEA (2020) A stepwise optimization strategy to formulate in situ gelling formulations comprising fluconazole-hydroxypropyl-beta-cyclodextrin complex loaded niosomal vesicles and Eudragit nanoparticles for enhanced antifungal activity and prolonged ocular delivery. Asian J Pharm Sci 15:617–636. https://doi.org/10.1016/j.ajps.2019.09.003

Fadlallah A, Zhu H, Arafat S, Kochevar I, Melki S, Ciolino JB (2016) Corneal resistance to keratolysis after collagen crosslinking with Rose Bengal and green light. Invest Ophthalmol Vis Sci 57:6610–6614. https://doi.org/10.1167/iovs.15-18764

Fan Y, Li C, Peng X, Jiang N, Hu L, Gu L, Zhu G, Zhao G, Lin J (2020) Perillaldehyde ameliorates Aspergillus fumigatus keratitis by activating the Nrf2/HO-1 signaling pathway and inhibiting dectin-1-mediated inflammation. Invest Ophthalmol Vis Sci 61:51. https://doi.org/10.1167/iovs.61.6.51

Figueiredo-Godoi LMA, Menezes RT, Carvalho JS, Garcia MT, Segundo AG, Jorge AOC, Junqueira JC (2019) Exploring the Galleria mellonella model to study antifungal photodynamic therapy. Photodiagnosis Photodyn Ther 27:66–73. https://doi.org/10.1016/j.pdpdt.2019.05.010

Fiori B, Posteraro B, Torelli R, Tumbarello M, Perlin DS, Fadda G, Sanguinetti M (2011) In vitro activities of anidulafungin and other antifungal agents against biofilms formed by clinical isolates of different Candida and Aspergillus species. Antimicrob Agents Chemother 55:3031–3035. https://doi.org/10.1128/AAC.01569-10

Freire F, de Barros PP, Pereira CA, Junqueira JC, Jorge AOC (2018) Photodynamic inactivation in the expression of the Candida albicans genes ALS3, HWP1, BCR1, TEC1, CPH1, and EFG1 in biofilms. Lasers Med Sci 33:1447–1454. https://doi.org/10.1007/s10103-018-2487-8

Fu T, Yi J, Lv S, Zhang B (2017) Ocular amphotericin B delivery by chitosan-modified nanostructured lipid carriers for fungal keratitis-targeted therapy. J Liposome Res 27:228–233. https://doi.org/10.1080/08982104.2016.1224899

Gabriel MM, Mcanally C, Bartell J (2018) Antimicrobial efficacy of multipurpose disinfecting solutions in the presence of contact lenses and lens cases. Eye Contact Lens 44:125–131. https://doi.org/10.1097/ICL.0000000000000308

Gao L, Jiang S, Sun Y, Deng M, Wu Q, Li M, Zeng T (2016) Evaluation of the effects of photodynamic therapy alone and combined with standard antifungal therapy on planktonic cells and biofilms of Fusarium spp. and Exophiala spp. Front Microbiol 7:617. https://doi.org/10.3389/fmicb.2016.00617

Garcia-Perez JE, Mathé L, Humblet-Baron S, Braem A, Lagrou K, van Dijck P, Liston A (2018) A framework for understanding the evasion of host immunity by Candida biofilms. Front Immunol 9:538. https://doi.org/10.3389/fimmu.2018.00538

Gholami L, Shahabi S, Jazaeri M, Hadilou M, Fekrazad R (2022) Clinical applications of antimicrobial photodynamic therapy in dentistry. Front Microbiol 13:1020995. https://doi.org/10.3389/fmicb.2022.1020995

Ghosh S, Humera Khathun AH, Athulya GS, Vignesh P, Mathan L, Mudaraddi N, Narendran S, Lalitha P, Venkatesh Prajna N (2023) Host cell-type and pathogen-specific immunomodulatory functions of macrophage migration inhibitory factor (MIF) in infectious keratitis. Exp Eye Res 236:109669. https://doi.org/10.1016/j.exer.2023.109669

Gonzalez-Ramirez AI, Ramirez-Granillo A, Medina-Canales MG, Rodriguez-Tovar AV, Martinez-Rivera MA (2016) Analysis and description of the stages of Aspergillus fumigatus biofilm formation using scanning electron microscopy. BMC Microbiol 16:243. https://doi.org/10.1186/s12866-016-0859-4

Gopinathan U, Sharma S, Garg P, Rao GN (2009) Review of epidemiological features, microbiological diagnosis and treatment outcome of microbial keratitis: experience of over a decade. Indian J Ophthalmol 57:273–279. https://doi.org/10.4103/0301-4738.53051

Gower EW, Keay LJ, Oechsler RA, Iovieno A, Alfonso EC, Jones DB, Colby K, Tuli SS, Patel SR, Lee SM, Irvine J, Stulting RD, Mauger TF, Schein OD (2010) Trends in fungal keratitis in the United States, 2001 to 2007. Ophthalmology 117:2263–2267. https://doi.org/10.1016/j.ophtha.2010.03.048

Gray TB, Cursons RT, Sherwan JF, Rose PR (1995) Acanthamoeba, bacterial, and fungal contamination of contact lens storage cases. Br J Ophthalmol 79:601–605. https://doi.org/10.1136/bjo.79.6.601

Greco G, di Piazza S, Chan J, Zotti M, Hanna R, Gheno E, Zekiy AO, Pasquale C, de Angelis N, Amaroli A (2020) Newly formulated 5% 5-aminolevulinic acid photodynamic therapy on Candida albicans. Photodiagnosis Photodyn Ther 29:101575. https://doi.org/10.1016/j.pdpdt.2019.10.010

Gu LW, Li C, Peng XD, Lin H, Niu YW, Zheng HR, Zhao GQ, Lin J (2022) Flavopiridol protects against fungal keratitis due to Aspergillus fumigatus by alleviating inflammation through the promotion of autophagy. ACS Infect Dis 8:2362–2373. https://doi.org/10.1021/acsinfecdis.2c00427

Gunasekaran R, Lalitha P, Megia-Fernandez A, Bradley M, Williams RL, Dhaliwal K, Prajna NV, Mills B (2020) Exploratory use of fluorescent SmartProbes for the rapid detection of microbial isolates causing corneal ulcer. Am J Ophthalmol. https://doi.org/10.1016/j.ajo.2020.06.014

Gunasekaran R, Chandrasekaran A, Rajarathinam K, Duncan S, Dhaliwal K, Lalitha P, Prajna NV, Mills B (2023) Rapid point-of-care identification of Aspergillus species in microbial Keratitis. JAMA Ophthalmol 141:966–973. https://doi.org/10.1001/jamaophthalmol.2023.4214

Guo H, Wu X (2009) Innate responses of corneal epithelial cells against Aspergillus fumigatus challenge. FEMS Immunol Med Microbiol 56:88–93. https://doi.org/10.1111/j.1574-695X.2009.00551.x

Guo Y, Karimi F, Fu Q, Qiao GG, Zhang H (2020) Reduced administration frequency for the treatment of fungal keratitis: a sustained natamycin release from a micellar solution. Expert Opin Drug Deliv 17:407–421. https://doi.org/10.1080/17425247.2020.1719995

Guridi A, Sevillano E, De La Fuente I, Mateo E, Eraso E, Quindos G (2019) Disinfectant activity of A portable ultraviolet C equipment. Int J Environ Res Public Health 16:4747. https://doi.org/10.3390/ijerph16234747

Gurzadyan GG, Görner H, Schulte-Frohlinde D (1995) Ultraviolet (193, 216 and 254 nm) Photoinactivation of Escherichia coli strains with different repair deficiencies. Radiat Res 141:244–251. https://doi.org/10.2307/3579001

Halili F, Arboleda A, Durkee H, Taneja M, Miller D, Alawa KA, Aguilar MC, Amescua G, Flynn HW, Parel JM (2016) Rose Bengal- and Riboflavin-Mediated Photodynamic Therapy to Inhibit Methicillin-Resistant Staphylococcus aureus Keratitis Isolates. Am J Ophthalmol 166:194–202. https://doi.org/10.1016/j.ajo.2016.03.014

Han F, Wang L, Shen L, Liu W, Li Y, Ma H, Wu X (2023) A20 ameliorates Aspergillus fumigatus keratitis by promoting autophagy and inhibiting NF-κB signaling. Int J Biol Macromol 253:127640. https://doi.org/10.1016/j.ijbiomac.2023.127640

Hassan I, Dorjay K, Anwar P (2014) Pentoxifylline and its applications in dermatology. Indian Dermatol Online J 5:510–516. https://doi.org/10.4103/2229-5178.142528

He M, You J, Liu X, Peng X, Li C, Yang S, Xu Q, Lin J, Zhao G (2022) Perillaldehyde protects against Aspergillus fumigatus keratitis by reducing fungal load and inhibiting inflammatory cytokines and LOX-1. Curr Eye Res 47:1366–1373. https://doi.org/10.1080/02713683.2022.2093382

Hicks PM, Singh K, Prajna NV, Lu M-C, Niziol LM, Greenwald MF, Verkade A, Amescua G, Farsiu S, Woodward MA (2022) Quantifying clinicians' diagnostic uncertainty when making initial treatment decisions for microbial keratitis. Cornea 10:1097. https://doi.org/10.1097/ICO.0000000000003159

Hiraoka T, Kaji Y, Wakabayashi T, Nanbu PN, Okamoto F, Oshika T (2007) Comparison of micafungin and fluconazole for experimental Candida keratitis in rabbits. Cornea 26:336–342. https://doi.org/10.1097/ICO.0b013e31802cd8a8

Hockberger PE (2002) A history of ultraviolet photobiology for humans, animals and microorganisms. Photochem Photobiol 76:561–579. https://doi.org/10.1562/0031-8655(2002)0760561AHOUPF2.0.CO2

Hoffman JJ, Burton MJ, Leck A (2021) Mycotic keratitis-A global threat from the filamentous fungi. J Fungi (Basel) 7:273. https://doi.org/10.3390/jof7040273

Hoffman JJ, Arunga S, Mohamed Ahmed AHA, Hu VH, Burton MJ (2022a) Management of filamentous fungal keratitis: a pragmatic approach. J Fungi (Basel) 8:1067. https://doi.org/10.3390/jof8101067

Hoffman JJ, Yadav R, Sanyam SD, Chaudhary P, Roshan A, Singh SK, Arunga S, Hu VH, Macleod D, Leck A, Burton MJ (2022b) Diagnosis of fungal keratitis in low-income countries: evaluation of smear microscopy, culture, and in vivo confocal microscopy in Nepal. J Fungi (Basel) 8:955. https://doi.org/10.3390/jof8090955

Hohl TM, Feldmesser M (2007) Aspergillus fumigatus: principles of pathogenesis and host defense. Eukaryot Cell 6:1953–1963. https://doi.org/10.1128/EC.00274-07

Hosny KM, Rizg WY, Alkhalidi HM, Abualsunun WA, Bakhaidar RB, Almehmady AM, Alghaith AF, Alshehri S, El Sisi AM (2021) Nanocubosomal based in situ gel loaded with natamycin for ocular fungal diseases: development, optimization, in-vitro, and in-vivo assessment. Drug Deliv 28:1836–1848. https://doi.org/10.1080/10717544.2021.1965675

Hossain P (2021) Microbial keratitis-the true costs of a silent pandemic? Eye (Lond) 35:2071–2072. https://doi.org/10.1038/s41433-020-01360-6

Hu J, Wu F, Huang Z, Ma S, Zhang J, Yang J, Han X, Xu G (2016) Raman spectroscopy analysis of the biochemical characteristics of experimental keratomycosis. Curr Eye Res 41:1408–1413. https://doi.org/10.3109/02713683.2015.1127393

Hu J, Chen F, Kan T, Zhuang H, Zhang J, Han X (2017) Inhibition of Fusarium solani infection in murine keratocytes by Lactobacillus salivarius ssp. salivarius JCM1231 culture filtrate in vitro. Curr Eye Res 42:1339–1347. https://doi.org/10.1080/02713683.2017.1317816

Hua X, Chi W, Su L, Li J, Zhang Z, Yuan X (2017) ROS-induced oxidative injury involved in pathogenesis of fungal keratitis via p38 MAPK activation. Sci Rep 7:10421. https://doi.org/10.1038/s41598-017-09636-w

Huang MC, Shen M, Huang YJ, Lin HC, Chen CT (2018) Photodynamic inactivation potentiates the susceptibility of antifungal agents against the planktonic and biofilm cells of Candida albicans. Int J Mol Sci 19:434. https://doi.org/10.3390/ijms19020434

Hung JH, Lee CN, Hsu HW, Ng IS, Wu CJ, Yu CK, Lee NY, Chang Y, Wong TW (2021) Recent advances in photodynamic therapy against fungal keratitis. Pharmaceutics 13:2011. https://doi.org/10.3390/pharmaceutics13122011

Ibrahim MM, Vanini R, Ibrahim FM, Fioriti LS, Furlan EM, Provinzano LM, de Castro RS, Sousa SJ, Rocha EM (2009a) Epidemiologic aspects and clinical outcome of fungal keratitis in southeastern Brazil. Eur J Ophthalmol 19:355–361. https://doi.org/10.1177/112067210901900305

Ibrahim YW, Boase DL, Cree IA (2009b) Epidemiological characteristics, predisposing factors and microbiological profiles of infectious corneal ulcers: the Portsmouth corneal ulcer study. Br J Ophthalmol 93:1319–1324. https://doi.org/10.1136/bjo.2008.151167

Imamura Y, Chandra J, Mukherjee PK, Lattif AA, Szczotka-Flynn LB, Pearlman E, Lass JH, O'donnell K, Ghannoum MA (2008) Fusarium and Candida albicans biofilms on soft contact lenses: model development, influence of lens type, and susceptibility to lens care solutions. Antimicrob Agents Chemother 52:171–182. https://doi.org/10.1128/AAC.00387-07

Ishihara K, Shi X, Fukazawa K, Yamaoka T, Yao G, Wu JY (2023) Biomimetic-engineered silicone hydrogel contact lens materials. ACS Appl Bio Mater 6:3600–3616. https://doi.org/10.1021/acsabm.3c00296

ISO14729:2001 (2010) Ophthalmic optics – contact lens care products – microbiological requirements and test methods for products and regimens for hygienic management of contact lenses [Online]. International Organization for Standardization, Geneva. Available: https://www.iso.org/standard/25382.html. Accessed June 2024

ISO18259:2014 (2014) Ophthalmic optics – contact lens care products – method to assess contact lens care products with contact lenses in a lens case, challenged with bacterial and fungal organisms [Online]. International Organization for Standardization, Geneva. Available: https://www.iso.org/standard/61901.html. Accessed June 2024

Jeng BH, Gritz DC, Kumar AB, Holsclaw DS, Porco TC, Smith SD, Whitcher JP, Margolis TP, Wong IG (2010) Epidemiology of ulcerative keratitis in Northern California. Arch Ophthalmol 128:1022–1028. https://doi.org/10.1001/archophthalmol.2010.144

Johnson CJ, Cabezas-Olcoz J, Kernien JF, Wang SX, Beebe DJ, Huttenlocher A, Ansari H, Nett JE (2016) The extracellular matrix of Candida albicans biofilms impairs formation of neutrophil extracellular traps. PLoS Pathog 12:e1005884. https://doi.org/10.1371/journal.ppat.1005884

Jordao CC, Viana De Sousa T, Inez Klein M, Mendonca Dias L, Pavarina AC, Carmello JC (2020) Antimicrobial photodynamic therapy reduces gene expression of Candida albicans in biofilms. Photodiagnosis Photodyn Ther 31:101825. https://doi.org/10.1016/j.pdpdt.2020.101825

Kackar S, Suman E, Kotian MS (2017) Bacterial and fungal biofilm formation on contact lenses and their susceptibility to lens care solutions. Indian J Med Microbiol 35:80–84. https://doi.org/10.4103/ijmm.IJMM_16_273

Kanafani ZA, Perfect JR (2008) Resistance to antifungal agents: mechanisms and clinical impact. Clin Infect Dis 46:120–128. https://doi.org/10.1086/524071

Kasım B, Koçluk Y (2020) Long-term outcomes of therapeutic penetrating keratoplasty for microbial keratitis in a tertiary care center in Turkey. Int Ophthalmol 40:3513–3519. https://doi.org/10.1007/s10792-020-01538-7

Katragkou A, Kruhlak MJ, Simitsopoulou M, Chatzimoschou A, Taparkou A, Cotten CJ, Paliogianni F, Diza-Mataftsi E, Tsantali C, Walsh TJ, Roilides E (2010) Interactions between human phagocytes and Candida albicans biofilms alone and in combination with antifungal agents. J Infect Dis 201:1941–1949. https://doi.org/10.1086/652783

Kaur S, Singh S (2014) Biofilm formation by Aspergillus fumigatus. Med Mycol 52:2–9. https://doi.org/10.3109/13693786.2013.819592

Keay LJ, Gower EW, Iovieno A, Oechsler RA, Alfonso EC, Matoba A, Colby K, Tuli SS, Hammersmith K, Cavanagh D, Lee SM, Irvine J, Stulting RD, Mauger TF, Schein OD (2011) Clinical and microbiological characteristics of fungal keratitis in the United States, 2001–2007: a multicenter study. Ophthalmology 118:920–926. https://doi.org/10.1016/j.ophtha.2010.09.011

Kernien JF, Snarr BD, Sheppard DC, Nett JE (2017) The interface between fungal biofilms and innate immunity. Front Immunol 8:1968. https://doi.org/10.3389/fimmu.2017.01968

Khan SA, Shahid S, Mahmood T, Lee CS (2021) Contact lenses coated with hybrid multifunctional ternary nanocoatings (Phytomolecule-coated ZnO nanoparticles: Gallic Acid: Tobramycin) for the treatment of bacterial and fungal keratitis. Acta Biomater 128:262–276. https://doi.org/10.1016/j.actbio.2021.04.014

Khan M, Mcdonald M, Mundada K, Willcox M (2022) Efficacy of ultraviolet radiations against coronavirus, bacteria, fungi, fungal spores and biofilm. Hygiene 2:120–131. https://doi.org/10.3390/hygiene2030010

Khor WB, Aung T, Saw SM, Wong TY, Tambyah PA, Tan AL, Beuerman R, Lim L, Chan WK, Heng WJ, Lim J, Loh RS, Lee SB, Tan DT (2006) An outbreak of Fusarium keratitis associated with contact lens wear in Singapore. Jama 295:2867–2873. https://doi.org/10.1001/jama.295.24.2867

Khor W-B, Prajna NV, Garg P, Mehta JS, Xie L, Liu Z, Padilla MDB, Joo C-K, Inoue Y, Goseyarakwong P, Hu F-R, Nishida K, Kinoshita S, Puangsricharern V, Tan A-L, Beuerman R, Young A, Sharma N, Haaland B, Mah FS, Tu EY, Stapleton FJ, Abbott RL, Tan DT-H (2018) The Asia Cornea Society infectious keratitis study: a prospective multicenter study of infectious keratitis in Asia. Am J Ophthalmol 195:161–170. https://doi.org/10.1016/j.ajo.2018.07.040

Kim H, Lee YR, Jeong H, Lee J, Wu X, Li H, Yoon J (2023) Photodynamic and photothermal therapies for bacterial infection treatment. Smart Mol 1:e20220010. https://doi.org/10.1002/smo.20220010

Knutsson KA, Iovieno A, Matuska S, Fontana L, Rama P (2021) Topical corticosteroids and fungal keratitis: a review of the literature and case series. J Clin Med 10. https://doi.org/10.3390/jcm10061178

Kodali S, Khan B, Zong AM, Moon J-Y, Shrivastava A, Daily JP, Gibralter RP (2024) Prognostic indicators of corneal ulcer clinical outcomes at a tertiary care center in the Bronx, New York. J Ophthalmic Inflamm Infect 14:18. https://doi.org/10.1186/s12348-024-00392-3

Kuzman T, Kutija MB, Juri J, Jandrokovic S, Skegro I, Olujic SM, Kordic R, Cerovski B (2014) Lens wearers non-compliance – is there an association with lens case contamination? Cont Lens Anterior Eye 37:99–105. https://doi.org/10.1016/j.clae.2013.08.004

Kwiatkowski S, Knap B, Przystupski D, Saczko J, Kedzierska E, Knap-Czop K, Kotlinska J, Michel O, Kotowski K, Kulbacka J (2018) Photodynamic therapy – mechanisms, photosensitizers and combinations. Biomed Pharmacother 106:1098–1107. https://doi.org/10.1016/j.biopha.2018.07.049

Lalitha CS, Nayak B, Hallur V, Kharolia A (2020) Invasion of the therapeutic contact lens by a rare fungal species presenting as the acute red eye: a case report. Indian J Ophthalmol 68:2548–2550. https://doi.org/10.4103/ijo.IJO_1634_20

Leck A (2009) Taking a corneal scrape and making a diagnosis. Community Eye Health 22:42–43

Lee MJ, Liu H, Barker BM, Snarr BD, Gravelat FN, Al Abdallah Q, Gavino C, Baistrocchi SR, Ostapska H, Xiao T, Ralph B, Solis NV, Lehoux M, Baptista SD, Thammahong A, Cerone RP, Kaminskyj SG, Guiot MC, Latgé JP, Fontaine T, Vinh DC, Filler SG, Sheppard DC (2015) The fungal exopolysaccharide galactosaminogalactan mediates virulence by enhancing resistance to neutrophil extracellular traps. PLoS Pathog 11:e1005187. https://doi.org/10.1371/journal.ppat.1005187

Lee JK, Park S, Kim YM, Guk T, Lee MY, Park SC, Lee JR, Jang MK (2022) Candidacidal and antibiofilm activity of PS1-3 peptide against drug-resistant Candida albicans on contact lenses. Pharmaceutics 14. https://doi.org/10.3390/pharmaceutics14081602

Li Y, Yang H, Wu X (2013a) Pretreatment with TLR2 and TLR4 ligand modulates innate immunity in corneal fibroblasts challenged with Aspergillus fumigatus. Invest Ophthalmol Vis Sci 54:4261–4270. https://doi.org/10.1167/iovs.12-11504

Li Z, Jhanji V, Tao X, Yu H, Chen W, Mu G (2013b) Riboflavin/ultravoilet light-mediated crosslinking for fungal keratitis. Br J Ophthalmol 97:669–671. https://doi.org/10.1136/bjophthalmol-2012-302518

Li C, Zhao G, Che C, Lin J, Li N, Hu L, Jiang N, Liu Y (2015) The role of LOX-1 in innate immunity to Aspergillus fumigatus in corneal epithelial cells. Invest Ophthalmol Vis Sci 56:3593–3603. https://doi.org/10.1167/iovs.14-15989

Li C, Li C, Lin J, Zhao G, Xu Q, Jiang N, Wang Q, Peng X, Zhu G, Jiang J (2020) The role of autophagy in the innate immune response to fungal keratitis caused by Aspergillus fumigatus infection. Invest Ophthalmol Vis Sci 61:25. https://doi.org/10.1167/iovs.61.2.25

Lian H, Fang X, Li Q, Liu S, Wei Q, Hua X, Li W, Liao C, Yuan X (2022) NLRP3 inflammasome-mediated pyroptosis pathway contributes to the pathogenesis of Candida albicans Keratitis. Front Med (Lausanne) 9:845129. https://doi.org/10.3389/fmed.2022.845129

Liang Z, Zhang Z, Yang J, Lu P, Zhou T, Li J, Zhang J (2021) Assessment to the antifungal effects in vitro and the ocular pharmacokinetics of solid-lipid nanoparticle in rabbits. Int J Nanomed 16:7847–7857. https://doi.org/10.2147/ijn.S340068

Liang Z, Zhang Z, Lu P, Yang J, Han L, Liu S, Zhou T, Li J, Zhang J (2023) The effect of charges on the corneal penetration of solid lipid nanoparticles loaded Econazole after topical administration in rabbits. Eur J Pharm Sci 187:106494. https://doi.org/10.1016/j.ejps.2023.106494

Liesegang TJ (2002) Physiologic changes of the cornea with contact lens wear. CLAO J 28:12–27

Liu Z, Pflugfelder SC (2000) The effects of long-term contact lens wear on corneal thickness, curvature, and surface regularity. Ophthalmology 107:105–111. https://doi.org/10.1016/s0161-6420(99)00027-5

Liu Y, Qin R, Zaat SAJ, Breukink E, Heger M (2015) Antibacterial photodynamic therapy: overview of a promising approach to fight antibiotic-resistant bacterial infections. J Clin Transl Res 1:140–167. https://doi.org/10.18053/jctres.201503.002

Liu G, Li K, Wang H, Ma L, Yu L, Nie Y (2020) Stable fabrication of zwitterionic coating based on copper-phenolic networks on contact lens with improved surface wettability and broad-spectrum antimicrobial activity. ACS Appl Mater Interfaces 12:16125–16136. https://doi.org/10.1021/acsami.0c02143

Liu S, le Mauff F, Sheppard DC, Zhang S (2022) Filamentous fungal biofilms: conserved and unique aspects of extracellular matrix composition, mechanisms of drug resistance and regulatory networks in Aspergillus fumigatus. NPJ Biofilms Microb 8:83. https://doi.org/10.1038/s41522-022-00347-3

Lochenie C, Duncan S, Zhou Y, Fingerhut L, Kiang A, Benson S, Jiang G, Liu X, Mills B, Vendrell M (2024) Photosensitizer-amplified antimicrobial materials for broad-spectrum ablation of resistant pathogens in ocular infections. Adv Mater 36:e2404107. https://doi.org/10.1002/adma.202404107

Luan J, Peng X, Lin J, Zhang Y, Tian X, Zhan L, Zhao G (2022) The therapeutic potential of chondroitin sulfate in Aspergillus fumigatus keratitis. Mol Immunol 147:50–61. https://doi.org/10.1016/j.molimm.2022.04.007

Luensmann D, Jones L (2012) Protein deposition on contact lenses: the past, the present, and the future. Cont Lens Anterior Eye 35:53–64. https://doi.org/10.1016/j.clae.2011.12.005

Makdoumi K, Mortensen J, Sorkhabi O, Malmvall BE, Crafoord S (2012) UVA-riboflavin photochemical therapy of bacterial keratitis: a pilot study. Graefes Arch Clin Exp Ophthalmol 250:95–102. https://doi.org/10.1007/s00417-011-1754-1

Malinovska Z, Conkova E, Vaczi P (2023) Biofilm formation in medically important candida species. J Fungi (Basel) 9:955. https://doi.org/10.3390/jof9100955

Mallela LS, Sharma P, Rao TSR, Roy S (2021) Recombinant IL-22 promotes protection in a murine model of Aspergillus flavus keratitis and mediates host immune responses in human corneal epithelial cells. Cell Microbiol 23:e13367. https://doi.org/10.1111/cmi.13367

Mallet JD, Rochette PJ (2013) Wavelength-dependent ultraviolet induction of cyclobutane pyrimidine dimers in the human cornea. Photochem Photobiol Sci 12:1310–1318. https://doi.org/10.1039/C3PP25408A

Mannan M, Nabeela S, Mishra R, Uppuluri P (2024) Host immune response against fungal biofilms. Curr Opin Microbiol 81:102520. https://doi.org/10.1016/j.mib.2024.102520

Marasini S, Mugisho OO, Swift S, Read H, Rupenthal ID, Dean SJ, Craig JP (2021) Effect of therapeutic UVC on corneal DNA: safety assessment for potential keratitis treatment. Ocular Surf 20:130–138. https://doi.org/10.1016/j.jtos.2021.02.005

Marasini S, Dean SJ, Swift S, Perera J, Rupenthal ID, Wang T, Read H, Craig JP (2022) Preclinical confirmation of UVC efficacy in treating infectious keratitis. Ocular Surf 25:76–86. https://doi.org/10.1016/j.jtos.2022.05.004

Marchetti O, Moreillon P, Glauser MP, Bille J, Sanglard D (2000) Potent synergism of the combination of fluconazole and cyclosporine in Candida albicans. Antimicrob Agents Chemother 44:2373–2381. https://doi.org/10.1128/aac.44.9.2373-2381.2000

Martins SA, Combs JC, Noguera G, Camacho W, Wittmann P, Walther R, Cano M, Dick J, Behrens A (2008) Antimicrobial efficacy of riboflavin/UVA combination (365 nm) in vitro for bacterial and fungal isolates: a potential new treatment for infectious keratitis. Invest Ophthalmol Vis Sci 49:3402–3408. https://doi.org/10.1167/iovs.07-1592

Martins M, Uppuluri P, Thomas DP, Cleary IA, Henriques M, Lopez-Ribot JL, Oliveira R (2010) Presence of extracellular DNA in the Candida albicans biofilm matrix and its contribution to biofilms. Mycopathologia 169:323–331. https://doi.org/10.1007/s11046-009-9264-y

Mcanally C, Walters R, Campolo A, Harris V, King J, Thomas M, Gabriel MM, Shannon P, Crary M (2021) Antimicrobial efficacy of contact lens solutions assessed by ISO standards. Microorganisms 9. https://doi.org/10.3390/microorganisms9102173

Mccarty MF, O'keefe JH, Dinicolantonio JJ (2016) Pentoxifylline for vascular health: a brief review of the literature. Open Heart 3:e000365. https://doi.org/10.1136/openhrt-2015-000365

Meek KM, Hayes S (2013) Corneal cross-linking–a review. Ophthalmic Physiol Opt 33:78–93. https://doi.org/10.1111/opo.12032

Meijome TE, Wozniak R, Kang L, Azzouz L, Niziol LM, Johnson WL, Kriegel M, Woodward MA (2022) Assessment of keratitis severity using quantitative image analysis in an in vivo murine model of Staphylococcus aureus bacterial keratitis. Transl Vis Sci Technol 11:12. https://doi.org/10.1167/tvst.11.11.12

Mela EK, Anastassiou ED, Gartaganis SP, Christofidou M (2015) Fungal isolation from disinfectant solutions of contact lens storage cases among asymptomatic users. Eye Contact Lens 41:87–90. https://doi.org/10.1097/ICL.0000000000000069

Mellon JE, Cotty PJ, Dowd MK (2007) Aspergillus flavus hydrolases: their roles in pathogenesis and substrate utilization. Appl Microbiol Biotechnol 77:497–504. https://doi.org/10.1007/s00253-007-1201-8

Mendonca JR, Dantas LR, Tuon FF (2023) Activity of multipurpose contact lens solutions against Staphylococcus aureus, Pseudomonas aeruginosa, Serratia marcescens and Candida albicans biofilms. Ophthalmic Physiol Opt 43:1092–1099. https://doi.org/10.1111/opo.13189

Mills B, Radhakrishnan N, Karthikeyan Rajapandian SG, Rameshkumar G, Lalitha P, Prajna NV (2021) The role of fungi in fungal keratitis. Exp Eye Res 202:108372. https://doi.org/10.1016/j.exer.2020.108372

Mimouni M, Tam G, Paitan Y, Kidron D, Segev F (2014) Safety and efficacy of intrastromal injection of 5% natamycin in experimental fusarium keratitis. J Ocul Pharmacol Ther 30:543–547. https://doi.org/10.1089/jop.2014.0004

Mohammed I, Said DG, Dua HS (2017) Human antimicrobial peptides in ocular surface defense. Prog Retinal Eye Res 61:1–22. https://doi.org/10.1016/j.preteyeres.2017.03.004

Mohammed I, Mohanty D, Said DG, Barik MR, Reddy MM, Alsaadi A, Das S, Dua HS, Mittal R (2021) Antimicrobial peptides in human corneal tissue of patients with fungal keratitis. Br J Ophthalmol 105:1172–1177. https://doi.org/10.1136/bjophthalmol-2020-316329

Moledina M, Roberts HW, Mukherjee A, Spokes D, Pimenides D, Stephenson C, Bassily R, Rajan MS, Myerscough J (2023) Analysis of microbial keratitis incidence, isolates and in-vitro antimicrobial susceptibility in the East of England: a 6-year study. Eye (Lond) 37:2716–2722. https://doi.org/10.1038/s41433-023-02404-3

Monod M, Capoccia S, Léchenne B, Zaugg C, Holdom M, Jousson O (2002) Secreted proteases from pathogenic fungi. Int J Med Microbiol 292:405–419. https://doi.org/10.1078/1438-4221-00223

Morelli KA, Kerkaert JD, Cramer RA (2021) biofilms: Toward understanding how growth as a multicellular network increases antifungal resistance and disease progression. PloS Pathogens 17:e1009794. https://doi.org/10.1371/journal.ppat.1009794

Morgan PB, Efron N (2006) A decade of contact lens prescribing trends in the United Kingdom (1996–2005). Cont Lens Anterior Eye 29:59–68. https://doi.org/10.1016/j.clae.2006.02.008
Moussa G, Hodson J, Gooch N, Virdee J, Penaloza C, Kigozi J, Rauz S (2021) Calculating the economic burden of presumed microbial keratitis admissions at a tertiary referral centre in the UK. Eye 35:2146–2154. https://doi.org/10.1038/s41433-020-01333-9
Mowat E, Lang S, Williams C, Mcculloch E, Jones B, Ramage G (2008) Phase-dependent antifungal activity against Aspergillus fumigatus developing multicellular filamentous biofilms. J Antimicrob Chemother 62:1281–1284. https://doi.org/10.1093/jac/dkn402
Mpakosi A, Kaliouli-Antonopoulou C (2024) Immune mechanisms of filamentous fungal keratitis. Cureus 16:e61954. https://doi.org/10.7759/cureus.61954
Mukherjee PK, Chandra J, Kuhn DM, Ghannoum MA (2003) Mechanism of fluconazole resistance in Candida albicans biofilms: phase-specific role of efflux pumps and membrane sterols. Infect Immun 71:4333–4340. https://doi.org/10.1128/IAI.71.8.4333-4340.2003
Mukherjee PK, Chandra J, Yu C, Sun Y, Pearlman E, Ghannoum MA (2012) Characterization of fusarium keratitis outbreak isolates: contribution of biofilms to antimicrobial resistance and pathogenesis. Invest Ophthalmol Vis Sci 53:4450–4457. https://doi.org/10.1167/iovs.12-9848
Murphy PJ, Patel S, Marshall J (2001) The effect of long-term, daily contact lens wear on corneal sensitivity. Cornea 20:264–269. https://doi.org/10.1097/00003226-200104000-00006
Naranjo A, Arboleda A, Martinez JD, Durkee H, Aguilar MC, Relhan N, Nikpoor N, Galor A, Dubovy SR, Leblanc R, Flynn HW Jr, Miller D, Parel JM, Amescua G (2019) Rose Bengal photodynamic antimicrobial therapy for patients with progressive infectious keratitis: a pilot clinical study. Am J Ophthalmol 208:387–396. https://doi.org/10.1016/j.ajo.2019.08.027
Nejabat M, Yaqubi N, Khosravi A, Zomorodian K, Ashraf MJ, Salouti R (2016) Therapeutic effect of intrastromal voriconazole, topical voriconazole, and topical natamycin on fusarium keratitis in rabbit. J Ophthalmol 2016:8692830. https://doi.org/10.1155/2016/8692830
Netea MG, Joosten LA, Van Der Meer JW, Kullberg BJ, Van De Veerdonk FL (2015) Immune defence against Candida fungal infections. Nat Rev Immunol 15:630–642. https://doi.org/10.1038/nri3897
Nett JE (2016) The host's reply to Candida biofilm. Pathogens 5. https://doi.org/10.3390/pathogens5010033
Ngo J, Khoo P, Watson SL (2020) Improving the efficiency and the technique of the corneal scrape procedure via an evidence based instructional video at a quaternary referral eye hospital. Curr Eye Res 45:529–534. https://doi.org/10.1080/02713683.2019.1676910
Niu L, Liu X, Ma Z, Yin Y, Sun L, Yang L, Zheng Y (2020) Fungal keratitis: pathogenesis, diagnosis and prevention. Microb Pathog 138:103802. https://doi.org/10.1016/j.micpath.2019.103802
Norren DV, Vos JJ (1974) Spectral transmission of the human ocular media. Vis Res 14:1237–1244. https://doi.org/10.1016/0042-6989(74)90222-3
O'day DM, Head WS, Robinson RD, Clanton JA (1986) Corneal penetration of topical amphotericin B and natamycin. Curr Eye Res 5:877–882. https://doi.org/10.3109/02713688609029240
Oliva M, Schottman T, Gulati M (2012) Turning the tide of corneal blindness. Indian J Ophthalmol 60:423–427. https://doi.org/10.4103/0301-4738.100540
Oliveira Dos Santos C, Kolwijck E, van Rooij J, Stoutenbeek R, Visser N, Cheng YY, Santana NTY, Verweij PE, Eggink CA (2020) Epidemiology and clinical management of Fusarium keratitis in the Netherlands, 2005–2016. Front Cell Infect Microbiol 10:133. https://doi.org/10.3389/fcimb.2020.00133
Ong HS, Fung SSM, Macleod D, Dart JKG, Tuft SJ, Burton MJ (2016) Altered patterns of fungal Keratitis at a London ophthalmic referral hospital: an eight-year retrospective observational study. Am J Ophthalmol 168:227–236. https://doi.org/10.1016/j.ajo.2016.05.021
Onurdag FK, Ozkan S, Ozgen S, Olmus H, Abbasoglu U (2011) Candida albicans and Pseudomonas aeruginosa adhesion on soft contact lenses. *Graefes Arch Clin Exp Ophthalmol* 249:559–564. https://doi.org/10.1007/s00417-010-1595-3
Oyama J, Fernandes Herculano Ramos-Milare AC, Lopes Lera-Nonose DSS, Nesi-Reis V, Galhardo Demarchi I, Alessi Aristides SM, Juarez Vieira Teixeira J, Gomes Verzignassi

Silveira T, Campana Lonardoni MV (2020) Photodynamic therapy in wound healing in vivo: a systematic review. Photodiagnosis Photodyn Ther 30:101682. https://doi.org/10.1016/j.pdpdt.2020.101682

Park M, Do E, Jung WH (2013) Lipolytic enzymes involved in the virulence of human pathogenic fungi. Mycobiology 41:67–72. https://doi.org/10.5941/myco.2013.41.2.67

Paziani MH, Tonani L, de Menezes HD, Bachmann L, Wainwright M, Braga GUL, Von Zeska Kress MR (2019) Antimicrobial photodynamic therapy with phenothiazinium photosensitizers in non-vertebrate model Galleria mellonella infected with Fusarium keratoplasticum and Fusarium moniliforme. Photodiagnosis Photodyn Ther 25:197–203. https://doi.org/10.1016/j.pdpdt.2018.12.010

Peiqian L, Xiaoming P, Huifang S, Jingxin Z, Ning H, Birun L (2014) Biofilm formation by Fusarium oxysporum f. sp. cucumerinum and susceptibility to environmental stress. FEMS Microbiol Lett 350:138–145. https://doi.org/10.1111/1574-6968.12310

Pereira R, dos Santos Fontenelle RO, de Brito EHS, de Morais SM (2021) Biofilm of Candida albicans: formation, regulation and resistance. J Appl Microbiol 131:11–22. https://doi.org/10.1111/jam.14949

Pereira AR, Braga DFO, Vassal M, Gomes IB, Simões M (2023) Ultraviolet C irradiation: a promising approach for the disinfection of public spaces? Sci Total Environ 879:163007. https://doi.org/10.1016/j.scitotenv.2023.163007

Petrillo F, Sinoca M, Fea AM, Galdiero M, Maione A, Galdiero E, Guida M, Reibaldi M (2023) Candida biofilm eye infection: main aspects and advance in novel agents as potential source of treatment. Antibiotics-Basel 12:1277. https://doi.org/10.3390/antibiotics12081277

Pham DQ, Bryant SJ, Cheeseman S, Huang LZY, Bryant G, Dupont MF, Chapman J, Berndt CC, Vongsvivut JP, Crawford RJ, Truong VK, Ang ASM, Elbourne A (2020) Micro- to nano-scale chemical and mechanical mapping of antimicrobial-resistant fungal biofilms. Nanoscale 12:19888–19904. https://doi.org/10.1039/d0nr05617k

Poomany Arul Soundara Rajan YA, Anitha V, Meenakshi R, Kasi M (2023) Therapeutic contact lens-related infection by a novel pathogenic fungus Coniochaeta hoffmannii-a case report. Indian J Med Microbiol 44:100361. https://doi.org/10.1016/j.ijmmb.2023.02.004

Prajna NV, Krishnan T, Mascarenhas J, Rajaraman R, Prajna L, Srinivasan M, Raghavan A, Oldenburg CE, Ray KJ, Zegans ME, Mcleod SD, Porco TC, Acharya NR, Lietman TM, Mycotic Ulcer Treatment Trial Group, F. T (2013a) The mycotic ulcer treatment trial: a randomized trial comparing natamycin vs voriconazole. JAMA Ophthalmol 131:422–429. https://doi.org/10.1001/jamaophthalmol.2013.1497

Prajna NV, Srinivasan M, Lalitha P, Krishnan T, Rajaraman R, Ravindran M, Mascarenhas J, Oldenburg CE, Ray KJ, Mcleod SD, Acharya NR, Lietman TM (2013b) Differences in clinical outcomes in keratitis due to fungus and bacteria. Jama Ophthalmol 131:1088–1089. https://doi.org/10.1001/jamaophthalmol.2013.1612

Prajna NV, Krishnan T, Rajaraman R, Patel S, Srinivasan M, Das M, Ray KJ, O'brien KS, Oldenburg CE, Mcleod SD, Zegans ME, Porco TC, Acharya NR, Lietman TM, Rose-Nussbaumer J (2016) Effect of oral voriconazole on fungal keratitis in the mycotic ulcer treatment trial II (MUTT II): a randomized clinical trial. JAMA Ophthalmol 134:1365–1372. https://doi.org/10.1001/jamaophthalmol.2016.4096

Prajna VN, Prajna L, Muthiah S (2017) Fungal keratitis: the Aravind experience. Indian J Ophthalmol 65:912–919. https://doi.org/10.4103/ijo.IJO_821_17

Prajna NV, Radhakrishnan N, Lalitha P, Austin A, Ray KJ, Keenan JD, Porco TC, Lietman TM, Rose-Nussbaumer J (2020) Cross-linking-assisted infection reduction: a randomized clinical trial evaluating the effect of adjuvant cross-linking on outcomes in fungal keratitis. Ophthalmology 127:159–166. https://doi.org/10.1016/j.ophtha.2019.08.029

Prajna NV, Radhakrishnan N, Lalitha P, Rajaraman R, Narayana S, Austin AF, Liu Z, Keenan JD, Porco TC, Lietman TM, Rose-Nussbaumer J (2021) Cross-linking assisted infection reduction (CLAIR): a randomized clinical trial evaluating the effect of adjuvant cross-linking on bacterial keratitis. Cornea 40:837–841. https://doi.org/10.1097/ICO.0000000000002510

Prajna V, Prajna L, Sharma S, de Freitas D, Hofling-Lima AL, Varnado N, Abdelrahman S, Cavallino V, Arnold B, Lietman T, Rose-Nussbaumer J (2024) A double-masked, sham-controlled trial of rose bengal photodynamic therapy for the treatment of fungal and acanthameoba keratitis: Rose Bengal Electromagnetic Activation with Green Light for Infection Reduction (REAGIR) Study. Res Sq. https://doi.org/10.21203/rs.3.rs-4165312/v1

Price MO, Tenkman LR, Schrier A, Fairchild KM, Trokel SL, Price FW Jr (2012) Photoactivated riboflavin treatment of infectious keratitis using collagen cross-linking technology. J Refract Surg 28:706–713. https://doi.org/10.3928/1081597X-20120921-06

Qi P, Liu X, Li C, Xu Q, Hu L, Duan H, Zhao G, Lin J (2024) Progranulin protects against Aspergillus fumigatus Keratitis by attenuating the inflammatory response through enhancing autophagy. ACS Infect Dis. https://doi.org/10.1021/acsinfecdis.4c00236

Qin Q, Hu K, He Z, Chen F, Zhang W, Liu Y, Xie Z (2022) Resolvin D1 protects against Aspergillus fumigatus keratitis in diabetes by blocking the MAPK-NF-kappaB pathway. Exp Eye Res 216:108941. https://doi.org/10.1016/j.exer.2022.108941

Radhakrishnan N, Pathak N, Subramanian KR, Das DR, Ningombam R, Khaitan I, Gandhi N, Rahul R, Prajna NV (2022) Comparative study on costs incurred for treatment of patients with bacterial and fungal keratitis – a retrospective analysis. Indian J Ophthalmol 70:1191–1195. https://doi.org/10.4103/ijo.IJO_2176_21

Raeiszadeh M, Adeli B (2020) A critical review on ultraviolet disinfection systems against COVID-19 outbreak: applicability, validation, and safety considerations. ACS Photonics 7:2941–2951. https://doi.org/10.1021/acsphotonics.0c01245

Rajaraman R, Bhat P, Vaidee V, Maskibail S, Raghavan A, Sivasubramaniam S, Namperumalsamy VP (2015) Topical 5% natamycin with oral ketoconazole in filamentous fungal keratitis: a randomized controlled trial. Asia Pac J Ophthalmol (Phila) 4:146–150. https://doi.org/10.1097/apo.0000000000000035

Rajendran R, Mowat E, Mcculloch E, Lappin DF, Jones B, Lang S, Majithiya JB, Warn P, Williams C, Ramage G (2011) Azole resistance of biofilms is partly associated with efflux pump activity. Antimicrob Agents Chemother 55:2092–2097. https://doi.org/10.1128/Aac.01189-10

Ramani R, Chaturvedi V (2011) Evaluations of shorter exposures of contact lens cleaning solutions against Fusarium oxysporum species complex and Fusarium solani species complex to simulate inappropriate usage. Antimicrob Agents Chemother 55:2265–2275. https://doi.org/10.1128/AAC.00634-09

Ramirez-Granillo A, Bautista-Hernandez LA, Bautista-de Lucio VM, Magana-Guerrero FS, Dominguez-Lopez A, Cordova-Alcantara IM, Perez NO, Martinez-Rivera MLA, Rodriguez-Tovar AV (2021) Microbial warfare on three fronts: mixed biofilm of Aspergillus fumigatus and Staphylococcus aureus on primary cultures of human limbo-corneal fibroblasts. Front Cell Infect Microbiol 11:646054. https://doi.org/10.3389/fcimb.2021.646054

Ranjith K, Sontam B, Sharma S, Joseph J, Chathoth KN, Sama KC, Murthy SI, Shivaji S (2017) Candida species from eye infections: drug susceptibility, virulence factors, and molecular characterization. Investig Ophthalmol Vis Sci 58:4201–4209. https://doi.org/10.1167/iovs.17-22003

Ranjith K, Kalyana Chakravarthy S, Adicherla H, Sharma S, Shivaji S (2018) Temporal expression of genes in biofilm-forming ocular Candida albicans Isolated from patients with keratitis and orbital cellulitis. Invest Ophthalmol Vis Sci 59:528–538. https://doi.org/10.1167/iovs.17-22933

Ranjith K, Nagapriya B, Shivaji S (2022) Polymicrobial biofilms of ocular bacteria and fungi on ex vivo human corneas. Sci Rep 12:11606. https://doi.org/10.1038/s41598-022-15809-z

Rao NA, Riggio DW, Delmage JM, Calandra AJ, Evans S, Lewis W (1985) Adherence of Candida to corneal surface. Curr Eye Res 4:851–856. https://doi.org/10.3109/02713688509095252

Rasoanirina BNV, Lassoued MA, Miladi K, Razafindrakoto Z, Chaâbane-Banaoues R, Ramanitrahasimbola D, Cornet M, Sfar S (2020) Self-nanoemulsifying drug delivery system to improve transcorneal permeability of voriconazole: in-vivo studies. J Pharm Pharmacol 72:889–896. https://doi.org/10.1111/jphp.13265

Rasool BK, Salmo HM (2012) Development and clinical evaluation of clotrimazole-β-cyclodextrin eyedrops for the treatment of fungal keratitis. AAPS Pharm Sci Tech 13:883–889. https://doi.org/10.1208/s12249-012-9813-4

Ratitong B, Pearlman E (2021) Pathogenic Aspergillus and Fusarium as important causes of blinding corneal infections—the role of neutrophils in fungal killing, tissue damage and cytokine production. Curr Opin Microbiol 63:195–203. https://doi.org/10.1016/j.mib.2021.07.018

Redd TK, Prajna NV, Srinivasan M, Lalitha P, Krishnan T, Rajaraman R, Venugopal A, Acharya N, Seitzman GD, Lietman TM, Keenan JD, Campbell JP, Song X (2022a) Image-based differentiation of bacterial and fungal keratitis using deep convolutional neural networks. Ophthalmol Sci 2:100119. https://doi.org/10.1016/j.xops.2022.100119

Redd TK, Prajna NV, Srinivasan M, Lalitha P, Krishnan T, Rajaraman R, Venugopal A, Lujan B, Acharya N, Seitzman GD, Rose-Nussbaumer J, Lietman TM, Campbell JP, Keenan JD, Corneal Ulcer Image Interpretation Study, G (2022b) Expert performance in visual differentiation of bacterial and fungal keratitis. Ophthalmology 129:227–230. https://doi.org/10.1016/j.ophtha.2021.09.019

Retuerto MA, Szczotka-Flynn L, Ho D, Mukherjee P, Ghannoum MA (2012) Efficacy of care solutions against contact lens-associated Fusarium biofilms. Optom Vis Sci 89:382–391. https://doi.org/10.1097/OPX.0b013e31824cb754

Roberts D, Thomas J, Salmon J, Cubeta MA, Stapelmann K, Gilger BC (2024) Cold atmospheric plasma inactivates Aspergillus flavus and Fusarium keratoplasticum biofilms and conidia in vitro. J Med Microbiol 73. https://doi.org/10.1099/jmm.0.001858

Robertson DM, Cavanagh HD (2008) The clinical and cellular basis of contact lens-related corneal infections: a review. Clin Ophthalmol 2:907–917. https://doi.org/10.2147/opth.s3249

Rohira H, Shankar S, Yadav S, Shah SG, Chugh A (2021) Enhanced in vivo antifungal activity of novel cell penetrating peptide natamycin conjugate for efficient fungal keratitis management. Int J Pharm 600:120484. https://doi.org/10.1016/j.ijpharm.2021.120484

Saglk A, Uçakhan ÖÖ, Kanpolat A (2013) Ultraviolet A and riboflavin therapy as an adjunct in corneal ulcer refractory to medical treatment. Eye Contact Lens 39:413–415. https://doi.org/10.1097/ICL.0b013e3182960fdf

Said DG, Otri M, Miri A, Kailasanathan A, Khatib T, Dua HS (2011) The challenge of fungal keratitis. Br J Ophthalmol 95:1623. https://doi.org/10.1136/bjophthalmol-2011-301148

Said DG, Elalfy MS, Gatzioufas Z, El-Zakzouk ES, Hassan MA, Saif MY, Zaki AA, Dua HS, Hafezi F (2014) Collagen cross-linking with photoactivated riboflavin (PACK-CXL) for the treatment of advanced infectious keratitis with corneal melting. Ophthalmology 121:1377–1382. https://doi.org/10.1016/j.ophtha.2014.01.011

Saluja G, Sharma N, Agarwal R, Sharma HP, Singhal D, Kumar Maharana P, Sinha R, Agarwal T, Velpandian T, Titiyal JS, Satpathy G (2021) Comparison of safety and efficacy of intrastromal injections of voriconazole, amphotericin b and natamycin in cases of recalcitrant fungal keratitis: a randomized controlled trial. Clin Ophthalmol 15:2437–2446. https://doi.org/10.2147/opth.S301878

Sarayar R, Lestari YD, Setio AAA, Sitompul R (2023) Accuracy of artificial intelligence model for infectious keratitis classification: a systematic review and meta-analysis. Front Public Health 11:1239231. https://doi.org/10.3389/fpubh.2023.1239231

Sav H, Rafati H, Oz Y, Dalyan-Cilo B, Ener B, Mohammadi F, Ilkit M, van Diepeningen AD, Seyedmousavi S (2018) Biofilm formation and resistance to fungicides in clinically relevant members of the fungal genus Fusarium. J Fungi (Basel) 4:16. https://doi.org/10.3390/jof4010016

Sengupta J, Saha S, Khetan A, Sarkar SK, Mandal SM (2012) Effects of lactoferricin B against keratitis-associated fungal biofilms. J Infect Chemother 18:698–703. https://doi.org/10.1007/s10156-012-0398-3

Sepulveda-Beltran PA, Levine H, Altamirano DS, Martinez JD, Durkee H, Mintz K, Leblanc R, Tothova JD, Miller D, Parel JM, Amescua G (2022) Rose Bengal photodynamic antimicrobial therapy: a review of the intermediate-term clinical and surgical outcomes. Am J Ophthalmol 243:125–134. https://doi.org/10.1016/j.ajo.2022.08.004

Setlow RB (1966) Cyclobutane-type pyrimidine dimers in polynucleotides. Science 153:379–386. https://doi.org/10.1126/science.153.3734.379

Sha X, Chan L, Fan X, Guo P, Chen T, Liu L, Zhong J (2022) Thermosensitive tri-block polymer nanoparticle-hydrogel composites as payloads of natamycin for antifungal therapy against Fusarium Solani. Int J Nanomed 17:1463–1478. https://doi.org/10.2147/ijn.S332127

Shah H, Radhakrishnan N, Ramsewak S, Chiu S, Joseph S, Rose-Nussbaumer J, Prajna NV (2019) Demographic and socioeconomic barriers and treatment seeking behaviors of patients with infectious keratitis requiring therapeutic penetrating keratoplasty. Indian J Ophthalmol 67:1593–1598. https://doi.org/10.4103/ijo.IJO_1821_18

Shao Y, Yu Y, Pei CG, Tan YH, Zhou Q, Yi JL, Gao GP (2010) Therapeutic efficacy of intracameral amphotericin B injection for 60 patients with keratomycosis. Int J Ophthalmol 3:257–260. https://doi.org/10.3980/j.issn.2222-3959.2010.03.18

Sharma N, Agarwal P, Sinha R, Titiyal JS, Velpandian T, Vajpayee RB (2011) Evaluation of intrastromal voriconazole injection in recalcitrant deep fungal keratitis: case series. Br J Ophthalmol 95:1735–1737. https://doi.org/10.1136/bjo.2010.192815

Sharma N, Chacko J, Velpandian T, Titiyal JS, Sinha R, Satpathy G, Tandon R, Vajpayee RB (2013) Comparative evaluation of topical versus intrastromal voriconazole as an adjunct to natamycin in recalcitrant fungal keratitis. Ophthalmology 120:677–681. https://doi.org/10.1016/j.ophtha.2012.09.023

Sharma N, Sahay P, Maharana PK, Singhal D, Saluja G, Bandivadekar P, Chako J, Agarwal T, Sinha R, Titiyal JS, Satpathy G, Velpandian T (2019) Management algorithm for fungal keratitis: the TST (topical, systemic, and targeted therapy) protocol. Cornea 38:141–145. https://doi.org/10.1097/ICO.0000000000001781

Sharma A, Sharma R, Chander J, Nirankari VS (2023) In vitro antimicrobial efficacy of riboflavin, ultraviolet-A radiation, and combined riboflavin/ultraviolet-A radiation on ocular pathogens. Taiwan J Ophthalmol 13:21–27. https://doi.org/10.4103/tjo.tjo_28_21

Shi H, Li J, Zhang H, Zhang J, Sun H (2016) Effect of 5-aminolevulinic acid photodynamic therapy on Candida albicans biofilms: an in vitro study. Photodiagnosis Photodyn Ther 15:40–45. https://doi.org/10.1016/j.pdpdt.2016.04.011

Shi H, Li J, Peng C, Xu B, Sun H (2021) The inhibitory activity of 5-aminolevulinic acid photodynamic therapy (ALA-PDT) on Candida albicans biofilms. Photodiagnosis Photodyn Ther 34:102271. https://doi.org/10.1016/j.pdpdt.2021.102271

Shibuya K, Paris S, Ando T, Nakayama H, Hatori T, Latgé JP (2006) Catalases of Aspergillus fumigatus and inflammation in aspergillosis. Nihon Ishinkin Gakkai Zasshi 47:249–255. https://doi.org/10.3314/jjmm.47.249

Shin H, Price K, Albert L, Dodick J, Park L, Dominguez-Bello MG (2016) Changes in the eye microbiota associated with contact lens wearing. mBio 7:e00198. https://doi.org/10.1128/mBio.00198-16

Shivaji S, Jayasudha R, Sai Prashanthi G, Kalyana Chakravarthy S, Sharma S (2019) The human ocular surface fungal microbiome. Invest Ophthalmol Vis Sci 60:451–459. https://doi.org/10.1167/iovs.18-26076

Shivaji S, Jayasudha R, Prashanthi GS, Arunasri K, Das T (2022) Fungi of the human eye: culture to mycobiome. Exp Eye Res 217:108968. https://doi.org/10.1016/j.exer.2022.108968

Shoham S (2015) Host/pathogen interactions in fungal keratitis. Curr Fungal Infect Rep 9:52–56. https://doi.org/10.1007/s12281-014-0214-x

Silva MP, dos Santos TA, de Barros PP, de Camargo Ribeiro F, Junqueira JC, Jorge AO (2016) Action of antimicrobial photodynamic therapy on heterotypic biofilm: Candida albicans and Bacillus atrophaeus. Lasers Med Sci 31:605–610. https://doi.org/10.1007/s10103-016-1876-0

Simmons RB, Buffington JR, Ward M, Wilson LA, Ahearn DG (1986) Morphology and ultrastructure of fungi in extended-wear soft contact lenses. J Clin Microbiol 24:21–25. https://doi.org/10.1128/jcm.24.1.21-25.1986

Skaat A, Zadok D, Goldich Y, Varssano D, Berger Y, Ezra-Nimni O, Avni I, Barequet IS (2014) Riboflavin/UVA photochemical therapy for severe infectious keratitis. Eur J Ophthalmol 24:21–28. https://doi.org/10.5301/ejo.5000330

Sliney DH (2002) How light reaches the eye and its components. Int J Toxicol 21:501–509

Soe H, Junthip J, Chamni S, Chansriniyom C, Limpikirati P, Thanusuwannasak T, Asasutjarit R, Pruksakorn P, Autthateinchai R, Wet-Osot S, Loftsson T, Jansook P (2023) A promising synthetic citric crosslinked β-cyclodextrin derivative for antifungal drugs: solubilization, cytotoxicity, and antifungal activity. Int J Pharm 645:123394. https://doi.org/10.1016/j.ijpharm.2023.123394

Soliman GM, Attia MA, Mohamed RA (2014) Poly(ethylene glycol)-block-poly(ε-caprolactone) nanomicelles for the solubilization and enhancement of antifungal activity of sertaconazole. Curr Drug Deliv 11:753–762. https://doi.org/10.2174/1567201811666140605151923

Somerville TF, Herbert R, Neal T, Horsburgh M, Kaye SB (2021) An evaluation of a simplified impression membrane sampling method for the diagnosis of microbial keratitis. J Clin Med 10:5671. https://doi.org/10.3390/jcm10235671

Somerville TF, Mdala S, Zungu T, Gandiwa M, Herbert R, Everett D, Corless CE, Beare NA, Neal T, Horsburgh MJ (2024) Microbial keratitis in Southern Malawi: a microbiological pilot study. BMJ Open Ophthalmol 9. https://doi.org/10.1136/bmjophth-2024-001682

Song C, Wen R, Zhou J, Zeng X, Kou Z, Li Y, Yun F, Wu R (2022) UV C light from a light-emitting diode at 275 nanometers shortens wound healing time in bacterium- and fungus-infected skin in mice. Microbiol Spectr 10:e03424–e03422. https://doi.org/10.1128/spectrum.03424-22

Sosnin EA, Stoffels E, Erofeev MV, Kieft IE, Kunts SE (2004) The effects of UV irradiation and gas plasma treatment on living mammalian cells and bacteria: a comparative approach. IEEE Trans Plasma Sci 32:1544–1550. https://doi.org/10.1109/TPS.2004.833401

Sourlis C, Seitz B, Roth M, Hamon L, Daas L (2022) Outcomes of severe fungal keratitis using in vivo confocal microscopy and early therapeutic penetrating keratoplasty. Clin Ophthalmol 16:2245–2254. https://doi.org/10.2147/opth.S358709

Spoerl E, Terai N, Scholz F, Raiskup F, Pillunat LE (2011) Detection of biomechanical changes after corneal cross-linking using ocular response analyzer software. J Refract Surg 27:452–457. https://doi.org/10.3928/1081597X-20110106-01

Stapleton F, Keay L, Edwards K, Naduvilath T, Dart JK, Brian G, Holden BA (2008) The incidence of contact lens-related microbial keratitis in Australia. Ophthalmology 115:1655–1662. https://doi.org/10.1016/j.ophtha.2008.04.002

Sun Y, Chandra J, Mukherjee P, Szczotka-Flynn L, Ghannoum MA, Pearlman E (2010) A murine model of contact lens-associated fusarium keratitis. Invest Ophthalmol Vis Sci 51:1511–1516. https://doi.org/10.1167/iovs.09-4237

Sun Y, Ogawa R, Xiao BH, Feng YX, Wu Y, Chen LH, Gao XH, Chen HD (2020) Antimicrobial photodynamic therapy in skin wound healing: a systematic review of animal studies. Int Wound J 17:285–299. https://doi.org/10.1111/iwj.13269

Tang Q, Che C, Lin J, He H, Zhao W, Lv L, Zhao G (2019) Maresin1 regulates neutrophil recruitment and IL-10 expression in Aspergillus fumigatus keratitis. Int Immunopharmacol 69:103–108. https://doi.org/10.1016/j.intimp.2019.01.032

Tavakoli N, Taymouri S, Saeidi A, Akbari V (2019) Thermosensitive hydrogel containing sertaconazole loaded nanostructured lipid carriers for potential treatment of fungal keratitis. Pharm Dev Technol 24:891–901. https://doi.org/10.1080/10837450.2019.1616755

Thew MR, Todd B (2008) Fungal keratitis in far north Queensland, Australia. Clin Exp Ophthalmol 36:721–724. https://doi.org/10.1111/j.1442-9071.2008.01879.x

Thiel MA, Zinkernagel AS, Burhenne J, Kaufmann C, Haefeli WE (2007) Voriconazole concentration in human aqueous humor and plasma during topical or combined topical and systemic administration for fungal keratitis. Antimicrob Agents Chemother 51:239–244. https://doi.org/10.1128/aac.00762-06

Tian X, Peng X, Lin J, Zhang Y, Zhan L, Yin J, Zhang R, Zhao G (2021) Isorhamnetin ameliorates Aspergillus fumigatus keratitis by reducing fungal load, inhibiting pattern-recognition receptors and inflammatory cytokines. Invest Ophthalmol Vis Sci 62:38. https://doi.org/10.1167/iovs.62.3.38

Ting DSJ, Galal M, Kulkarni B, Elalfy MS, Lake D, Hamada S, Said DG, Dua HS (2021) Clinical characteristics and outcomes of fungal keratitis in the United Kingdom 2011–2020: a 10-year study. J Fungi (Basel) 7:966. https://doi.org/10.3390/jof7110966

Ting DSJ, Chodosh J, Mehta JS (2022) Achieving diagnostic excellence for infectious keratitis: a future roadmap. Front Microbiol 13:1020198. https://doi.org/10.3389/fmicb.2022.1020198

Tuft S, Somerville TF, Li JO, Neal T, De S, Horsburgh MJ, Fothergill JL, Foulkes D, Kaye S (2022) Bacterial keratitis: identifying the areas of clinical uncertainty. Prog Retin Eye Res 89:101031. https://doi.org/10.1016/j.preteyeres.2021.101031

Turicea B, Sahoo DK, Allbaugh RA, Stinman CC, Kubai MA (2024) Novel treatment of infectious keratitis in canine corneas using ultraviolet C (UV-C) light. Vet Ophthalmol. https://doi.org/10.1111/vop.13265

Uchio E, Saeki Y, Tsukahara-Kawamura T, Kadonosono K, Ozaki H (2019) Clinical outcome after air-assisted manual deep anterior lamellar keratoplasty for fungal keratitis poorly responsive to medical treatment. Clin Ophthalmol 13:1913–1919. https://doi.org/10.2147/opth.S211099

Uddaraju M, Mascarenhas J, Das MR, Radhakrishnan N, Keenan JD, Prajna L, Prajna NV (2015) Corneal cross-linking as an adjuvant therapy in the management of recalcitrant deep stromal fungal keratitis: a randomized trial. Am J Ophthalmol 160:131–4.e5. https://doi.org/10.1016/j.ajo.2015.03.024

Ung L, Acharya NR, Agarwal T, Alfonso EC, Bagga B, Bispo PJ, Burton MJ, Dart JK, Doan T, Fleiszig SM, Garg P, Gilmore MS, Gritz DC, Hazlett LD, Iovieno A, Jhanji V, Kempen JH, Lee CS, Lietman TM, Margolis TP, Mcleod SD, Mehta JS, Miller D, Pearlman E, Lalitha P, Prajna NV, Seitzman GD, Shanbhag SS, Sharma N, Sharma S, Srinivasan M, Stapleton F, Tan DT, Tandon R, Taylor HR, Tu EY, Tuli SS, Vajpayee RB, van Gelder RN, Watson SL, Zegans ME, Chodosh J (2019) Infectious corneal ulceration: a proposal for neglected tropical disease status. Bull World Health Organ 97:854–856. https://doi.org/10.2471/BLT.19.232660

Urwin L, Okurowska K, Crowther G, Roy S, Garg P, Karunakaran E, Macneil S, Partridge LJ, Green LR, Monk PN (2020) Corneal infection models: tools to investigate the role of biofilms in bacterial keratitis. Cells 9:2450. https://doi.org/10.3390/cells9112450

Wainwright M, Maisch T, Nonell S, Plaetzer K, Almeida A, Tegos GP, Hamblin MR (2017) Photoantimicrobials—are we afraid of the light? Lancet Infect Dis 17:e49–e55. https://doi.org/10.1016/S1473-3099(16)30268-7

Walsh TJ, Petraitis V, Petraitiene R, Field-Ridley A, Sutton D, Ghannoum M, Sein T, Schaufele R, Peter J, Bacher J, Casler H, Armstrong D, Espinel-Ingroff A, Rinaldi MG, Lyman CA (2003) Experimental pulmonary aspergillosis due to Aspergillus terreus: pathogenesis and treatment of an emerging fungal pathogen resistant to amphotericin B. J Infect Dis 188:305–319. https://doi.org/10.1086/377210

Walther G, Stasch S, Kaerger K, Hamprecht A, Roth M, Cornely OA, Geerling G, Mackenzie CR, Kurzai O, von Lilienfeld-Toal M (2017) Fusarium Keratitis in Germany. J Clin Microbiol 55:2983–2995. https://doi.org/10.1128/JCM.00649-17

Wang F, Zhao L, Song F, Wu J, Zhou Q, Xie L (2021) Hybrid natural hydrogels integrated with voriconazole-loaded microspheres for ocular antifungal applications. J Mater Chem B 9:3377–3388. https://doi.org/10.1039/d1tb00263e

Wang Y, Yin M, Gu L, Yi W, Lin J, Zhang L, Wang Q, Qi Y, Diao W, Chi M, Zheng H, Li C, Zhao G (2023) The therapeutic role and mechanism of 4-Methoxycinnamic acid in fungal keratitis. Int Immunopharmacol 116:109782. https://doi.org/10.1016/j.intimp.2023.109782

Wang D, Zeng N, Li C, Li Z, Zhang N, Li B (2024) Fungal biofilm formation and its regulatory mechanism. Heliyon 10:e32766. https://doi.org/10.1016/j.heliyon.2024.e32766

Wei Z, Wang S, Wang Z, Zhang Y, Chen K, Gong L, Li G, Zheng Q, Zhang Q, He Y, Zhang Q, Chen D, Cao K, Pang J, Zhang Z, Wang L, Ou Z, Liang Q (2023) Development and multi-center validation of machine learning model for early detection of fungal keratitis. EBioMedicine 88:104438. https://doi.org/10.1016/j.ebiom.2023.104438

Whitcher JP, Srinivasan M, Upadhyay MP (2001) Corneal blindness: a global perspective. Bull World Health Organ 79:214–221

Whitcher JP, Srinivasan M, Upadhyay MP (2002) Prevention of corneal ulceration in the developing world. Int Ophthalmol Clin 42:71–77. https://doi.org/10.1097/00004397-200201000-00010

Willcox MD, Carnt N, Diec J, Naduvilath T, Evans V, Stapleton F, Iskandar S, Harmis N, de La Jara PL, Holden BA (2010) Contact lens case contamination during daily wear of silicone hydrogels. Optom Vis Sci 87:456–464. https://doi.org/10.1097/OPX.0b013e3181e19eda

Wilson LA, Sawant AD, Simmons RB, Ahearn DG (1990) Microbial contamination of contact lens storage cases and solutions. Am J Ophthalmol 110:193–198. https://doi.org/10.1016/s0002-9394(14)76991-0

Wollensak G, Spoerl E, Seiler T (2003) Stress-strain measurements of human and porcine corneas after riboflavin-ultraviolet-A-induced cross-linking. J Cataract Refract Surg 29:1780–1785. https://doi.org/10.1016/s0886-3350(03)00407-3

Wong A, Fallon M, Celiksoy V, Ferla S, Varricchio C, Whitaker D, Quantock AJ, Heard CM (2022) A composite system based upon hydroxypropyl cyclodextrins and soft hydrogel contact lenses for the delivery of therapeutic doses of econazole to the Cornea, in vitro. Pharmaceutics 14. https://doi.org/10.3390/pharmaceutics14081631

Wu YT, Zhu H, Harmis NY, Iskandar SY, Willcox M, Stapleton F (2010) Profile and frequency of microbial contamination of contact lens cases. Optom Vis Sci 87:E152–E158. https://doi.org/10.1097/OPX.0b013e3181cf86ee

Wu YT, Willcox M, Zhu H, Stapleton F (2015a) Contact lens hygiene compliance and lens case contamination: a review. Cont Lens Anterior Eye 38:307–316. https://doi.org/10.1016/j.clae.2015.04.007

Wu YT, Willcox MD, Stapleton F (2015b) The effect of contact lens hygiene behavior on lens case contamination. Optom Vis Sci 92:167–174. https://doi.org/10.1097/OPX.0000000000000477

Xie L, Zhai H, Shi W (2007) Penetrating keratoplasty for corneal perforations in fungal keratitis. Cornea 26:158–162. https://doi.org/10.1097/01.ico.0000248381.24519.0d

Xie Z, Thompson A, Sobue T, Kashleva H, Xu H, Vasilakos J, Dongari-Bagtzoglou A (2012) Candida albicans biofilms do not trigger reactive oxygen species and evade neutrophil killing. J Infect Dis 206:1936–1945. https://doi.org/10.1093/infdis/jis607

Xu R, Lin J, Zhao GQ, Li C, Che CY, Xu Q, Liu M (2018) Production of interleukin-1β related to mammalian target of rapamycin/Toll-like receptor 4 signaling pathway during Aspergillus fumigatus infection of the mouse cornea. Int J Ophthalmol 11:712–718. https://doi.org/10.18240/ijo.2018.05.02

Yang Z, Klionsky DJ (2010) Eaten alive: a history of macroautophagy. Nat Cell Biol 12:814–822. https://doi.org/10.1038/ncb0910-814

Yang J-H, Wu U-I, Tai H-M, Sheng W-H (2019a) Effectiveness of an ultraviolet-C disinfection system for reduction of healthcare-associated pathogens. J Microbiol Immunol Infect 52:487–493. https://doi.org/10.1016/j.jmii.2017.08.017

Yang X, Zhao G, Yan J, Xu R, Che C, Zheng H, Zhu G, Zhang J (2019b) Pannexin 1 channels contribute to IL-1beta expression via NLRP3/Caspase-1 inflammasome in Aspergillus fumigatus keratitis. Curr Eye Res 44:716–725. https://doi.org/10.1080/02713683.2019.1584321

Yang H, Wang Q, Han L, Yang X, Zhao W, Lyu L, Wang L, Yan H, Che C (2020) Nerolidol inhibits the LOX-1/IL-1β signaling to protect against the Aspergillus fumigatus keratitis inflammation damage to the cornea. Int Immunopharmacol 80:106118. https://doi.org/10.1016/j.intimp.2019.106118

Yang RB, Wu LP, Lu XX, Zhang C, Liu H, Huang Y, Jia Z, Gao YC, Zhao SZ (2021) Immunologic mechanism of fungal keratitis. Int J Ophthalmol 14:1100–1106. https://doi.org/10.18240/ijo.2021.07.20

Ye Y, He J, Wang H, Li W, Wang Q, Luo C, Tang X, Chen X, Jin X, Yao K, Zhou M (2022) Cell wall destruction and internal cascade synergistic antifungal strategy for fungal keratitis. ACS Nano 16:18729–18745. https://doi.org/10.1021/acsnano.2c07444

Yeniad B, Yigit B, Issever H, Kozer Bilgin L (2003) Effects of contact lenses on corneal thickness and corneal curvature during usage. Eye Contact Lens 29:223–229. https://doi.org/10.1097/01.icl.0000086494.50288.70

Yi J, Sun Y, Zeng C, Kostoulias X, Qu Y (2023) The role of biofilms in contact lens associated fungal keratitis. Antibiotics (Basel) 12. https://doi.org/10.3390/antibiotics12101533

Yike I (2011) Fungal proteases and their pathophysiological effects. Mycopathologia 171:299–323. https://doi.org/10.1007/s11046-010-9386-2

Yoon KC, Jeong IY, Im SK, Chae HJ, Yang SY (2007) Therapeutic effect of intracameral amphotericin B injection in the treatment of fungal keratitis. Cornea 26:814–818. https://doi.org/10.1097/ICO.0b013e31806c791e

Zarnowski R, Westler WM, Lacmbouh GA, Marita JM, Bothe JR, Bernhardt J, Lounes-Hadj Sahraoui A, Fontaine J, Sanchez H, Hatfield RD, Ntambi JM, Nett JE, Mitchell AP, Andes DR (2014) Novel entries in a fungal biofilm matrix encyclopedia. *mBio* 5:e01333–e01314. https://doi.org/10.1128/mBio.01333-14

Zarnowski R, Noll A, Chevrette MG, Sanchez H, Jones R, Anhalt H, Fossen J, Jaromin A, Currie C, Nett JE, Mitchell A, Andes DR (2021) Coordination of fungal biofilm development by extracellular vesicle cargo. Nat Commun 12:6235. https://doi.org/10.1038/s41467-021-26525-z

Zhang X, Sun X, Wang Z, Zhang Y, Hou W (2012) Keratitis-associated fungi form biofilms with reduced antifungal drug susceptibility. Invest Ophthalmol Vis Sci 53:7774–7778. https://doi.org/10.1167/iovs.12-10810

Zhang S, Meng P, Liu G, Liu K, Che C (2018) ATF4 involvement in TLR4 and LOX-1-induced host inflammatory response to Aspergillus fumigatus Keratitis. J Ophthalmol 2018:5830202. https://doi.org/10.1155/2018/5830202

Zhang R, Peng X, Lin J, Zhang Y, Zhan L, Tian X, Yin J, Zhao G (2021) The role of SREC-I in innate immunity to Aspergillus fumigatus keratitis. Invest Ophthalmol Vis Sci 62:12. https://doi.org/10.1167/iovs.62.9.12

Zhang L, Hu Q, Zhang Y, Wang Y, Liu N, Liu Q (2023) Rapid inactivation of mixed biofilms of Candida albicans and Candida tropicalis using antibacterial photodynamic therapy: based on PAD Plus. Heliyon 9:e15396. https://doi.org/10.1016/j.heliyon.2023.e15396

Zhao G, Li S, Zhao W, He K, Xi H, Li W, Zhou Q, Wang Y (2012) Phage display against corneal epithelial cells produced bioactive peptides that inhibit Aspergillus adhesion to the corneas. PLoS One 7:e33578. https://doi.org/10.1371/journal.pone.0033578

Zhao GQ, Qiu XY, Lin J, Li Q, Hu LT, Wang Q, Li H (2016) Co-regulation of Dectin-1 and TLR2 in inflammatory response of human corneal epithelial cells induced by Aspergillus fumigates. Int J Ophthalmol 9:185–190. https://doi.org/10.18240/ijo.2016.02.01

Zhao G, Hu M, Li C, Lee J, Yuan K, Zhu G, Che C (2018) Osteopontin contributes to effective neutrophil recruitment, IL-1beta production and apoptosis in Aspergillus fumigatus keratitis. Immunol Cell Biol 96:401–412. https://doi.org/10.1111/imcb.12010

Zhou Q, Chen H, Qu M, Wang Q, Yang L, Xie L (2011) Development of a novel ex vivo model of corneal fungal adherence. Graefes Arch Clin Exp Ophthalmol 249:693–700. https://doi.org/10.1007/s00417-010-1601-9

Zhou C, Peng C, Shi C, Jiang M, Chau JHC, Liu Z, Bai H, Kwok RTK, Lam JWY, Shi Y, Tang BZ (2021) Mitochondria-specific aggregation-induced emission luminogens for selective photodynamic killing of fungi and efficacious treatment of keratitis. ACS Nano 15:12129–12139. https://doi.org/10.1021/acsnano.1c03508

Zhu CC, Zhao GQ, Lin J, Hu LT, Xu Q, Peng XD, Wang X, Qiu S (2015) Dectin-1 agonist curdlan modulates innate immunity to Aspergillus fumigatus in human corneal epithelial cells. Int J Ophthalmol 8:690–696. https://doi.org/10.3980/j.issn.2222-3959.2015.04.09

Ziental D, Mlynarczyk DT, Czarczynska-Goslinska B, Lewandowski K, Sobotta L (2021) Photosensitizers mediated photodynamic inactivation against fungi. Nanomaterials (Basel) 11:2883. https://doi.org/10.3390/nano11112883

Zieske JD, Gipson IK (1986) Protein synthesis during corneal epithelial wound healing. Invest Ophthalmol Vis Sci 27:1–7

The manufacturer's authorised representative in the EU is Springer Nature Customer Service Centre GmbH, Europaplatz 3, 69115 Heidelberg, Germany. If you have any concerns regarding our products, please contact ProductSafety@springernature.com

Printed and bound by CPI Group (UK) Ltd, Croydon, CR0 4YY
30/09/2026
02225591-0015